STRATIGRAPHY *and Life History*

STRATIGRAPHY
and Life History

MARSHALL KAY

Professor of Geology, Columbia University

EDWIN H. COLBERT

Curator of Vertebrate Paleontology, The American Museum of Natural History;
Professor of Vertebrate Paleontology, Columbia University

with a chapter on "The Origin of the Earth"
by DEAN B. McLAUGHLIN
Professor of Astronomy, University of Michigan

John Wiley & Sons, Inc.   *New York*   *London*   *Sydney*

Maps by Jean Paul Tremblay

Section and fossil drawings by Genevieve Shimer

Charts and diagrams by George Kelvin

Preface

The reader who is about to turn the pages of this book may wish to know the ends toward which it is directed. The first word of the title may need some explanation. In its simplest sense, stratigraphy is the placing of events in their sequence in time; thus it would seem that the subject is one of history, physical and organic. This might lead to a plan of discussing the progression of events and life through time. However, the span of geologic time is very long, several billion years, and the area of the surface of the earth or even of North America is very great. Even if the history of this entire time and area were well known (which it is not) and did not require lengthy explanations, the material would be much too great to contain in a single book, perhaps even within a library of books. And the question of whether a student would find great value in having such a compendium, except as a reference source, might properly be raised. Nevertheless, the scholar may be interested in and concerned with how some of the interpretations about the history of the earth have been gained. So we will deal with the principles of stratigraphy, as applied to earth history, and with some aspects of the development of life.

The study of the stratigraphy of North America entails the review of much literature in geology, for there are few articles in this field that do not concern the arrangement of events in their order of sequence in time. *Stratigraphy and Life History* presents the principles, as well as giving a summary of some of the main events in the history of North America. Historical geology concerns the records through a succession of time intervals, somewhat accidentally assembled into systems and smaller time-stratigraphic units. For each span, such as of a system, the details are provincial—of interest particularly to those who are near to them. What is of greatest interest to the student in one region will be little appreciated by someone in a distant place, where geographic separation reduces the pertinence of the material and the ability of the individual to comprehend it. There are reference articles that will be most useful for students living in each of the many parts of North America; the instructor will be familiar with these.

With the progress of research, there has been a succession of publications in the field of geology—each one of these building on its predecessors. A number of general references present the principles of stratigraphy, and guide the student in locating further material on the several phases of the science. These principles will be gained progressively through the pages of this book and summarized in a closing chapter. Principles are also discussed in: C. O. Dunbar and John Rodgers, *Principles of Stratigraphy* (John Wiley and Sons, 1959);

W. C. Krumbein and L. L. Sloss, *Stratigraphy and Sedimentation* (second edition, W. H. Freeman and Co., 1963); and J. M. Weller, *Stratigraphic Principles and Practice* (Harper and Brothers, 1961). *Paleocurrents and Basin Analysis* by Paul Potter and F. J. Pettijohn (Academic Press, 1963) presents aspects of interpreting primary sedimentary structures that bear on stratigraphy.

Few books systematically discuss the stratigraphic history of North America. The United States Geological Survey is preparing a series of atlas volumes presenting lithic data for each system, with maps and sections and a wealth of references; some have been published. A series of volumes is being published on *The Precambrian* (K. Rankama, Editor, John Wiley and Sons). A few books contain papers summarizing the stratigraphy in limited areas. For example, many authors contributed information on the Cambrian System to *El Sistema Cambrico* (International Geological Congress, Mexico, 1956) and on the *Pennsylvanian System in the United States* (American Association of Petroleum Geologists, 1962).

Jurassic Geology of the World by W. J. Arkell (Hafner Publishing Co., 1956) is one of a very few books treating the record of a span of time throughout the earth. *The Quaternary Era* by J. K. Charlesworth (Edward Arnold, 1957), *Glacial and Pleistocene Geology* by R. F. Flint (John Wiley and Sons, 1957), and *The Quaternary* (K. Rankama, Editor, John Wiley and Sons, 1964) summarize the last million years of geologic time. *History of the Earth* by Bernhard Kummel (W. H. Freeman and Co., 1961) is the best source of general information on the whole earth. Comprehensive information on the stratigraphy of each system in North America is contained in the *Correlation Charts*, prepared by committees of the National Research Council and published by the Geological Society of America at intervals through many years.

Excellent books treat the structural history of the continent. *The Geological Evolution of North America* by T. H. Clark and C. W. Stearn (Ronald Press, 1960) is an historical geology textbook progressing by regions. *Structural Geology of North America* by A. J. Eardley (second edition, Harper and Brothers, 1961) has colored structural maps of past times; and the *Evolution of North America* by P. B. King (Princeton University Press, 1960) is a lucid presentation of the story of the development of the continent. An exceptionally fine summary of the *Atlantic and Gulf Coastal Province of North America* is by G. E. Murray (Harper and Brothers, 1960). One of the authors of the present book prepared a study of *North American Geosynclines* (M. Kay, Geological Society of America, Memoir 48, 1951) from which many illustrations have been adapted for this book.

The great source of information on systematic invertebrate paleontology is the *Treatise on Invertebrate Paleontology* (R. C. Moore, Editor, Geological Society of America and the Kansas University Press), published at intervals in many volumes, each treating one taxonomic division of organisms. Similarly, a useful source of information on systematic vertebrate paleontology is to be found in several volumes of the *Traité de Palaeontologie,* Jean Piveteau, Editor, Masson et Cie, Paris). *Vertebrate Paleontology* by Alfred S. Romer (University of Chicago Press, 1945) is the most comprehensive single volume devoted to this subject, while *Evolution of the Vertebrates* by Edwin H. Colbert (John Wiley and Sons, 1961) is a presentation for the general reader.

The illustrations made from photographs have been credited to the photographer or to the contributor, except for those made by the authors. We appreciate the help of geologists who have contributed suitable pictures or directed our search for them; although some do not receive this acknowledgment in the captions, where the pictures were taken by others. Sections and maps are difficult to acknowledge properly. All have been prepared for this book, and most have been drawn from originals by Kay. Some of these are based in part on illustrations published by others, in which cases we have recognized the sources. But as much of the information on such illustrations is gained from data gathered by a host of geologists, it is not possible to trace adequately all of their origins. The word "after" is intended to show either that the drawing is prepared from one that has been published by the person named, or that the illustration has been prepared from data contained in his publications. Many of the drawings are based on those in Kay's *North American Geosynclines* and Colbert's *Evolution of the Vertebrates,* but some of these were compiled from earlier sources. The fossil illustrations are after those in many publications; some are adapted from those in volumes of the *Treatise on Invertebrate Paleontology.* They were drawn by Genevieve Shimer, who also drafted the sections. Charts and diagrams were made by George Kelvin. The writers are grateful for the assistance of Inez Kay in editing, and in arranging the index. John Tremblay gave valued advice as well as technical skill in the preparation of the maps.

New York City
October, 1964

Marshall Kay
Edwin H. Colbert

Contents

The investigation of the geology of all the places visited was far more important, as reasoning here comes to play. On first examining a new district, nothing can appear more hopeless than the chaos of rocks; but by recording the stratification and nature of the rocks and fossils at many points, always reasoning and predicting what will be found elsewhere, light soon begins to dawn on the district, and the structure of the whole becomes more or less intelligible.

Charles Darwin,

Journal of Researches into the Geology and Natural History of the Various Countries Visited by H. M. S. Beagle, 1839

1
Introduction

Restoration of an ancient sea bottom—Silurian (The Smithsonian Institution).

1 *Introduction*

The physical history of the earth is a subject of intrinsic interest, and its interpretation presents the origin and the development of the rocks and the landscapes that surround us. The history of life on the earth is a subject equally interesting. It reveals the evolution of animals and plants through the ages in countless aspects—some strange and bizarre, others familiar and commonplace.

Stratigraphy, the field of geology that concerns the physical history of the earth, implies in its name a description of stratified or layered rock. However, the treatment of the nature of sedimentary rocks and the interpretation of their origin have come to be contained in the science of sedimentology or sedimentary petrology; and the biological aspects of the rocks fall specifically within the domain of paleontology. The term stratigraphic has come to be associated in particular with the determined chronological order of formation of the rocks and their positions in the scale of progressing time. A standard time classification has been developed from experience, based on strata in which the relative positions and fossil contents were determined. The recent advent of geochemical dating has given additional means of determining relative ages and has shown that the earth's history extends through billions of years. Stratigraphic relations, then, refer to the time of deposition or emplacement of one rock relative to another.

The rocks and their surface expressions in landforms are the records from which the interpretations of geologic history must be obtained. The significance of the conclusions about them depends on the skill and judgment of the geologist; but the processes and techniques that he employs need to be understood to evaluate the validity of his conclusions.

The discussion of stratigraphy will begin with a development of the principles from recorded observations of rock formations in specific areas. The intention is to learn how knowledge of the history has been acquired, and incidentally what some of the facets of this story have been. The basic events in the development of the continent of North America are presented, so that the principles can be learned from the successive systems of rocks formed in geologic time. As the methods are introduced, they will be applied and reviewed in succeeding systems. In the chapters on the later stages of the history of the continent, the text will tend to follow the more conventional practice of describing the earth's history as it is gained from the application of principles.

The story is one of deposition and erosion of surfaces that are progressively changing in relative elevation, forming the lands and directing the courses and plans of the troughs and seas. These continually varying patterns can be summarized in several kinds of maps and in vertical plans of restored structure sections, although with the enormity of geologic time and the vastness of the area even of a continent, only meagre but intendedly representative and most informative selections can be portrayed. An aim in stratigraphy as in structural geology is the advancing of one's skill in gaining mental images of the three-dimensional forms that are developing progressively.

The treatment of organisms is intended to give an understanding of their principal attributes and their most significant applications to stratigraphy and biology; their multiplicity and varied geographic and stratigraphic distributions make it practical to present only the barest essentials. The records of vertebrate fossils have the greatest esthetic interest and popularity, and they demonstrate important principles of vertebrate evolution.

The stratigraphic record is presented before treatment of the origin of the earth, a topic having great philosophic interest. The theories of origin involve applications of physical laws to postulates reflecting astronomical observations. The characters of the main organic groups will be summarized and the principles of stratigraphy and their applications reviewed in the last chapters in the book.

The presentation of historical geology can be made in several ways: preferences in treatment are matters of taste. There are excellent books that treat the history of the earth, emphasizing the different regions of North America and of the world. Some textbooks give fullest account of the physical development, or the stratigraphic sections of the continent through time, or the succession of organisms. We present aspects of each of these topics, developing principles by induction, then applying those principles derived from relations in specific examples to the general knowledge of earth history.

Geology, like all sciences, involves the classification of substances, properties and events, and each scheme of classification has its own terminology. Certainly, in order to converse, there must be a language. Learning this language is not an end but a means by which the scholar can convey impressions and concepts to the reader and auditor. Some technical words present ideas that would otherwise require many words. The language of historical geology is largely that of physical geology, though some of the terms are acquired from the biological sciences, and others are peculiar to the considerations of time. We present the definitions of terms that are likely to be novel, but the reader is not denied the privilege of occasionally referring to a dictionary or glossary.

The fund of facts and details of geologic history are virtually unlimited, so a selection of examples has to be made to best develop principles. Errors of fact are probably here; we hope that they are few. The book's differences from some views that are conventional are matters of subjective judgments, and authorities do not agree on many details. Alternative views are presented in cases where differences in opinion bear on the bases for interpretation—but not in other instances where the differences are semantic or not particularly pertinent to the subject of the book. The interested student must go to original references or textbooks more directly concerned with such problems. The varying interpretations emphasize the fact that there are prospects for many contributions to stratigraphic knowledge and needs for methods that can improve the chances of solving some of the problems. Stratigraphy is a field in which conclusions are continually subject to review and refinement.

2
The Physical Features
of the North American
Continent

Lavas of the Columbia Plateau in eastern Washington, at the falls of the Palouse River
(A. Devaney).

2 *The Physical Features of the North American Continent*

The principles of stratigraphy are to be applied, for the most part, to the rocks of the continent of North America. The main physical features are presented briefly, in order that the student may have a better realization of the nature of the continent whose history is to be understood. The character of the rocks at the surface can be gained by direct observation and then expressed on maps. We will be concerned, too, about the characters that are revealed by the penetration of the crust of the continent directly, and the interpretation of data that can be recorded on the surface because of effects produced by the properties of rocks that lie at depth.

The principal political divisions, the states and provinces of the United States, Canada, and Mexico, must be familiar, for they are the means of locating the geologic features that will concern us. A map of these political units (Fig. 2–1) shows the shores of the oceans and the largest lakes and the courses of principal streams, as well as the arbitrary lines that men have selected to form some of the boundaries.

The physical geography of the continent can be represented on maps of several kinds. In a standard atlas, the elevations are commonly shown by shades or colors, lines connecting points of equal elevation, contours, at given levels above the sea separating the successively higher divisions. The maps are very small-scale topographic maps having large contour intervals; such a simple elevation-line map or hypsometric map of the continent is shown in Fig. 2–2. A student of geology gains greater understanding from physiographic maps that symbolically represent the form and character of the landforms as though seen obliquely from above, rather than from those having only the elevations; such maps showing plains, plateaus, and mountains for an area as large as a continent must be extremely generalized and symbolic (Fig. 2–3).

Geologic maps are of several kinds, for the rocks at each place can be classified by such attributes as their colors, textures, compositions, and structures, or into such categories as sedimentary, igneous, and metamorphic. Moreover, each rock has an age or several ages, such as the time of deposition, of deformation or of alteration. Commonly the term geologic map is applied to one having colors or patterns representing the progressive ages of the mapped units; these are maps of interpretations rather than of facts. It will serve our purposes better in the beginning if we consider the continent more objectively in respect

Fig. 2–1. Map of North America showing the provinces and territories of Canada, the states of the United States, and the states of Mexico. The map is on a sinusoidal projection—one having distances to scale on the central longitude and on all latitudes, but introducing strong angular distortion in the upper corners of the map area.

Fig. 2–2 (p. 8). Map of North America showing the general elevations by means of shading. On a map of such a small scale, it is not possible to portray the intricate pattern that the areas would have on a map of very large scale. The map is on an azimuthal equal-area projection.

Fig. 2–3 (p. 9). Physiographic map of North America (one showing surface features symbolically as though viewed from above and northward). The principal physiographic provinces are distinguished by surface forms that are dependent on factors of rock structure, elevation, and dissection.

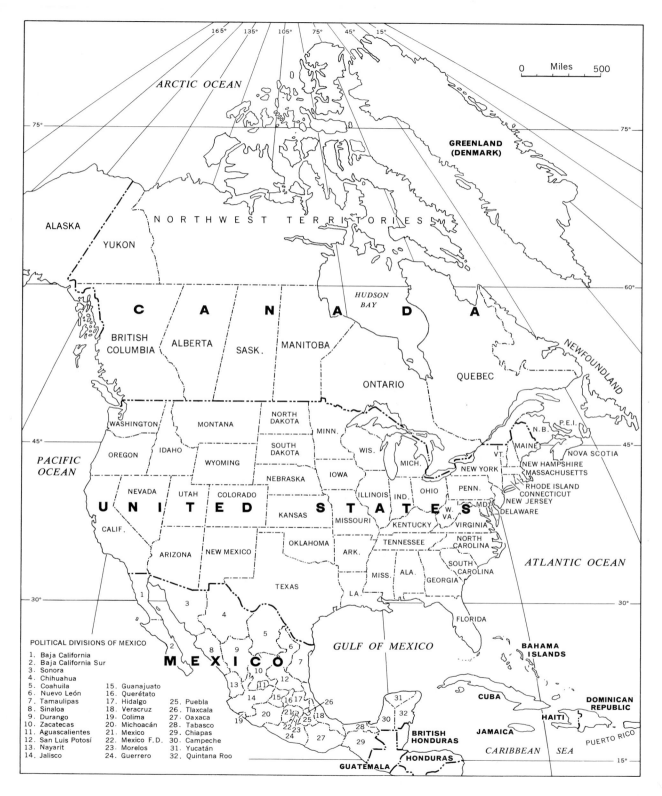

POLITICAL DIVISIONS OF MEXICO

1. Baja California
2. Baja California Sur
3. Sonora
4. Chihuahua
5. Coahuila
6. Nuevo León
7. Tamaulipas
8. Sinaloa
9. Durango
10. Zacatecas
11. Aguascalientes
12. San Luis Potosí
13. Nayarit
14. Jalisco
15. Guanajuato
16. Querétato
17. Hidalgo
18. Veracruz
19. Colima
20. Michoacán
21. Mexico
22. Mexico F.D.
23. Morelos
24. Guerrero
25. Puebla
26. Tlaxcala
27. Oaxaca
28. Tabasco
29. Chiapas
30. Campeche
31. Yucatán
32. Quintana Roo

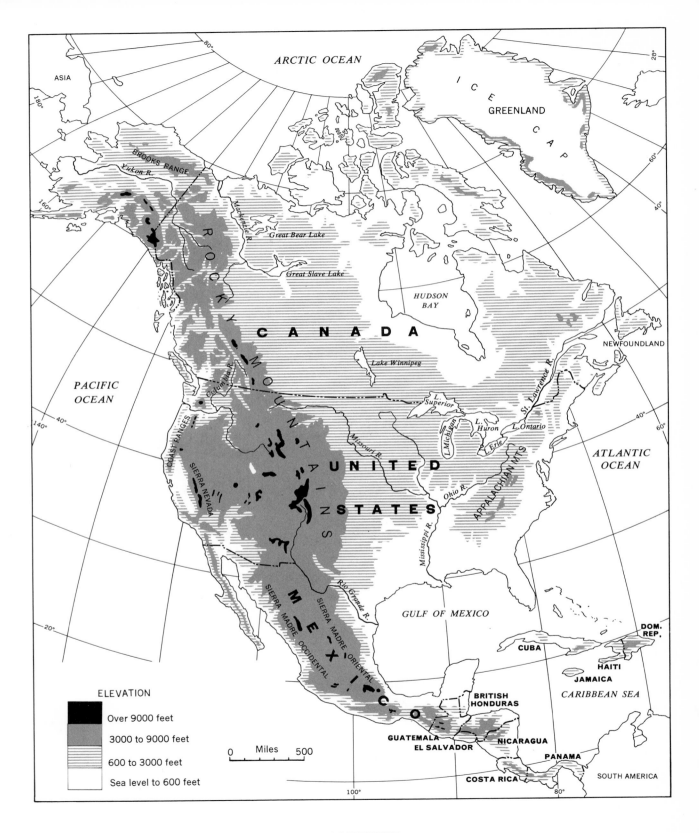

ARCTIC OCEAN

ASIA

GREENLAND

ICE CAP

BROOKS RANGE

Yukon R.

Mackenzie R.

Great Bear Lake

Great Slave Lake

HUDSON
BAY

CANADA

NEWFOUNDLAND

PACIFIC
OCEAN

R O C K Y   M O U N T A I N S

Columbia R.

Lake Winnipeg

L.
Superior

St. Lawrence R.

L. Huron

L. Ontario

L. Erie

ATLANTIC
OCEAN

COAST RANGES

SIERRA NEVADA

U N I T E D

Missouri R.

L. Michigan

APPALACHIAN MTS.

S T A T E S

Ohio R.

Mississippi R.

SIERRA MADRE OCCIDENTAL

SIERRA MADRE ORIENTAL

M E X I C O

Rio Grande R.

GULF OF MEXICO

DOM.
REP.

CUBA

HAITI

JAMAICA

CARIBBEAN SEA

BRITISH
HONDURAS

GUATEMALA
EL SALVADOR

NICARAGUA

PANAMA

COSTA RICA

SOUTH AMERICA

ELEVATION

Over 9000 feet

3000 to 9000 feet

600 to 3000 feet

Sea level to 600 feet

Miles

0   500

ARCTIC OCEAN

ASIA

QUEEN
ELIZABETH
ISLANDS

ICELAND

GREENLAND ICECAP

BROOKS RANGE

Yukon R.

BAFFIN
ISLAND

Great Bear
Lake

Mackenzie R.

Great Slave
Lake

HUDSON
BAY

NEWFOUND-
LAND

CANADIAN   SHIELD

Lake
Winnipeg

Great Lakes

St. Lawrence River

PACIFIC
OCEAN

COAST RANGE

SIERRA NEVADA

BASIN RANGES

Snake R.

Colorado R.

Rio Grande R.

Missouri R.

Mississippi R.

Ohio R.

APPALACHIAN MOUNTAINS

ATLANTIC
OCEAN

COASTAL   PLAIN

GULF OF MEXICO

CUBA

HISPANIOLA

JAMAICA

CARIBBEAN SEA

SOUTH
AMERICA

0   Miles   1000

to the kinds of rocks that are known at the surface and within the limits of observation at depth.

One must recognize the limitations of generalizations. First, there is the matter of scale. The maps of largest scale are those showing greatest detail, because each ground area has a relatively large map area. In the United States and Canada, the most common detailed maps are of about 1 inch to 1 mile which is 1 inch to 63,360 inches (1:63,360 scale); United States maps commonly are 1:62,500 scale because that is an even division, one-sixteenth of 1:1,000,000. State road maps generally are of scales of about 20 miles to an inch—1 inch representing somewhat more than a million inches, but they vary with the density of habitation in the regions they portray. A continent is a very large area, and must be represented on a very small-scale map—that is, one that has a very small distance to represent the true dimensions, if the map is to be of moderate size.

The elevation lines, contours, on a large-scale map can be very elaborate in an area with complex stream systems. However, the detail is so intricate that it must be lost in making a contour map of a large area on a smaller scale—as a small-scale map of a state or particularly of a continent.

There is similar difficulty in mapping the relationships among rocks represented on geologic maps. In some instances, rocks of different kinds are in sharp contrast at their contact. A dike of lava may cut directly across bedded sediments; or consolidated gravels, conglomerate, may lie in direct contact on gneiss or granite at an unconformity. On a large-scale map, these contacts can be represented by sharp lines. But the lines can become quite complicated in pattern on a large-scale map of an area of deformed rocks; the generalization must destroy much of this detail when placed on a continental map of small scale. In addition to the problems in portraying rocks having intricate though well-defined contacts, other problems arise because some contrasts are not sharp. Thus an intrusive igneous rock may so alter and penetrate the surrounding sedimentary rocks that a gradual progression results from igneous through metamorphic to sedimentary rock. These phases must be distinguished by lines, but even on large scale maps they are somewhat arbitrary. Therefore, a map showing the kinds of rocks on the surface of the continent (Fig. 2–4) is one having colors or symbols representing some of the many rocks that may be found in each locality.

THE KINDS OF ROCKS

The continent first can be described in terms of the principal kinds of rocks that lie at the surface: (1) soils and unconsolidated sediments,

Fig. 2–4.   Map of North America showing the distribution of the principal kinds of rocks, other than the veneer of surficial sediments of recent origin. As in the case of all small scale maps, the distribution is generalized, so that what is shown is a representation of the principal rock types within each part of the continent.

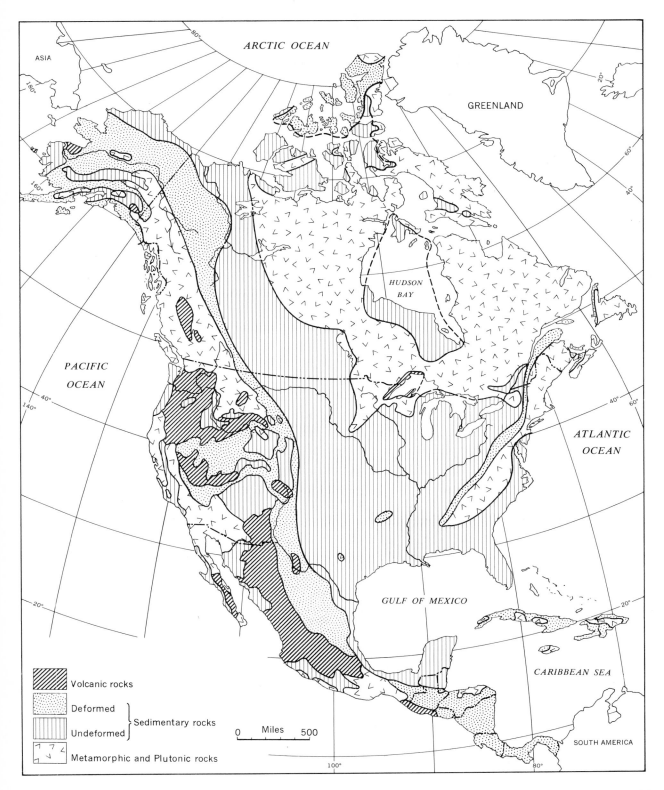

ARCTIC OCEAN

ASIA

GREENLAND

PACIFIC
OCEAN

HUDSON
BAY

ATLANTIC
OCEAN

GULF OF MEXICO

CARIBBEAN SEA

SOUTH AMERICA

Volcanic rocks

Deformed

Undeformed

Sedimentary rocks

Metamorphic and Plutonic rocks

0   Miles   500

11

(2) lavas and other rocks of surficial volcanic origin, (3) sedimentary rocks, some of them folded and faulted, and (4) more indurated, solid metamorphic sedimentary rocks and intrusive igneous rocks.

1. Solid rock fortunately is not universally present at the surface, for we are quite dependent on the soils and loose surface sediments for our agricultural economy. Some soils are but weathered residues of the rocks that lie below them, and this alteration decreases with depth; such surface soils really are formed from original surface rocks. In other large areas in the northern part of the continent, surface materials are the tills and fluvial glacial sediments moved from their source rocks as fragments frozen in the glacial ice and released by the melting of the ice (Fig. 2–5). Floors of valleys and shores of lakes and seas have variable thicknesses of water-laid clays, sands, and gravels. If all of these loosely consolidated materials were removed from the surface of the earth, there would be little change in land form, for generally they are but a shallow veneer concealing the solid rock below. Special kinds of geologic maps and soil maps show the distribution of surficial materials. Geologic maps of large areas generally disregard them in their generalizations. So the commonest geologic maps show the kind of rock that would be present at the surface if the loosely consolidated soils and surficial rocks were removed. This is the character of the map showing the kinds of rocks (Fig. 2–4) and their geologic ages (Fig. 2–6).

2. Basalts and other volcanic rocks lie below large areas in western North America from the Aleutian Islands and the peninsula of Alaska to central Mexico. They form the plateaus cut by the Columbia and Snake rivers, and the high, dormant volcanoes of the Cascades, such as Mount Shasta and Mount Rainier. Some of the scenic features of Yellowstone Park and the Craters of the Moon in Idaho are made of these rocks. Other volcanic peaks are in coastal Alaska and Mexico, the most recent being the spectacular Paricutín that began in a field and shortly developed a cone hundreds of feet high. These lavas and associated fragmental sediments are from a few feet to thousands of feet thick, concealing rocks like those in other regions, though the latter were penetrated by the deeply buried conduits through which the molten lava welled from the earth's interior. The limits of the volcanic rocks are not invariably abrupt, for often the flows end as wedges penetrating the sediments that abound in adjoining regions (Fig. 2–7). These sediments may in themselves contain large proportions of detritus gained from the volcanism. Thus the great lava fields of the Cascades pass westward into the sediments of western Oregon and Washington.

3. Sedimentary rocks underlie most of the United States and much of western Canada. Wells have penetrated to their base through large

Fig. 2–5 (right).   Air photograph northwestward over a part of Ohio where a veneer of glacially deposited sedimentary materials conceal the solid rocks below (J. Muench). The striking pattern of rectangular fields is systematically arranged within the roads enclosing each square mile.

Fig. 2–6 (*see inside of jacket*).   Map of North America showing the distribution of kinds of rocks by their ages of deposition, for the sedimentary and metamorphic rocks derived from sediments, and the age of intrusion or extrusion of igneous rocks or their metamorphic products.

Fig. 2–7 (right).   Hoover Dam and the lower Colorado River with California on the left and Arizona on the right in a northward view (A. Devaney). The rock distribution is complex, with rather flat-lying lavas partly concealing sedimentary and igneous rocks, as along Lake Mead.

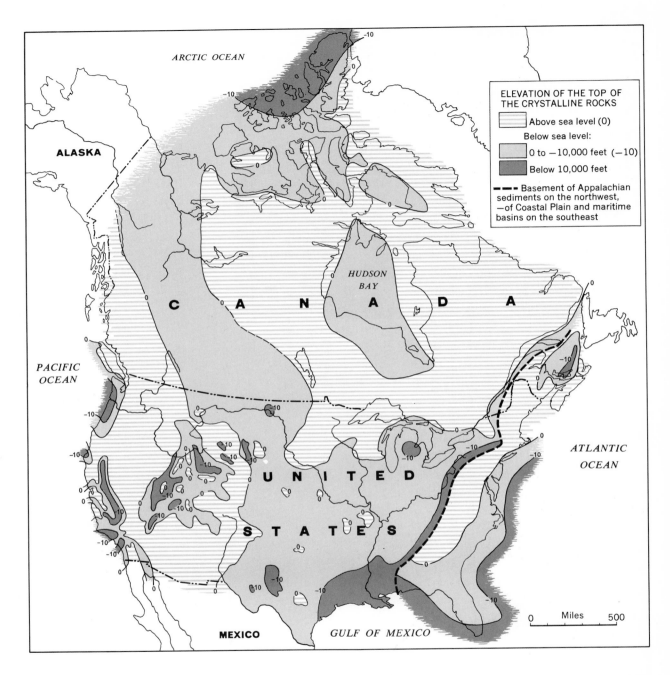

ELEVATION OF THE TOP OF
THE CRYSTALLINE ROCKS

Above sea level (0)

Below sea level:

0 to −10,000 feet (−10)

Below 10,000 feet

── ── Basement of Appalachian
sediments on the northwest,
—of Coastal Plain and maritime
basins on the southeast

Fig. 2–8. Map showing the elevation of the basement of metamorphic and igneous rocks in North America by structure contours (lines of equal elevation). Inasmuch as sedimentary rocks become metamorphosed in some regions, the separation of the two is not abrupt; thus in the Appalachian region, the broken line represents a band wherein sedimentary rocks become so metamorphosed that they become effectively the crystalline basement. The map is very generalized and incomplete in areas of complex structures, but it shows the depths of sedimentary rocks in such regions as the Atlantic and Gulf coastal plains, the central states, and the Prairie Provinces.

Fig. 2-9 (above). The Mackenzie Plain, Northwest Territories, extending into the distance southeastward from the broad anticlinal nose of the Paleozoic rocks of the Canyon Range, with the Carcajou River in the eastern flank (Royal Canadian Air Force). The highway to Canol, an oil field on the Mackenzie River, is in the lower right.

Fig. 2-10 (right). Appalachian Mountains; view southeastward near Warm Springs, in the western anticlines of the mountains in Virginia (Fairchild Aerial Surveys). The sedimentary rocks were folded and then eroded, the more resistant rocks, principally quartzites, forming the linear ridges, and the less resistant limestones and shales forming the intervening valleys.

areas, and in extensive regions the depth can be ascertained by geophysical methods, such as the study of the reflection and refraction of waves created by artificial explosions. The sediments differ not only in constituents but also in structure—most are so nearly flat that their inclinations are scarcely apparent, but in other areas the sediments form sharp folds, and have been broken into separated suites by faults. The manner and time of origin of such structures will form one of the principal objects of historical geology. We will first outline the broadest features of the distribution of the sedimentary rocks in North America.

A broad belt of sediments lies along the Atlantic Coast from New York to Florida, where it swings westward to skirt the northern border of the Gulf of Mexico into Texas and northeastern Mexico. Wells penetrating this belt along the Atlantic show it thickening seaward from its western termination to several thousand feet at the shore of the Atlantic, and submarine geophysical studies indicate that it continues with increasing depth of base into the Atlantic. The rocks beneath are like the metamorphic and igneous rocks that are exposed on the north and west. Along the Gulf Coast, thickness increases southward as determined from geophysical records, approaching ten miles along the shore in Louisiana and Texas. Thus, if we could strip away all the sediment from the Atlantic and Gulf Coasts (Fig. 2–8), we would have a surface that increased in depth beneath the margins of the ocean and the gulf.

An even greater expanse of sediments spreads across the interior of the United States from the Blue Ridge and New England to western Nevada and Idaho. To the north, a continuing belt narrows northward from the international border of the prairie provinces and easternmost British Columbia to the lower Mackenzie (Fig. 2–9), then westward along the Arctic slope of Alaska. Through most of the interior, the rocks are almost horizontal. These sediments lie on a gently undulating surface having basins as much as four miles deep and broadly domed areas with little or no sediment cover. If the sediments could be stripped away, a basin nearly three miles deep would be centered in lower Michigan, rising to sea level on the west, north, and east, and approaching it on the south. On the other hand, the surface in the Ozark region of Missouri and Arkansas would be above sea level, descending gradually into depressions, the deepest on the south.

The flat-lying sediments pass into belts having folded and faulted rocks with steeper dips. On the east, such a belt extends from western Newfoundland and Gaspé through the Appalachian Mountains (Fig. 2–10) to Alabama. Another great belt of folded and faulted sediments extends from north of the Yukon in Alaska through the Canadian

Fig. 2–11. Folded Paleozoic sedimentary rocks in the Kananaskis Range of the Rocky Mountains, Banff National Park, Alberta; view to the northward (Royal Canadian Air Force).

Fig. 2–12. Canadian Shield west of Hudson Bay near Artillery Lake, Northwest Territories (Royal Canadian Air Force). The nearly flat surface on Precambrian crystalline rocks has lakes trending northwestward parallel to the gneissic structure of the metamorphic rocks.

Rockies (Fig. 2–11) to western Montana and south to Nevada and California. In the southern course, it is separated by a broad plateau with gently dipping sediments from another branch of the fold belt passing through Colorado into New Mexico. There are additional smaller areas of deformed strata, such as that of the Ouachita Mountains in southern Arkansas and southeastern Oklahoma. Similar rocks continue southwards beneath the base of the previously described sediment wedge passing under the Gulf. The limits of the sedimentary belts and regions are not abrupt, for the change from sediment to metamorphosed sediment is one of degree. However, there are other belts in which little is sediment and much is metasediment and igneous intrusion.

4. Finally, there are great areas of the continent having a predominance of intrusive igneous rocks, such as granites and gabbros, but with varying proportions of metasedimentary and metavolcanic rocks, the so-called crystalline rocks, though really virtually all rocks have mineral crystals. For the most part, these are regions of severely folded rocks that have been invaded and soaked by magmatic matter. The great region of metamorphic and igneous rocks is the Canadian Shield (Fig. 2–12), forming an arc surrounding Hudson Bay and eastward to Labrador (Fig. 2–13), with extensions into Minnesota, Wisconsin, Michigan, and New York (Fig. 2–4). A second belt of similar rocks extended within the Pacific Coast from Baja California, Western Mexico to Alaska, the Sierra Nevada of California and Coast Range of British Columbia being among the larger mountainous areas within it. The third great area extends within the Atlantic Coast from central Newfoundland through the Maritime Provinces and New England (Fig. 2–14) to the Blue Ridge and Piedmont of the southern States.

Fig. 2–13. The southeastern margin of the Canadian Shield (G. Hunter). View northward of the Laurentide Mountains on the north shore of the St. Lawrence River near Port Cartier, Quebec; the railroad serves the iron mines of Labrador, hundreds of miles to the north.

There are considerable areas of sediments within each of the three. A large area south of Hudson Bay has a rather thin veneer of sediments. But they are preserved to a thickness of several miles in narrow bands within both the Pacific and Atlantic belts. Additional small areas of metamorphic and igneous rocks rise within the preponderantly sediment-surfaced interior. Many of these are in the ranges of the Rocky Mountains of the United States, particularly in central Colorado, but smaller areas appear in the Black Hills, in the St. Francis Mountains of Missouri and in central Texas. And wherever the sediments have been penetrated, the drill encounters rocks of metamorphic and igneous types. If we could remove all surficial volcanic and sedimentary rocks from the crust of North America, we would have a surface of metamorphic and igneous rocks with elevations rising to the height of the highest present mountains (Fig. 2–8). Such peaks

as Mount McKinley in Alaska (20,300 feet), Mount Logan in southwest Yukon (19,850 feet, Fig. 2–15), Mount Whitney in the Sierra Nevada of California (14,495 feet), Pikes Peak in Colorado (14,110 feet), Mount Mitchell in North Carolina (6684 feet) and Mount Washington in New Hampshire (6288 feet), are made of metamorphic and igneous rocks. A few of the higher elevations are volcanic, as Popocatepetl (17,887 feet) in Mexico and Mount Shasta (14,162 feet) in northern California, and some such as Mount Robson (12,972 feet) and others in the Canadian Rockies and Mount Wheeler (13,058 feet) in eastern Nevada are in the belt of folded and faulted sediments. The elevations on the surface of metamorphic and igneous rocks would pass miles below sea level, perhaps to the deepest level along the Gulf Coast.

THE AGES OF ROCKS

Each of the regions of metamorphic and igneous rocks has had somewhat similar history, but the events took place at different times, in different places. It is our purpose to learn how these events progressed, and how long ago they took place. Perhaps it will be well to consider the general problem of age. One of the questions most commonly asked is "How old are the mountains?" This is not as easy to answer as "What is the elevation of its top?" for there are various ages for different phases in the record of most mountains.

The surface of the earth is undergoing change through continuous reduction of exposed surface by agencies of erosion. The surfaces developed in rocks that were raised to their high level by preceding events. A few have materials that have been raised in recent times, such as the volcanic Parícutin. This mountain is only a few years of age, and it has been built by the accumulation of lava flows and explosive volcanic fragments since the first eruptions on a flat plain in 1943. But the rocks in most mountains are not made of materials that have just been brought to their present form.

The rocks have high elevation because they have been raised to that level, through a short time or a longer span in the past, or in deformation that is still progressing. The crustal bending that raised the rocks is epeirogeny. The present elevation of the rock, then, has an age—measured from the end of the time that the mountain rose or stopped rising. As the word mountain in common language is applied to something high, perhaps this is the most proper age to apply, but more commonly the reference is to the antiquity of the rocks that are in the mountains.

Plutonic, igneous, and metasedimentary rocks bearing structures

Fig. 2–14. The top of Mount Washington, New Hampshire, from the northwest, a mountain of metamorphosed sediment and some intrusive igneous rocks (New Hampshire Division of Economic Development). The surface of crystalline rocks slopes gently southeastward toward the Atlantic Ocean.

Fig. 2–15. St. Elias Mountains, Yukon Territory, on the border of Alaska, from the east (Royal Canadian Air Force). The face of the Kluane Mountains in the foreground may be a fault scarp. The nearer mountains are of sedimentary rocks with some lavas. The more distant higher ranges are little explored, but they are considerably of igneous and metamorphic rocks. Their highest peak, Mount Logan (19,850 feet) is to the south, left of the view; the peak on the extreme right is in Alaska.

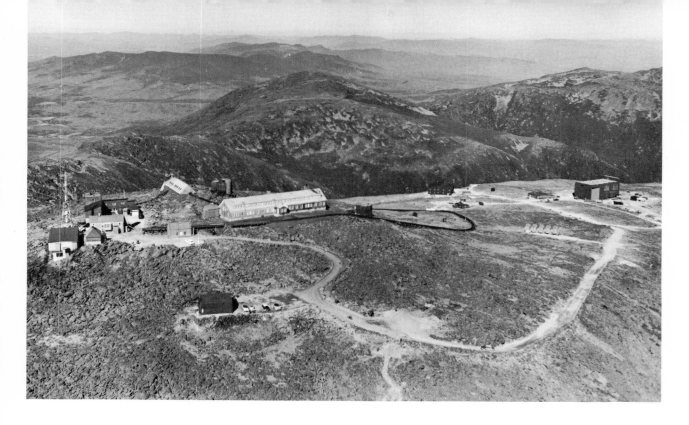

formed by strong deformation are the prevalent materials in most of our highest mountains, not only in North America, as has been stated, but elsewhere in the world. This is true of such peaks as Everest (29,141 feet) in the Himalayas, and Mont Blanc (15,781 feet) in the Alps; but some of the highest, particularly those in the Andes, such as Aconcagua in western Argentina (23,081 feet), are volcanic. The plutonic rocks, such as granites, invaded the surrounding rocks at one time or through a span of time, perhaps metamorphosing them through the high temperature and introduction of magmatic fluids and gases. Thus we might say that the granite of the mountain has an age. As we shall see, geochemical methods sometimes permit determination of the approximate age of the formation of such plutonic rocks. The meta-sedimentary rocks may have folds and other structures that were formed at the time of the intrusion, or by earlier orogeny (literally "mountain-making"), and these structures would have an age. And the rocks must have been laid by streams or under marine waters at a still earlier time, which is the age as it is commonly shown on a colored geologic map (Fig. 2–6).

A mountain range has many ages. The surface is of present origin, the product of continuing erosion. The rocks were raised to their present elevation at a past time. There are sediments laid at one time, deformed at another, and invaded by plutonic intrusions of lava at another. And there may have been more than one such event in history. Consideration has been given to the rocks at the surface of the continent and to the depth of the sedimentary rocks that cover the crystalline rock basement. The rocks at the surface give little insight into the earth as a whole. Wells have been drilled to a depth of more than four miles, and the deepest mines descend to more than half that depth, but this is but a very slight part of the 4000 miles to the center of the earth. Nothing on the surface suggests that there should be a crust of different properties than the interior. To learn the nature of this crust, methods have been developed which make use of physical phenomena that penetrate far below the wells and mines.

DEPTH OF ROCKS

Rocks of the outer part of the earth can be described in terms of the velocity with which compressional seismic waves pass through them— a velocity of about three-and-a-half miles a second—a characteristic similar in this respect to that of the average granite. Such velocities apply to the crystalline rocks that have been described as forming the surface over large areas and passing below the sediments in others.

When seismographs record waves at a distance from the source of

Fig. 2–16. Sections of the crust of the North American continent. The upper drawing has the crust drawn approximately to scale, the base, the Mohorovičić Discontinuity or Moho, lying some thirty miles below the surface under the continent but only a few miles beneath the oceans. The lower diagram shows the crust with exaggeration of scale, so that its thickness is ten times as great as the scale of the horizontal; sea level is shown as a horizontal line instead of with the curvature of the earth. The very thin layers shown in the upper part are those of the sedimentary rocks that underlie the surface of much of the continent, but which form only a thin veneer on the crust, even when the scale is so exaggerated. The density of rocks in the crust generally increases with depth, density then increasing rather abruptly at the Moho separating the crust from the mantle.

propagation, whether of an earthquake or an artificial explosion, the first waves arrive not in the time required by passage through crustal rocks of granitic character but somewhat faster. The waves follow a path through denser rocks by descending some twenty miles, traveling through rocks having properties that transmit waves about five miles a second, velocities comparable to those in basic igneous rocks, somewhat greater than that of ordinary basalt. Hence it has been determined that beneath the continents there is a change in velocity of transmission of compression shock waves, a discontinuity, at a depth of twenty miles or so; this is the Mohorovicic* Discontinuity or Moho (see Fig. 2–16). When one sees a sector of the earth drawn to scale, the limited relative depth of the crust is the more apparent. The rocks above the Moho are of lighter mass in a given volume, less dense than those below the discontinuity; the upper, lighter rocks are called sialic, because they are relatively rich in silica and alumina, and those below the discontinuity, simatic, rich in magnesium. In the principal oceanic areas, the sialic crust is lacking; the bottom of the ocean is underlain by some three or four miles of basaltic rock transmitting waves at a velocity somewhat less than those below the underlying Moho discontinuity. Thus continents are distinguished from ocean basins by more fundamental differences than simply the absence or presence of marine waters. The nature of the mantle beneath the Mohorovicic Discontinuity and the core of the earth lies outside the field of concern of historical geology, but it has been of such scientific interest that it is the object of deep drilling.

We will next return to the consideration of rocks, concentrating our attention on one area in northern Minnesota with the view of learning the methods by which the succession of events and their relative ages are determined.

* Alternate spelling with diacritical marks, Mohorovičić—after the Yugoslavian geophysicist, A. Mohorovičić, who discovered the contrast in the early years of this century.

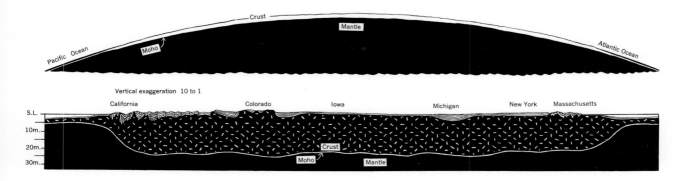

23

3

The Determination
of Stratigraphic Sequences
from Physical Evidence

Open-pit iron mine on the Mesabi Range; the Missabe Mountain Mine near Hibbing, Minnesota (Standard Oil Company, New Jersey).

3 The Determination of Stratigraphic Sequences from Physical Evidence

North America, as we have seen, can be divided into broad regions: those with a preponderance of lava flows and volcanic rocks just beneath the surface, those having sedimentary rocks, and those having metamorphic and plutonic igneous rocks. The ages of the rocks in a range of mountains in a region of metasediments and intrusive magmas involve phases of a long, complex history. We will consider the methods by which such a sequence of events can be determined by examining the record in the iron ranges of Minnesota and western Ontario (Fig. 3–1).

LAKE SUPERIOR IRON RANGES

Duluth and Two Harbors, Minnesota, and Superior, Wisconsin are cities at and near the western tip of Lake Superior, the northwestern-most of the Great Lakes. The waterfront in each of these ports (Fig. 3–2) has huge, high piers. Trains of cars filled with iron ore run out on these piers, so that the ore can be dumped into long, low ships and carried down the Great Lakes and overland to the iron and steel furnaces of Pennsylvania, Ohio, Indiana, and Illinois. About fifty miles to the north of western Lake Superior is the Mesabi Range, a rather inconspicuous belt of hills near Hibbing and Eveleth, Minnesota. Enormous pits in this range yield iron ore which is stripped by giant power shovels and loaded in the cars that form the trains (Fig. 3–3). The Mesabi Range has for a half-century yielded four-fifths or more of the iron ore mined in the United States, ore that has made the United States the world's greatest producer of iron and steel. For many years, tens of millions of tons have been mined in the Mesabi, the total output being nearly one hundred million tons a year during World War II. Scores of the long ships carry loads of ore in the seven- or eight-month season when the lakes and canals are free of ice. Thus the Mesabi Range has been of great importance in the national economy. In recent years, other regions have come to be relatively large producers, but new methods of beneficiation of low-grade Mesabi ores assure the continued importance of the range as a source of siliceous iron carbonates or taconite. But the range will also serve our immediate purpose as a fine source of examples of the application of stratigraphic methods to the reconstruction of geologic history.

Suppose that we were to go to some cliffs of rock between Hibbing and Eveleth. We might see a brown-surfaced ledge of iron-rich siliceous

Fig. 3–1. Geologic map of the north-west shore of Lake Superior in Minnesota and Ontario (after C. K. Leith, R. J. Lund, and A. Leith). Duluth lies at the western end of the lake, Port Arthur and Fort William are to the northeast, and Isle Royale National Park, Michigan, is on an island in the lake. The Mesabi Range is shown on the map by the east-trending, somewhat irregular band of outcrop of the iron formation and associated sedimentary rocks. The Vermilion Range lies to the northeast across the international boundary between Minnesota and Ontario.

ONTARIO

Lake Superior

MINN.

WIS.

MICH.

Lake Michigan

Lake Huron

ONTARIO

IRON FORMATION

ROCKS OLDER THAN

KEWEENAWAN-DULUTH GABBRO

KEWEENAWAN VOLCANIC AND SEDIMENTARY ROCKS

MINNESOTA

Virginia

Hibbing   Eveleth

Grand Rapids

LAKE SUPERIOR

"HURONIAN" SLATE

■ Iron formations and associated rocks

Two Harbors

0   MILES   50

Duluth

Superior

KEWEENAWAN VOLCANIC AND SEDIMENTARY ROCKS

WISCONSIN

Fig. 3–2. Ore dock at Duluth, Minnesota (Standard Oil Company of New Jersey). Trains of railroad cars containing iron ore run out on the pier. Their loads are dumped into bins, passing through chutes into the holds of the ore-carrying lake ships. The principal iron range, the Mesabi Range, lies about fifty miles north of Duluth.

carbonate rock; where this has been deeply weathered, it yields the ore that is mined. If we were to wander about the area, we might find exposures of some other rocks, such as a quartzite (composed of grains of quartz sand) or a slate (strongly cleaved argillaceous rock). The first stratigraphic problem is that of the relationship of the iron-bearing carbonate rock, the "iron formation" to other kinds of rocks. We might start by trying to find all the places where the iron-bearing rock is exposed, marking them on a map; they would extend along a curving band or belt of outcrop (Fig. 3–1). And in one of the mines (the Embarrass Mine) to the east of Eveleth, the iron formation can be seen lying with gentle inclination of bedding on the south side of the mine. Directly above is a rather sharp contact with slaty argillaceous rock, the slate in Fig. 3–4.

LAW OF SUPERPOSITION

The most elementary principle applied in determining relative ages of rocks is that younger beds of sedimentary rock will lie upon older beds of sedimentary rock; this is called the *law of superposition*. The higher rock must have been laid after the lower one, for it is hard to imagine a situation in which the upper bed could be laid and the lower one laid *under* it. It is the rare exception when beds that were originally deposited one on the other have been so deformed that their relative positions are reversed; it is possible, but the prospect is so small that it is commonly disregarded. Means that are available to enable recognition of the uncommon relationship will be described presently. Of course, the contact between the two might not be one of deposition—there might have been faulting. These are the questions that the geologist must answer by his critical study of the exposures, but normally the younger bed lies on top.

So the evidence in the Embarrass Mine strongly suggests that the slate formation is younger than the iron formation. If we were to go along the belt of outcrop of the iron formation to Gunflint Lake, we might find a place where the iron formation is adjacent to a different rock, a quartz-sandy formation that can be called the quartzite. By methods like those we applied to the slate on iron formation, it can be determined that the quartzite is the older, for it lies below the iron formation. Thus we have belts of quartzite, iron formation, and slate formation arranged in that order on a map, with the quartzite the oldest, and the slate the youngest. The relative ages have been judged from their superposition of the deposited sedimentary rocks.

Fig. 3–3. Iron mine in the Mesabi Range near Virginia, Minnesota. The iron ore has been concentrated naturally by the alteration of iron-bearing siliceous and carbonate-bearing rocks; the incomplete alteration accounts for the irregular form of the mine. The iron ore once was loaded into trains of cars pulled by steam locomotives; diesel engines are still used in some of the mines. Great trucks have become principal movers, bringing the ore to conveyer belts that carry it to upper levels; loaded trains proceed to the ore docks in Duluth and Superior. These mines are in soft, naturally concentrated ores. In recent years mines in the eastern Mesabi Range yield their iron-bearing sedimentary rock, which is carried to mills that concentrate the iron content in what is called beneficiation; the treated iron concentrates are shipped to the steel furnaces rather than the raw ore.

Fig. 3–4. Virginia slate overlying the Biwabik iron-formation on the south face of the Embarrass Mine in the eastern part of the Mesabi Range, near Eveleth, Minnesota. The iron formation is composed of ferruginous and siliceous carbonate rocks that are mined where leaching has concentrated the iron as oxide, hematite; the Virginia Slate, lying conformably on the Biwabik, is of consolidated ferruginous argillite. Inasmuch as the sequence seems in normal order, the Virginia is a younger deposit than the Biwabik. The bluff in the distance has sands and gravels of quite recent origin.

But along Gunflint Lake the quartzite is not just pure quartz sand. It has pebbles of other kinds of rocks, particularly of coarse pink granite that is very like granite that is exposed nearby, as though the granite had been exposed in the past, and pebbles of its kind had been carried to the place where the sands that became the quartzite were being deposited. The pebbles need not have come from this particular granite, but as they look like it, and the granite seems to lie along the base of the quartzite, this source seems plausible. As a rule, *a bed is younger than rocks like those contained in its pebbles*—younger than the source rocks of the pebbles.

The particular granite that yielded the pebbles had to have cooled and to have been "unroofed" before it could be eroded, for it was originally an intrusive at depth. The granite must have been so many years older than the sediments above it that erosion removed the rocks which deeply buried the granite when it was formed at depth in the earth's crust. On this basis, the granite, or some other granite very much like it, is older than the quartzite formation. And the grains of sand must have come also from a rock older than the quartzite. We have, then, a succession of granite, quartzite with granite pebbles, iron formation that is mined where altered, and a slate formation that can be seen to overlie it. There was an erosional period between the intrusion of the granite and the deposition of the quartzite with pebbles; they are separated by an unconformity.

UNCONFORMABLE CONTACTS

Suppose that instead of the sediments lying on the granite (that is, above it), the contact of the two had been a vertical one; we would still know the sediments to be younger, because *sediments in depositional contact with plutonic igneous rocks are younger;* the granite had to be unroofed before it could have sediment laid on it. We would need to be sure that the contact was an unconformity; the geologist would need to consider carefully the evidence that the contact is not one of a granite intruded into the sediment or faulted against it. So the presence of the pebbles in the quartzite is an important piece of evidence. It is the sort of information that was needed by the first geologists who came to study the iron ranges, and it is through the analysis of such evidence that the geologist reconstructs the sequence of events. The geologist is continually reaching judgments from his analysis of evidence.

Let us study this granite further. As it is exposed near Vermilion Lake, it has been called the Vermilion Granite so that we can distinguish it from other granites. And the quartzite is known as Pokegama

Fig. 3–5. *Dikes in Precambrian rocks, younger than the intruded rocks.* *A.* Near Rainy Lake, Kenora, western Ontario. The dike beneath the hammer cuts the one higher in the view, and thus is younger; the older dikes in turn invaded the older gneiss.

B. Dikes in a block of Precambrian rock in the Medicine Bow Range west of Laramie, Wyoming. The horizontal dikes are of coarse-textured granite, a pegmatite. They cut darker bands of gneiss that may be partly a product of the invasion of the dark schist by magmatic fluids. The order of relative ages can be determined.

Fig. 3–6. A Precambrian intrusive igneous rock near Ely, Minnesota containing blocks or xenoliths (pronounced *zeno*-liths) of older rocks that have been broken from the walls of the conduit; such inclusions are older than the enclosing rocks. In some instances blocks are but phases of a magma that crystallized at an earlier stage.

(Po-kee′-ga-ma) Quartzite, the iron formation as the Biwabik (Bee-wa′-bik) Formation—the Indian word for a piece of iron—after the old Biwabik Mine where it was exposed, and the slate, the Virginia Slate from the name of a city in the region. The iron formation at Biwabik Mine, a distinct stratified rock unit contrasting with those adjoining it, is called the Biwabik, the Biwabik Formation. It might seem simpler to give numbers or letters to such formations, but it is the practice to use the names of places, which carry a designation of where the rock can be seen, giving distinction to the name. There can be no question of which iron formation is at Biwabik Mine, for that is the type locality of the Biwabik Formation, a specific rock or lithic stratigraphic unit. Such rock stratigraphic names are defined under the principles of a code of stratigraphic nomenclature.

When we explore around the area having granite exposures we find some other rocks. Here and there are exposures of schist, a rock that resulted from the metamorphism of a sediment, or perhaps of a volcanic rock. Exposures are not very good where they adjoin those of the granite. There is the possibility that the granite cuts through the schist, gneiss, or slates (Fig. 3–5A and B) from which they might have been altered. Certainly there must have been some rock for the granite to enter, for it is an intrusive rock. *A plutonic rock is younger than any rock that it intrudes, cuts across, or alters.* Perhaps there is a dike-like extension of the granite entering into the schist. Perhaps the mineral composition of the granite changes as it approaches the schist, so that it more and more resembles the schist, because it actually has been formed by fluids altering, soaking into the schist. Perhaps there are fragments of the schist suspended in the granite—fragments of the walls of an intrusive that fell into the fluid magma (Fig. 3–6), or possibly parts of the schist that have not been so altered that the original appearance has become unrecognizable. The younger intrusive in some way affects the older intruded rock. The schist is older than the granite.

It will be rather difficult to determine relative ages if there are two granites in contact with each other. But if there were a distinct contact, the younger should alter the older and be affected by it. The geologist might find in the field or in a laboratory study of thin sections of the rock that one becomes finer and finer-textured—comes to have smaller and smaller crystals—toward the contact; this would indicate that the border had been chilled, for the more slowly the magma cools, the larger its crystals will be, and the cooling will be most swift near the cold rock surrounding it. The presence of large crystals grading into fine crystals means that the fluid was chilled at its border and the crystals did not have time to grow as large as those in the part of the intrusive

A

B

Fig. 3–7. *Conglomerate in the Precambrian Knife Lake Group in northern Minnesota (Minnesota Geological Survey).*
A. Beds near Ogishke Lake. The pebbles of igneous and metamorphic rocks, many of them well-rounded, were derived from lands eroded during the time of deposition and are thus representative of rocks older than the Knife Lake Group. As the exposed surface is horizontal, the rocks are standing with vertical dip; determination of younger beds depends on examination of primary depositional features and structures formed during subsequent deformation.

B. Conglomerate north of Ely Minnesota. The pebbles are of granite like the Saganaga Granite; in the deformation of the formation, the pebbles have been "stretched," compressed and elongated. They are of course made of rocks older than the sediments that contain them.

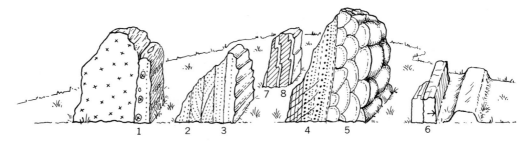

Fig. 3–8. Criteria for determining the tops of beds of sedimentary and volcanic rocks. The sketches are of beds having tops to the right, successive beds thus being younger in that direction. Such evidence must be found in order to enable learning the structure of the rocks and the chronological order of sequence of the lithic units.

Primary structures that are illustrated are: (1) unconformity with pebbles from older bed; (2) cross-stratification; (3) ripple marks; (4) graded bedding; (5) pillows in a lava flow; and (6) linear sole marks, expressed on the bottom of a bed that has fallen away. Secondary structures include: (7) axial-plane cleavage; and (8) drag folding.

that was farther away from the cold surrounding rock. The principle that the intruding rock is the younger is so obvious that it does not need much comment, but the evidence by which we learn that one is the intruder is not always so easily gained. And there is little to indicate *how much younger* the intruding rock is, or whether the intrusive is just a late phase of the same rock as the rock that is intruded.

DEPOSITIONAL STRUCTURES

To return to the Vermilion Granite, we have learned that it is younger than some schists or slates. But it also cuts across some conglomerates (Fig. 3–7) that must thus be older than the granite. As these older rocks are exposed along Knife Lake, we will refer to them as the Knife Lake Group of sedimentary rocks in order to distinguish them from other rock units. How are we to learn whether the conglomerates are older or younger than associated slates? They are found near together, and they are in bands of exposure that lie parallel to each other. Where the two are found close together at one locality, we can examine the depositional structures in the conglomerate (Fig. 3–8). We find that there are sandy beds which have cross-stratification; the laminae are concave upward, and are truncated by overlying laminae as though in minor angular unconformity. Some beds fill channels in underlying layers. Other beds of a foot or so have larger fragments on one side and smaller on another; the finer texture is upward, for only with decreasing velocity would a stream or submarine current leave successively finer fragments. In some poorly sorted sandy sediments laid in deep water, each graded layer, with the coarsest at the base, represents the deposit of a mass of mud and water that flowed down the bottom slope as a turbulent density current, from which the coarsest particles settled first to the bottom as the velocity decreased. Rippled sand beds show pointed crests toward the younger rocks (Fig. 3–9). All these criteria and other primary sedimentary structures will show whether a particular band of slate is on the younger or older side of a band of conglomerate or quartzite. In the slates, we can study the cleavage: suppose that where the dip of the beds is away from the conglomerate, the angle of dip of the cleavage is a little steeper than the bedding; this will indicate that it is not overturned, for the cleavage commonly nearly parallels the axis of a fold, and the bedding diverges from the axis (Fig. 3–10). Relations such as these shown in vertical beds enable us to determine the younger side and thus the one toward the syncline. Many criteria are useful in determining the relative ages of folded beds, and these are the methods employed by the structural geologist.

Fig. 3–9 (above). Ripple marks exposed on Precambrian quartzite in the Baraboo Range, in southern Wisconsin. Such ripple marks are useful in determining which side of a bed was originally uppermost; in this case, the top is facing southward toward the observer.

Fig. 3–10. Van Hise Rock, an exposure of Precambrian quartzite and argillaceous quartzite northwest of Baraboo, Wisconsin along the highway and the Chicago and Northwestern Railway. The bedding is vertical and cleavage inclined northward, to the left. To reconstruct a syncline having an axial plane, inclined as is the cleavage, requires that the syncline be asymmetrical and that the trough be to the right, as shown in the sketch. So the quartzite bed is younger than the cleaved argillaceous quartzite —tops are to the right. The rock was dedicated to the memory of a distinguished geologist, Charles Richard Van Hise (1857–1918) who recognized the structure of the range (R. F. Mueller).

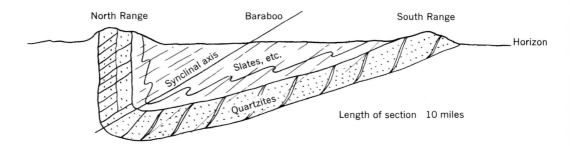

North Range   Baraboo   South Range

Horizon

Synclinal axis   Slates, etc.

Quartzites

Length of section   10 miles

ORDER OF SUCCESSION

The conglomerates are of about the age of the slates, for they are found as beds within the slates; the whole assemblage forms the Knife Lake Group. The conglomerates have pebbles from a number of kinds of rocks, particularly of granite (Fig. 3–7), and of altered lava (called greenstone), and of iron formation. The granite pebbles cannot be of the Vermilion Granite, for we have seen that the Vermilion is intrusive into the conglomerate; and the iron formation must also be different than the one that we studied near the iron mines, for it was stratigraphically above the Vermilion. There must be an older granite as well as the Vermilion. Conglomerates are found to unconformably overlie such a granite near Saganaga (Sa′-ga-na-gaw′) Lake. So we are arranging quite a number of rock units in their order of age, even though we have not converted this into history. The Saganaga Granite is much like the Vermilion Granite, but the latter intrudes conglomerate lying unconformably on the Saganaga and containing pebbles of granite like the Saganaga.

Some other granite than the Saganaga might have produced the boulders. If there were two granites that were identical, we might have been unable to determine which had yielded the pebbles. The boulders could have come from one, but this is not to say that they did. Geology is constantly involved in judgments that are not susceptible to absolute proof; evidence has to be weighed. We have to reason whether the evidence is fairly conclusive or rather weak. We know that the Saganaga Granite is older than the conglomerate, because it underlies it, and that the pebbles in the conglomerate are like the Saganaga Granite; but we cannot prove that the pebbles came from the Saganaga; it is only very probable that they did.

OTHER CRITERIA

There must have been a land surface exposing not only the Saganaga Granite but lavas that yielded the greenstone pebbles and iron formation that was eroded to produce pebbles of that sort in the conglomerate. But what is their order of age? The same methods that we have already applied determine that the lavas (Fig. 3–11) exposed near Ely, and the iron formation (Fig. 3–12), found near Soudan Mine, are older than the Saganaga Granite, and that the iron formation overlies the greenstone. Lavas have additional structures that are often useful in determining top sides of beds. When they flow into water, they harden into elongated structures called "pillows" having rounded, loaf-like upper surfaces and pointed processes below (Fig. 3–8); and there may be bubbles of increasing size upward in the lava. Many of these

Fig. 3–11. *Pillow lava in the Soudan Formation in northern Minnesota.* A. Pillows in an exposure in the town of Ely. Although pillows can be useful in determining the tops of lava flows, those in the view illustrate the difficulty in applying the method, for few of these pillows have projections penetrating between convex tops of older layers. The top is probably to the right, south, as the pillow above the hammer and to the right seems to be indenting the pillows to its left. Such pillows form where lava flows into a body of water, a lake or the sea.

B. Surface cutting similar pillow lavas near Saganaga Lake, in which the top of the flows seems toward the top of the view, away from the observer. Note the convex upper surface of pillows below and above the hammer, and the projections toward the bottom of the view (Minnesota Geological Survey).

Fig. 3–12. Banded red chert, or jasper, in the Precambrian Soudan Formation near Soudan, Minnesota (Minnesota Geological Survey). Jasper is iron-rich and is commonly associated with other iron-bearing rocks, such as iron formations, and with volcanic rocks, such as pillow lavas. The silica is attributed to settling of extremely fine particles or colloidal aggregates of silica, perhaps with reorganization after deposition and before consolidation; the bands are interlayered with other fine terrigenous sediments of silty texture. Such rocks are readily deformed, and thus develop intricate drag folds and flow structures under pressure.

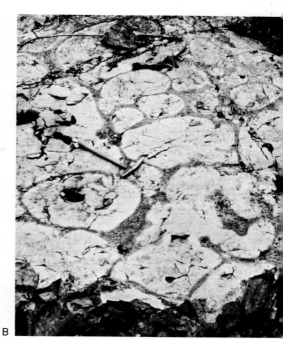

A

B

criteria were first recognized as significant by experience; the explanations of causes were subsequently made. This is the whole order: Ely Greenstone, Soudan Formation, Saganaga Granite, unconformity, Knife Lake Group, including slates, Vermilion Granite, unconformity, Pokegama Quartzite with pebbles of granite, Biwabik Formation of the iron mines, and the Virginia Slate.

The term stratigraphic sequence is applied to the arrangement of rock units in successive age. A number of stratigraphic methods have been discussed—methods of determining the sequence of rocks and events. First is the law of superposition, that younger sedimentary rocks are deposited above older sedimentary rocks. The younger sediments may contain fragments of older rocks as pebbles and grains. However, the simple depositional succession may become disturbed so that the bed that was originally the upper one must be recognized; thus we can use primary sedimentary structures such as cross-stratification, ripple marks, and graded bedding to determine the top of each bed. There are also structures in lava flows that permit determination of original top, such as the form of pillows, and the presence of gas bubbles, filled as amygdules, and surface flow features. When there are contacts with intrusive igneous rocks, the intrusions may alter as well as cut across the older rocks, and the cooler country rock may chill the margin of the intrusive so as to give it a finer grain size. There are also structural criteria. In the simplest form, as a corollary of the law of superposition, we can say that older rocks form the cores of anticlines, younger rocks the centers of synclines. But in folded and faulted rocks, the sediments may become overturned, so that the principle of superposition is not sufficient. The most useful means of determining the form of structures is the use of axial-plane cleavage and drag folds, where axial planes parallel that of the larger structures (Fig. 3–10). Thus the principles of structural geology become a necessary basis for determining stratigraphic sequence in deformed rocks. And at the same time, determined stratigraphic units permit the mapping and interpretation of structures.

INTERPRETING HISTORY

The sequence of rock units gives a basis for interpreting the regional history and present structure (Fig. 3–13). The oldest rock, the Ely Greenstone, records the outpouring of great streams of lava. The Soudan iron formation is water-laid sediment, iron and silica carried in suspension or solution in streams from its source in deeply weathered volcanic terrane to deposition, where the fresh water mixed with the salt water of a sea. These rocks were deeply buried, for Saganaga

Fig. 3–13. A structure section constructed through the Vermilion and Mesabi Ranges north of Lake Superior and east of Duluth, Minnesota. The section shows the relative ages of the several rock units represented and the nature of their subsurface extensions.

Fig. 3–14. *Keweenawan lavas along the shore of Lake Superior east of Duluth, Minnesota. A.* East dipping-lavas forming the points with the more readily eroded amygdaloidal phases in the coves (A. E. Sandberg).

B. Several flows with interbedded tuffs at Sugarloaf Point, with the top of a flow in the foreground (F. F. Grout).

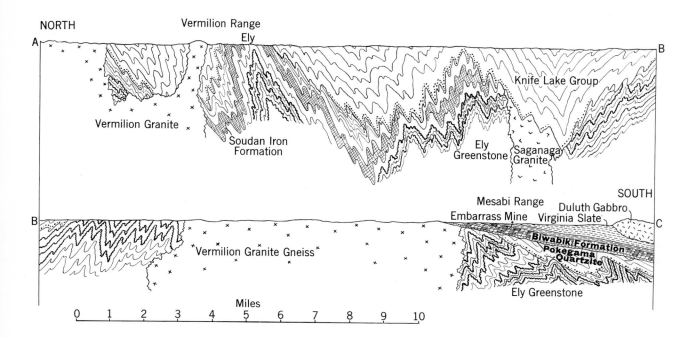

NORTH

Vermilion Range
Ely

A ——— B

Knife Lake Group

Vermilion Granite

Soudan Iron
Formation

Ely
Greenstone

Saganaga
Granite

SOUTH

B ——— C

Mesabi Range
Embarrass Mine

Duluth Gabbro
Virginia Slate

Vermilion Granite Gneiss

Biwabik Formation
Pokegama
Quartzite

Ely Greenstone

Miles

0   1   2   3   4   5   6   7   8   9   10

A

B

Granite intruding them crystallized below the earth's surface. The intrusion was accompanied by folding—mountain making or orogeny that deformed the lavas and sediments and raised them.

The revolution ended, and with erosion, pebbles of all the older rocks, lavas, sediments, and granite, were laid in the conglomerate at the base of the next sedimentary sequence, the Knife Lake Group. Deposition of other sediments succeeded. Then a second revolution interrupted with intrusion of the Vermilion Granite.

These intrusions and all the older rocks were then to yield pebbles to layers in the sedimentary sequence of Pokegama Quartzite, Biwabik Formation and Virginia Slate. This is as far as our preceding descriptions went. Later formations have been intruded by basaltic lavas; and near Duluth a sequence of more than five miles of lava flows is beautifully exposed along Lake Superior's rocky shores (Fig. 3–14). Although these were originally laid in fairly horizontal layers, they have been tilted and eroded during the long time that has elapsed since they formed, so that all can now be seen.

The history of the Lake Superior region has three large divisions, the first terminated with intrusion of the Saganaga Granite and its uplift and erosion, the second by similar events culminating in the invasion of the Vermilion Granite, and the third included the later occurrence of sedimentation and volcanism. The unraveling of the history is initially dependent on the ledges in the field, and particularly the recognition and correct interpretation of exposures that contain evidence critical to the determination of the rock sequence. Now that we have established the order of relative ages, we can proceed to arrange the interpretations of the successive rocks and their interrelations into a chronological order, giving a story of the events in the region of immediate investigation. We must then fit this history into that of many other regions in order to gain the history of successive times over larger regions and the continent. Such integration requires the application of somewhat different principles and techniques.

4
Some Methods of Correlation by Physical Characters

Folded Precambrian metasedimentary rocks near Gordon Lake, Northwest Territories (Royal Canadian Air Force).

4 *Some Methods of Correlation by Physical Characters*

The preceding chapter dealt with the means of determining the relative ages of rock units, and applying them to a description of a succession of events. We will turn now to the problems of comparing rocks and events in one region with those in another. The word correlate is applied to comparisons; the greater the likeness between things, the better they correlate. So in geology, two rock units that are correlated are alike in some way or ways. Whether the likeness is significant is a matter of opinion and judgments, for there may be many reasons why two things are alike or resemble each other to a degree. Generally in stratigraphic geology, correlation connotes comparable age, a decision that is dependent on the judgment that similarities in some one or more characters seem significant of age.

We developed the sequence of events in the Mesabi Range in northeastern Minnesota and its immediate region. We shall next concern ourselves with the correlation and classification of some of the rocks. The Biwabik Formation has many distinctive characters. It is nearly one thousand feet thick and is composed largely of beds of iron-bearing siliceous carbonate rock, frequently called "cherty iron carbonate." But there is neither uniform nor disorderly distribution of these elements in its composition—some parts are more siliceous, others more ferruginous, others quite argillaceous. There are iron-rich beds that have the mineral hematite, iron oxide, in little granules of spheroidal structure with layers like the "skins" of an onion. We can divide the formation roughly into four parts—a lower more cherty division, a lower argillaceous or slaty member, an upper cherty, and an upper slaty (Fig. 4–1). That does not tell much about iron content. If we were mining, we would not be taking out the iron formation in its ordinary condition, but only where it had been deeply altered so as to have had the silica leached away. With the new methods of beneficiation that are now under development, the geologists will be interested in finding the part of the unweathered formation having the highest iron-oxide content. Beneficiation is a process by which the iron-bearing part of the ore is separated partly from the other constituents, so that a higher proportion of iron is concentrated. A graph illustrates the differences in percentage of iron from base to top of the Biwabik iron formation; some parts are much better sources than others. Thus, when we speak of the Biwabik Formation, we are referring to a term that covers a great many details of lithology.

Fig. 4–1. *Sections of the Biwabik Formation, an iron-formation in the Mesabi Range northwest of Lake Superior in Minnesota (after D. A. White). A. The columnar section is at a representative locality and shows that the formation is composed of several kinds of rocks in its sequence. If sections are similarly plotted at other localities, they differ appreciably in the thicknesses of rock types but are generally composed of the same kinds of rocks.*

B. Figure showing analyses of the iron-content of the sediment from a succession of samples in each of three localities. To be economical, the iron must be further concentrated either through natural processes or by artificial beneficiation.

42

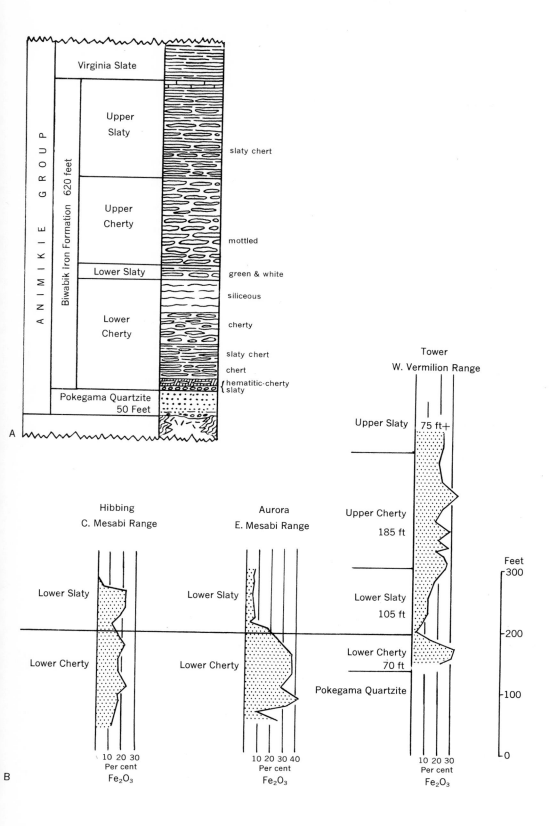

43

LITHOLOGY IN CORRELATION

The simplest and most positive sort of correlation would be to start at Biwabik Mine and walk along on continuous exposure of the same kind of rock—but we cannot do this, for there are soils and glacial drift and swamps and lakes that conceal the bedrock. However, the iron-formation is high in magnetic iron, so that if we fly a plane over the area and plot the magnetic attraction, we can make a map that shows the continuity almost as though we were walking along on the iron formation (Fig. 4–2).

If the two ends of the map had exposures of iron-bearing rock, we could say that they look alike. If the succession of kinds of rock in each succession were similar, or if both attracted the magnet more than the surrounding rocks, those would be other forms of likeness. And if the magnetic intensity continued between the two in an unbroken band, we would be warranted in concluding that all of the exposures are to be classified in the Biwabik Formation. The larger the number of attributes or characters that the one has in common with the other, the greater the likeness, the greater is the degree of correlation in a *statistical* sense—it is not that the rocks in various sections do or do not correlate, but that they correlate to a greater or lesser degree.

The Biwabik iron formation at Biwabik Mine lies between exposures of consolidated quartz sandstone or quartzite on the north and dark argillite or slate on the south; those are demonstrable facts. We might say that these are the Pokegama Quartzite and the Virginia Slate. That is an interpretation—we believe that the quartzite is like that exposed at Pokegama Lake, and the slate is like that found near the city of Virginia; we are correlating the rocks with those at the respective type localities. However, we have introduced one kind of supporting evidence; in each locality there is similar sequence, which adds to the likeness of the three.

CORRELATION IN TIME

The term correlation, when used with rock units having locality names, carries more meaning than simple likeness. We might know a quartzite in Sweden or Australia that is nearly identical in its lithic characters to that at Pokegama Lake, but we would not call it the Pokegama, even though it correlates strongly or even perfectly in the statistical sense. Correlation in the stratigraphic sense carries the additional connotation of continuity or synchroneity—that is, the rocks that we examine are not only correlated with those that we believe originally to have been continuously of the same kind or laid at about

Fig. 4–2. Geologic and aeromagnetic maps of a part of the Mesabi Range; the position of the area can be recognized from Figure 3–1 (after the United States Geological Survey). The areas from which iron ore has been removed are shown by the black areas that are mines. The aeromagnetic map was made from planes by equipment measuring the strength of the magnetic field at an elevation of about 1000 feet from the ground and along parallel routes about one mile apart. The magnetic Biwabik iron-formation is marked by prominent magnetic attraction that decreases gradually toward the south as the beds decline at about a 15-degree dip beneath the Virginia Slate. Note the similarity of trend of the magnetic lines to the trend of the iron formation, and the rather constant magnetic field over the older rocks to the north. The variations within the trends are not readily explained; they are caused by such factors as destruction of the magnetite in the alteration of the rock in its enrichment to form ore.

Fig. 4–3. Diagram showing the changing age of shore sands of a sea advancing on an eroded surface. The sections on the blocks show the progressively younger sands that formed on the margin of a sea advancing on an erosion surface, forming an unconformity. The lower, earlier stage represents conditions at the time of deposition of the surface at a time indicated at the isochron line in the upper diagram. The sands are of the same age only along the trends of shore lines of a particular time.

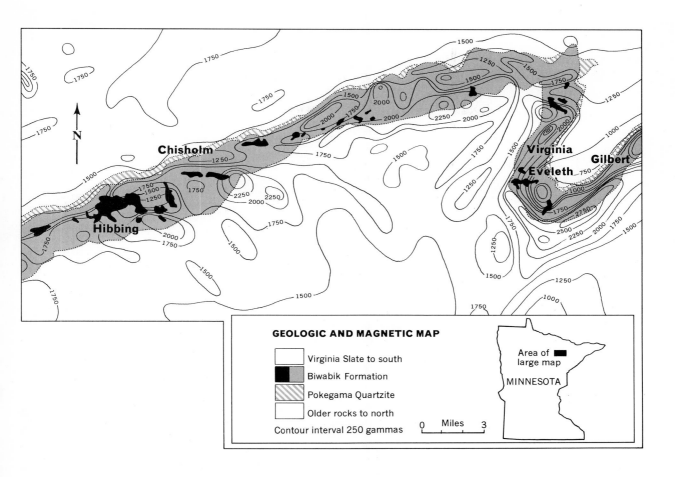

GEOLOGIC AND MAGNETIC MAP

Virginia Slate to south

Biwabik Formation

Pokegama Quartzite

Older rocks to north

Contour interval 250 gammas

0   Miles   3

Area of
large map

MINNESOTA

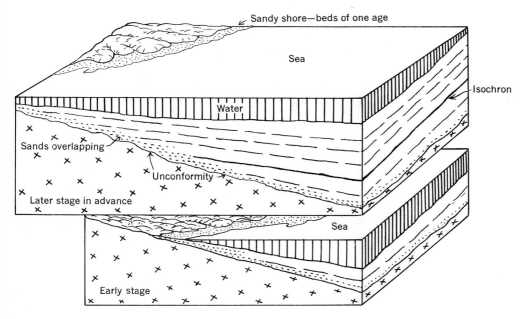

Sandy shore—beds of one age

Sea

Isochron

Water

Sands overlapping

Unconformity

Later stage in advance

Sea

Early stage

45

46   SOME METHODS OF CORRELATION BY PHYSICAL CHARACTERS

Fig. 4–4. Gently dipping, almost flat-lying Keweenawan lavas and underlying sedimentary rocks northeast of Port Arthur and north of Lake Superior (Geological Survey of Canada). The Trans-Canada Highway is in the foreground and Nipigon River is in the middle distance.

Fig. 4–5 (left). Keweenawan lavas and sedimentary rocks on north-central Isle Royale, in the National Park in northwestern Lake Superior, Michigan. (United States Geological Survey). The air view is north-northwestward; Chickenbone Lake in the center is about three miles long. The more resistant gently south-dipping lava flows form ridges separated by valleys in more easily eroded beds. The angle of dip is twenty degrees or so; thousands of feet of section are displayed. A depression has been eroded along a northeast trending fracture zone or fault extending northeastward into McCargoe Cove; lateral or vertical displacement is a few hundred feet. The Keweenawan Series is the youngest principal division of the Precambrian.

Fig. 4–6 (right). Geologic map of the south shore of western Lake Superior in Wisconsin and Michigan, to the south of the area shown in Fig. 3–1 (after C. K. Leith, R. J. Lund, and A. Leith). A principal iron range is that west of Marquette, but folded belts of sediments that are iron-bearing are scattered through northern Michigan and Wisconsin.

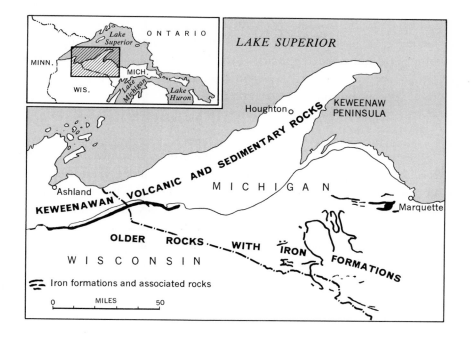

the same time, or both; they are also correlated with all rocks of the same general *time* of formation.

These matters require judgment in analysis. Logically, rocks of the same kind in two regions would be of the same age, synchronous, if that particular kind of rock were formed but once, if the conditions that resulted in its formation were never repeated. However, we have already seen that rocks of the same sort are repeated; in the Mesabi Range there were iron-formations and conglomerates and granites. Perhaps the iron-bearing rock in each of the different iron-formations has some peculiarity of composition or structure that does make it distinguishable and unique; but experience shows us that kinds of rocks are repeated.

WERNERIAN DOCTRINE

Geologists of the eighteenth century did believe that the earth was made of layers of successive differing rock units. This was the widely accepted theory of Abraham Gottlieb Werner (1750–1817), professor of mining at Freiberg in Saxony. He thought all the rocks to be precipitates from an original "chaotic paste," whatever that might be. But Werner did not travel far and generalized from the knowledge of the rocks in the limited area near his home. Some of his students went to Vesuvius and the ancient volcanoes in southern France, and decided that the supposed "precipitates" included real lava flows. The views of Werner seemed fantastic when men endeavored to fit their fuller knowledge into this theory. But there were proponents so ardent that they concluded that the trouble was with the explorers: "the recent custom among geologists of cutting up and subdividing seems to be upon the point of ruining the simplicity of the Wernerian arrangement." Correlations are not generally warranted on an assumption that a kind of rock appears but once in geologic time.

ANIMIKIE GROUP

Geologists about 1879 studied the region around Thunder Bay, which has the Indian name Animikie (A-ni-mi-kee'), along Lake Superior northeast of Port Arthur and Fort William, Ontario. The Mesabi Range and Duluth lie to the southwest (Fig. 3–1). They found hundreds of feet of somewhat metamorphosed sedimentary rocks in a sequence of quartzite, iron formation, and slate, to which they gave local names, just as were later to be given in the Mesabi Range after discovery of iron ore in the late eighties. The rocks of the three formations were put together into the Animikie Group. The formations of the Animikie Group are very like those bearing the iron ore of the Minnesota

Fig. 4–7. A table showing the correlation of the successions of rock units in several districts. Rocks having the same relative ages are arranged so that they lie along a horizontal line in the diagram; absence of deposits of a time is indicated by the vertical closely spaced lines. Diagonal lines show lack of information, as when rocks on the top of the succession may have been eroded away. Names in quotation marks are from outside the Lake Superior region.

Such a diagram is known as a correlation table.

Million years			Ontario–Minnesota		Wisconsin	Michigan
			Vermilion–Port Arthur	Mesabi Range	Gogebic	Marquette
0						
70		CENOZOIC				
		MESOZOIC		Cretaceous cgl.		
220						
		PALEOZOIC				
600						Cambrian Sandstone
	PROTOZOIC	Orogeny			sediments	
1000		Keweenawan System			gabbro and granite flows and sediments	
			Logan Sills	Duluth Gabbro		
			Osler Flows			
			Sibley Sediments		sediments	
1700	PRECAMBRIAN	Penokean Orogeny				granite
	ANIMIKIAN	"Huronian" System	Rove Slate	Virginia Slate	slate	sediments and volcanics
			Gunflint Iron f.	Biwabik Iron fm.	Ironwood Iron f.	Negaunee Iron f.
			quartzite	Pokegama Qu.	quartzite	slate quartzite
					dolomite quartzite	slate quartzite dolomite
2500		Algoman Orogeny	Vermilion Gr.	granite		
		Timiskamian System	Knife Lake seds.	Knife Lake seds.		
2700+		Saganagan Orogeny	granites		granite	intrusive rocks
		"Keewatinian" System	Soudan Iron fm. Ely Greenstone	schists	volcanic and sedim. schists	schists

49

ranges, are in the same order of succession, and are not far away. This leads to the judgment that the Pokegama Quartzite, Biwabik Formation, and Virginia Slate are Animikian; that is, they are about the same age as the formations of the Animikie Group. The quartzite at the base in each area lies on an unconformity—sand of a sea shore advancing over a land of metasedimentary rocks, altered lavas, and intrusive granites. The granite is called Algoman and the most ancient lavas, the Keewatin. As the sea spreads over land, beaches are not found everywhere at once, but only along the narrow band of the shore. When the sea advances, the shore sands are continuous with those laid earlier, but the earlier sands become seaward relative to the later—they come to lie beneath the sea as it advances (Fig. 4–3). Thus the apparently identical sands are continuous but except in a trend, not exactly synchronous, not of the same age. The sequence in the Mesabi area is about the same as the Animikie Group, but the base would be exactly the same age only if it formed along the shore line at the same time, and we do not now have a way of determining whether it did form at the precise time. Thus the correlation of the succession in the Mesabi Range with that west of Port Arthur implies something more than mere likeness; it carries a time connotation, but the correlated rock units are probably not exactly synchronous.

The geologic sequence near Thunder Bay has thousands of feet of sediments and lava flows overlying the Animikian and forming the rock of Thunder Cape (Fig. 4–4). Very similar lavas are exposed in Isle Royale, Michigan (Fig. 4–5), a National Park in Lake Superior to the south, as well as along the shore of the lake east of Duluth in Minnesota (Fig. 3–14). Similar lavas form the long peninsula, Keweenaw Point, jutting into Lake Superior on the Michigan shore. On Keweenaw Point, several miles of red sandstones, shales, and lavas have conglomerate layers containing metallic copper that was mined by the Indians before the arrival of the whites. Then it became the most important source of copper. The mines were carried down the dip of the beds for thousands of feet until the cost of removing the copper ore became so great that the mines were abandoned. The sediments and lavas have long been called the Keweenawan. They are so much like the sequences on Thunder Bay, along the shore east of Duluth, and on Isle Royale that all have been called Keweenawan, correlated on their strong lithic similarities and their positions in sequence above the iron formations and associated sediments in each area.

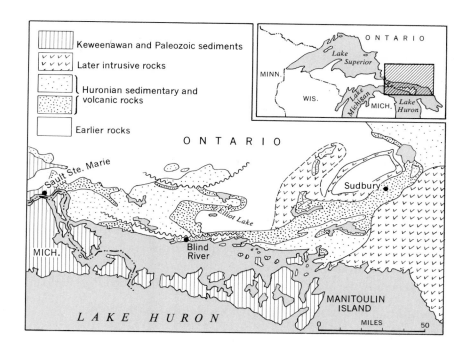

Fig. 4–8. Geologic map of the North Shore of Lake Huron east of Sault Sainte Marie, at the southeastern end of Lake Superior. Sudbury lies just south of the oval Sudbury Basin in the northeastern part of the map.

Legend:
- Keweenawan and Paleozoic sediments
- Later intrusive rocks
- Huronian sedimentary and volcanic rocks
- Earlier rocks

Fig. 4–9. Ripple marks on the surface of Huronian quartzite on the highway east of Sault Sainte Marie, Ontario. The beds are dipping southward, toward the observer, and are right side up—for the troughs of the ripples are concave, and the crests of the ripples form ridges. They seem like symmetrical, oscillation ripples formed beneath waves that moved on the surface of the sea and that produced to and fro movements of water along the top of the sediment. The younger overlying bed on the left has ripples oriented nearly at right angles to the older ones. Thus the directions are not constant; only a large number of observations might shed light on the dominant directions of the waves, or the currents responsible for the structures.

51

CLASSIFICATION OF TIME

The largest spans of geologic time are called *eras*. The rocks that we have been examining have few fossils, though fossil plants—calcareous algae—have been found in some places. The oldest rocks having abundant fossil animals are called the Cambrian from exposures in Wales—Roman Cambria—that we will discuss. In Minnesota, south of the west end of Lake Superior, Cambrian rocks unconformably overlie the Keweenawan sedimentary rocks; thus the Keweenawan, being older than the Cambrian, is *pre*-Cambrian. And as the Keweenawan are the youngest of the rocks along the lake, all of the rocks that we have been discussing are *pre*-Cambrian rocks. A century ago, Adam Sedgwick, a British geologist, defined the Cambrian Period of time as the earliest part of the Palaeozoic, or as we now spell it, the Paleozoic Era, and he called *pre*-Cambrian time the Protozoic Era. But generally the pre-Paleozoic time and rocks are called Precambrian.

CORRELATION TO SOUTH SHORE OF LAKE SUPERIOR

The principles of correlation can be illustrated further by considering other mining districts in Wisconsin and Michigan south of Lake Superior (Fig. 4–6); none of them is as productive of iron ore as the Mesabi Range. The Marquette district is just east of Keweenaw Point. The iron ore comes from a formation, the Negaunee, very like the Biwabik of the Mesabi Range, and it is underlain by quartzite and overlain by slate, each having a local name (Fig. 4–7). Similar sequences pertain in other ranges, such as in the Gogebic Range along the Wisconsin border to the west. The iron formation has very similar characters, and can be divided into lower cherty, lower slaty, upper cherty, and upper slaty like the Biwabik. Hence it is logical to correlate across Lake Superior and classify these iron formations and associated quartzite and slates as Animikian. But though they lie on granitic intrusive rocks in some places, other areas have an intervening sequence of quartzite, carbonate rock, and slate. If our preceding correlation is correct, there are deposits on the south shore without corresponding formations on the north (Fig. 4–7).

The slates overlying the Negaunee iron formation, correlated with the Virginia Slate of the north shore, have hundreds of feet of volcanic surficial rocks, tuffs, agglomerates, and flows, in the sections in the Marquette region. Although the sequences north and south of Lake Superior have similarities that are so great as to warrant correlation of some of the units, other units seem limited to the south shore, or are considerably different than on the north.

Let us now consider the region north of Lake Huron, extending

Fig. 4-10 (columnar section diagram)

Feet thick

SERIES: HURONIAN SERIES

Cobalt Group
- Lorrain Quartzite — 6000 ft
- Gowganda Formation — 800 ft

Bruce Group
- Serpent Formation — 900 ft
- Espanola Formation — 750 ft
- Mississagi Formation — 2600 ft
- Nordic Formation — 350 ft
- Matinenda Formation uranium-rich — 500 ft
- Granite

Fig. 4–11. Precambrian Huronian boulder-bearing graywacke, or lithified mudstone near Blind River, north of Lake Huron in Ontario (Ontario Department of Mines). This sedimentary rock, in the Gowganda Formation, has a matrix of fine-textured material containing suspended fragments of several kinds of igneous and metamorphic rocks, some of them well-rounded, others quite angular; the boulders are of rocks older and different than the matrix. The boulder beds are associated with laminated sediments having graded bedding. Such associations are commonly attributed to the slump of shallow-laid sediments into deeper water, transported and settling from turbid currents denser than the overlying water; deposits of such origin commonly have masses of sediment similar to the matrix. The Gowganda on the other hand has boulders of older and varied igneous rocks, and seems to lie in some places on striated and grooved underlying surfaces; hence the Gowganda is considered to be a consolidated glacial till, a tillite, with associated sediments that might be likened to varves. The tillite is evidence of the rigorous climates that prevailed a billion and a half years ago in Huronian time.

Fig. 4–10. Columnar section of the type Huronian Series in the uranium-mining district near Blind River on the North Shore of Lake Huron (after F. M. Roscoe). The Mississagi Quartzite contains uranium in sufficient quantity to be an important reserve of ore. Although the rocks have been thought of about the same age as the Animikian Series northwest of Lake Superior, comparison with those sections will reveal that close lithic similarities are few. The section also shows the manner in which rock-units are named and classified.

into western Quebec and Ontario (Fig. 4–8). This will give familiarity with another important region of Precambrian rocks and will permit further application of the principles of age determination and correlation. The region north of Lake Huron was first studied by Canadian geologists a century ago. They started their work along the north shore of Lake Huron, extending it northeastward (Fig. 4–7). The Sudbury region has mines that have produced about three-fourths of the world's nickel as well as large quantities in copper. To the north, the gold mines in a belt running eastward from Timmins, Ontario into western Quebec are among the most productive in North America.

THE TYPE HURONIAN

Along the north shore of Lake Huron, the first geologists found thousands of feet of sedimentary rocks that they called Huronian, quite appropriately. These have basal quartzite and conglomerate, car-

Fig. 4–12. *Preparation of a geologic map of a small area east of Sudbury, Ontario. A.* Air photograph of an area of about 40 square miles along the Wanapitei River. Only a limited part has rock exposures (Royal Canadian Air Force).

B. A large-scale geologic map of a small area near the base of the air photo, which is in part of the shaded area on the inset map. The limits of exposures are shown, and the field record of the principal rock types and their structures, originally plotted on an air photograph. The structure was interpreted as having two east-plunging anticlines and an intervening syncline southeast of the road. The numbers refer to several kinds of gneisses, without regard to their relative ages. The geologic map of the whole Sudbury Basin has been prepared and generalized from details of this sort, too small to be shown on a small scale map (after the Ontario Department of Mines).

bonate rocks, slates, and other quartzites (Fig. 4–9) in a lower sequence, the Bruce Group (Fig. 4–10) succeeded by regional unconformity by an upper sequence of sedimentary rocks, the Cobalt Group. The Mississagi Quartzite in the basal Bruce is the source of the uranium-bearing ores of the Blind River district, among the most productive on the continent. The lower part of the Cobalt has some peculiar rocks having pebbles of considerable variety suspended in a matrix of solidified mud and associated with laminated slates—the Gowganda Formation. The rock is thought to be glacial till that has been indurated, a tillite (Fig. 4–11). Northward in the gold belt, Cobalt tillite lies directly on granite such as underlies the Bruce Group on the north shore of Lake Huron. Though these Huronian rocks have some similarity to the Animikian of Lake Superior, some two hundred miles away, there is no iron formation of any consequence in this part of Ontario, and the sequence is far from identical.

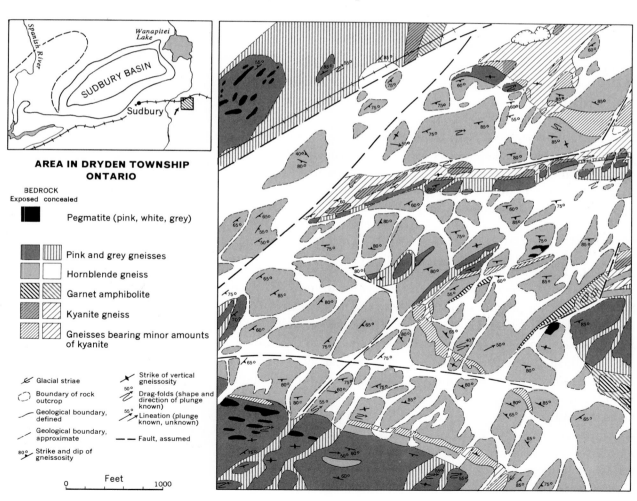

AREA IN DRYDEN TOWNSHIP ONTARIO

BEDROCK
Exposed concealed

Pegmatite (pink, white, grey)

Pink and grey gneisses

Hornblende gneiss

Garnet amphibolite

Kyanite gneiss

Gneisses bearing minor amounts of kyanite

Glacial striae

Boundary of rock outcrop

Geological boundary, defined

Geological boundary, approximate

Strike and dip of gneissosity

Strike of vertical gneissosity

Drag-folds (shape and direction of plunge known)

Lineation (plunge known, unknown)

Fault, assumed

Feet
0   1000

In the gold mining region, the granite beneath the Gowganda tillite, the Lorain Granite, intrudes sedimentary rocks (the Timiskaming Group) and volcanic rocks. The Timiskaming sediments in some localities lie unconformably on lava flows and tuffs, but in others volcanic rocks are interbedded in the sediments, as are the sediments with the lavas. The Timiskaming conglomerates have pebbles of granite, so there are granites older than the Timiskaming—and others that are younger than Timiskaming but older than the Huronian.

The Huronian rocks throughout the region are intruded by basic dikes and sills. It has become conventional to classify the rocks in the following divisions:

> "Keweenawan" intrusions
> Huronian sediments—the Bruce and Cobalt groups
> *erosion interval*
> Algoman intrusions, such as the Lorain Granite
> Timiskamian sediments and volcanics
> "Keewatin" lavas and "Laurentian" granites

Normally the "Laurentian" is placed as younger than the "Keewatin," but the relations are not clearly thus. The names in quotations are from distant places and involve correlation, the Keewatin from lavas in western Ontario, the Laurentian from mountains north of Ottawa and Montreal in Quebec, and Keweenawan from the south shore of Lake Superior. The term Killarnian has been given to granite interpreted as intrusive into the "Keweenawan," a disposition that is disputed. It is tempting to match this sequence with the one northwest of Lake Superior—to correlate the rocks and events.

CORRELATION WITH LAKE SUPERIOR SECTIONS

The first conclusion regarding correlation that we can reach with assurance is that the events which transpired at the western end of Lake Superior were not just the same as those to the north of Lake Huron, for the rocks in each region have their distinctive peculiarities. The Huronian sediments are enough like the Animikian and the associated pre-Animikian sediments south of Lake Superior that all of these rocks have been called Huronian in the past by geologists who have studied them. However, the similarities in lithologies are not great enough to make the correlation one of certainty rather than probability. The history in the western Lake Superior region is separable into three occurrences of sedimentation and volcanism interrupted by two "revolutions," times of folding and intrusion of granites preceding and following the deposition of the Knife Lake Group. A similar division of the record

Fig. 4–13. Mine and smelter producing from nickel-copper ore along the southeast margin of the Sudbury Basin at Falconbridge, Ontario (Ontario Department of Mines). In the distance to the west are other smelters near Sudbury, along the south margin of the basin at Coniston. The Nickel Eruptive forms low hills that surround the basin but are not apparent in a general aerial view of the region. The air photograph (Fig. 4–12) has the mine in the extreme upper left or northwest corner.

Fig. 4–14. Rhyolite breccia in the Onaping Formation southeast of the Sudbury Basin, northeast of Lake Huron in Ontario (Ontario Department of Mines). The rock has been interpreted as containing volcanic fragments that were deposited from hot gaseous clouds and explosive eruptions coming from volcanic vents that surrounded the Sudbury Basin; the associated rocks are consolidated tuffs.

north of Lake Huron can be made, as granites older than the Timiskaming are contained in pebbles in the Timiskamian Series and later granite that intruded the series. Thus we correlate the Timiskamian with the Knife Lake, and the Huronian with the Animikian on the basis of similar position in sequence and history; there is perhaps an unwarranted assumption that the events affecting the one region were sufficiently universal to have affected the other region at the same time.

CLASSIFICATION OF PRECAMBRIAN

Precambrian time is very long, as the multiplicity of events suggests, so it is advantageous to divide it into parts. Through the years, the terms Archeozoic and Proterozoic came into use for earlier and later Precambrian spans of time. Some included the Timiskamian rocks in the former, others in the latter. Not only is there doubtful basis for making a two-fold separation, but perhaps the Great Lakes region is not fully representative of Precambrian times. As stated, Protozoic has been applied to the whole of the pre-Paleozoic, and more recently the term Cryptozoic has come into use, but the time and rocks are commonly called Precambrian.

SUDBURY REGION INTERPRETED

Relations between rock units cannot always be determined with assurance. Exposures are not continuous—soil, glacial drift, lakes and swamps, and other objects frequently conceal the places that should show the critical relationships; or the contact between rock types, when it is seen, has been disturbed by faulting or other alteration (Fig. 4–12). Mines may reveal data obscured at the surface. But interpretations can vary greatly even in areas as thoroughly studied as the great nickel-copper mining district around Sudbury, Ontario, lying within the region north of Lake Huron. The most conspicuous surface feature there is a low range of hills (Fig. 4–13) of intrusive rock, the Sudbury Eruptive, that forms an ellipse about thirty miles in length in a northeastward trend and fifteen miles across. The "eruptive" is of dark, granular igneous rock varying in composition from basic-like basalt or gabbro to acidic syenite and granite. Within the interior of the ellipse, in the Sudbury Basin, are bedded sedimentary rocks—graywackes, and water-laid volcanic rocks such as tuffs and agglomerates (Fig. 4–14), the Whitewater Group, gently dipping, with a few simple folds (Fig. 4–15). On the outside of the ellipse are rocks of several kinds, some of them mineralized by the ore-bearing solutions where they approach the Sudbury Eruptive. The Wanapitei Formation consists of thousands of feet of well-bedded quartzite; the Stobie

Fig. 4–15. *Gently dipping late Precambrian argillite and arkose of the Whitewater Group in the middle of the Sudbury Basin, near Sudbury, Ontario. A.* The sediments are gently folded and little altered and are the youngest Precambrian sediments in the area; they are cut by faults and by later basaltic dikes (Ontario Department of Mines).

B. The air photo shows that they are displaced by faults as well as folded (Royal Canadian Air Force). The scale is about 1.5 miles to one inch.

A

B

Formation has pillow lavas, water-laid sediments, tuffs. Granites, such as the Creighton Granite, intrude these as well as the Sudbury Eruptive, and there are also other kinds of intrusives.

The relations among the many rocks have been interpreted in different ways. There are two dominant views. The conventional and long-prevalent interpretation has been that the Wanapitei Quartzite and some associated sedimentary rocks are Huronian, like the Bruce Group of the North Shore of Lake Huron. The Stobie Formation, lavas and associated volcanic rocks and sediments, is considered to be older and is compared to the Keewatin. The Whitewater Group is considered to have been laid on a great regional unconformity that cut across the Stobie, Wanapitei, and associated formations. It has been correlated with the slates above the Animikian iron formations or, by some geologists, with the Keweenawan Series. The Sudbury Eruptive is considered to have been intruded on a pre-Whitewater unconformity as a rather

Fig. 4–16. *The structure of the Sudbury Basin, northeast of Lake Huron in Ontario. A. In an alternative interpretation, the eruptive was intruded as a sheet along an unconformity between older rocks that now lie outside the basin and younger rocks that now lie within it. Subsequently, the eruptive was folded to form the present synclinal form. In this interpretation, the rocks within the basin are younger than those outside it, though both are antecedent to the intrusion. The view is eastward, and the section has north on the left. The numbers show the rock units in order of relative ages.*

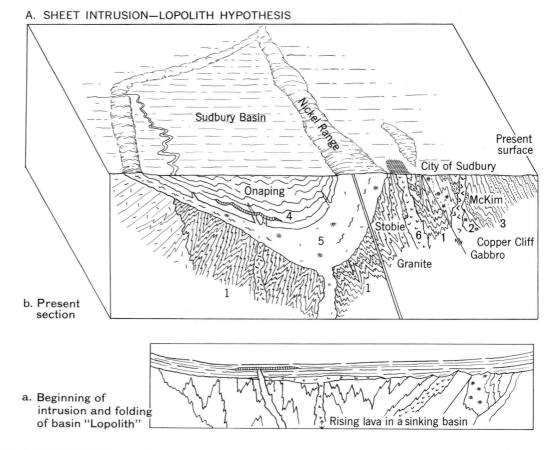

A. SHEET INTRUSION—LOPOLITH HYPOTHESIS

flat sill in which a lower, more basic phase has crystals that formed first in the melt and settled to the base of the fluid; and the upper, more siliceous phase has the later residues. The intrusive sill was then folded into the basin form, having the Whitewater Group in the center. Later granites such as the Creighton were intruded, as well as basic basaltic dikes. And the whole has been eroded to the present surface. Under this scheme (Fig. 4–16A), the Stobie, Wanapitei, Whitewater, and the Sudbury Eruptive were in that order, with erosion preceding the Whitewater and a time of folding followed the emplacement of the Sudbury Eruptive.

In another interpretation (Fig. 4–16B), the Stobie outside the basin is considered the same age as the agglomerates in the Whitewater (Fig. 4–14) on the inside. The Sudbury Eruptive invaded an elliptical fracture system that had earlier been the site of a ring of explosive volcanoes that produced the volcanic tuffs and fragmental rocks of the

B. The interpretation given is that the igneous rock of the Sudbury "eruptive" invaded the rocks of the region as a steeply dipping dike-like ring shown in the upper section, in the site of earlier explosive volcanoes shown in the lower block, the rocks within and outside the basin being older than the eruptive. North is on the left.

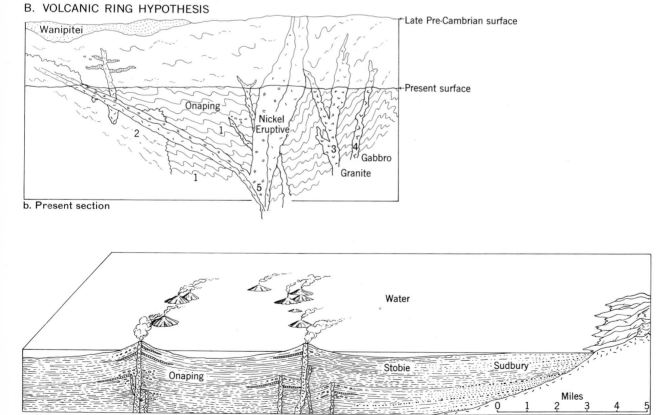

B. VOLCANIC RING HYPOTHESIS

b. Present section

a. Volcanism in early stages

Whitewater and Stobie, volcanoes of the Pelean type from which volcanic debris was transported by *nuée ardente* (hot gaseous clouds). There are rhyolite masses that are considered to be a dike complex, having an outer basic intrusion and an inner more silicic one. The Wanapitei Quartzite is part of an older floor and not Huronian, for the resemblance to the Mississagi is not great, and the Wanapitei lacks the uranium mineralization so distinctive of the Mississagi. In this interpretation, the sequence is Wanapitei and associated rocks, volcanoes and Whitewater-Stobie volcanic and sedimentary rocks, the Sudbury Eruptive, and granite and other later intrusives.

We may wonder why there should be any uncertainty about such relationships. There is little question about the composition of the rocks, for they have been surveyed for a century—and with particular intensity in some parts of the area because of their economic importance. However, in many localities, the relation of one type of rock to another is concealed or obscured by faults, fracture zones, and intrusives. For example, in the Sudbury Eruptive we have yet to learn whether it does in fact form the floor of a basin or continue dike-like into depth. Perhaps it will be possible to learn this from future drilling, or from the effect of the magnetic properties on instruments measuring magnetism on the surface, or the effect that the form of the surface of the eruptive has on the reflection of explosion waves generated on the surface and returned to seismographs. The relationships among some of the rocks in the Sudbury district are clearly shown on surface exposures and in mines; but there is a dispute among geologists as to whether the eruptive has continuous gradation from basic to acidic phases or consists of two separate intrusives. Some have maintained that the fragmental volcanic rocks have quartzite rather than rhyolite blocks.

Thus the determination of the history of such a region entails the making of many observations and interpreting the significance of some of the facts; solutions, at best, are but an approach to the truth. One promising method of analysis, the geochemical study of radioactive isotopes, offers the prospect of solving some of the problems through the dating of the rocks (as will be discussed in the next chapter). Some results of this sort are contributing to our understanding of the history at Sudbury. Another phenomenon, paleomagnetism, discussed in Chapter 19, depends on the fact that when a lava cools, it retains a record of the earth's magnetic field in its minerals. If the Sudbury Eruptive was a flow, later folded, the poles of remanent magnetism on the two limbs (Fig. 4–16*A*) should have changed from original parallelism to divergence. Recent studies do show divergence but not as much as postulated. Possibly some additional information can be derived from

Fig. 4–17. Precambrian sedimentary and volcanic rocks, dipping steeply westward and trending northward in the Belcher Islands, Northwest Territories, in the eastern part of Hudsons Bay (Royal Canadian Air Force). The succession, thousands of feet thick, has sedimentary rocks very similar to those of the Huronian of the Great Lakes region, and with some iron formation that has not been economical to mine.

63

aeromagnetic maps of the Sudbury area (such as have already been illustrated for the Mesabi Range) or from seismic refraction records (such as will be discussed in Chapters 18 and 22). These studies, like any field relationships, are constantly subject to critical review.

VALIDITY OF CORRELATIONS

We have been concerned with correlations in the region of the Upper Great Lakes, Huron and Superior. Within limited areas, rock characters seemed to warrant time correlation; the iron formation of the Mesabi is so like that near Thunder Bay and those to the south of Lake Superior that there is little hesitation in believing them synchronous. The Belcher Islands (Fig. 4–17), in eastern Hudson Bay and a belt in Labrador, Newfoundland and Quebec have similar iron formations, a thousand miles away; yet north of Lake Huron, only a few hundred miles away, such rock is not recognized. In the past as in the present, marine sediments of one kind were laid in some areas and belts, but other regions having differing conditions had different sediments. If there are difficulties in carrying correlations for a few hundred miles, they must increase in considering rocks as distant as those in the Grand Canyon, or on another continent. Yet, to understand the history of regions and continents, rocks and events must be compared in time among the many areas of exposure. The geologist endeavors to establish time correlations with recognition of the limitations that records and methods impose.

If rock types were universal at a particular time and appeared but once, identification of the rock would also classify it in time. In fact in the eighteenth and early nineteenth centuries, a widely accepted hypothesis was that all the rocks of one kind were universal and synchronous. But experience has shown that rocks of one kind within the limits of the means of identification are not widely distributed at one time, as can be seen in the present deposits; and rocks of one kind appear more than once in many successions. In nearby places, similar rocks are likely to be found at one time, particularly along belts and trends. To this degree, lithology is a means of correlation. The validity of correlations by lithologies and rock sequences is generally greatest within a small regional scope and tends to decrease with distance. Other methods must be recognized and applied in correlations among distant regions and continents. In Precambrian rocks physical methods must be used, for fossils are virtually absent. Later eras can use biological evidence as well.

5
Geochemistry
in Precambrian Correlations

Precambrian sedimentary rocks of the Belt Group along Going-to-the-Sun Highway, in Glacier National Park, northwestern Montana (A. Devaney).

5 *Geochemistry in Precambrian Correlations*

The events in each region of Precambrian rocks have been arranged in time by determining the physical relationships among the rocks that represent their history. And the events in one region have been correlated with those in another on the basis that they are relatively of the same age, but not of a known absolute age. In a somewhat analogous way, we might speak of the destruction of Pompeii as having been during the reign of the Roman Emperor Titus, without knowing that the date was A.D. 79. In the Western world, we use the method of classifying events in terms of the time before or after a standard date of reference thought by those who devised our calendar to be the beginning of the year following the birth of Christ. Thus two events that are of the same date are correlated in time. There are means of dating rocks that also make use of determinations of absolute time. Conditions at a mineral locality will illustrate the principles.

URANIUM-LEAD RATIOS

Wilberforce is a small village in southern Ontario, Canada, about one hundred miles northeast of Toronto. Nearby are several small mines and quarries, none of great economic importance, but places that are a delight to mineral collectors and to geochemists who are concerned with the dating of rocks (Fig. 5–1). A representative pit is in a pegmatite dike (Fig. 5–2) that has a breadth of a few tens of feet or less, dips quite steeply, and is bordered by metamorphic rocks, principally schists and marbles of the Grenville Formation (Fig. 5–3). The pegmatites are granite-like rocks with large crystals of quartz, feldspar, and muscovite mica, as well as many other minerals, including small crystals of the radioactive mineral uraninite, composed dominantly of oxides of uranium. Radium was first extracted from a form of the mineral known as pitchblende, and the application of the substance to the production of energy has brought the present atomic age. But Wilberforce uraninite is of scientific rather than economic importance. The analyses that must be made are carried forward in geochemical laboratories. The substances are present in extremely small amounts, and the apparatus must thus be very delicate to detect and measure the substances (Fig. 5–4).

Uranium is, of course, an element that disintegrates; in fact there are two forms of isotopes, each of which disintegrates. The uranium isotopes are forms that have the same number of protons in their nuclei, 92 being the atomic number, but that have differing numbers of neutrons there, so that their atomic weights are 235 and 238. The latter is

Fig. 5–1. *The abandoned Wilberforce uraninite mine in Ontario. A.* This narrow pit in a dike of pegmatite was the source of the oxide of uranium, uraninite, from which radium was extracted in the years after the first world war. This uraninite was the subject of study to evaluate the uranium-lead isotope method of determining the ages of rocks; that in the pegmatite was found to be about 1,000,000,000 years old. The pit further illustrates the tremendous change in the interest in uranium, for it was considered to be a significant possible source of radium at a time when the use of uranium as a source of atomic energy had not been considered.

B. The deposit is a band of coarse crystalline feldspar, fluorite, calcite, and other minerals within banded gneisses. The pegmatite is thought to be formed from the precipitation of the minerals from fluids intruded at high temperature, though it is possible that the band is a replacement of an original limestone bed in sediments that have become matamorphosed to gneisses. In the detail, the light rock is calcite, in which there are darker crystals of hornblende, tourmaline and magnetite, some light-colored feldspar and a few crystals of uraninite and other minerals.

Fig. 5–2. Section through the uraninite-bearing pegmatite (shown with a pattern with black bars) at Wilberforce, Ontario, that yielded radioactive minerals that were useful in calculating that the intrusion formed somewhat more than one billion years ago (after R. B. Rowe).

A

B

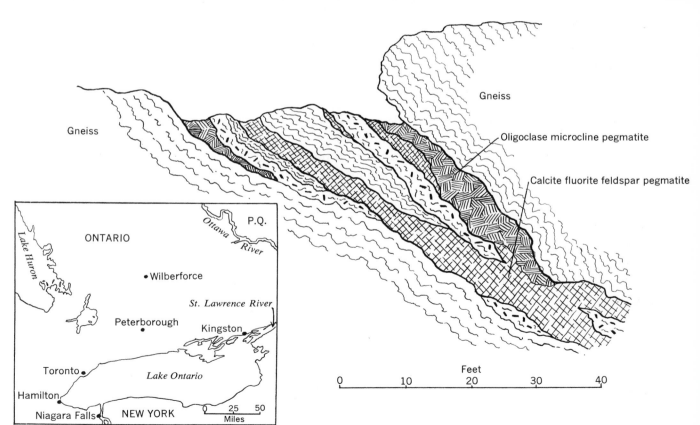

Gneiss

Gneiss

Oligoclase microcline pegmatite

Calcite fluorite feldspar pegmatite

ONTARIO

Lake Huron

Ottawa River

P.Q.

• Wilberforce

St. Lawrence River

• Peterborough

Kingston •

• Toronto

Lake Ontario

Hamilton •

Niagara Falls •

NEW YORK

25   50

Miles

Feet

0   10   20   30   40

about 140 times as abundant as the former at the present time. Uranium 238 has been found to disintegrate through the loss of eight alpha particles (helium atoms, each of atomic weight 4) and six beta particles (electrons) to an isotope of lead, atomic number 82, having an atomic weight of 206 (Fig. 5–5). Now consider a principle. If a crystal of uraninite formed, was composed of uranium only of the isotope 238, and lacked any lead, the disintegration of this primary uranium would produce an increasing amount of lead through time. And if the rate at which the uranium disintegrated were known, and known to be constant, the ratio of lead to uranium would be a measure of the time that elapsed since the formation of the uraninite. It has been determined that if we had a mass of uraninite at a given time, one-half of the atoms will have decayed to lead in about four and one-half billion years or 4500 million years. Hence a method is available that should give the time since the formation of the uraninite.

The uraninite from Wilberforce also contains uranium 235 and the disintegration product lead 207 that comes from its decay, the parent uranium having a half-life of about 700 million or .7 billion years. And it has thorium of atomic weight 232 that decays to lead of atomic weight 208, with half-life of about 14 billion years. Geochemists are able to differentiate the several isotopes by means of such instruments as the mass spectrograph. In addition to the three isotopes of lead that are produced by disintegration from the two isotopes of uranium and one of thorium, there is another lead isotope, atomic weight 204, that is primary lead. The mass spectrograph (Fig. 5–6) operates on the principle that when a beam of atoms is deflected in a magnetic field, the atoms of least atomic weight or mass will be deflected most easily and farthest, those of greatest mass, the least. So it is possible to determine the proportions among the several isotopes and compute the probable age of each parent substance. Many determinations have been made on the uraninite in the Wilberforce district, and most measurements are very nearly one billion and fifty million years or 1.05 billion years.

One of the most useful observations has been that the proportions of uranium 235 and 238 vary with time. Thus the ratio between the two will be the same for all minerals formed one billion years ago, but the present ratio will have changed from the original one because the uranium 235 will have decayed to lead more rapidly than the uranium 238, their half lives being 0.7 and 4.5 billion years, respectively. The use of this ratio has the advantage that it would remain constant regardless of the loss of some of the parents or decay products, such as lead, during the time of decay.

Fig. 5–3. Precambrian Grenville Marble near Tweed, southeastern Ontario. The contorted marble is associated with other metamorphosed sedimentary rocks that are widely distributed in southeastern Ontario and the Adirondack region of New York in the Grenville Precambrian province, in which principal intrusions are a little more than one billion years old. The province is separated from that in which Huronian rocks are found, as at Sudbury, by a sharp line of contrast called the Grenville Front, perhaps a principal zone of faulting.

Fig. 5–4. Spectrometer used for argon-40 determinations relating to the determination of age of potassium-bearing rocks. The equipment is a gas mass spectrometer in the Geochemistry Laboratory, Lamont Geological Observatory, Columbia University.

The rocks at Wilberforce contain other elements that are radioactive. Studies of their disintegration should give further means of verifying the validity of the lead ages. Potassium (K) has an atomic number of 19, but has several isotopes—the one with atomic weight 39 is preponderant. About 1 per cent of potassium is commonly of the radioactive isotope with atomic weight 40, which disintegrates to argon 40 and calcium 40 (Fig. 5–7). A number of very sensitive techniques permit the geochemist to determine the ratios of these substances, and the half-life of about 1.3 billion years enables the scientist to determine the time of crystallization of the mineral. We will refer to this method later in the discussion of earth history. The Wilberforce pegmatites contain mica that has given ages comparable to those from the lead methods. Moreover, the element rubidium (Rb) has two isotopes of atomic weight 85 and 87, with atomic number of 17, the heavier isotope forming about a quarter of the present element. Rubidium 87 is radioactive, changing to strontium 87 with a half-life of some 6 billion years. This promises to be a further useful means of determining the age of ancient rocks.

A basic assumption in the determinations of ages by means of radioactive element decay is that the disintegration has been at a constant rate through time. Within the limits of laboratory experiment, it does not seem that rates are changed through introduction of varying temperatures and pressures. And within the very limited spans of time available in the laboratory, the rate seems constant. Further support for the assumption's truth lies in the consistent ages determined by several methods from several radioactive elements or isotopes in single minerals formed at one time.

PLEOCHROIC HALOES

The disintegration of elements such as uranium and thorium produces alpha particles, helium atoms. These are projected from the site of the disintegrating element or isotope into the surrounding mineral, forming a spheroid of damaged mineral that in section is a small "halo" (Fig. 5–8), a discolored or less translucent area in cross section. At one time, ages of rocks and minerals were crudely determined by "helium ratios," based on the fact that the ratio of helium to parent element will increase with time, just as do the lead isotopes that are other disintegration products. The method proved subject to error because of the likely escape of the gas, helium, from the parent mineral.

The minerals around Wilberforce have been studied by many methods over many years. Experience shows that the results are consistent for most of the analyses, but as can be anticipated, some of the ratios are

Fig. 5–5. Stages in the disintegration of uranium isotope 238 to lead isotope 206 through the emission of alpha particles, helium atoms, and beta particles. The particles are emitted with differing energies from stage to stage, and the half-life, or span of time in which one-half of each distintegrates, differs for each stage.

Fig. 5–6. Diagram of the principle of the mass spectograph, an instrument that distinguishes among the isotopes of elements according to their masses. Atoms of gas are admitted on the left into a virtual vacuum in the tube. They are ionized at the ion source and accelerated by electrodes into a beam of ionized atoms. The magnetic field deflects the ions, those of the greatest mass the least; the amount of deflection can also be altered by changing the strength of the magnetic field. A collecting mechanism on the right determines the proportions of the atoms of differing mass in the ion stream.

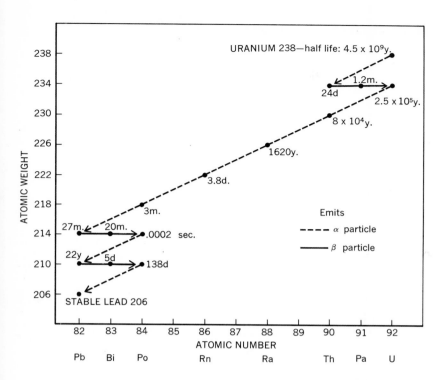

URANIUM 238—half life: 4.5 x 10⁹y.

STABLE LEAD 206

Emits
- - - - α particle
———— β particle

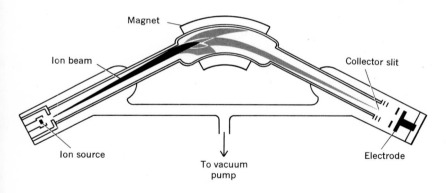

Magnet

Ion beam

Collector slit

Ion source

To vacuum pump

Electrode

Fig. 5–7 (right). Disintegration of potassium 40 (after J. L. Kulp). Potassium (K), of atomic number 19, has three isotopes: one of atomic weight 39, forming 93 per cent of common potassium; one of 41, nearly 7 per cent; and the rare potassium 40, forming only .012 per cent of common potassium. K^{40} is unstable, with a half-life of 1.35 billion years, and disintegrates principally to calcium (Ca^{40}). The disintegration to Ca^{40} atomic weight 20 is by beta emission, loss of a nuclear electron. The geochemist is more interested in some 12 per cent of K^{40} that disintegrates to argon (A^{40}). The latter is by capture of an electron in an outer shell (k-shell) of the atom of K^{40}, and prompt gain of an electron in the nucleus with gamma-ray emission so as to lower the atomic number to that of argon, atomic weight 18. Age determination depends on knowledge of the rate of disintegration and of proportions of disintegration to the respective product. With this information, the ratio of argon 40 to potassium 40 gives a measure of the time involved since the potassium-bearing mineral crystallized.

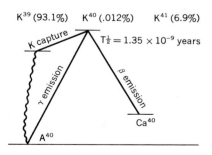

K^{39} (93.1%) K^{40} (.012%) K^{41} (6.9%)

K capture

γ emission

β emission

$T_{\frac{1}{2}} = 1.35 \times 10^{-9}$ years

A^{40}

Ca^{40}

variants because of such factors as alteration and loss of parent elements or decay products and analytical errors. The knowledge that the age is a little more than one billion years simply tells us that the metasediments of the Grenville (Figs. 5–3, 5–9) are older than that. Metamorphic and igneous rocks along the Atlantic Coast (Figs. 2–4, 5–10) were thought to be Precambrian, but many are known to be younger. Until one has the absolute ages of the rocks that were classified in order of succession in the region north of Lake Huron or west of Lake Superior, the age has little geologic importance. Information on these rocks is being acquired rapidly, and soon it will be possible to judge the relative age and correlation of the Wilberforce pegmatites with events in the other regions.

AGES OF ROCKS ALONG UPPER GREAT LAKES

The determinations of the ages of rocks along the north shore of Lakes Superior and Huron and in the Sudbury district on the disintegration of isotopes of other series than that of uranium have been reported in the past few years, and are still in progress. The ages are based on studies of the ratios of potassium 40 to argon 40, and of strontium 87 to rubidium 87. They suggest that the Animikian rocks in the Minnesota-Ontario sections northwest of Lake Superior are of about the same age as the rocks of the type Huronian; the base of the Huronian has the uranium-bearing Mississagi Quartzite, so it is older than the calculated age of that rock. The age of the Animikian is controlled by knowledge that it is older than dated intruding granite. The definition of ages is such that the Huronian might be younger than the Animikian, even though of the same approximate age.

The Wanapitei and other sediments of the "Sudburian Series" are older than the Huronian as had been surmised from the field relations. The Ely and Soudan formations of Minnesota are quite old, as old as the Sudburian if not older. So it may be that the Precambrian has a succession of sedimentary series such as the Sudburian and Huronian or Animikian, separated by orogenies having the granites and other intrusive rocks that were placed in sequence in their respective stratigraphic relations. All the sediments have been called Huronian at one time or other in the past because of their lithic similarities to rocks of the typical Huronian of the north shore of lake Huron; see Table I.

The problems of correlation with distant places can be shown by discussing the Precambrian rocks of the Grand Canyon region, Arizona, and those of the Glacier Park region, Montana.

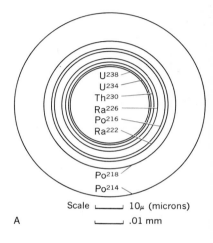

U^{238}
U^{234}
Th^{230}
Ra^{226}
Po^{216}
Ra^{222}
Po^{218}
Po^{214}

Scale ⊢——⊣ 10μ (microns)
⊢——⊣ .01 mm

A

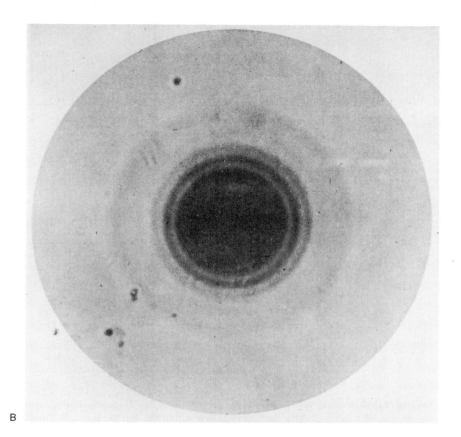

Fig. 5–8. *Pleochroic haloes.* In the disintegration of radioactive elements such as uranium, a minute crystal of the mineral uraninite in biotite releases alpha particles with energies that penetrate the biotite for varying distances. *A. Idealized section in biotite.* The successive rings are those in the disintegration of U^{238}, U^{234}, Th^{230}, Ra^{226}, Po^{216}, Ra^{222}, Po^{218}, and Po^{214}; see Fig. 5–5.

B. Photograph of haloes in a section of biotite. Inasmuch as the rings are sections of spherical surfaces, their relative distances from the center of the section will vary depending on whether the section passes directly through the central emitting source of radioactivity (D. E. Kerr-Lawson).

B

Fig. 5–9. Precambrian Hastings Conglomerate near Kaladar, along the Toronto-Ottawa Highway in southeastern Ontario. The conglomerate has pebbles of quartzite and of igneous rocks that have been compressed and elongated by tectonic forces; the formation is associated with the Grenville marbles of uncertain age, but older than the Wilberforce intrusives which cut through them.

TABLE I. SUMMARY OF PRECAMBRIAN ROCKS IN THE GREAT LAKES REGION

		Million Years Ago
Beginning of Paleozoic Era		600

———————————————— Lipalian interval[1] ————————————————

Precambrian		
Upper Keweenawan System		
	"Grenville" ("Wilberforce") orogeny[2]	1050
Lower Keweenawan System		
	Penokean orogeny[3]	1700
Huronian-Animikian System		
	Algoman orogeny	2400
Timiskamian System		
	"Saganagan" orogeny[4]	(?)2600
Keewatinian[5] System		

[1] Although Paleozoic rocks very commonly lie unconformably on Precambrian rocks, the latter are generally deformed as a result of orogenies within the Precambrian rather than at the close of the era. The uppermost Keweenawan in Wisconsin and Minnesota seems to have been warped but not severely folded prior to Cambrian time. The time wherein the erosion surface beneath the Cambrian rocks formed has been called the Lipalian interval. It is Precambrian—in the sense that it lies beneath Cambrian rocks; but since these are not the oldest possible Cambrian rocks, and since the erosion continued into Cambrian time, it is not entirely in the Precambrian Era.

[2] The deformation of Grenville rocks preceded the intrusion of the Wilberforce pegmatite in southeastern Ontario. The orogeny is known to be severe and extensive in a region south and east of that containing Huronian rocks. The name Grenville, though widely used, is not very apt, for it is the orogeny that deformed Grenville sediments, which are of undetermined older age.

[3] We have read that Animikian rocks were deformed prior to Keweenawan deposition. In areas in Michigan and Wisconsin, intrusions through Animikian rocks that are unconformably succeeded by Keweenawan rocks have been dated as about 1700 million years old, and important granites in southwestern Minnesota are of this age. The name is from an iron-formation-bearing mountain range in northern Michigan, the Penokee Range. The orogeny has also been called Hudsonian.

[4] It is also known as the Kenoran orogeny, of about 2500 million years.

[5] Keewatin has long been applied to the ancient volcanic rocks at this stratigraphic position. The name Ontarian has been used and would be better, except that it has been applied for more than a century to a system of younger rocks that include those exposed at Niagara Falls.

The terminology in this table is partially informal.

Fig. 5–11. Grand Canyon of the Colorado River in northwestern Arizona: the view is northward from the South Rim. The river is flowing in Precambrian sedimentary rocks that are overlain unconformably by the Cambrian sandstone that forms the prominent platform below the middle of the view; late Paleozoic Permian sediments form the plateau at the skyline (J. Muench).

Fig. 5–10. Fordham Gneiss in The Bronx, New York City; an example of a metamorphic rock that has been thought to be Precambrian.

A belt of metamorphic rocks extends along the Atlantic Coast from Newfoundland to Georgia (Fig. 2–4); the cities of New York, Philadelphia, and Baltimore are built largely on these rocks, and they form such mountains as the White Mountains in New Hampshire. Because they are crystalline metamorphic rocks, they were thought to be very old and generally were called Precambrian. In a few cases, as in New Hampshire, recognizable fossils found in the gneisses and schists showed them to be extremely metamorphosed sedimentary rocks much younger than the Precambrian. In other instances, the determination of the age of minerals by radioactive isotope analyses has shown that the alteration took place during Paleozoic time, though such ages are but the youngest possible limit to the age of the deposited rocks. In some places, age determinations from the minerals do give Precambrian age. Research has shown that degree of metamorphism of a rock is not a direct evidence of its age, and that the metamorphic rocks along the Atlantic Coast include Precambrian and Paleozoic rocks that can be dated and distinguished occasionally by radiometric and faunal evidence.

In New York City, the Fordham Gneiss is the oldest formation. It is succeeded by Inwood Marble and Manhattan Schist, the latter to be seen in the parks in the city. They are intruded by granitic rocks, some of them dikes that are quite undeformed and are dated as Paleozoic rather than Precambrian.

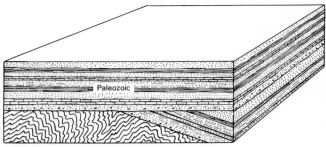

(5) Paleozoic: after deposition of sedimentary rock sequence

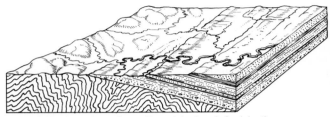

(4) Early Paleozoic: erosion surface on Precambrian in early Cambrian time

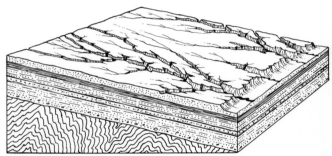

(3) Precambrian: tilting and beginning of erosion of Grand Canyon Group

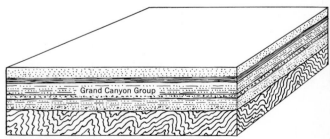

(2) Precambrian: after deposition of the Grand Canyon Group

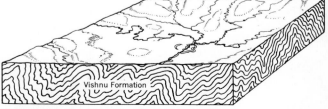

(1) Precambrian: after deformation of the Vishnu and erosion

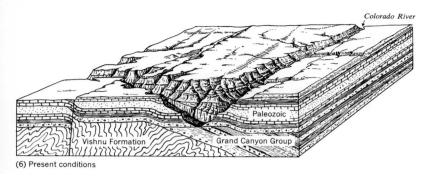

Colorado River

Paleozoic

Vishnu Formation — Grand Canyon Group

(6) Present conditions

Fig. 5–12. A succession of block diagrams illustrating the history of the Grand Canyon of the Colorado River in Arizona (after Douglas Johnson).

CORRELATION OF ARIZONA SECTIONS

A traveler standing on the brink of the Grand Canyon of the Colorado River looks across a great stream valley at an orderly succession of bedded rocks miles away on the far wall of the great chasm (Fig. 5–11). As we follow the layered rock down we come to a formation of quartz sandstone that is classified as Cambrian for reasons that will be discussed in the next chapter. If we accept the placement of the sandstone as Cambrian, the rocks below it are pre-Cambrian on the principle of superposition—younger rocks overlie older rocks. There is an erosion surface below the Cambrian sandstones. From Museum Point on the south rim, one can see below the flat Cambrian sediments overlying other bedded sediments that are tilted eastward (Fig. 5–12). This succession, the Grand Canyon Group, forms a wedge of rocks that comes to a point westward, for it is sharply limited by an unconformity at its base. The rocks of the Granite Gorge of the Colorado—actually not of granite—lie below the Grand Canyon Group on the right (or east) and below the Cambrian beyond the end of their wedge on the left (or west).

The Grand Canyon Group consists of thousands of feet of sediments —quartz sandstones, shales, and thick carbonate formations, with a lava flow at one level. The base is beautifully exposed, a rather flat surface cutting across the deformed rocks of the underlying Vishnu Formation of the Granite Gorge. The Vishnu is made largely of micaceous schist, such as might form by metamorphism of argillaceous sediment, but it has a few quartzite beds, and there are intrusive igneous rocks, granite dikes cutting through the steeply dipping metasediments. We can readily reconstruct the history. Miles of sediment were laid on a subsiding floor of undetermined character, the sediments of the Vishnu Formation. These were invaded by plutonic rocks, metamorphosed and folded in a great period of mountain-making. The mountains were eroded to a rather level surface, a peneplane, which came to be buried beneath the sediments of another period of sedimentation on a subsiding floor, resulting in the Grand Canyon Group. The sediments were then tilted prior to their erosion by the close of the Protozoic or Precambrian Era.

Consider the possible classification of these greater rock units, the Vishnu and the Grand Canyon. The Vishnu is perhaps most like the Timiskamian rocks of the Great Lakes region, and the Grand Canyon Group is most similar to the typical Huronian of the north shore of Lake Huron. The two series in the Grand Canyon have a strong likeness in their lithologies to the Timiskamian and Huronian; they have a high degree of similarity. There is further likeness in that they are separated by an unconformity. In Ontario there is a similar great unconformity between the Timiskamian and Huronian, representing the Algoman Revolution. Before putting too much confidence in the similarities as an indication of age, let us outline the implications that are basic to its postulates.

Classification of the Vishnu as Timiskamian and the Grand Canyon as Huronian assumes that there is time significance in the similarity between lithologies in regions as far distant as Lake Huron and Arizona. It connotes acceptance of a principle that rocks of the same types are distributed extensively at one time and are distinctive of that time and no other. We have already seen that such an assumption is tenuous. The great orogeny and erosion represented by the unconformity between the two would be classed as Algoman, if the correlation could be accepted. This implies that there was synchronous orogeny in distant regions. The alternatives might be that although conditions of deposition were similar in Arizona in the time of deposition of the Grand Canyon Group to those north of Lake Huron when the Huronian sediments were laid, these events may have transpired at different times. Though there was a great thickness of muddy and sandy sediment laid in the Timiskamian time in Ontario, the deposition of similar rocks in Arizona need not have been at the same time. It is surely hard to conceive of the latter having been laid universally, for some part of the continent would need to be a source while other parts would have to be subsiding to receive the detritus. The idea that the earth has shells of differing kinds was a theory widely held in the late eighteenth century by the followers of the German mineralogist Werner, but it was later disputed.

Fig. 5 – 13. Diagrammatic section through the Precambrian rocks in central Arizona, the interrelations of the several units giving their respective relative ages. The later intrusions in the region seem to be about one and one-half billion years old from the study of the potassium-argon and rubidium-strontium isotope ratios in a few samples.

Fig. 5–14. Precambrian sedimentary rocks of the Belt Group near Logan Pass in Glacier National Park, northwestern Montana; the limestones and argillites have been very little metamorphosed (A. Devaney). The Belt sediments are in a crustal block that was thrust eastward over the Cretaceous rocks of the Great Plains on the Chief Mountain fault in the eastern face of the mountains; thus the thrust formed in the post-Cretaceous Laramian Orogeny.

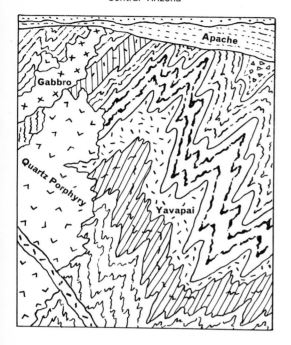

Central Arizona

Apache

Gabbro

Quartz Porphyry

Yavapai

In central Arizona, south of the Grand Canyon, rocks very like the Vishnu schists and intruding plutonic rocks are overlain unconformably by thousands of feet of quartzites, lithified quartz sandstone (Fig. 5–13). They were intruded by granitic rocks, which in turn are unconformably succeeded by other sediments older than Cambrian. In this area there is record of two orogenies within the Protozoic, which might be correlated with the Saganagan and Algoman—but if this is a correct assumption, the Vishnu would need to be older than the Timiskamian that it so closely resembled. The principal later granites have been given absolute ages of about one and a half billion years on the basis of a limited number of potassium-argon and rubidium-strontium age determinations; further measurements will test the validity of these ages. There would be little to suggest that the thousands of feet of sediments in the Grand Canyon are of any particular age, so correlation depends ultimately on geochemical methods.

CLASSIFICATION OF MONTANA PRECAMBRIAN

Another of the National Parks, Glacier Park in Montana, has a sequence of thousands of feet of sedimentary rocks much like the Grand Canyon Group (Fig. 5–14). Similar rocks are widely distributed through western Montana and eastern Idaho, continuing northward in the Selkirk Range of British Columbia. The sequence has been named the Belt Group from the Little Belt Mountains in central Montana; in the region of Glacier Park, the Belt has as much as six miles of sedimentary rocks, considerably of limestone, with a few volcanic flows. The limestones in Glacier Park have the finest display of Precambrian algae known (Fig. 5–15), but similar ones are found in the Huronian rocks of Michigan and Ontario (Fig. 5–16) and the Grand Canyon Group. The Belt lies with great unconformity on metasedimentary and intrusive rocks in western Montana. Because of the slight deformation and metamorphism of the Belt rocks, they have been considered generally to be very young Precambrian, perhaps Keweenawan, though their lithologies more closely resemble Animikian and Huronian rocks. Similar rocks in the mountains of northern Idaho have been correlated with the typical Belt. There have been a few age determinations made on intrusives into these latter "Belt" sediments in the Coeur d'Alene mining district, Idaho, that give an age of more than 1.5 billion years. On the other hand, potassium-argon ages on shales in Montana are about 0.7 million years. Thus if the Belt Group is of latest Precambrian age—the beginning of the Cambrian seems to have been about 600 million years ago—the age of the Idaho intrusives have been erroneously

A

Fig. 5–15. *Algal beds in the Precambrian rocks of Glacier National Park, Montana. A. Collenia; B. Conophyton* (R. Rezhak, United States Geological Survey).

The algae are the simplest plants, in which many cells are not differentiated to perform distinct functions. There are a great many kinds living today in many environments, such as in the soil and in fresh and salt water. They are the green scum on a pond or the seaweeds of the sea. Some of these organisms cause calcium carbonate to precipitate around them, forming mound structures, *stromatolites*, known from Precambrian rocks in many places. Though the rocks are not clearly organic, they can be attributed to an organic cause. The best known of these calcareous algae in the Precambrian in North America are those in Glacier National Park, where they are exposed along the highway just east of Logan Pass. They grew in such abundance as to form significant parts of limestones in these Precambrian rocks of the Belt Group. The structures are thought to have been formed in water, doubtfully marine, probably by blue-green algae (Cyanophyta).

B

made, or the correlation by physical methods with the typical Belt Group of Montana is in error, or the latter ages are too low.

Physical stratigraphy permits determination of the relative ages of rocks in a single area, and gives probable correlations within limited regions. Correlations between distant places are tenuous by such methods. The geochemical analysis of radioactive isotopes of elements such as uranium, thorium, potassium and rubidium supplement the physical methods, and give a better means of comparing the events in different parts of continents and of the world.

THE EARLIEST ROCKS

A matter of interest is the absolute age of the earliest rocks. It would be most fortuitous if the geologist should discover the oldest rock, and within each region he would search for it by the stratigraphic methods that have been discussed. Having determined which of the rocks is relatively the oldest, the age of minerals within the rock could be computed from any one of the radioactive elements. But if the rock were an intrusive, it would be younger than the rocks intruded. If it were a metamorphic rock, the age might be that of alteration rather than of the deposition of the primary sediment. The oldest rocks known on the earth are stony meteorites, which have given ages of 4500 to 5000 million years —4.5 to 5 billion years—by ratios of lead isotopes and by potassium-argon ratios. In North America pegmatites in southeastern Manitoba have uraninite that has been analyzed for uranium-lead and thorium-lead ages, lepidolite mica that has given rubidium-strontium ages, and feldspar that has given potassium-argon ages. There is little difference among the age determinations, which average about 2700 million years. Rocks of similar age have been determined in southern Montana and in the Black Hills of western South Dakota. Age determinations by rubidium-strontium and potassium-argon ratios of sedimentary rocks in the Cambrian range to nearly 600 million years.

Thus if the oldest rocks are about 2700 million years, and the oldest of the Cambrian rocks is of about 600 million years, the rocks of the Precambrian span about 2100 million years, more than three times the length of all post-Protozoic or post-Precambrian time. And if the earth has been in existence as long as the stony meteorites that reach it, the rocks record only a little more than half of the time since the earth was formed.

The divisions of time in the post-Protozoic eras are universally recognized, and the classification of eras and periods to be used in the succeeding chapters is given in Table II.

Fig. 5–16. Calcareous algae from the Precambrian Animikian rocks northwest of Lake Superior in Minnesota and Ontario. The fossils are from the Gunflint and Biwabik iron formations, some 1.7 billion years old; they are the silicified remains of calcareous algae that formed mounds or reefs on the floor of the Huronian seas. The specimens are (A) from west of Port Arthur, Ontario (W. W. Moorhouse and F. W. Beales) and (B) from the Mesabi Range north of Duluth, Minnesota (specimen from R. Heller); each is somewhat reduced in scale.

B

TABLE II. CLASSIFICATION OF PALEOZOIC AND LATER ERAS

Youngest	*Million Years to Beginning*[1]	
Cenozoic Era	65	
Pleistocene period		1
Tertiary Period		65
Mesozoic Era	230	
Cretaceous Period		135
Jurassic Period		180
Triassic Period		230
Paleozoic Era	600	
Permian Period		280
Carboniferous Period[2]		345
Devonian Period		405
Silurian Period		425
Ordovician Period		500
Cambrian Period		600
Precambrian (Protozoic[3] Era)		

Oldest

[1] Dates after Kulp, 1961.
[2] In North America the rocks of the Carboniferous System are placed in two subsystems,
Mississippian (older) and Pennsylvanian, that are generally treated as systems.
[3] Also called Cryptozoic.

6
Classification and Faunal Correlation in the Cambrian System

Middle Cambrian shale along a street in St. John, New Brunswick.

6 Classification and Faunal Correlation in the Cambrian System

In the summer of 1831, the Reverend Adam Sedgwick (1785–1873), professor of geology in Cambridge University, England, arrived in Bangor, North Wales (Fig. 6–1), with a young assistant and recent graduate, Charles Darwin (1809–1882), who in late December of the same year was to begin his great voyage on the *Beagle*. Sedgwick entered a region where no geologist had carried on significant studies, to determine the structure and the succession of lithologies. These were the transition rocks which had been thought indeterminate by his predecessors. In the course of a few years, Sedgwick was able to arrange the sediments in stratigraphic order from the oldest exposed beds in the core of the great anticlinal Harlech Dome to the rocks on the borders of North Wales (Fig. 6–2). He thought it appropriate that the succession of strata be called the Cambrian System, inasmuch as the Romans had called the region Cambria.

ORIGIN OF THE NAMES OF SYSTEMS

In the same year, Roderick Murchison (1792–1871) started an excursion into central Wales. The base of the known succession of sedimentary rocks was the Old Red Sandstone, a conglomeratic formation that had been made famous by the discovery that it contained well-preserved fish remains, popularized in articles by the Scotch quarryman and stone-mason, Hugh Miller (1802–1856). So Murchison followed a logical course in beginning his studies at the base of that formation along the river Wye and assembling information on successively older beds. In 1835, having built an orderly succession and recorded not only the kinds of rocks but their contained fossils, he proposed that the name Silurian System be given to the sequence, the Silures having been ancient inhabitants there. Subsequently, he prepared a large volume with full description of the rocks and fossils, maps of the areal geology and sections of the structure.

Sedgwick and Murchison had for years been the closest of friends and they assumed that their labors had been directed toward different parts of the rocks of Wales. The uppermost rocks in Sedgwick's Cambrian were exposed in the Berwyn Hills, in eastern Wales near the borders of England. He and Murchison visited the area, and at the time Sedgwick understood that Murchison considered the rocks to be older than those contained in his Silurian System. But when Murchison returned to London and examined his collections, he reached the

Fig. 6–1. Lower Cambrian slates in the quarries near Bangor in North Wales, where Adam Sedgwick began his studies of the Cambrian System in 1831 (Her Majesty's Geological Survey).

Fig. 6–2. Geologic map of Wales, below the center of the map, and surrounding parts of the British Isles. The boundary between Scotland and England, not shown, extends northeastward from the bay near to the Isle of Man in the northern Irish Sea. The Cambrian System was named for Wales or Roman Cambria, and the Ordovician and Silurian gained names from early inhabitants of the area. Though the Cambrian was named by Sedgwick after work in North Wales, the base is not exposed there; but it can be seen in Pembrokeshire on one of the southwest points of land. When Murchison studied the Silurian System, he started from the base of the previously known Old Red Sandstone along the River Wye in eastern Wales.

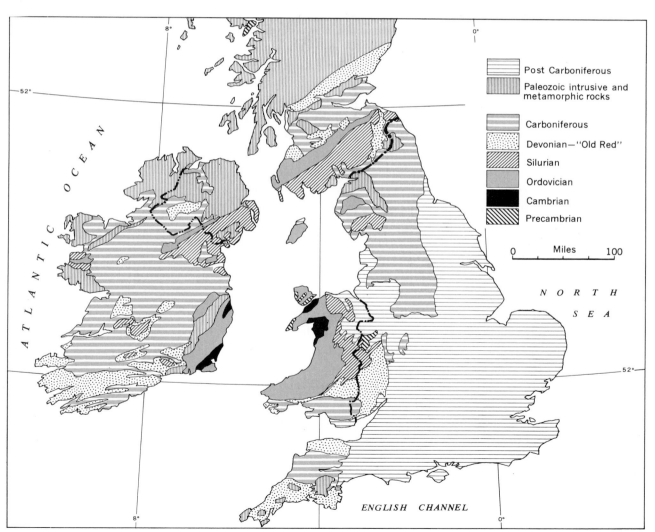

▭	Post Carboniferous
▥	Paleozoic intrusive and metamorphic rocks
▤	Carboniferous
⦂	Devonian—"Old Red"
◪	Silurian
▨	Ordovician
■	Cambrian
◩	Precambrian

0   Miles   100

ATLANTIC OCEAN

NORTH SEA

ENGLISH CHANNEL

opposite conclusion that the upper beds of Sedgwick's Cambrian were indeed the lower part of his Silurian System. Murchison became the Director of the Geological Survey, and perhaps because he had prepared such a fine summary of his Silurian System the original upper Cambrian became the lower Silurian, to the dismay of Sedgwick and his students.

It was not until after the death of the two antagonists, that another scientist, Charles Lapworth (1842–1920), professor of geology at Birmingham, proposed that the controversial beds be removed from both systems and placed in an intermediate Ordovician System, named from an aboriginal tribe that dwelt in Wales. Thus the names of the Cambrian, Ordovician, and Silurian systems came into being.

In the eighteenth century and the earliest years of the nineteenth, a classification had developed that was based largely on the consolidation and induration of the rocks. The granite and the most consolidated were the Primary, followed by Secondary and Tertiary. These names became applied in time to a roughly sequential arrangement, as most of the granites and similar rocks were found to underlie the less consolidated sedimentary rocks. For rocks that were not granite, but slates and hardened sandy argillaceous rocks, graywackes, the name Transition came to be used. The Old Red Sandstone for a time was the base of the Secondary, and the Cambrian and Silurian systems were developed from the Transition. At this stage the importance of organisms as constituents of the rocks came to be recognized, as we shall shortly discuss. Hence it was proposed by Sedgwick that the Cambrian and some succeeding periods of time, corresponding to the systems, be included in a Paleozoic Era, and that the time prior to the Paleozoic become the Protozoic Era, the latter a name that was little used because of continuing controversy as to whether the Precambrian rocks do contain evidence of "primitive life."

Thus by the end of the fourth decade of the nineteenth century a system of rocks called the Cambrian had been described from northern Wales and placed in the first or Cambrian Period in the Paleozoic Era of time. The Silurian System had been described quite fully, but some of the strata were found to be contained in both the Cambrian and Silurian systems. It was not until 1879 that the latter were removed to form the Ordovician System. Most significant in this extended discussion of the source of the names is the observation that systems were not devised to contain some successions of rocks that have limits with distinctive and peculiar characteristics. They were names given to successions of rocks found in certain areas for the purpose of having a term of reference. It would be too much to expect each system to be of equal

Fig. 6–3. Columnar section of the Cambrian rocks of the Harlech Dome and nearby areas in North Wales, where the system was first studied and described. It is only in recent years that fossils have been found below the beds containing the trilobite *Paradoxides*, since the lower Cambrian section is quite barren of fossils.

				Garth Sandstone		

Garth Sandstone

ORDOVICIAN SYSTEM

Tremadocian Series — *Ceratopyge*, *Dictyonema*

1000 ft

Ctenopyge

Dolgelly Shale — *Parabolina spinulosa*, *Orusia lenticularis*

Festiniog Sandstone — *Lingulella davisi*

Homagnostus obesus

Olenus truncatus

Maentwrog Formation — *Glyptagnostus reticulatus*

4600 ft

Clogan Slate — 300 ft *Paradoxides*

CAMBRIAN SYSTEM — **Harlech Group**

Gamlan Sandstone

Barmouth Sandstone

Manganese Shale

Rhinog Sandstone — *Myopsolenus*

Llanbedr Slate

Dolwen Sandstone — 5200 ft plus

Base of exposure

Fig. 6-4. Lower Cambrian impure sandstones exposed in the cliffs along the shore of the Irish Sea at Hell's Mouth, Cardigan Bay, North Wales (D. A. Bassett). Similar graywackes, impure sandstones, in the Harlech Dome have graded beds suggesting that they were laid in waters with a depth of hundreds of feet.

length to all others or that each begin with or be terminated by some phase of unusual significance. To a degree the defined limits have been shifted; but many of the proposed alternate boundaries have not been accepted as having compelling advantages, so most systems have been continued with essentially their original limits.

TABLE III. CATEGORIES OF STRATIGRAPHIC UNITS

Time units and corresponding *Time-rock* units:

Era	"rocks of the . . . Era"
Period	System
Epoch	Series
"age of the . . . Stage"	Stage

Rock units:

Group
Formation
Member of "the . . . Formation"

TIME AND TIME-ROCK CLASSIFICATIONS

Eras came to be divided into periods of time with corresponding systems of rock. These were in turn separated into smaller divisions, epochs of time and series of rocks. There are three series in the Cambrian Period, each with a geographic name. Whether or not this is a natural division, it is convenient to have these smaller reference units. The English language is perhaps handicapped by having the terms early, medial, and late, and lower, middle, and upper, but not having words to refer to the quarters or fifths or sixths. Some have formalized the three-fold terminology, but if the periods and systems are in themselves rather arbitrary units, it follows logically that there should not be just three subdivisions having peculiar properties. The divisions of series are the stages, the rocks deposited within a part of the epoch (Table III). It is common to speak of the smaller lithologic divisions of rocks as formations, thus a formation might represent a stage, or be larger or smaller; one (the stage) is defined by rock laid through time, the other (the formation) by some rock character. In summary, the geologic time is divided into successively smaller categories, eras, periods, and epochs. The rocks deposited during the time of periods are systems, during epochs are series, and within smaller spans of time, stages; these are time-rock stratigraphic units. Formations are rock units, normally laid through time of the duration of a stage or less; several formations together are referred to as a group; subdivisions of formations are members of these units. The rules that con-

Fig. 6–5. *Trilobites and animal classification.* Illustrated is a trilobite, *Crassifimbria*, showing the cephalon (Ce), eye (E), glabella (Gl), free cheek (F.C.), thorax (Th) with its many segments, and pygidium (Py). The living animal had other structures, such as appendages that are rarely preserved. Trilobites, like many other anthropods, shed their carapaces as they grew, so that many fragments preserved in rocks are of the moults rather than of dead animals.

Trilobites are the most distinctive organisms found in the rocks of the Cambrian System, though they are common in many Paleozoic rocks. They are rather flat or gently convex animals of oval outline, commonly an inch or so long, but occasionally nearly a foot long. The external skeleton was of chitin, a horny substance of carbon, hydrogen, oxygen, and nitrogen that is quite resistant but which in fossils commonly is altered or replaced. The body of the trilobite is strongly three-lobed, a longitudinal middle axial lobe being flanked by lobes of similar width. The animals have, from front to back, a head or cephalon, a many-segmented thorax and a tail or pygidium. Exceptionally well-preserved specimens, such as some found in shales in British Columbia, preserve many legs, antennae, and other parts. The form and position of the eyes are significant in the classification.

Animals are separable into those with vertebrae (Vertebrata) and those without (Invertebrata). They are classified into a number of categories, discussed in conventional order in Chapter 26. Forms with greatest likenesses are species having two names, such as *Olenellus thompsoni*, written as though in Latin. This binomial system of species classification is sometimes called Linnean because it was first used by the Swedish naturalist Carl Linneaus in 1785 for animals and in 1753 for plants. Organisms such as fossils that have no common names are given distinctive names, and as all are written in Latin, they are the same to scientists who speak any language. *Olenellus thompsoni* is the name of a kind of trilobite first found in Cambrian rocks in Vermont by a man named Thompson, and, described and named after him by a paleontologist James Hall in 1842 (Fig. 6–14). The genus *Olenellus*, the first half of the binomial, contains trilobites having rather close similarities and differing from those placed in other genera; minor differences among the forms of *Olenellus* lead to their being placed in several species, *Olenellus thompsoni* being one, *Olenellus gilberti*,

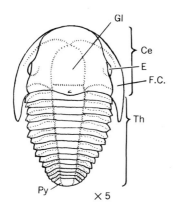

another, for example. Properly, the name of the paleontologist who named the species is written after the specific name—thus *Olenellus thompsoni* (Hall); the author's name is placed in parentheses if he assigned the species name to some other genus when he first described it, as was done in this case.

The successively higher categories in the classification of organisms are species, genus, family, order, class, and phylum. The trilobites belong in arthropods, the phylum Arthropoda, which includes such common present-day animals as Crustacea—lobsters and crabs, Myriapoda—centipede worms, insects, and Arachnidna—spiders. Trilobites are in the phylum Arthropoda and in the class Trilobita, though that is sometimes called a subclass of the class Crustacea. There are several orders, and *Olenellus* has such resemblance to the genus *Mesonacis* (pronounced me-so-ná-sis) that it is placed in the family Mesonacidae. The science that concerns the arrangement of organisms into their biological classification is called taxonomy.

The progressive developments of individual organisms is known as ontogeny, the successive stages from birth to death. In several places exceptional conditions resulted in the destruction, burial, and preservation of many trilobites of one kind in all stages of growth from the larvae to the adults, so that the ontogeny of several species of trilobites has been determined (Fig. 15–6). The comparison of forms that are believed, from their similarities, to be descendents in successively younger beds is known as phylogeny; phylogenetic successions have been interpreted also for some Cambrian trilobites.

Fig. 6–6. The hills east of Barmouth, North Wales, composed of Cambrian sandstones and shales surrounding the Harlech Dome, where the older beds of the Cambrian System were first observed and described (Aero Pictorial, Limited).

trol usage are in the published code of the American Commission on Stratigraphic Nomenclature. We shall concern ourselves only incidentally with the smallest categories of stratigraphic division.

CAMBRIAN SYSTEM IN WALES

The name Cambrian came to denote a sequence of argillaceous sandstones and shales exposed in North Wales; their base was revealed in the ledges exposed in the center of a structural dome, the Harlech Dome, and their top was defined by the placement of the base of the Ordovician System, whose rocks had once been contained in the Cambrian (Fig. 6–3). These are the typical Cambrian rocks. Rocks called Cambrian in other regions are believed to be of the same age, are correlated with the Cambrian rocks of North Wales, and are placed in the Cambrian System because it is thought that they were deposited in the Cambrian Period of time. To some degree this correlation might be based on the criteria that were applied to the Protozoic or Precambrian rocks, but to a larger extent it is dependent on the interpretation of the fossils, remains of organisms preserved in the rocks.

TRILOBITE SUCCESSION

The sediments of North Wales are not very fossiliferous. A traveler might spend many days in traversing much of the section without noticing any organic relic at all. But here and there, fossils are frequent,

Fig. 6–7. *Cambrian Trilobita.* The trilobites that are shown are distinctive of successive times within the Cambrian period in Scandinavia, Britain, and the Atlantic provinces of Canada. The genera appear in the following order. Lower Cambrian: (1) *Callavia* and (2) *Holmia,* closely related to *Olenellus,* but with differences that can be readily recognized by comparing the drawings; (3) *Protolenus,* found at the top of the lower Cambrian. Middle Cambrian: (4) *Paradoxides* and (5) *Centropleura.* Upper Cambrian: (6) *Agnostus;* (7) *Olenus;* (8) *Ctenopyge;* (9) *Peltura,* and (10) *Niobe.* The great differences among the forms indicate that they are unrelated or distantly related, not successive descendants of each other. In most of the illustrations, cephalon, thorax and pygidium is shown for each type; but the others, only the cephalon and pygidium; in the specimen of *Centropleura,* the cephalon lacks the free cheeks. Most are about their natural size, though *Agnostus* is several times enlarged.

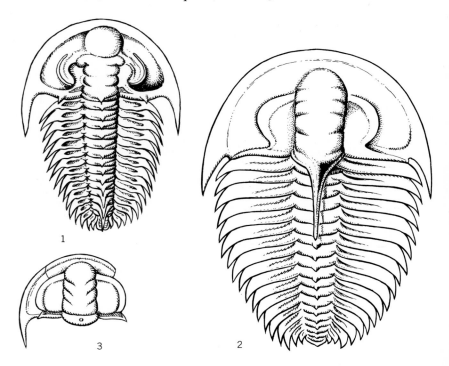

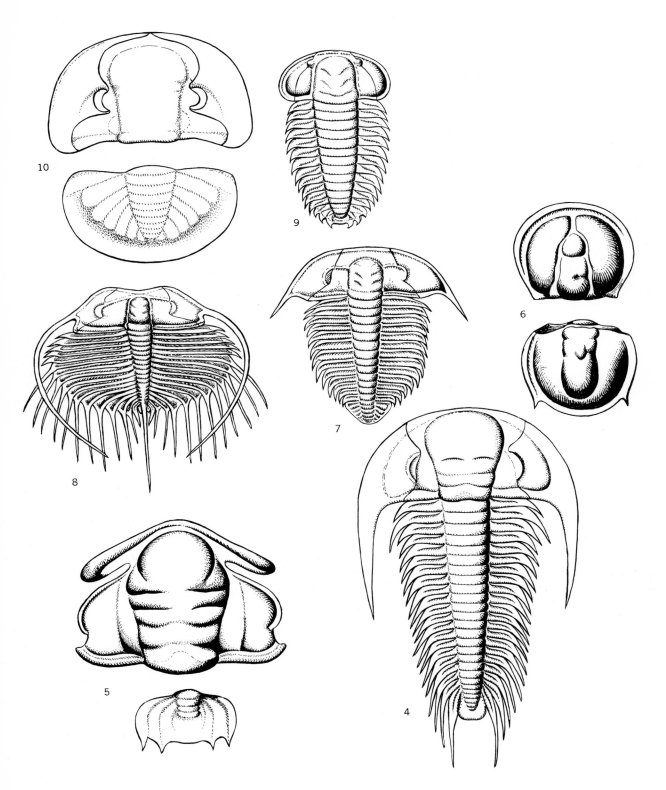

and in a few layers, quite abundant. For example, on the shore of Cardigan Bay (Fig. 6–4) and in the hills near Tremadoc, argillaceous and calcareous shales contain trilobites (Fig. 6–5), brachiopods, and graptolites. The organisms differ from bed to bed, and have been placed in genera, each genus including forms having close similarity in construction. The fossils are distributed in a distinctive order of succession; we may find several kinds of trilobites and brachiopods in some layers at one level in the succession, and quite different forms in other localities having higher and lower layers. Very few fossils are known from the thousands of feet of sandy beds in the Harlech Dome (Fig. 6–6); until recently none had been found in the lowest five thousand feet of the sequence. Occasionally, we will find a specimen or two of the trilobite *Paradoxides* far above the base. Succeeding beds are more fossiliferous. The brachiopod "*Lingula*" becomes quite common in some higher quartzites. Then, quite a large number of forms appear in some argillaceous beds exposed near Tremadoc, among them in order *Olenus, Orusia, Peltura, Niobe,* and *Dictyonema* (Fig. 6–7). Thus the Cambrian of North Wales is a sequence of rocks which in some parts contain a sequence of fossils.

SUCCESSIONS IN OTHER PLACES

The names of most of the genera of fossils were given long before geologists first explored North Wales. A Swedish naturalist, Carl Linnaeus (1707–1778), had described many of them that he had found, or that had been brought to him in his native country. Other scientists collected these and other forms among the rocks of the western and southern part of Sweden (Fig. 6–8). And when they determined the sequence, a few hundred feet of argillaceous rocks, they found that there are *Paradoxides, Olenus, Orusia, Peltura, Niobe,* and *Dictyonema* beds in the same order as in North Wales; and below are quartz arenaceous beds containing another trilobite, *Holmia* lying in turn on Precambrian gneisses and schists.

Returning to Great Britain, the coast of Pembrokeshire, the southern prong of the peninsula of Wales, quartz sandstones with *Holmia* lie on an unconformity on the Precambrian rocks below; higher beds contain *Paradoxides.* In western England in Shropshire and the Malvern Hills of Herefordshire, *Holmia, Paradoxides, Olenus, Orusia, Peltura,* and *Dictyonema* are known to occur in that order. We can cross the Atlantic Ocean to the eastern part of Newfoundland (Fig. 6–9) and find many of the forms (Fig. 6–10) in argillites along Conception Bay (Fig. 6–11) and Trinity Bay (Fig. 6–12) west of St. John's. Most again are found in argillaceous rocks in the same sequence on Cape Breton Island, in

Fig. 6–8. Upper Cambrian dark, bituminous shales underlying lower Ordovician limestone in a quarry near Skovde, Vastergotland, southwestern Sweden; a glauconite-bearing sandstone lies disconformably on the *Peltura*-bearing Cambrian shales, which are so rich in carbonaceous matter that they are burned in kilns, to calcine blocks of the overlying limestone in producing lime. The locality is also one in which uranium has been extracted from pitchblende in the shales to determine the age by uranium-lead ratios of several isotopes. The limestone, only the lower part of which is shown in the view, is a principal source for the cement industry in Sweden.

Fig. 6–10. *Paradoxides*-bearing middle Cambrian limestone from Trinity Bay, eastern Newfoundland (H. B. Whittington). Though illustrations are of specimens showing the cephalon, thorax, and pygidium of most forms, such specimens are generally not found. Trilobite fossils are commonly the dismembered moults of the living organisms, or of dead animals that have been broken and transported by the currents in the marine waters. The slab of limestone illustrates the more common appearance of rocks having an abundance of fossil trilobites.

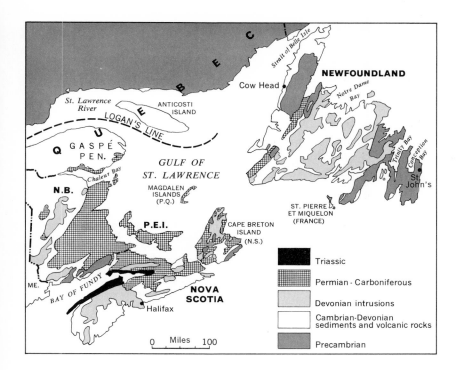

Fig. 6–9. Geologic map of the Atlantic Provinces of Canada (after E. W. R. Neale, W. A. Nash, R. R. Potter, and W. H. Poole). The Cambrian outcrops only in several small areas. Rocks similar in fauna to those in Europe are exposed along Trinity and Conception Bays in eastern Newfoundland, on the islands of the French colony of St. Pierre and Miquelon, in eastern Cape Breton Island, Nova Scotia, and near St. John in New Brunswick. Faunas of a different sort are present in sandstones and limestones in western Newfoundland.

northeastern Nova Scotia, and in the vicinity of St. John, New Brunswick (Chapter 6 opening).

DISCOVERY OF TRILOBITES AT BOSTON

There are many additional genera of trilobites and other organisms that are consistent in their position in the same sequence in many localities. Long ago, in the vicinity of Boston, a geologist came upon a slate doorstep having a fine specimen of *Paradoxides,* though it was some time before a quarry in Braintree was found to be the source (Fig. 6–13). Other forms, such as *Holmia* have been found in the Boston area, but unfortunately not in exposures showing how they are related in stratigraphic position to each other; but it is supposed that the *Paradoxides* beds are younger than those with *Holmia* from analogy with the sequence in the other regions.

SIGNIFICANCE OF SUCCESSIONS

We know that the rocks in North Wales are Cambrian because that is the succession to which the name was assigned. The succession in North Wales has some fossils in an order of sequence. The fossils had been described earlier from Sweden, where they were found to be in the same order, with some additions; and the sequence is repeated essentially in South Wales, western England, eastern Newfoundland, Cape Breton Island of Nova Scotia, and southeastern New Brunswick. Hence when a paleontologist, a student of fossils, found a fossil like one known in the Cambrian sequence in uncertain stratigraphic position, near Boston, he judged that it belonged in the same place as in the other sections.

Such occurrences might seem due to the organisms evolving through time—some being the descendants of others. But that is not the reason that the beds containing *Paradoxides* are judged to be younger than those with *Holmia* and older than those with *Olenus* and *Peltura.* Experience simply shows that they occur only in that order within the limits of our experience. Further experience sometimes alters our interpretation; the sequence of trilobites in northwestern Argentina, though very nearly the same as in Sweden, differs to the degree that some genera that are always found separately in Sweden occur together in Argentina. The conclusion that beds with *Paradoxides* are younger than beds with *Holmia* is reached from observation. It is very doubtful that many of the forms were very closely related biologically, hence that they are descendants of each other; as you can see in the illustrations, the differences are so great that they are unlikely to have been related at all. Their ancestors, if they were at all like them, lived in

Fig. 6–11. *Cambrian sedimentary rocks on Manuels Brook, west of St. John's in eastern Newfoundland. A.* Precambrian rocks unconformably overlain by lower Cambrian conglomerate. The boulder beds lie on an irregular, eroded surface of Precambrian lavas and sediments and dip gently northwestward. Sediments directly above the conglomerates have yielded *Callavia,* a trilobite found in lower Cambrian rocks in Europe.

B. Middle Cambrian shales on Manuels Brook are sparsely fossiliferous but have a few specimens of the trilobite *Paradoxides,* such as is present in middle Cambrian beds in Europe.

Fig. 6–12. Lower Cambrian shales along Chapel Arm, Trinity Bay, eastern Newfoundland, from the south. The red and green argillites in the headland lie with unconformity on Precambrian sedimentary and volcanic rocks in the foreground; and the contact is beyond the tree-bearing slope near the center of the view. Lower Cambrian rocks, having a maximum thickness of about 1000 feet in the area, are sparsely fossiliferous.

A

B

places that are not known to us. The ones we find are genera that for reasons we cannot clearly understand lived in profusion over wide areas for limited times, then were superseded by others that found conditions to their liking, only in turn to be succeeded by still others. The order of succession of Cambrian fossils is based on their observed presence in beds whose stratigraphic order is determined by the same means as were applied to the Precambrian rock formations in the Mesabi Range. We shall see that in some later times, the forms in one bed seem to have descended from those in others, but the Cambrian does not have facts that demonstrate such relationships among the distinctive genera. Insofar as one can see, the successive fossils could have evolved from unknown ancestors, or they could have been created in successive times.

BEGINNINGS OF STRATIGRAPHY

It is surprising that the significance of fossils in the classification of sedimentary rocks was so long in being recognized. William Smith (1769–1839) was an Englishman from Oxfordshire who attended the village school, but gained no further formal education. He learned surveying, and in the last years of the century became an engineer in the construction of canals and roads in southern England. His occupation gave him the opportunity to travel widely and to observe critically. He became able to arrange the strata in systematic order, and made the great discovery that fossils found in a single stratum are distinctive of it and of no other. He skillfully prepared maps showing the distribution of the fossiliferous beds of southern England as a product of his observations. At about the same time, in the first decade of the nineteenth century, Georges Cuvier (1769–1832), distinguished in the court of Napoleon as well as in anatomy, and Alexandre Brongniart (1870–1847), a naturalist and the son of the architect who designed the Bourse, while working out the sequence of layers in the Paris region also discovered that the fossils differed from one bed to another. Though others may have made similar observations, these scientists of a century and a half ago found the key to the unravelling of the history of the earth, that fossils are guides to time. Stratigraphy, the arrangement of events in sequence, came into being with the recognition of this principle.

OTHER SUCCESSIONS OF TRILOBITES

Fossils are distinctive in that they have progressive differences, those that abound at one time virtually do not reappear unchanged at a later time. We may find rock formations in two very different parts of the rock succession that seem identical, as though laid under exactly the

Fig. 6–13. Specimen of the trilobite *Paradoxides* on a slab of slate from a quarry in Braintree, Massachusetts, near Boston (H. B. Whittington). A similar specimen from an unknown locality was the first Cambrian fossil recognized in America, described by a physician in Boston in 1834; more than twenty years elapsed before specimens were observed in place in Braintree slates. *Paradoxides* is representative of the middle Cambrian Acadian Series of the Atlantic Coast.

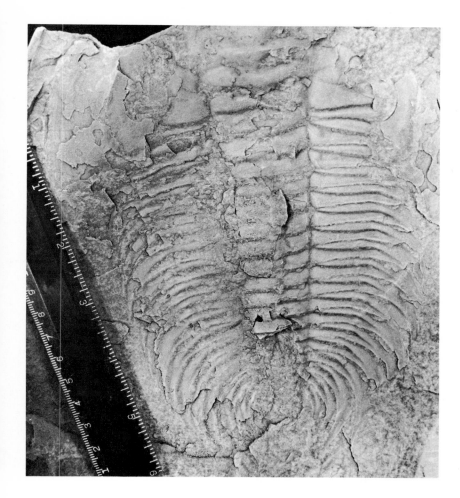

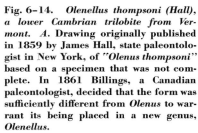

Fig. 6–14. *Olenellus thompsoni (Hall),
a lower Cambrian trilobite from Vermont. A.* Drawing originally published
in 1859 by James Hall, state paleontologist in New York, of *"Olenus thompsoni"*
based on a specimen that was not complete. In 1861 Billings, a Canadian
paleontologist, decided that the form was
sufficiently different from *Olenus* to warrant its being placed in a new genus,
Olenellus.

B. Later on better specimens were
found such as those illustrated in 1886
by Walcott of the United States Geological Survey; at the time the species was
thought to be younger than *Paradoxides,*
but Walcott soon decided that it was older
on the basis of similar trilobites, *Callavia,* lying below *Paradoxides* in Newfoundland. *Olenellus* and *Paradoxides*
came to be thought of as distinctive of
lower and middle Cambrian faunas.

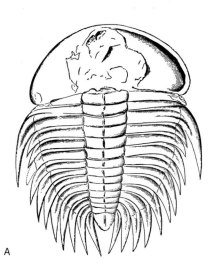

A

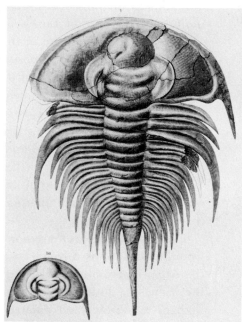

B

same conditions of depth, current, and other factors. But however alike the beds may be, however much it seems that they were laid under like environments of deposition, the fossils in the one will differ measurably from those in the other. Nor are the fossils like those that we have been discussing as characteristic of the Cambrian, from Sweden to Massachusetts and Argentina, found all over the world. We need only go a little to the west in North America to find successions in which they are lacking.

More than a century ago, along the strait between Newfoundland and Labrador, explorers found quartz sandstones and limestones lying on granites and gneisses, and containing a variety of fossils generally similar to those we have been describing but not identical. The principal trilobite was called *Olenellus*, which can be seen to be very like *Holmia* but yet is not the same, for it lacks the inner spines along the posterior edge of the cephalon (Fig. 6–14). The overlying beds, exposed on the west coast of Newfoundland, had a whole succession of forms, of which very few if any are like the ones that are found in the eastern part of the province. Instead, their similarities are to fossils found through much of the rest of the continent, whether in Vermont, Alabama, Nevada, or British Columbia. Rocks containing these fossils are also called Cambrian. Their fossils are probably more like those found in the true Cambrian of Wales than those in any part of the succession of fossiliferous beds. Perhaps it is more pertinent that in a number of regions, these so-called Cambrian fossils lie in beds directly below the rocks having fossils like those found overlying the Cambrian rocks of Europe. The rocks that have fossils that are somewhat similar to those in the typical Cambrian thus occupy the same relative stratigraphic position as the typical Cambrian. A few fossils found in the western of the American sequences are like those in the eastern or Atlantic sequences; for example in western Vermont near St. Albans, a trilobite closely related to *Paradoxides, Centropleura* (Fig. 6–7), is present in some argillites lying above the beds in which *Olenellus* was first described and below beds bearing a great number of the forms characteristic of the upper part of the western sequence. We have been endeavoring to understand the manner in which fossils occur, and why they are interpreted as evidencing the age of the beds that enclose them.

There must have been a reason why the faunas in the Cambrian rocks from Massachusetts through eastern Newfoundland to Sweden along the Atlantic, are much alike, yet quite unlike those thought to be of the same age found farther west and characteristically American. Perhaps a narrow land entered from the Atlantic, passed through central Newfoundland and New England, separating the two kinds of organisms. Or the fossils of the two kinds might represent different times, so that *Paradoxides,* for instance, lived after the time of some of the American province forms and before that of others. Or maybe the organisms of the Atlantic Province were living in colder water than those in the American, or warmer, or in seas having muddier bottom, or greater depths, thus gaining organisms adapted to this environment. These are the sorts of explanations that have been offered. Likely the cause is one or more of these factors.

SIGNIFICANCE OF DIFFERING SUCCESSIONS

The hypotheses to explain the contrasting faunas in the Atlantic and the "Pacific" or American provinces are of three general categories: (1) differences in time of living, (2) differences by isolation through prevention of migration by geographic barriers, or (3) differences due to conditions at the place of deposition. These are the effects of time, province, and environment, and the latter two are closely related; geographic barriers are those with environment not hospitable to the organisms. The first explanation would have the successive organisms living for a time on one side of the line, then on the other, and so forth. It would require changes in earth structure causing seas to occupy alternately one side and then the other of a narrow land in a rather implausible fashion. The second explanation would require that the organisms of one province were prevented from mingling with those of the other through the presence of a continuous environmental barrier, such as a narrow land, or a band of such depth of water as to prevent bottom-dwelling organisms and their larvae from crossing; thus there would be independent development of the faunas in the two provinces. Such an extensive narrow and sinuous land barrier is hardly analogous to conditions that we know on the earth today. The third explanation is one of environmental controls, encouraging some organisms to live in one area and preventing others that could not tolerate the conditions from doing so. There is suggestion that such factors as would be contained in the environment have controlled the faunas, for not only are the Atlantic province faunas in prevalently argillaceous and silty sediments, while the American forms are in principally sandy and calcareous ones, but some of the Atlantic-type forms appear in the more argillaceous rocks even in the western part of North America, and with continuing study mixed assemblages are being found. The conditions must have been widely distributed through long periods of time to account for such marked contrasts in the two faunal provinces. The organisms lived only through certain spans of time, and during these stages, only in certain environments, spreading only as far as the environments that they could tolerate persisted. There were different provinces, evidenced by the faunal distribution, but the reasons for the contrasts are subject to several interpretations.

The fossils of the Cambrian are useful in distinguishing the successive stages of time in the period. Contrasting assemblages of fossils have geographic distributions that may have been related to differing environments, one prevalent along both sides of the Atlantic, from Sweden to Massachusetts and Argentina, and the other characteristic in the interior of North America.

The introduction of a variety of organisms in the early Cambrian, including such complex forms of the arthropods as the trilobites, is surprising. The commonest and most widely distributed fossils known in the Precambrian are calcareous algae or similar structures. Other reported fossils are questionable organisms, or those from rocks that are doubtfully or barely Precambrian (Fig. 6–15).

THE EARLIEST FOSSILS

The introduction of abundant organisms in the record would not be so surprising if they were simple. Why should such complex organic forms be in rocks about six hundred million years old and be absent or unrecognized in the records of the preceding two billion years? If

organisms evolved, it should have taken a long time for them to have developed into forms such as arthropods. Many suggestions have been made. They may have had soft bodies, structures that were not resistant enough to be preserved as fossils. As we will see, remarkably well-preserved records of soft-bodied Cambrian organisms are known in Alberta—animals that otherwise are not known through the geologic record because they have so rarely been preserved. Perhaps fossils are not preserved in older rocks because they have been destroyed by metamorphism. Most Precambrian rocks are quite metamorphosed, but some are as unaltered as later rocks in which fossils are well-preserved or at least recognizable. It has been thought that organisms in Precambrian time lived in such an environment as the oceanic depths, where the sediment record has not been recovered. Some have even suggested that the rocks that we study in the Precambrian are not of marine origin, and hence are from environments unfavorable to the marine life that we know from later eras. The ancestors of the Cambrian fossils must have lived in some place, if they evolved through a long span of time. Life of the Cambrian is known from the fossils of many kinds found in widely scattered parts of the earth, whereas records in earlier rocks are sparse and controversial. If there has been evolution of life, the absence of the requisite fossils in the rocks older than the Cambrian is puzzling.

Fig. 6–15. *Ancient fossils from sandstones in the Ediocara Hills, South Australia. A, Dickinsonia, and C, Spriggina, are segmented and worm-like; B, Parvancorina, resembles no other known organism. The illustrations are a little reduced* (M. F. Glaessner).

The specimens are of organisms that have been found in some abundance in bedded sandstones in the southern part of South Australia. They lie some five hundred feet, apparently in a conformable succession, below beds with characteristic Cambrian trilobites. They are Precambrian, if the system is defined as having its base at the first appearance of Cambrian trilobites; others consider beds conformably below those with Cambrian fossils to lie within the system. The Australian fauna has many forms of organisms, some of them of uncertain biological relationships.

B

C

7
The Tectonic Elements
of the Cambrian Continent

Upper Cambrian Potsdam Sandstone lying unconformably on Precambrian gneiss along the highway near Elgin, southeastern Ontario. In western New England, beyond the Adirondack Mountains to the east, several thousand feet of lower and middle Cambrian rocks intervene between the Precambrian and upper Cambrian.

7 The Tectonic Elements
of the Cambrian Continent

The stratified rocks containing fossils like those in the Cambrian System of the type region of Wales, or occupying the same stratigraphic position, are classified as in the Cambrian. And the series and stages, the time-rock divisions within the Cambrian System, can be correlated and classified on the basis of their organisms. We should, then, be able to compare the history of one part of the Cambrian Period with that of another, and from the distribution of rocks in each stage, learn something of the early Paleozoic character of the North American continent. It will be convenient to divide the rocks into three series, and the corresponding time into three epochs. One of the regions where the Cambrian has been best studied is in the Canadian Rockies of Alberta and British Columbia. The lowest part of the two miles or more of Cambrian in that area has specimens of the trilobite *Olenellus* (Fig. 6–14). These rocks, correlated with *Olenellus*-bearing strata in southeastern California near Death Valley, have been called Waucoban, from springs of that name in the desert ranges of California.

BASE OF THE CAMBRIAN SYSTEM

How far down does the lowest part of the Cambrian, the Waucoban Series, go? This was discussed somewhat in the preceding chapter. At best, fossils are not very common in these rocks. The presence of a trilobite such as *Olenellus* or of its common associates, is accepted as evidence of Cambrian age. Perhaps rocks older than those having trilobites are not Cambrian but Precambrian. The richest fauna in so-called Precambrian rocks is one from South Australia having coelenterates resembling modern jellyfish and seapens, annelid worms, and other strange creatures (Fig. 6–15). Quite similar fossils have been found on quartzites in western England, and recently on Axel Heiberg Island in Arctic Canada. The beds are in a continuous succession upward into those containing trilobites, so they are Precambrian only if we define the base of the system as being the horizon of the appearance of trilobites. Under this definition, the rocks of the Harlech Dome in North Wales would be Precambrian, for they lack known fossils in the lowest mile or so of thickness; or perhaps it is better to say that they seemed to lack them until one hundred and twenty years after the work of Sedgwick when trilobites were discovered in a bed about half way to the base from the previously known forms. The Australian fossils do lie below beds having an early Cambrian fauna, whereas the beds

Fig. 7–1. *Middle and upper Cambrian succession of over 9000 feet, partially shown in Mount Robson, British Columbia, along the Canadian National Railways west of Jasper Park, Alberta.* A. View of Mount Robson from the southwest. The oldest rocks, Lower Cambrian, are exposed on the mountain above the sign on the left. In Mount Robson, Middle Cambrian is exposed to the top of the light band of Arctomys Limestone, the Upper Cambrian Lynx Limestone forming the upper slopes of the mountain. The peak (12,972 feet) is the highest in the Canadian Rockies (Canadian National Railways). B. A section of the Cambrian rocks in the vicinity of Mount Robson (E. Mountjoy, Geological Survey of Canada).

Fig. 7–2. Mount Eisenhower (9390 feet), a peak of Cambrian sedimentary rocks in Banff National Park, in the Canadian Rockies of Alberta (National Film Board, Canada). The southwest side of the mountain, on the left, facing the Banff—Lake Louise Road, has more than one thousand feet of lower Cambrian quartzites and overlying shales, overlain at the base of the principal cliff by 2000 feet of middle Cambrian extending to the top of the peak; the rocks are dominantly limestones, with a shale-forming terrace breaking the cliff. The Upper Cambrian has nearly 1500 feet of shale and limestone on the distant skyline. The fossils from the middle Cambrian of this region are typical of the Albertan Series.

A

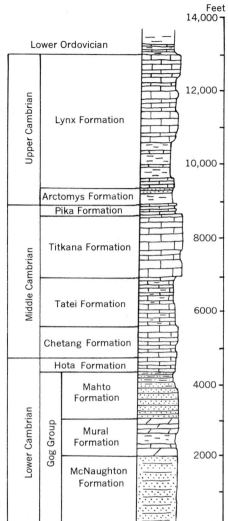

B

in North Wales were below rocks having trilobites known not to be as old. Aside from the "nearly Cambrian" fossils that come from rocks in rather continuous stratigraphic succession, not far below trilobite-bearing beds, as in Australia, well-authenticated fossil invertebrates are not known in older rocks.

It may seem better to classify as Cambrian the rocks bearing the trilobite faunas, as well as those that lie essentially conformably beneath them. This would retain the oldest rocks in the type section in North Wales, and would place the base of the system as the prominent unconformity lying below the trilobite-bearing sediments in South Wales or Sweden. Occasionally this, too, leads to contradictions. For example, in western Vermont a quartzite a thousand feet or so thick has *Olenellus* near its top; the base lies on an angular unconformity on argillaceous sandstone—graywacke and limestone that would be called "Precambrian." Farther north in Vermont and in southern Quebec, quartzite like that overlying the unconformity lies quite conformably on the correlative graywacke and interbedded limestone—hence the "Precambrian" becomes "Cambrian." These latter graywackes overlie thousands of feet of altered lavas, which are variably called Cambrian and Precambrian.

The position of the lowest fossil find is bound to be somewhat fortuitous; more careful search or a more fortunate exposure may lead to finding lower and older fossiliferous beds. The beds containing the oldest trilobite found in southern British Columbia were mapped to the international border; to the south in northeastern Washington, in ledges thousands of feet lower down, the lowest trilobite fragments were discovered. So the "Precambrian" in Canada became the "Cambrian" in the United States. No method is known by which a moment in earth history marking the beginning of the Cambrian can be identified universally. With present methods of classification, we only approximate a consistent placement of the base of the Cambrian System and the beginning of the Cambrian Period. Because of the inadequate basis in defining the bottom of the Cambrian System, we must be cautious in assuming that young "Precambrian" rock at one locality is not of the same age as old "lowest Cambrian" at another. Perhaps geochemical methods will enable us to define time more closely in numbers of years.

CONTRASTING AMERICAN FAUNAS

Above the Waucoban Series in the Canadian Rocky Mountains of Alberta (Figs. 7–1 and 7–2) are other rocks containing several stages, each with distinctive trilobites whose names need not concern us particularly, but rather the fact that they are like trilobites found in

Fig. 7–3. Distribution of thicknesses of Cambrian rocks in the western geosynclinal belt (after M. Kay). The thicknesses of lower Cambrian, total Cambrian, and of middle and upper Cambrian combined, are shown in the maps from left to right. The size of the spot at each locality is a record of thickness. The lower Cambrian map also has lines connecting points of approximate equal thickness, known as isopachs. The maps are stratigraphic, for they represent a span of geologic time within which the rocks were deposited.

Fig. 7–4. Restored sections of the Cambrian sediments on the western and eastern margins of the relatively stable craton of central North America. The sections show the thicker successions in the geosynclines margining the craton, including lower and middle Cambrian beds that are absent or sparsely represented in the interior.

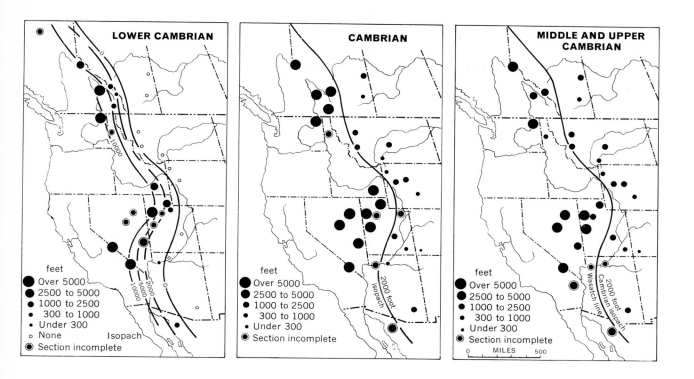

LOWER CAMBRIAN

feet
- ● Over 5000
- ● 2500 to 5000
- ● 1000 to 2500
- ● 300 to 1000
- • Under 300
- ○ None
- ◉ Section incomplete

Isopach

CAMBRIAN

feet
- ● Over 5000
- ● 2500 to 5000
- ● 1000 to 2500
- ● 300 to 1000
- • Under 300
- ◉ Section incomplete

2000 foot isopach

MIDDLE AND UPPER CAMBRIAN

feet
- ● Over 5000
- ● 2500 to 5000
- ● 1000 to 2500
- ● 300 to 1000
- • Under 300
- ◉ Section incomplete

0   MILES   500

Wasatch line

2000 foot Cambrian isopach

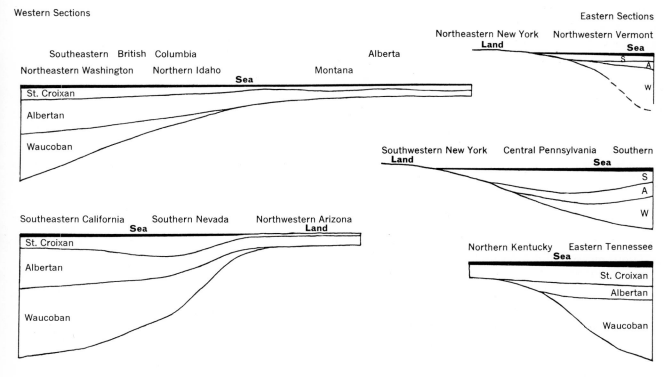

Western Sections

Southeastern British Columbia
Northeastern Washington   Northern Idaho   Alberta   Montana

Sea

St. Croixan

Albertan

Waucoban

Southeastern California   Southern Nevada   Northwestern Arizona

Sea   Land

St. Croixan

Albertan

Waucoban

Eastern Sections

Northeastern New York   Northwestern Vermont
Land   Sea

S
A
W

Southwestern New York   Central Pennsylvania   Southern
Land   Sea

S
A
W

Northern Kentucky   Eastern Tennessee
Sea

St. Croixan

Albertan

Waucoban

109

exposures as far away as western Newfoundland, Alabama, and northwestern Mexico; and they appear in the same order. This second series was called the Albertan from the very region that we are discussing and at the top of that is a succession of limestones containing other genera of trilobites first found by explorers a century ago along the St. Croix River valley on the Wisconsin-Minnesota boundary; these have been called the St. Croixan Series. So the Cambrian System of the American province is divided into the lower or Waucoban, the middle or Albertan and the upper or St. Croixan series, each recognized by its distinctive genera of fossils, particularly trilobites. The first, Waucoban, is thought to be about equivalent to the *Holmia*-bearing rocks of the Atlantic province, the second to most of the *Paradoxides* beds, and the third to the younger Cambrian. The correlations between the two faunally contrasting successions is based on some mixing of representative trilobites and the presence of distinctive forms of one province in stratigraphic successions dominated by forms of the other; for example, *Centropleura* (Fig. 6–7) normally found with highest *Paradoxides* in Europe, appear in shale in northwestern Vermont above beds with Waucoban trilobites, the type *Olenellus* in fact, and below those with St. Croixan genera.

THICKNESS DISTRIBUTIONS

The Cambrian rocks are not of equal thickness and identical character from place to place. The nature of the differences should bear on Cambrian history. Let us consider the thickness of the system in the states to the south of Alberta and British Columbia. Cambrian rocks have been found in mountain ranges scattered over much of the area, in Washington, Idaho, Montana, Wyoming, Nevada, Utah, Colorado, California, Arizona, and Sonora. We can summarize the thicknesses by placing circles on a map, larger circles representing thicker sections of rocks containing fossils that have been considered as Cambrian types (Fig. 7–3). The thickest sections are in the western localities (Fig. 7–4), where a maximum of about three miles of Cambrian has been measured near Death Valley, and more than two miles in several ranges from British Columbia and northeastern Washington to eastern Nevada, western Utah (Fig. 7–5), and southeastern California. On the other hand, sections measured in wells under the plains of Alberta, in the mountains of Montana, Wyoming (Fig. 7–5B), and Colorado, and in wells in southeastern Utah are less than two thousand feet. And though there are only a thousand feet or so of Cambrian rocks in the Grand Canyon section of northwestern Arizona (Fig. 7–6), nearly two miles is

Fig. 7–5. *Cambrian sections in western Utah and western Wyoming. A.* Cambrian sediments in the western face of the House Range, western Utah. Lower Cambrian quartzites outcrop in the lower left and are intruded by sheets of granite; the whole of the succession of about a mile of middle and upper Cambrian crops out in the slopes and cliffs of the range, Ordovician overlying it on Notch Peak, some 9000 feet in elevation nearly a mile above the valley in the foreground. The white patches in the middle distance are along the shore of Lake Bonneville, a lake that spread over much of western Utah in rather recent, Pleistocene time. The Cambrian rocks of the House Range were discovered nearly a century ago, and although the range is quite isolated, the fossil succession was one of the first that was determined in the Cambrian of the West. The section contrasts with those of a few hundred feet that are found to the east in Colorado and Wyoming outside the Cordilleran geosynclinal belt.

B. Cambrian sediments lying on the Precambrian along Bull Lake and Dinwoody Canyon on the east side of the Wind River Range in western Wyoming. The Cambrian is composed of about one thousand feet of sandstone, shale, and limestone, the latter forming the bright cliffs in the distance, above the stripped Precambrian peneplane. The rocks are middle and upper Cambrian, their thickness in this section in the continental interior or craton contrasting with thickness of thousands of feet to the west in Idaho. The Cambrian limestone is succeeded by a hundred feet or so of Ordovician limestone, and that by Carboniferous.

A

B

Fig. 7–6 (below). A restored section of the stratigraphy of the Cambrian rocks exposed along the Grand Canyon of the Colorado River in Arizona (after E. D. McKee). Thin and distinctive rock units and fossil units can be traced along the canyon walls, or correlated among the many exposed sections. The rock units can be seen to disappear laterally as tongues within other lithologies, thus serving as the means of determining the interrelations of the strata in time. Time-planes or isochronous surfaces lie within the limits of lithic tongues. The compiled data show that the beds at the base of the section, lying unconformably on the Precambrian, are overlapping; the base is becoming younger eastward and the basal sand is rising in age toward the eastward; see Fig. 3–10. Sections in mountain ranges a few miles to the westward and northwestward come to have much greater thickness of lower Cambrian, its basal part still older than the beds in the western Grand Canyon, as will be seen in the next figure.

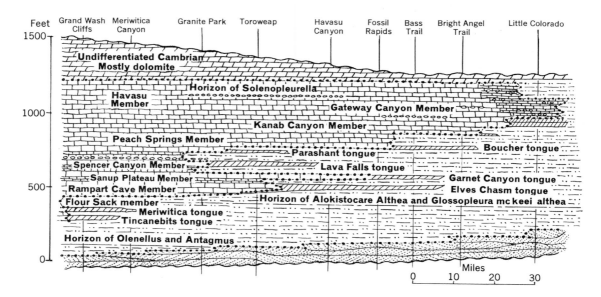

preserved in the Pioche mining district about one hundred miles northwest.

One might draw lines on the map connecting points of equal thickness. Such a line is an *isopach*. The two-thousand-foot isopach of the Cambrian—the line connecting all points having two thousand feet of Cambrian—passes from western Alberta through western Montana into the corner of Wyoming, thence across Utah through western Arizona into Sonora, Mexico. Sections west of this isopach are thicker, and those to the east are thinner.

Before explaining the differing thicknesses of the Cambrian, let us record some other facts. Instead of taking the whole of the Cambrian System, we can make similar maps for each one of the three Cambrian series (Fig. 7–3B). The distribution of the lower Cambrian or Waucoban Series shows the same thinning of sections eastward as for the whole of the system, but the Waucoban Series *disappears* along a line approximating that we previously drew as the two thousand foot isopach of the whole Cambrian. In other words, the Cambrian is less than two thousand feet in sections that do not have lower Cambrian—which might lead us to assume that the great thicknesses are there only because the Waucoban at the base is thick. But the middle and upper Cambrian series each have sections of a mile or more in the west and diminishing thickness east of the two-thousand-foot isopach line that was drawn for all Cambrian; the upper Cambrian is thousands of feet thick in eastern Nevada (Fig. 7–7) but only a few hundred feet in central Wyoming (Fig. 7–8) and Colorado (Fig. 7–9). We can conclude that the Cambrian as a whole, or any series of the Cambrian, decreases in thickness east of a flexure passing through eastern Utah—hence called the Wasatch Line, the lower Cambrian disappearing entirely where the whole is less than two thousand feet.

RESTORED SECTIONS

Let us see something of the kinds of rocks that are represented. The Cambrian rocks are invariably quartz sandstones in the lowest part, regardless of the age of the beds; thus the early Cambrian (Waucoban) in Nevada is quartzite, and so is the base of the medial Cambrian (Albertan) in western Montana and Wyoming, and of the late Cambrian (St. Croixan) in eastern Wyoming, Colorado, and South Dakota. Where it has been possible to follow strata characterized by a certain fossil assemblage and believed of the same age from westward to eastward, they pass from the limy beds into the sandy. Hence *restored sections,* sections such as would appear on the side of a trench cut through the rocks after their deposition but with horizontal scale very greatly

Fig. 7–7. Cambrian and younger Paleozoic rocks exposed on the west face of the Egan Range 35 miles south of Ely, Nevada. The Egan Range, in the Basin and Range region of western United States, rises about a mile above the intermontane basin in the foreground. The relatively great thickness of Paleozoic rocks in the non-volcanic geosynclinal belt in western North America is exemplified by this magnificent section of regularly east-dipping rocks. The mountain slope at the left has Upper Cambrian limestone and dolomite exposed through thickness of 2350 feet, with the lowest 200 feet of Ordovician extending to the first peak. Ordovician Canadian and Chazyan limestones totaling 4600 feet in thickness extend through the lower cols and the two hogbacks to a disconformity at the base of the light-colored band of quartzite on the higher mountain on the right of center; the Eureka quartzite, more than 500 feet thick is upper Trentonian, and is succeeded by about 500 feet of dark-colored Cincinnatian dolomite. More than 1000 feet of lighter-colored Silurian dolomite forming part of the mountain face above the quartzite is succeeded by Devonian limestone and dolomite nearly a mile thick in the higher part of the mountain and the ranges farther to the right. The section continues through thousands of feet of Carboniferous. Adding lower and middle Cambrian rocks exposed elsewhere in the range, the Paleozoic succession has a thickness approaching four miles. The lower Paleozoic systems have many times the thickness found in the regions to the east outside the geosyncline.

Fig. 7–8. *Upper Cambrian sandstones overlying Precambrian rocks in Wyoming. A.* Hills west of Rawlins, south-central Wyoming. The unconformable contact lies below the bedded rocks gently dipping southeast on the right of the hill in the background. The presence of upper Cambrian of a thousand-foot thickness or less in Wyoming is in marked contrast to the thousands of feet of beds, including earlier Cambrian Series, in the mountain ranges of the states to the west.

B. Upper Cambrian sandstone overlying Precambrian gneiss in the canyon of the Platte River, east-central Wyoming; the site of the photograph now lies beneath the waters of the Alcova Reservoir (A. Devaney).

A

B

113

reduced, show the rocks of a stage thinning eastward and passing into beach sands at the eastern termination (Fig. 7–10). This is not surprising, as the Cambrian lies unconformably on Precambrian rocks (Fig. 7–8), and the earliest beds must have been laid by a sea advancing over the terrane of older rocks, with a beach belt along its margin; evidently the beaches migrated from western positions in earlier time to eastern areas at a later date. We might show on a map successive lines eastward representing successive positions in the advance of the beaches of the Cambrian (Fig. 7–11). The beds above these lower sandstones or quartzites are commonly carbonate rocks, such as might represent the accumulation of shell fragments in the waters seaward from the beach. There are exceptions—most famous is the argillaceous shale of Mount Stephen, Alberta, which contains well-preserved impressions of the most delicate soft parts of organisms (Fig. 15–2). With few exceptions, Cambrian rocks that overlie the basal beds are carbonates, principally limestones.

PALEOGEOGRAPHIC MAPS

Maps showing the kinds of sediments forming the sea bottom from stage to stage show the sandy bottoms to the east passing into the calcareous bottoms to the west (Fig. 7–12). Such maps are *paleogeographic maps* of a particular kind, either showing the kind of sea bottom at a moment of time or lithologic phases (*lithofacies*) through a span of time. We might also show depth of water at a time in another type of paleogeographic map, for the water would be deepening westward from the sandy beach. But such a map would entail interpretation, conversion of facts about the rock and its fossils into depths of water at time of deposition. The several sections and maps show that from time to time the region was one in which lands were to the east, margined by sandy beaches passing seaward into regions having limy bottoms; that they were shallow is dependent on interpretation of the depths to which the organisms can live in analogy with present related forms, and the significance of rock structures and compositions, again largely depending on the comparisons with modern conditions for the interpretation of those of the past.

INTERPRETATION OF DISTRIBUTIONS

There are orderly changes in the thicknesses and lithologies. But how might they have happened? Let us consider some methods; those that now may seem rather ridiculous have in the past been thought excellent, and quite likely some that we accept as valid will prove to have been similarly fallacious. We have learned that the basal beds are invariably

Fig. 7–9. Upper Cambrian sandstones overlying Precambrian rocks along Glenwood Canyon of the Colorado River, followed by Highway 40 and the Denver and Rio Grande Railway in central Colorado. The bedded Cambrian consolidated sandstones, orthoquartzites, form the upper half of the prominent bluff to the right as well as the more distant mountain.

Fig. 7–10. Restored section of the Cambrian System from the geosynclinal belt in eastern Nevada and Utah to the thinner, less complete cratonal sections in Wyoming. Such a section interprets the relations as they may have been near the close of the Cambrian Period; the present rocks have been deformed and altered by events in the later history of the region.

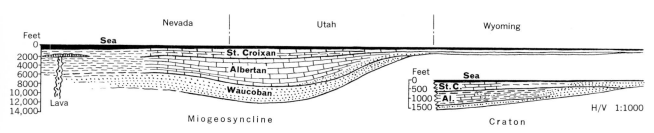

Nevada | Utah | Wyoming

Feet

0

2000
4000
6000
8000
10,000
12,000
14,000

Sea

St. Croixan

Albertan

Waucoban

Lava

Miogeosyncline

Feet

0
500
1000
1500

Sea

St. C.

Al.

H/V 1:1000

Craton

Length of section   800 miles

Horizontal to vertical   1:300

quartz arenites such as might have been beach sands; and that as much as fifteen thousand feet of Cambrian sediment lies above the early Cambrian base in some western sections in North America.

We might first assume that the original relief at the beginning of the Cambrian Period in this region was that of a land surface rising gradually eastward from the sea in Nevada to the crest of a level plateau of Precambrian rocks in Colorado at an elevation of about three miles (Fig. 7–13A). The water in the sea might then gradually rise, advancing eastward until by the close of Cambrian time it had surmounted the top of the plateau. This would be an *eustatic movement*—a change in level of the sea, the earth's surface remaining constant throughout the time. In such a case, the streams should have flowed westward from the plateau with such velocity as to produce very poorly sorted sediments; whereas the quartz sands of the basal Cambrian to the east in Colorado are very well sorted. They should have eroded deep valleys in their earlier stages, such as would produce local thick sections lying in low trenches in the basement; but the thicknesses are quite constant

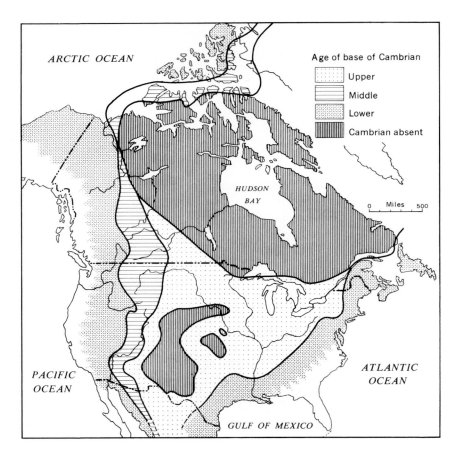

Fig. 7–11. Outline map showing the age of the rocks that lie at the base of the Cambrian System in the interior of North America. The shaded area seems to have remained as land, or at least retained no sediments of Cambrian age. Late Cambrian (St. Croixan) sediments extended beyond those of the medial (Albertan) and early (Waucoban) series, though the latter two are not separated on parts of the map. There are of course Albertan and St. Croixan rocks above Waucoban where it is shown, for the map shows only the age of the basal rocks. The distributions of Cambrian rocks on the eastern and western margins of the continent are omitted.

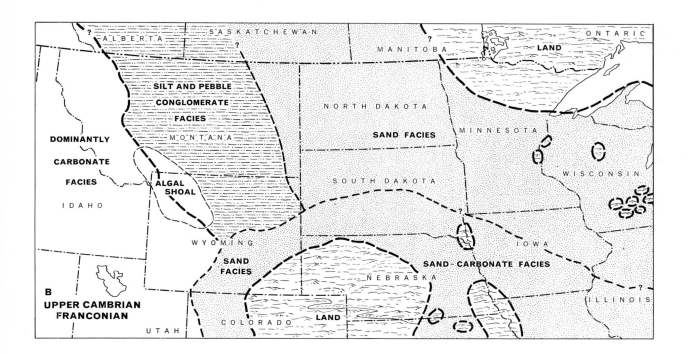

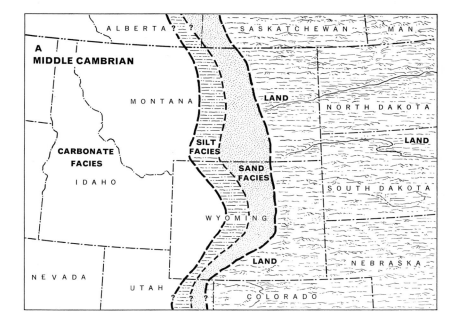

Fig. 7-12. Paleolithologic maps, showing the distribution of types of sediments
on the floor of the seas at two stages in
Cambrian history in north-central United
States. The upper map is of the late Cambrian Franconian stage, the lower is
medial Cambrian. As the boundaries of
the sediment types are generally gradu-
tional, the lines on the map, called isoliths, connect points of similar properties, symbolically separating contrasting
sediments on opposite sides. The interpretation of such paleogeographic maps
is that lands on the east were bordered
by deepening seas in which organically
formed carbonates accumulated on some
shoals (C. Lochman).

in Colorado and the surface beneath the unconformity more like a peneplane. Moreover, it is doubtful that a plateau with elevation such as we have assumed could have remained for even a small part of such a long period of time as elapsed in the Cambrian—several tens of millions of years; erosion would have reduced it to a region of rugged relief by the time the upper Cambrian seas covered it. Such an hypothesis as this was abandoned long ago as untenable. It is an extreme hypothesis of eustatic movement—change in sea level.

A second hypothesis (Fig. 7–13B) can assume that the surface of the region was one formed by long-continued erosion of the Precambrian rocks by the beginning of the Cambrian Period. The surface came to subside more on the west than on the east, so that seas encroaching from some unknown source of marine waters, probably the ocean basins, first flooded the part that first subsided in Nevada. During early Cambrian (Waucoban) this western region eastward to the Wasatch line continued to subside as the seas advanced gradually eastward; and by the close of the epoch a surface that had been nearly horizontal had been so depressed that it lay a mile or two below the sea on the west, but it remained at sea level on the east, the shore following a sinuous line from Alberta through Utah to Sonora (Fig. 7–11). In medial Cambrian, the sea advanced gradually eastward, particularly into Wyoming and Montana, and continued to do so in late Cambrian; but the differential movement continued, for the top of the lower Cambrian is more than a mile below the top of the upper Cambrian in the western sections, but the full section is only a few hundred feet in Wyoming, Colorado, and eastward. Thus by the close of the Cambrian, a surface originally a virtual plain (or peneplane) had been deformed so that it was as much as three miles below the sea in central Nevada but virtually at sea level or a little above in Colorado and the Dakotas. And all that time the land relief was so low that the sediments were carried from slowly disintegrating rocks to a beach where they were slowly sorted into clean sands, the present quartzites. This hypothesis attributes the entire change to *epeirogenic movement,* warping of the crust, with regions originally at the same distance from the center of the earth becoming nearer or farther from the center, subsiding or rising. Such a history seems to better account for the facts.

It may be that the gradual spread of the seas was really the result of a sea-level rise (eustatic movement), yet most of the changes in thickness are due to independent warping epeirogenic movement. The distribution of seas throughout geologic time are resultant of these two factors, universal changes in sea level and local regional or

Fig. 7–13. *Hypotheses on the accumulation of sediments through time* (compare to the restored section of Cambrian in western United States, Fig. 7–10). *A.* Sea level stands at S early in time, sediments having accumulated on the original surface of erosion. Land stands more than 12,000 feet higher on the coast to the east. A rise in sea level of 10,000 feet to S^1 will cause the sea to spread over the land until the latter has little more than 2000 feet of relief, and if the sediment supply permits, 10,000 feet of sediment might accumulate in the sedimentary basin. Such rise of sea level is *eustatic movement.* This hypothesis was conventional in the nineteenth century to explain the Cambrian.

B. Sea level stands at S early in the time, sediments having accumulated on the original surface of erosion. Land stands about 2000 feet higher on the coast to the east. Sediments to thickness of 10,000 feet can accumulate in the basin if the base of sedimentation P subsides to P^1, and erosion of the land to L' produces sufficient sediment to fill the basin. Such subsidence of the basin is through *epeirogenic movement,* crustal warping. If there is some rise of sea level through the same time, as to a level 2000 feet higher at S^2, there might be spread of seas over the margin of the lands, a more complex history that might be further complicated if there be differential warping of the land. Distributions of lands and seas are controlled by both eustatic and epeirogenic movements in varying degrees.

Fig. 7–14. Dells of the Wisconsin River at Kilbourn, Wisconsin, exposing shallow-laid upper Cambrian St. Croixan sandstones; some of the beds show cross-bedding, such as one with inclination toward the right just to the right of the old river boat (A. Devaney).

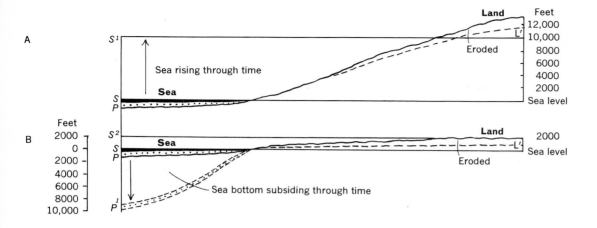

A

S^1 **Land**   Feet
12,000
10,000   L'
Sea rising through time   Eroded   8000
6000
4000
S **Sea**   2000
P Sea level

B

Feet
2000
0   S^2 **Sea**   **Land**   2000
2000 S L'   Sea level
4000 P Eroded
6000
8000   Sea bottom subsiding through time
10,000 P^1

119

continent-wide warping of the crust, to which there is the added effect of sediment filling the lower areas and so displacing marine shores.

THE WESTERN GEOSYNCLINE

The sections of the Cambrian rocks through the western part of the continent show a region of great subsidence on the west separated by a zone of bending, flexure, from a rather stable, little subsiding region to the east. A belt that subsides deeply as its contained rocks are laid is known as a *geosyncline*. So we find that a rather stable part of the interior of the continent was margined on the west by a belt of subsidence, a geosyncline. We can only surmise the Cambrian history of the region farther west from knowledge of younger rocks, for strata with Cambrian fossils are unknown. For the present, we will consider that the history is unknown, and that the western limit of the geosyncline is not determinable. But there is abundant evidence of the nature of more stable region to the east, for Cambrian rocks are known through a large part of the continental interior.

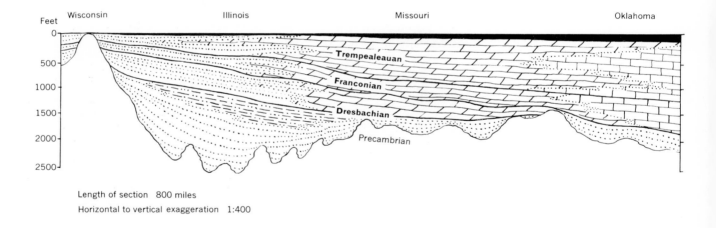

Fig. 7–15. Restored section of Cambrian rocks as at the end of Cambrian time from Wisconsin to southern Oklahoma; note great exaggeration of vertical scale. Only the St. Croixan Series is present along this line of section, represented by the three stages. Near-shore sands grade into offshore dolomite and limestones (calcitites). But thicknesses vary greatly locally because of the topographic relief on the land surface that the Cambrian seas covered, as well as through somewhat different rates of subsidence from place to place, differing rates of sedimentation with varying sediment supply, and compaction. The stages are defined by surfaces of synchrony judged from the presence of zones having successive fossil associations; these zonal subdivisions of the stages can also be distinguished.

THE CENTRAL CRATON

The Cambrian rocks preserved in the interior of North America are restricted to the area south of the Great Lakes. The rocks are wholly late Cambrian (St. Croixan), except in limited areas east of the sinuous limit of the lower Cambrian. Thickness is generally of a few hundred feet, though in some areas it is as much as a thousand feet. The sea lapped against hills on the Precambrian surface, so that successive stages of the upper Cambrian lie higher and higher on such monadnocks on the pre-Paleozoic peneplane. The thicknesses are constantly small compared to those in the geosyncline to the west. The sediments are almost wholly quartz sandstones in the states in the north from South Dakota to New York (Fig. 7–14), becoming limestones and dolomites overlying basal sandstones southward, as in the Ozark region of Missouri (Fig. 7–15). This entire area has much the same character as the relatively more stable belt east of the western geosyncline. It is the *craton* of Cambrian time, the broad stable continental interior.

Fig. 7–16. Chimney Rock, Tennessee, in the Great Smoky National Park. The sedimentary rocks of the Ocoee Group are extensively exposed in the Great Smoky Mountains of Tennessee and North Carolina. They lie stratigraphically below the fossiliferous Cambrian, but are in essentially continuous depositional sequence. Their classification as Precambrian or as Cambrian depends on the definition of the term Cambrian (Thornton Studio).

Eastward there is great change, like that to the west. A line can be drawn from the coast of Labrador along the east of the Adirondack Mountains in New York and the west of the Appalachian Mountains to Alabama. To the east of this line (Fig. 7–11) lower Cambrian is almost invariably present, thickening eastward. The problems of its base have been the subject of study for many years, the critical area being in and around the Great Smoky National Park in Tennessee and North Carolina. The *Olenellus* fauna that distinguishes lower Cambrian is known sparingly from some upper quartzites in a group of consolidated sands and clays (The Chilhowee Group) underlying thick carbonate rocks with middle Cambrian fossils. The quartzite succession is separated by thrust faults from other, conglomerate-bearing terrigenous sedimentary rocks, the Ocoee Group, widely preserved in Great Smoky Park. The Ocoee rocks (Fig. 7–16) were for a time thought to be the time equivalents of the Chilhowee quartzites and shales. As they contain coarser sediments, conglomerates, they were thought to evidence an eastern origin for the lower Cambrian, derivation from lands southeast of the geosyncline that subsided to contain their thousands of feet of shallow-laid rocks. Further study has led to the conclusion that the sediments equivalent to the Chilhowee are finer textured, and lie above the Ocoee; thus the Chilhowee sediments may overlap the Ocoee on the west, and both may have been derived from that western source prior to the deposition of fossiliferous lower Cambrian. The Chilhowee is lower Cambrian in its upper part because it contains characteristic trilobites there; whether its lower part is Cambrian or Precambrian depends on the manner of definition. The Ocoee is still older and is generally called Precambrian, though it seems to belong in the same tectonic setting as the rocks of the succeeding Cambrian.

So, rocks containing *Olenellus* are not only restricted to a belt east of the continental interior, but the pre-*Olenellus* sedimentary rocks thicken enormously southeastward in eastern Tennessee into North Carolina. The succeeding middle and upper Cambrian generally have thicknesses of a few thousands of feet in the southeast, and fossils very like those found in the western geosyncline as well. And all are dominantly carbonate rock above basal quartz arenites. To the west of this belt of thick Cambrian, as in New York and Ontario, the upper Cambrian lies unconformably on Precambrian (Chapter 7 opening) and has a few hundred feet of quartz sandstone (Fig. 7–17). The relatively stable part of North America was margined on the east, too, by a subsiding belt, a geosyncline, trending from western Newfoundland to Alabama. The region eastward is known to have the Cambrian

Fig. 7–17. *Upper Cambrian sediments in northern New York. A. Ausable Chasm, a gorge eroded in bedded sandstones of the Potsdam formation in northeastern New York; the Precambrian rocks lie a few hundred feet below. To the east, beyond the Champlain thrust fault along Lake Champlain in Vermont, the corresponding upper Cambrian sediments lie above thousands of feet of lower and middle Cambrian rocks. Ausable Chasm is on the margin of the deeply subsiding Cambrian geosyncline, from which the seas did not spread westward until late Cambrian time (New York State Museum).*

B. Wellesley Island, in the Thousand Islands of the St. Lawrence River, northwest of the Adirondack Mountains; a roadside cut in bedded upper Cambrian (Potsdam) sandstone with a lens of conglomerate; Precambrian is only a few feet below the exposure, and the sediments were laid very near the shore.

A

B

rocks of the Atlantic Provinces of Canada and of eastern Massachusetts, beds with *Holmia* and *Paradoxides* that we have discussed. They are so separated that original relations have not been determined across the intervening areas.

CRATON AND GEOSYNCLINES

To the south of the craton, the Cambrian is of the late Cambrian St. Croixan Series as far as it is accessible to view; south of Oklahoma and central Texas, the rocks are concealed by the coastal plains sediments. Northward, the Cambrian thins and disappears before it reaches the Hudson Bay region, which evidently remained until the close of the period as a source for the quartz sands that were laid in the seas of the states bordering the Great Lakes. The shores were irregular with peninsulas and embayments (Fig. 7–18). Thus this central region of the craton subsided least of all, standing as low land margined by shallow seas until the end of the period.

The margin of the craton was a flexure passing into the deeper subsiding geosynclinal belts on each side. Cambrian North America had a fairly simple structural plan. Compare this pattern of belts with that of the present structure discussed in an earlier chapter. The sedimentary rock regions of North America were divided into those with rather flat lying beds, and those having folded and faulted sediments. The lines drawn at the margin of the Cambrian craton, the region between the Cambrian geosynclines, about coincides with the lines separating the folded and faulted beds from the flat lying rocks of the interior on the present rock map (Fig. 2–4). The belts that subsided deeply during the Cambrian later were deformed, whereas the rather stable interior of Cambrian time has remained little deformed. Areas in which Cambrian rocks are folded and faulted are also those in which Cambrian sedimentary rocks are thicker. The character and distribution of Cambrian rocks is of interest not only in permitting interpretation of the state of North America more than a half-billion years ago, but in revealing significant relations with the present structure of the continent.

Fig. 7–18. Paleogeographic map showing directions of current flow on the floor of the late Cambrian (St. Croixan-Dresbachian) sea in Wisconsin, derived from the recording of the inclination of cross-bedding in the near-shore-laid sandstones. In addition, the map is stratigraphic, for it shows isopachs (lines of equal rock thickness). The small insert shows the records of plotted inclinations on which an average direction is indicated by an arrow; the numbers of observations in each small sector are indicated relatively by the extension of each sector (after W. K. Hamblin).

Fig. 7–19. Restoration of the bottom of the Cambrian sea based on the fossils in the middle Cambrian rocks of southeastern British Columbia (The Smithsonian Institution). Trilobites will be recognized readily, and there are other kinds of arthropods. The pipe-like structures on the right are sponges, and the floating disc with appendages is a jellyfish. The Burgess Shale of Mount Stephen, British Columbia, has yielded a great variety of fossils, some preserving the record of the soft parts; they will be described in Chapter 15.

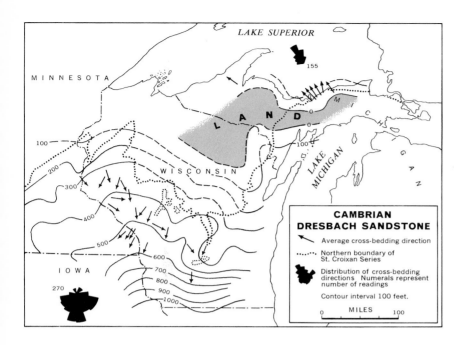

**CAMBRIAN
DRESBACH SANDSTONE**

→ Average cross-bedding direction

·······• Northern boundary of
St. Croixan Series

⬛ Distribution of cross-bedding
directions. Numerals represent
number of readings

Contour interval 100 feet.

MILES
0 _____ 100

8

The Facies Faunas in the Ordovician Marginal Volcanic Belts

Ordovician pillow lava, Willow Canyon, Toquima Range, central Nevada; the surface of a flow.

8 *The Facies Faunas in the Ordovician Marginal Volcanic Belts*

The name Ordovician was applied by Lapworth in 1879 to rocks in Wales that had been included in the upper part of the Cambrian by Sedgwick and in the lower Silurian by Murchison. The Welsh Ordovician is composed of thousands of feet of graywackes—quartz silty and sandy argillites—and volcanic rocks—flows, agglomerates, and tuffaceous sediments (Fig. 8–1). The Cambrian provided the basis for the judgment that rocks containing similar fossils should be classified as of about the same age. Hence fossils found in North America that are like those present in the typical Ordovician of Wales will be considered to be evidence sufficient to warrant placing the rocks that contain them in the Ordovician System.

GRAPTOLITES AND THEIR DISTRIBUTION

The Ordovician in Wales can be subdivided into series and stages having several kinds of graptolites—chitinous or horny-skeletoned colonial animals whose individuals live in minute cup-like chambers attached to a thread or ribbon-like structure (Fig. 8–2). The beds succeeding the *Dictyonema*-bearing Cambrian have many kinds that are distinctive of a succession of fossil zones; among them are forms resembling two intersecting leaves, *Phyllograptus*, others having four ribbon-like stipes extending from a center, *Tetratraptus*, and the lower Ordovician culminates in shales with tuning-fork-shaped forms, *Didymograptus*. The middle and higher parts of the system have other graptolites such as one with a sinuous thread having straight ribbons projecting at regular intervals from one side, *Nemagraptus*, and another ribbon-like with the little cups extending from both sides, *Climacograptus*. These are but a few among a great many genera present in the argillaceous rocks of the Ordovician System in several areas in the British Isles. A very similar succession of well-preserved graptolites is found in Norway and Sweden.

Lower Ordovician graptolites such as *Phyllograptus*, *Tetragraptus*, and *Didymograptus* have been found in dark argillites and graywackes in many places in North America (Fig. 8–3). They are widely distributed in Newfoundland and known in Cape Breton, Nova Scotia, in New Brunswick, and south of the St. Lawrence in Quebec (Fig. 8–4*A*), along the Hudson in New York (Fig. 8–4*B*), and southwest to Alabama, in the Ouachita Mountains of Arkansas and Oklahoma, and in west Texas. They are found in scattered places through the mountains of

Fig. 8–1. Snowdon Mountain (3560 feet), the highest peak in north Wales, composed of middle Ordovician volcanic and sedimentary rocks in the region from which the Ordovician System received its name (Aero Pictorial, Ltd.). The view from the west shows the railway line that ascends to the peak.

the West in central Nevada, Idaho (Fig. 8–5), and extreme northeastern Washington, and in a number of localities in eastern British Columbia, the Yukon, and southeastern Alaska. Another band of graptolite-bearing rocks adjoins the Arctic Ocean from the mouth of the Mackenzie River through Cornwallis Island to northwest Greenland. Lower Ordovician rocks are present, then, along the continental margins; but their graptolites are rare in the interior.

The graptolites are not easily found in many of these sections; careful search and good eyes are requisites to their discovery. Dark, shaly argillites are most likely sources, like those in which they appear in Britain and Scandinavia. The graptolites are thought to have been free-floating organisms that are preserved only where they settled into favorable environments, such as in deposits laid in quiet water. Other means will thus be needed to identify rocks of the same age laid under conditions unfavorable to graptolite life and preservation but containing other kinds of organisms.

VOLCANIC ROCKS AND GRAYWACKES

The sort of sequence that contains the graptolite shales has such beds in scattered horizons in thousands of feet of sedimentary rocks, whether in the Welsh Ordovician or sections in Newfoundland or Nevada. Very deep subsidence was needed to accommodate such great thickness of marine-laid sediments; either a deep trench was filled, or the base of the Ordovician subsided as deposition progressed. In some sections,

B

Fig. 8–2. *Graptolites. A.* Ordovician Graptolithina: the illustrations are of Ordovician graptolites that appear generally in the following sequence, though several are found together. (1) *Dictyonema*, a genus that ranges from the Cambrian to the Carboniferous—a dendroid graptolite. Lower Ordovician: (2) *Goniograptus*; (3) *Tetragraptus*; (4) *Phyllograptus*; (5) and (6), *Didymograptus* of two sorts, extensiform and pendant or "tuning-fork"-like. Middle and upper Ordovician: (7) *Nemagraptus*; (8) *Dicranograptus*; (9) *Dicellograptus*; and (10) the biserial *Climacograptus*, showing the whole stipe and detail in section; and (11) the similar *Amplexograptus*—*a.* part of the stipe with the alternating thecae, *b.* the growth of the thecae from the sicula within the stipe. Magnifications here vary from natural size to an enlargement of ten times.

B. Graptolite of the genus *Nemagraptus* from Ordovician Bolarian shale at Lusters Gate, Virginia (The Smithsonian Institution).

The figures illustrate individuals as normally observed, and as known from well-preserved and prepared specimens. Graptolites are colonial marine organisms known only from early Paleozoic rocks, particularly those of the Ordovician and Silurian. Each colony (rhabdosome) has one or more blades (stipes) bearing small cup-like individuals (theca, plural thecae). The colony grew from a single individual (the sicula) from which one or two buds developed; these in turn gave rise to succeeding thecae in varied and complex arrangements. The colonies were suspended from disk-like floats or from seaweeds; graptolites were free-floating on the surface of the sea, and drifted widely over the earth. They are found commonly as flat, carbonaceous films that resemble narrow saw blades having teeth, the thecae, on one side (uniserial) or both sides (biserial) of the central suspending filament (virgula). Well-preserved forms in three dimensions are found rarely, some retaining the original chitin, a horny resistant sub-

stance having complex molecules of carbon, hydrogen, oxygen and nitrogen; when prepared and clarified, these are beautifully transparent amber fossils preserving intricate structural details. In some cases, the chitin and internal matter have been replaced, as by pyrite, retaining the form of the original. The intricate manner of growth and arrangement of the stipes and thecae, constant within single species, and the rapidly changing forms through time, have made graptolites very useful fossils in classifying the rocks in which they are preserved.

The placement of graptolites in the organic classification has been disputed. Their many small cups give them resemblance to bryozoans and corals, and they have been placed in those phyla; however, the chitinous skeleton suggests that of the arthropods, such as the trilobites. The skeletal details are most similar to those in some of the types of protochordates—organisms that are commonly placed as a phylum of the most primitive vertebrates.

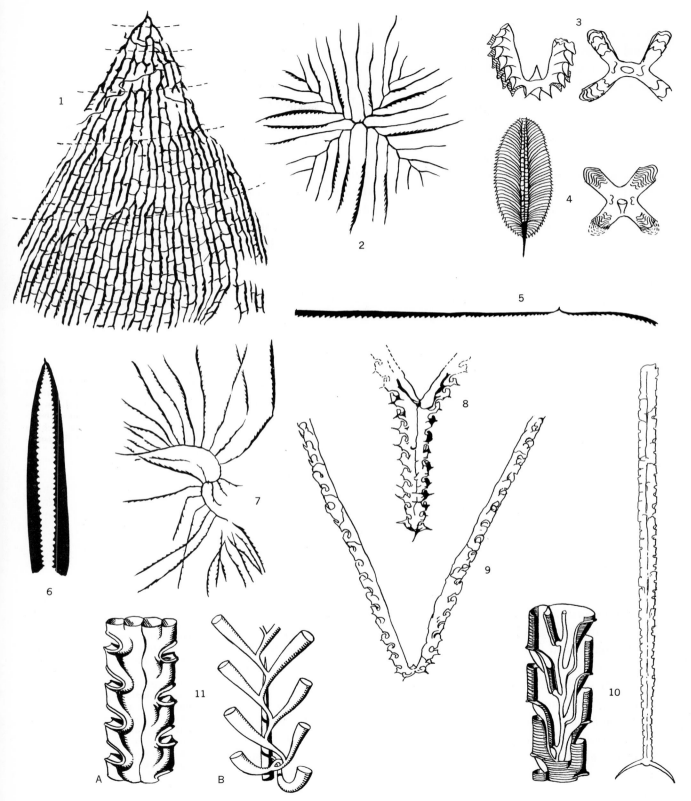

131

there are associated conglomerates containing pebbles of granite and of varied volcanic and sedimentary rocks (Fig. 8–6). Parts of the sections are of chert with delicate radiolaria, minute one-celled animals having siliceous skeletons. And thousands of feet are volcanic—basalt and andesite flows (Fig. 8–7 and Chapter 8 opening), agglomerates and tuffs, and fragmental volcanic deposits. Not only must there have been great subsidence but also there were tectonic or structurally raised lands as sources for the sediments that were carried to the subsiding sea floor. Graywackes are of particles of minerals and rocks that have been little sorted, quite in contrast to the well-sorted quartz arenites that came from the interior in the Cambrian. The poor sorting shows that deposition was at too great a depth to permit waves to winnow the detritus; the quiet waters of the bottom prevented destruction of the delicate graptolite skeletons that settled. And there were volcanoes in the same region, some in the area that was subsiding deeply. The sediments were eroded and transported by streams from nearby lands, both of older rocks and volcanoes, and perhaps some slumped into the deep water or were carried in turbid muddy flows.

The sections that contain the graptolites do not all have lava flows and fragmental volcanic rocks, but the known volcanic rocks are all in the graptolite-bearing graywacke belts. Some successions that seem Ordovician because they directly underlie fossiliferous Silurian do not seem to contain graptolites but have some volcanic rocks, as in the somewhat metamorphosed rocks of the White Mountains region of New Hampshire, and in the Klamath Mountains of northern California. So we can reconstruct the geography of the Atlantic and Pacific margins of North America as having deeply subsiding belts with adjoining highlands that yielded some of the constituents of argillite, graywacke, and conglomerate; volcanoes and volcanism were widely associated.

SHALY AND SHELLY FACIES

Rocks in the interior of the continent very rarely contain Ordovician graptolites and their classification against the graptolite succession is difficult. In some localities thin zones of graptolite-bearing shaly argillite have been found within sequences of limestone having fossils of shell-bearing animals of other classes, such as bivalved brachiopods (Fig. 8–8), snails, chambered cephalopods, and trilobites. Occasionally, graptolites are found in limestones; because these rocks do not compact as much as argillites, the fossils are frequently well preserved, and can be removed because the limestone is soluble in acids that do not affect the graptolites. In a very few places, large blocks of limestone with other fossils lie within graptolite-bearing shales, (Fig. 8–9); the

Fig. 8–3. Principal graptolite-bearing shale localities in North America. The triangular symbols are places where lower Ordovician graptolites (Arenigian and Llanvirnian), as well as younger forms, have been found in abundance; the crosses are localities where only the younger Ordovician (Llandeilan, Caradocian, and Ashgillian) forms have been found. Generally the fossils are common only in a few thin beds in such successions. Graptolites are also found occasionally in Ordovician rocks such as limestones in the interior of the continent.

Fig. 8–4. *Classic graptolite localities in eastern North America.* **A.** Lower Ordovician Arenigian shales along the St. Lawrence River at Levis, across the river from Quebec City; the beds dip to the left, southward. Graptolites such as *Tetragraptus* and *Didymograptus* were discovered and described at Levis more than a century ago; many of the distinctive forms were identified here before they were found in western Europe. **B.** Lower Ordovician shales along Deepkill, a stream north of Troy, New York, that has yielded many representative graptolites similar to those found at Levis, and in northwestern Europe. The beds dip gently eastward up the stream.

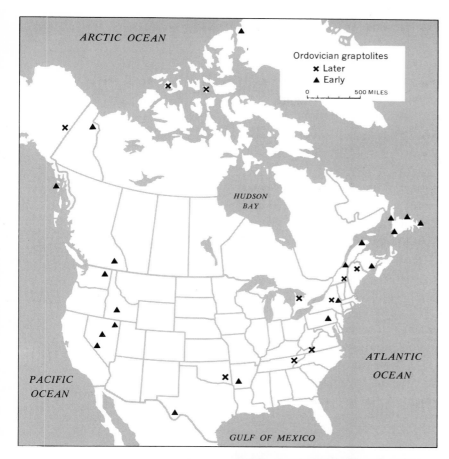

A

B

graptolites are there known to be younger than the fossils in blocks in the immediately underlying conglomerates. Thus in western Newfoundland, richly fossiliferous limestones yield fossils nearly of the age of the associated shales with indigenous graptolites. But at Levis, across the St. Lawrence from Quebec city, conglomerates interbedded in the Ordovician shales have much older, Cambrian, limestone boulders (Fig. 8–10), some that are scores of feet long.

FACIES CHANGES

In some instances, Ordovician limestones of changing lithologies and fossils can be traced through mapping of continuous distinctive lithic and faunal beds into shale with the graptolite faunas. These changes in synchronously laid beds from one lithology to another are changes in *facies;* there are the two principal contrasting facies, the "shaly" or argillitic with graptolites, and the calcareous and sandy with shells of a variety of other animals, the "shelly" facies.

Gradations from carbonate facies to graptolite shale facies or argillite facies can be traced through exposures in several regions. In New York south of the Adirondack Mountains, about two hundred feet of middle Trentonian limestones (Fig. 8–11) on the west were shallow-laid calcite sands—calcarenites—and silts with argillaceous partings, which contained an abundance of brachiopods: bryozoans, trilobites,

Fig. 8–5. Ordovician and Silurian graptolite-bearing shale along Trail Creek northeast of Sun Valley, central Idaho. The broad slope in the distance exposes nearly 1000 feet of dark shale between the west- and left-dipping quartzite beds in the tree-covered area on the right and the mountain top on the left. The succession contains graptolites ranging from the lower Ordovician forms like those at Deepkill, New York and Levis, Quebec on the right to Cincinnatian or Ashgillian forms coming in the upper light-colored slope near the trees on the left, where there are lower Silurian graptolites. The region lies in the graptolite facies, carbonate rocks of the same age appearing in ranges nearby to the east.

Fig. 8–6. Columnar section of the Ordovician rocks exposed near Girvan, Ayrshire, Scotland, showing the great variety represented (after A. Williams). There are a few graptolites in shales associated with the lavas low in the sequence. The upper beds of limestone and sandy shale contain principally shelly-facies faunas of brachiopods, trilobites, and other forms. The rocks, laid in a volcanic geosynclinal or eugeosynclinal belt, contain an unusual proportion of conglomerates; many of the boulders are of igneous rocks thought to have come from fault scarps that bordered the geosyncline. Tens of miles to the southeast, near Moffat, the rocks equivalent to those above the volcanics at Girvan are graptolite-bearing shaly, cherty argillites not much more than one hundred feet thick—a *condensed section* laid in a deep trough that gained very little terrigenous detritus.

Feet

ORDOVICIAN

Trentonian — Cascade — 400–610

Caradocian — Ardmilian — Ardwell Group — Assel

Tormitchell — –240 *–4500*

Balclatchie Group — *1000*

Bolarian — Barr Group — Benan — 180–2100

Superstes Shale — –125

Stinchar Limestone — –220

Confinis Sandstone — –150

Auchensoul Limestone — –60

Kirkland Conglomerate — –800

Ballantrae Volcanic rocks — Feet below surface

Fig. 8–7. *Pillow lavas in Newfoundland and Nevada. A.* Pillow lava in Ordovician rocks at Twillingate, on an island off the northeast coast of Newfoundland. The pillows are of basaltic lava, having a band of greenish epidote (calcium iron-aluminum silicate) beneath the surface, the result of alteration. The hammer lies on the convex top of a pillow that has the characteristic downward extension that permits determination of the orientation of beds. Pillow lavas are formed by lava flowing into marine or fresh water.

B. Pillow lava in Ordovician rocks in the Toquima Range, central Nevada; a view of the surface of a flow. Lavas of this sort are restricted to the volcanic geosynclinal belts along the margins of the continent, the eugeosynclinal belts. The associated rocks are argillites, probably formed from ash, and red, green, and black cherts containing siliceous organisms—radiolaria.

A

B

135

Fig. 8 – 8. *Ordovician Brachiopoda.*
A. The illustrations show representatives of the classes of brachiopods and of several of the orders that are present in the middle Ordovician rocks; they are of natural size or somewhat enlarged. Listed below are the localities from which the specimens came; most of the genera are known in many places, and the various species differ little from the examples. The drawings are after illustrations by G. A. Cooper, principally in *Chazyan and Related Brachiopoda* (The Smithsonian Institution, 1956).

The upper figures are of inarticulate brachiopods: (1) and (2) an atreme of the genus *Plectoglossa* from Oklahoma; (3) and (4) a neotreme, *Schizotreta* collected in Alabama.

The other figures show the great range in appearance among the articulate brachiopods belonging to each of several of the orders. (5–9) Five views from the brachial, lateral, posterior, brachial interior, and pedicle interior sides of an

orthid (order Orthoidea) of the genus *Hesperorthis* collected in Oklahoma. (10–14) A syntrophid, *Camarella*, from Tennessee. (15–22) Pentamerids, (15–18) are *Rostricellula*, from Minnesota, and (19–22) *Zygospira* from Tennessee. (23–24) A spiriferid, *Catazyga*, from Ontario. (25–27) A plectambonitid, *Sowerbyella* from Virginia. (28–30) A strophomenid, *Strophomena*, and (31–34) A dalmanellid, *Dalmanella*, both from Tennessee.

B. Some genera of Ordovician Brachiopoda that look very much like each other but are not closely related (homeomorphs).

The upper two rows have from left to right, views from the pedicle, posterior side, anterior, and brachial sides. The compared genera are (1–5) *Doleroides* and (6–10) *Pionodema*, dalmanellid and orthid brachiopods, respectively—the former having a non-punctate, fibrous shell structure, the latter a punctate shell. Below them are pedicle, side, posterior,

and brachial views of two other externally similar forms: (11–14) genera, *Chaulistomella*, an orthid and (15–18) *Strophomena*, a strophomenid; note the closed pedical opening of the latter. Brachiopods may have very close external similarities but have differing internal structures or shell materials that show that they do not have close biologic relationships. To assure correct classification they must be studied with care and skill.

Brachiopods. The brachiopods, of the phylum Brachiopoda, are marine animals having two shells that differ, though each is bilaterally symmetrical, the valves are called brachial and pedicle. The pedicle valve generally has an opening at the hinged end through which passed a fleshy appendage or pedicle by which the animal was attached. Brachiopod shells differ from pelecypod or clam shells in that the latter are a similar pair, left and right, but each is not symmetical. Brachiopods live today, but are uncommon. They were very widely distributed and varied in Paleozoic rocks, and are among the most useful organisms in the classification of Paleozoic strata in time. The fossils preserve in the interior a configuration showing the attachments of the muscles, and the calcareous structures pertaining to the hingement and the arms or brachidia that supported fleshy parts that directed food to the mouth. The brachiopods illustrate the manner of classification within a phylum.

The phylum is divided into two classes, the Inarticulata and Articulata, the former having valves connected only by muscles and other soft parts, whereas the latter have interlocking teeth in the hingement. The Class Inarticulata has two orders, Atremata having the pedicle opening passing between the valves, and Neotremata having the pedicle opening confined to the pedicle valve. One of the best known of modern brachiopods, *Lingula anatina* is an atreme, and there are a few modern atremes. The shells are either of chitin or of calcareous material, commonly of both in alternating layers; some have phosphatic composition. The Class Articulata has valves with teeth that lock in sockets in the opposite valve. The articulate brachiopods have shells of calcium carbonate, though fossils can have the original composition replaced by silica or other substances; some of the most attractive specimens are prepared by dissolving the limestone from around brachiopods whose skeletons have been silicified and are hence insoluble in most acids.

Articulate brachiopods have many distinctive attributes that permit their differentiation. They are generally oval, elliptical, or subcircular and less than an inch long, though some reach breadths of several inches, and true giants are known to have attained widths of a foot. They can be distinguished by their external outline, shape, and ornamentation. The shells are three-layered, the outermost organic layer not preserved in fossils. Of the other two layers, the external features are preserved in an outer lamellar layer, and the internal surface on a thicker prismatic layer. The inner layer on some forms has minute holes or punctae, others have fibrous shell, and in a third type, small spicules in the impunctate shell weather to simulate pits, so the shells are called pseudopunctate. Shells with the same type of shell structure differ widely in external shapes and in structures preserved in the interior, impressions of muscle attachments and internal organs, the supporting structures of soft parts, and the hingement teeth and sockets. Some shells having similar exteriors have markedly different shell structure and internal characters or both. They are judged not to be closely related, are placed in different genera; such mimics are known as *homeomorphs*.

Combinations of some of these characters lead to association of genera and families into many suborders. Among those in the Ordovician, the most important are the impunctate Orthacea, the pseudopunctate Strophomenacea and the puncatate Dalmanellacea. Among these, the orthid *Doleroides* mimics the punctate *Pionodema*, and the orthid *Campylorthis* resembles the strophomenid *Strophomena*; there are internal differences as well as those of shell structure. Some of the orthids, such as *Platystrophia*, resemble the Spiriferacea of later systems; the latter, also impunctate, have spirally coiled internal calcareous structures called brachidia, and in the Ordovician are represented by few forms, such as *Catazyga*. The Ordovician contains representatives of several other suborders, not here distinguished.

There were Cambrian brachiopods, and their descendants live rather uncommonly in modern seas—so they were long-lived as a phylum. They were most common and varied in the Paleozoic. With limited study, a student of fossils can come to know some of the more distinctive classes and genera, but only a trained specialist can know more than a few of the thousands of kinds of brachiopods that thrived in the Paleozoic Era and lived in diminishing numbers through succeeding eras.

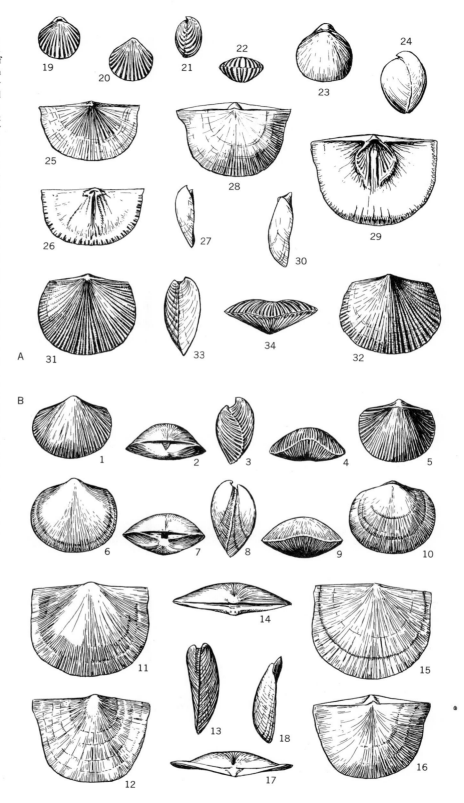

137

and molluscs. Eastward, the limestones pass into more than a thousand feet of black shaly graptolite-bearing argillite (Fig. 8–12) within a few miles. As the textures become finer and the argillaceous content rises, rather sparsely fossiliferous beds have proportional increase in the number of molluscs, particularly chambered cephalopods. If we continue eastward, the limy silts and clays—calcisiltites and calcilutites—become more argillaceous until they are shaly argillutites, thin-bedded consolidated argillaceous clays (Fig. 8–13), with frequent graptolites and a few trilobites unlike those farther west. If the change could not be followed progressively, the equivalence of the beds at the two ends would hardly be suspected, because the faunas are wholly different.

EFFECTS OF THRUST FAULTS

Commonly, later deformation has so disturbed and broken the continuity of rocks of the carbonate and argillite facies that their original relations are obscure. In central Nevada there is an example of the relations. Graptolite-bearing black argillites, graywackes, cherty argillites, and volcanic rocks lie on thrust faults above carbonate rocks that seem of the same age, the latter continuing eastward into Utah. The argillite-chert-volcanic facies should lie farther westward, but long after the Ordovician, the rocks of that belt were carried eastward as great thrust sheets that concealed the correlated carbonate rocks until erosion

Fig. 8–9. Ordovician Cow Head conglomerate on the west coast of Newfoundland. The light-colored rock in the foreground is a large block, more than ten feet long, that is associated with boulders and finer matrix in a steeply south-dipping band that extends into the cliff in the background. The boulders have many kinds of beautifully preserved trilobites. Conglomerate bands are interbedded with shaly argillites containing graptolites, as on the left of the distant cliff. Thus the trilobites are at least as old as the associated graptolites. The trilobites in one band seem of about the same age, and differ from those in other bands. It is believed that the limestone boulders are from sediment laid in shallower water at about the same time as the shales were laid in deeper water, and that the boulders were dislodged and slid into the deeper water. The conglomerates and associated shales give one of the best means of determining the relative ages of fossils that are not generally found in direct association because of the differing environmental preferences.

brought them to view (Fig. 8–14). When the large fault-bounded blocks are restored to their relative original positions, lower Ordovician-Canadian limestones are found to become argillaceous westward. Brachiopods, other than a few phosphatic forms, become quite uncommon, and trilobites, generally in fragments in carbonate rocks once east and now tectonically below, are complete more frequently; the water deepened so that the waves did not disturb and separate the segmented parts. In higher thrust slices there are graptolite-bearing cherts, shaly argillites, and volcanic rocks that originally lay farther westward.

Thrust-faulting is prevalent along such zones of changing facies. Argillitic and cherty rocks of the western Ouachita Mountains in Oklahoma have been thrust over rocks of carbonate facies, and similar relations pertain in the vicinity of Quebec city along the St. Lawrence, and in the northern end of the Taconic Range in eastern New York and southwestern Vermont (Figs. 8–15, 8–16). If we are to reconstruct the geography of the time, it is necessary that we move the tectonically higher, overthrust mass of argillitic rocks back to its relative original place. This creates a peculiar type of map in which the rocks of folded and faulted regions are "unfaulted" and "unfolded," so as to regain differences in distance and direction that they had when they were laid. Such a map is a *palinspastic* map, a "stretched-back" map.

Fig. 8 – 10. *Limestone conglomerates that are interbedded in graptolite-bearing lower Ordovician black shales at Levis, across the St. Lawrence River from the city of Quebec (F. F. Osborne).* *A.* An exposure of conglomerate with small slabs of limestone.
B. A quarry in a single block of middle Cambrian limestone tens of feet in diameter. Such blocks are thought to have traveled into the shale in great submarine slides, possibly dislodged from rising fault scarps.

A

B

CLASSIFICATION OF CARBONATE ROCK FACIES

These sorts of evidence make it possible to determine which fossils in the limestone sequences are of about the age of graptolites in specific stages in the argillite-graywacke successions. We can correlate one with the other, and demonstrate that the limestones with fossils of other sorts, such as trilobites and brachiopods, are Ordovician. When the latter are systematically arranged, they permit a separating of the carbonate rocks into many stages distinguished by fossils other than graptolites.

The rocks of the "shelly" carbonate facies in North America have been placed in five series, though the classifications differ. The terms adopted here are Canadian, Chazyan, Bolarian, Trentonian, and Cincinnatian. The first two correspond with the lower Ordovician of the graptolite facies nomenclature but the term "lower" is often restricted to the Canadian; the Bolarian and Trentonian are commonly called Mohawkian. The names are but convenient terms to use to distinguish events in successive Ordovician epochs. These are in turn divided into stages, two or more in a series. These terms will be usefully applied in discussing the examples of changing conditions.

GEOSYNCLINAL BELTS

Thicknesses of the Cambrian rocks were found to be greater in the belts east and west of the relatively little-subsiding interior cratonal area. The Ordovician has volcanic geosynclinal belts beyond the non-volcanic

Fig. 8–11. *Trentonian limestones at Trenton Falls, north of Utica, in central New York. A.* West Canada Creek at high water; the gorge exposes about two hundred and fifty feet of limestones, the middle Trentonian (Denmark) being overlain by the upper Trentonian (Cobourg) at the base of the upper High Falls in the middle distance.

B. Sherman Fall, in middle Trentonian (Denmark) limestones at the lower end of the gorge. The two re-entrants on the right, nine feet apart, are in clay beds, altered volcanic ash falls.

A

B

140 THE FACIES FAUNAS IN THE ORDOVICIAN MARGINAL VOLCANIC BELTS

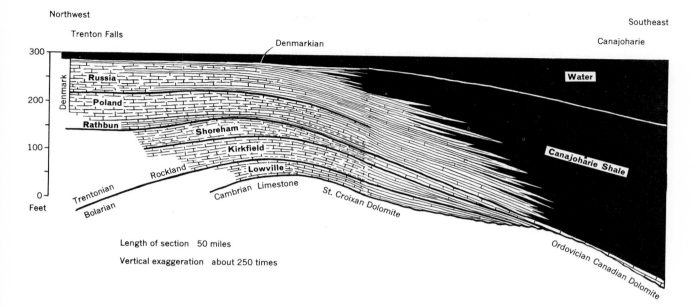

Northwest

Trenton Falls

Southeast

Canajoharie

Denmarkian

300

Russia

Water

200

Poland

Rathbun

Shoreham

Canajoharie Shale

100

Kirkfield

Lowville

Rockland

0

Trentonian

Cambrian   Limestone

St. Croixan Dolomite

Feet

Bolarian

Ordovician Canadian Dolomite

Length of section   50 miles

Vertical exaggeration   about 250 times

Fig. 8–12 (above).   Restored section of Ordovician rocks south of the Adirondack Mountains in New York, showing the changing facies in the middle Trentonian Denmarkian Stage from the richly fossiliferous limestones on the northwest, through barren argillaceous limestones with shale interbeds into the laminated Canajoharie shales, represented in black, on the right. The limestone beds below overlap southeastward on a pre-Bolarian erosion surface.

Graptolites are present in the black shales, with few other associated fossils. They are rarely found in the limestones, but the relations permit the comparison in time of fossils of quite different biological classes.

Fig. 8–13 (right).   The black shale of the middle Trentonian at Canajoharie, New York. These shales change facies into the limestones of the Trenton Falls gorge to the west and into the sandstones and shales eastward near Schenectady and Albany, thickness increasing greatly from west to east. The deep re-entrants or cavities between beds have been eroded into clays that are probably thin volcanic ash deposits. The Canajoharie shale has graptolites and a few trilobites.

geosynclines recognized in the Cambrian. Thus if we restore the Ordovician rocks from Colorado westward, we begin on the craton with a few hundred feet of late Ordovician limestones at the top of the section overlying a few feet of quartz sandstone. These rocks persist westward to central Nevada, thickening to a thousand feet or so from central Utah westward, becoming more argillaceous westward in central Nevada (Fig. 8–17). In Colorado these later Ordovician sediments lie on Cambrian or on thin lower Ordovician Canadian, but in central Utah the Canadian limestones thicken rapidly with younger beds, principally Chazyan, coming in at their top until they exceed three thousand feet as a somewhat variable but persistent plate through western Utah to central Nevada; then thrust faults, previously discussed, bring Ordovician volcanic-bearing argillites structurally above the carbonate facies, though originally they graded into each other. Thus we pass from the craton to a non-volcanic geosynclinal belt and then to a volcanic geosynclinal belt. Similar thickening takes place to the east of the Cambrian craton, particularly for lower Ordovician. The later Ordovician in the eastern belts became a time of increasing deformation culminating in the mountain making that will be discussed more fully in the next chapter.

In general, the Ordovician of the interior has a few hundred to a thousand feet or so of carbonate rocks with some quartz sandstones. Most were laid in shallow water, slowly so that they are of small

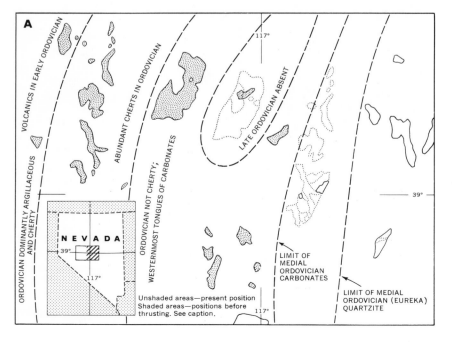

Fig. 8–14. *Palinspastic maps across zones of change in lithology in Ordovician rocks in central Nevada. A (left).* **Palinspastic map of an area in central Nevada; unshaded outlined areas have present positions; relative positions prior to thrusting are represented by similar shaped area in original place of deposition. The map is made on the interpretation of original paleogeography represented by the notations.**

B (right). **Restored section of the lower Paleozoic rocks in central Nevada, showing interpreted lateral relations prior to thrusting. The problems of thrusting will be discussed in Chapter 13.**

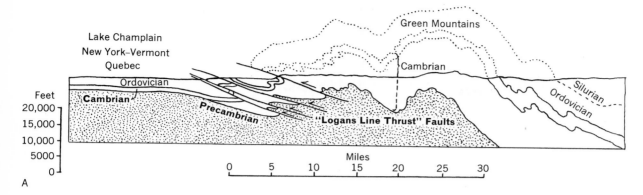

Feet
20,000
15,000
10,000
5000
0

Miles
0   5   10   15   20   25   30

A

Fig. 8–15. *Palinspastic map of Ordovician in the Adirondack–New England region.* *A.* Structure section along the international border eastward from Lake Champlain, showing thrust faults that reduce the original breadth of a surficial crustal segment; the rocks are restored as though preserved far above the present surface. The thrusts are thought to be of late Devonian age.

B. Map showing the lithologies dominant in the Ordovician upper Canadian Series within the region; note the anomalous shale area in the Taconic Mountains contrasting with carbonate rocks surrounding it.

C. An interpretation of the original distribution of the rocks in the region in which the fault blocks have been returned to the presumed relative original positions in Ordovician times, a *palinspastic* map (after W. M. Cady and M. Kay). The maps were made many years ago and are continually revised as knowledge is gained. The thrust around the Taconic Mountains is thought to be late Ordovician (Taconian) and those along Lake Champlain and the St. Lawrence River to be later Devonian (Acadian).

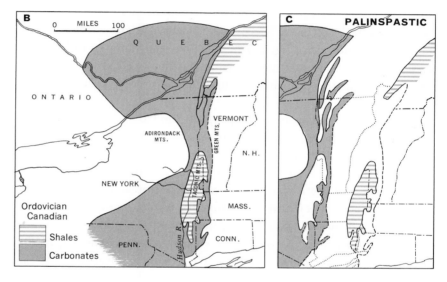

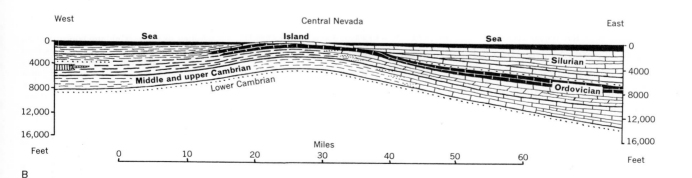

West   Central Nevada   East

Feet
0
4000
8000
12,000
16,000

Miles
0   10   20   30   40   50   60

B

143

thickness and with many interruptions in deposition and surfaces of erosion. The limestones have characters that enables the geologists to recognize them over wide areas, distinguish the lateral facies changes and judge their conditions of formation (Fig. 8–18). These rocks in the relatively stable cratonal areas thicken into geosynclinal belts having about the same position as the Cambrian belts of greater thickness and fuller depositional record. But beyond these geosynclines are the other belts that gained argillaceous sediments, graywackes, and some conglomerates from lands adjoining them and separating them into troughs; and within the belts were volcanoes and other sources of lava flows and volcanic detritus. The volcanic geosynclinal belts have been called the *eugeosynclinal* or "really" geosynclinal belts in contrast to the *miogeosynclinal* or "lesser" geosynclinal belts that are essentially lacking in local volcanism (Fig. 8–19). The paleogeography of the Ordovician volcanic belts must have been somewhat like that of the island arcs and associated submarine troughs of the present East Indies and West Indies.

The effects of volcanism were not wholly limited to the eugeosynclinal belts, however, for explosive volcanoes spread fine fragments to form ash deposits. If there had been but one great explosion, the ash bed would be an ideal datum for correlation of rocks throughout the region of dispersal, but there were many explosions spreading their ashes to differing extents. They have the distinctive feature that each represents a discrete fall over a wide region; beds in each section over a wide area must be exactly of the same age as some in other sections (Fig. 8–20). The problems are in determining which beds are equivalent. Beds can be correlated by normal methods using lithologies and contained fossils; a limited number of ash beds is found within seeming correlatives. When experience shows their number to be constant and certain beds to be consistent in thickness and lithic associations, the prospect of precise correlation becomes great. In central Pennsylvania, at least five volcanic ash beds persist in the upper Bolarian and eight or ten in lower and middle Trentonian. A few can be traced with fair confidence from New York to southern Virginia (Fig. 8–21) and Tennessee reaching thickness of more than a foot; distinguishable beds of an inch or so are known as far as west of the Mississippi River from Missouri to Minnesota. The volume of ash blown into the atmosphere and carried westward from volcanoes in the eastern volcanic belt during medial Ordovician is measured in scores of cubic miles. Fossils, progressively changing, are the best means of making approximate correlations, but within limited areas, precise correlation can be made only by lithic effects of physical or chemical changes that produced simultaneous

Fig. 8–16. The Champlain Fault, a thrust fault exposed on the shore of Lake Champlain north of Burlington, Vermont; Lower Cambrian dolomite has been thrust on middle Ordovician argillaceous shales. The stratigraphic throw, the thickness that would separate the rocks above the thrust and those below it in a normal succession, is probably a mile or so. But the fault plane developed on the shales below the dolomite, rising on an inclined break through a section of rather competent carbonate rocks to slide westward on middle Ordovician shales; the lateral movement on the thrust plane may have been several miles. The thrust is one of the principal breaks within the carbonate rocks of the geosynclinal belt that lies east of the Adirondack Mountains of New York. Although long attributed to the Taconian Disturbance late in the Ordovician, the thrust is now thought to have formed in the Acadian Orogeny in the late Devonian.

Fig. 8–17. Restored section of the Ordovician System from central Nevada to central Colorado. The thicker sections on the west are in more rapidly subsiding, geosynclinal belts, contrasting with the relatively stable cratonal area to the east. The carbonate rocks of the non-volcanic miogeosynclinal belt grade into graptolite-bearing argillites in the belt to the west, the eugeosynclinal belt having interbedded lavas and other volcanic rocks. Quartzites in the section were the product of the sorting and concentrating of the more resistant minerals gained from crystalline and earlier sedimentary rocks by streams on the cratonal lands and the currents of the marginal seas; carbonates accumulated in seas that terrigenous sediments did not reach; the sandy argillites in the west may have come from lands to the west. In late Ordovician, the argillite and carbonate belts were partly separated by narrow islands produced by crustal warping; see Fig. 8–14.

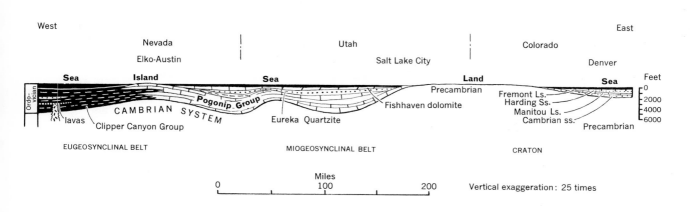

recognizable differences in rocks, or permitted introduction or caused destruction of fossil assemblages.

CORRELATION TABLES

Though the Ordovician rocks in the interior of the continent differ in thickness from place to place, they are never more than a thousand feet thick or so. When the lithologies and fossils are studied, the rocks can be correlated with the stages of the standard series. These can be shown graphically by placing the successive stages in vertical sequence along the left or right margin of a graph, the oldest at the bottom and the youngest at the top; time is represented as the ordinate on the graph. The content of sections in many localities can be plotted in successive columns from left to right, presence of beds being distinguished symbolically from absence of beds, so as to indicate which stages are represented by deposits in each section and which are lacking. Such a graph is known as a *correlation table* (Fig. 8–22). Such tables have been made for each system, recording interpretations of the relative ages of the thousands of formations in the innumerable exposures in North America.

When we take such a series as the Canadian and divide it into a number of parts, the plotted sections reveal that some stages are present in many sections and others in very few; if we made four divisions,

Fig. 8–18. *Ordovician limestones. A.* Limestones in a road cut on Highway 500, north of Bobcaygeon, central Ontario. Five feet of lower Trentonian limestone at the top of the cut overlies some thirty feet of Black River (Bolarian) limestone. Lithic units such as these are very persistent for long distances, with gradual changes in facies reflecting changing environments of deposition. Ten feet of limestone in the upper Black River gradually increases to nearly forty feet 200 miles to the east in New York—local sections are remarkably constant. Occasional metabentonites, volcanic ash beds, enable close time-correlation in the many sections.

B. Upper Ordovician limestone in the Franklin Mountains, southern New Mexico, near El Paso, Texas. The original deposit of calcium carbonate mud was partially replaced by silica during consolidation. Such distinctive lithologies persist over great areas in the Ordovician rocks in the continental interior. The illustrated limestone is one that was laid in the very extensive clear seas that spread from the Arctic in late medial and late Ordovician time. Environments were so different from the waters affected by the orogenics in the east that faunas, rich in corals, are difficult to correlate with those of the eastern standard (H. E. Howe).

Fig. 8–19. Paleogeographic map of North America in the middle Ordovician Chazyan Epoch (after M. Kay). The great central stable region or craton was margined by non-volcanic and volcanic geosynclinal belts in which sediments and igneous rocks accumulated to greater thickness than on the craton. The relationships between the belts were varied, and their sediments were the result of conditions in source areas, in transportation, and in the sites of deposition.

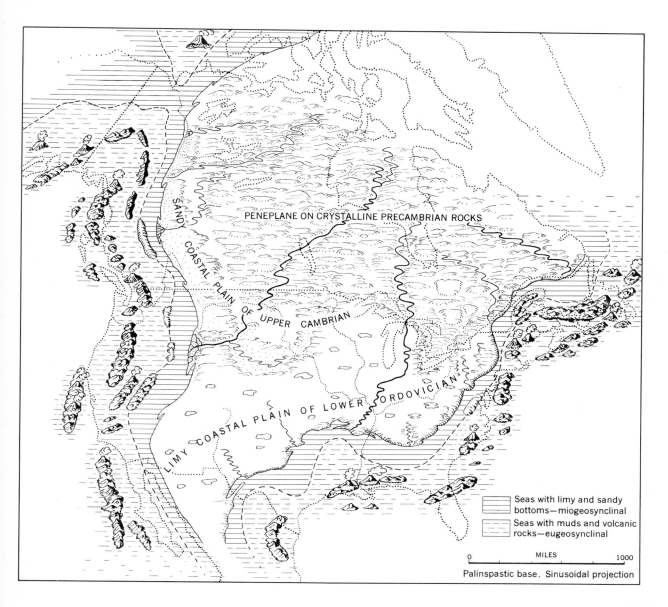

PENEPLANE ON CRYSTALLINE PRECAMBRIAN ROCKS

SANDY COASTAL PLAIN OF UPPER CAMBRIAN

LIMY COASTAL PLAIN OF LOWER ORDOVICIAN

| Seas with limy and sandy bottoms—miogeosynclinal |
| Seas with muds and volcanic rocks—eugeosynclinal |

0   MILES   1000

Palinspastic base. Sinusoidal projection

147

the first is generally present if any Canadian is preserved, the third is next most frequent, then the second, and the fourth is least present. As the rocks are rather shallow-water marine deposits, this shows that at two times the seas covered larger parts of the continent than at two other times. Advance of the sea may be due to a rise of sea level—an eustatic change—or to bending down of the part of the sea's surface—an epeirogenic or warping movement. The former should have the same effect universally, inasmuch as a rise in the level of the sea at one place must be expressed all over the earth at the same time. But warping movements may vary from place to place. Therefore, when the correlation table shows simultaneous introduction of deposits over great areas or similar absence of strata, the suggestion is strong that there have been rises and falls of the sea, such as local withdrawal in a time of general marine spread. Exceptions can be attributed to greater local uplift (epeirogeny) than the eustatic rise.

There was great marine withdrawal at the close of the Canadian Epoch; the surface of the Canadian rocks was eroded to a hill and valley topography in much of eastern United States. Subsequently, the sea returned, marine deposits spreading gradually over the interior of the craton. The shores gained sands that had been eroded from older Cambrian sandstones, and perhaps from still exposed Precambrian granites and metamorphosed sediments. These filled the valleys of the

Fig. 8–20 (right). *Altered volcanic ash beds or metabentonites in the middle Ordovician Trentonian in southwestern Virginia. A.* The clay lies on limestone that has been changed to chert at the top through the addition of silica that came from the alteration of the overlying ash. The metabentonite, being quite incompetent, has been sheared and deformed in the differential movements in the area by the folding of the Appalachian Mountains. The exposure is southwest of Clifton Forge. Volcanic ash beds are useful in giving precise correlations within the limits of the ash falls.

B. A volcanic ash bed such as is shown in detail in part *A*, near Abingdon.

Fig. 8–21 (below). Restored section showing the distribution of altered volcanic ash beds or metabentonites in middle Ordovician limestones in central Pennsylvania (after M. Kay). Their persistence enables determining precisely the thickness of sediment in successive spans of time, showing their convergence toward a submarine shoal to the southeast. Present thicknesses will have been reduced from the original ones because of the compaction and induration of the beds.

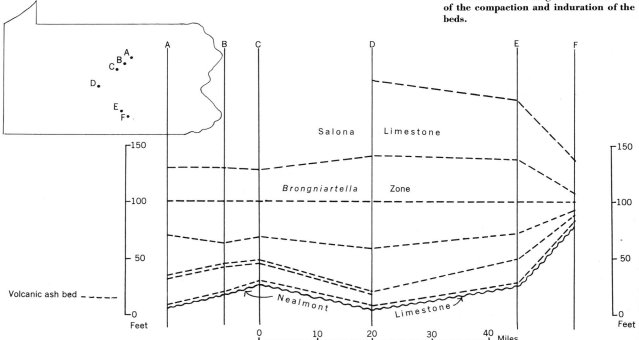

A

B

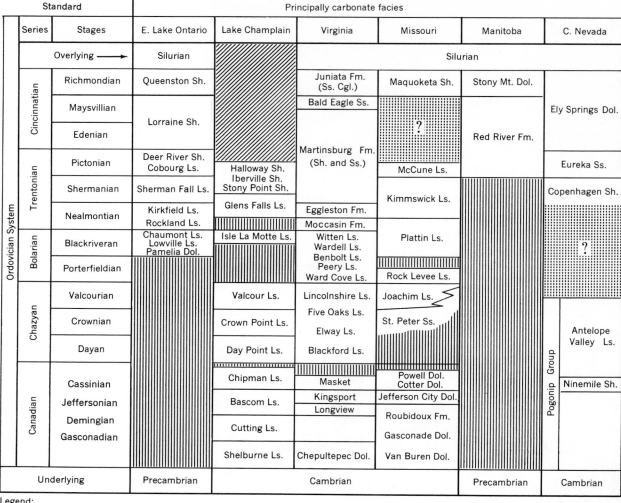

Legend:
- Not deposited / Unconformity
- Eroded out / Concealed / No information
- Changing facies— gradation

Fig. 8–22. A correlation table placing formations in several localities in comparison with a classification of the Ordovician period into series and stages.

Vertical lines in such a table indicate that the beds are absent because they were not deposited or were eroded out— there is unconformity, though not necessarily of angular nature. Diagonal lines indicate that information is lacking, as where the system is buried too deeply for observation, or where only a part of the original rock section remains preserved. A correlation table gives no indication of the relative thicknesses nor is there adequate indication of lithologies. It permits ready comparison of the

C. Nevada	C. Idaho	W. Texas	E. New York	Quebec	Zones	Series	System
(hatched)	Silurian	Devonian	*(hatched)*	*(hatched)*	← Overlying		
(hatched)	Phi Kappa Shale	Maravillas Chert	*(hatched)*	Quebec City Shale	15. *Dicello. anceps*	Ashgillian	Ordovician System
					14. *Dicello.complanatus*	Ashgillian	
					13. *Pleuro. linearis*	Caradocian	
		Woods Hollow Sh.			12. *Dicrano. Clingani*	Caradocian	
Sams Spring Fm. (Sh. Chert)					11. *Climaco. wilsoni*	Caradocian	
					10. *Climaco. peltifer*	Caradocian	
			Normanskill Shale		9. *Nema. gracilis*	Caradocian	
Petes Summit Fm. (Sh. and Ss.)		Fort Peña Fm. (Sh. Ls.)	*(hatched)*	*(hatched)*	8. *Glypto. teretiusculus*	Llandeilan	
					7. *Didymo. murchisoni*	Llanvirnian	
					6. *Didymo. bifidus*	Llanvirnian	
		Alsate Sh.	Deepkill Sh.	Levis Shale	5. *Didymo. hirundo*	Arenigian	
Charcoal Canyon Fm. (Sh., Chert, Tuff)	*(hatched)*				4. *Didymo. extensus*	Arenigian	
		Marathon Fm. (Sh. and Dol.)			3. *Tetra. approximatus*	Arenigian	
			Schaghti-coke Sh.		2. *Bryo. cambriensis*	Tremadocian	
					1. *Dictyonema socialis*	Tremadocian	
(hatched)	*(hatched)*	Cambrian			Underlying		

Left side: "Clipper Canyon Group" labels Sams Spring Fm., Petes Summit Fm., and Charcoal Canyon Fm.

supposed equivalent rocks in the several columns. The table is but a small part of one for the Ordovician, for the time-classification can be more refined, and the number of sections and their named stratigraphic units increased greatly. And when the similar charts for all other systems are added, the fund of information is such as to stress the enormity of details that are generally only of local or provincial importance.

The names of graptolites in the "Zones" column on the right have been abbreviated by deletion of the "-*graptus*" at the end of each generic name; thus *Dicellograptus* is *Dicello*.

erosional surface. The sand particles on the shore were drifted by winds, and as they struck one another they developed minutely chipped surfaces much like frosted glass; finally, they were laid on the beaches of the encroaching waters. The extremely pure sand deposits (St. Peter Sandstone) (Fig. 8–23) are extensive in the Mississippi Valley, where they are quarried as a source of silica for the making of glass and other products; their purity resulted from the winnowing and sorting by winds and the waves. Similar sands lie interruptedly at the base of later Ordovician sediments of varying ages as far west as the western miogeosynclinal belt through Utah and Nevada and northward to British Columbia. The retreat of the sea in the late Canadian and Chazyan epochs contrasts with their seeming spread and persistence through the Cambrian into the Canadian. The Cambrian and lower Ordovician are in a relatively continuous sequence, having a great unconformity at the base and generally interrupted at the top though overlain by later Ordovician sandstones. Physically the systemic bound-

Fig. 8–23. *The St. Peter Sandstone, a widespread middle Ordovician formation in the upper Mississippi Valley (Illinois Geological Survey). A.* A quarry in the sandstone at Ottawa, Illinois. The rock is composed of quartz sand that makes an excellent source of silica because it has little impurity and is readily removed, having little cement. The sand is used in the manufacture of glass.

B. Grains of sand from the St. Peter are very well rounded, and have a dull or frosted surface. They seem to have been transported by winds and laid on the shores of an advancing middle Ordovician sea. The grains have average diameter of about one-half millimeter.

A

B

ary at the top of the Cambrian in North America is not as distinct as that above the Canadian, emphasizing that the periods and systems are not great natural divisions of time and rock. If we were to start anew in dividing geologic time, we might have quite different limits to some of the divisions.

The record of the Ordovician adds materially to our knowledge of the early Paleozoic character of the continent in that it shows the presence of deeply subsiding volcanic belts—eugeosynclines—separated by the non-volcanic miogeosynclines that were recognized in the Cambrian from the rather stable interior, the craton. Within the interior, the carbonate rocks and quartz sandstones show by their distribution in time that the seas advanced and retreated several times, evidently as the result of rises and falls of the level of the sea. It also shows well the manner in which the sedimentary record evidences the progression of a great period of mountain making (of orogeny), the Taconian Revolution, which will be the subject of further discussion.

Fig. 8–24. Reconstruction of the bottom of the sea in late Ordovician as it might have been in southern Indiana (The Smithsonian Institution). The large conical animals are cephalopods; there are spheroidal colonial corals and branching masses of bryozoans, a trilobite below the center of the view, and a gastropod or snail in the foreground.

9
The Phases of Orogeny: The Taconian Revolution

Lower Devonian limestone lying with angular unconformity on folded Ordovician shales in quarry face in Becraft Mountain, east of the Hudson River at Hudson, New York, an expression of the Taconian Orogeny.

9 The Phases of Orogeny: The Taconian Revolution

Nearly a century ago, before the Ordovician System had been removed from the more inclusive Silurian, geologists observed in a number of localities from eastern Quebec to Pennsylvania that rocks with Silurian fossils lie with angular discordance on the Ordovician sediments (Fig. 9–1 and chapter opening illustration). The folding and erosion of the Ordovician rocks seemed to represent mountain making that came to be called the Taconian Orogeny or Taconian Revolution, from the Taconic Mountains that lie east of the Hudson River in eastern New York, western Massachusetts, and Vermont.

The Taconian Revolution is but the culmination of a progression of disturbances that extended through much of the Ordovician Period along the Atlantic margin of the continent. The history may seem of interest, but the manner by which the conclusions are reached is of greater importance. Generalizations have been given in preceding chapters, that early Paleozoic North America was separable into a central rather stable craton margined by geosynclinal belts, and that the latter had inner non-volcanic belts, miogeosynclines, with prevalent carbonate rocks and outer deeply subsiding geosynclines with associated volcanoes and tectonic islands, eugeosynclines that margined the present Atlantic and Pacific shores. The generalizations give too simple an impression of the Ordovician history along the Atlantic Coast. And study of the subsequent Silurian System, will show that the relatively stable craton, too, has some complexity of deformation.

The deformation that progressed through the Ordovician Period and culminated in the Taconian Revolution in eastern North America is determined from the study of the distribution of the thicknesses and kinds of rocks in successive stages and series, the records of spans of time within the system. The data are far too numerous to permit more than a presentation of examples that illustrate the means of gaining an understanding of the history. We can take the record in the latitude of the Great Lakes and New England as representative of the changes that took place across the belts. Discussion of parts of the system in the Appalachian Mountains region to the southeast emphasize that similar events transpired at different times from place to place. The whole is to show that highlands were restricted to islands in the eugeosynclinal belt along the coast in the early Ordovician, lands that produced sediments that did not spread far toward the interior. Tectonic movements raised lands in the eugeosynclinal and miogeosynclinal belts at inter-

Fig. 9–1. *Unconformity of the Taconian Orogeny in southeastern New York.* A. Silurian conglomerate lying unconformably on sandy shales of Ordovician age in a cut on the old railway at Otisville, New York, north of the northwestern corner of New Jersey. The Ordovician rocks had been tilted northwestward to dip about 15 degrees and then eroded before they were buried by the Silurian stream-laid quartz gravels.

B. Lowest Devonian Manlius limestone lying with unconformity on folded middle Ordovician Normanskill cherty shale on Becraft Mountain, a Devonian outlier east of the Hudson River south of Albany, New York and on the west side of the Taconic Mountains. The folding of the Ordovician rocks is attributed to the Taconian Orogeny, which thus took place prior to the Devonian on this evidence; but Silurian rocks lie on the folded Ordovician at other places.

Fig. 9–2. Ordovician Chazyan limestone on Isle Lamotte, northern Lake Champlain, Vermont. The rock in the mound is a reef consisting of intergrown alga-like organisms, sponges, bryozoans, and occasional corals, with associated lime mud derived from the wearing and breaking of other calcareous organisms. The reef is cut by eroded channels of darker calcarenite, calcite sandstone, in which fragments of echinoderms, crinoids, and cystids (see Chapter 26) are the principal constituents; the hammer lies on such a channel. The reefs grew on relatively little subsiding parts of the sea floor, finer lime muds accumulating in the deeper water of more rapidly subsiding areas; such reefs have been compared to the patch reefs of present seas.

156

A

B

157

vals through the period, so that by the end of the Ordovician, terrigenous sediments had spread far into the interior from lands that were raised along the cratonal margin. These developments can be illustrated in restored sections and in maps of thicknesses and lithofacies in successive times.

PALEOTECTONIC ELEMENTS

We have referred to the larger divisions, epochs, in the Ordovician Period, and their corresponding rocks in the series. During the earliest epochs, the Canadian and Chazyan (Fig. 8–24), the craton was bordered by gently subsiding submarine plains in which carbonate rocks, the detritus of calcareous organisms, accumulated (Fig. 9–2). They extended beyond the Cambrian to overlap the irregular surface of Precambrian rocks in New York and Ontario (Fig. 9–3). In the latitude of New England and eastern Canada, carbonate rocks, thin or absent on the craton, thickened to a few thousand feet in a miogeosynclinal belt (Fig. 9–4). Subsidence was greater to the east, and in some subsiding troughs the water was also deep, for sediment from the adjoining raised islands and volcanoes did not fill the belt as rapidly as it sank.

LIMESTONE PEBBLE CONGLOMERATES

The early Ordovician record of the relations of the limestone and argillite belts suggests instability that is but an harbinger of events to come. In the last chapter, the association of graptolite-bearing shales with limestone boulder conglomerates at Levis gave means of correlating the shaly and shelly facies of the Ordovician. Bands within the shales have large blocks of limestone (Figs. 8–9, 8–10) that reach lengths of scores of feet and seem much too large to have been carried by streams or ocean currents. Lower Ordovician shales with conglomerates of this sort are known for two hundred miles or so along the south shore of the St. Lawrence to the tip of Gaspé. The conglomerates are not present farther south in southeastern Quebec, though there are exposures of similar argillites. How were very large blocks of limestone with fossils from early Cambrian to early Ordovician emplaced in layers within lower Ordovician graptolite shales along a two-hundred-mile belt?

The limestone blocks seem to belong north of the containing shales, since neither limestones nor boulders are known farther southeast. Yet they are not exposed north of the river, for at Montmorency Falls (Fig. 9–5), middle Ordovician (Trentonian) limestone lies on Precambrian, and rocks of the age of the boulders are missing at the unconformity (Fig. 9–6). Cambrian and Ordovician like those in the

Fig. 9–3. Ordovician Black River (Bolarian) limestone with basal conglomerate overlapping Precambrian rocks in Ontario. A. Overlap on a knob of Precambrian granite along Highway 101 in Kingston, Ontario.

B. In the vicinity of Marmora, sandstones unconformably overlie the undulating surface on Precambrian magnetite-bearing rocks in Marmoraton Mine near a bench in the wall of the open-pit. The higher walls of the mine, seen to the left in the background, expose the limestones higher in the Black River Group and ten feet of lower Trentonian limestone which appear just beneath the ground surface at the top; the Black River rocks are more than 100 feet thick.

Fig. 9–4. Restored section of the Canadian and Chazyan Series across northern New York and northwestern Vermont. Canadian rocks thicken eastward through the divergence of stratigraphic units as well as the addition of beds that are missing through unconformity in western sections. The Chazyan overlaps westward from the miogeosyncline; patch reefs of stromatoporoids, corals, sponges, and bryozoans formed on shoals developed on structurally higher areas.

A

B

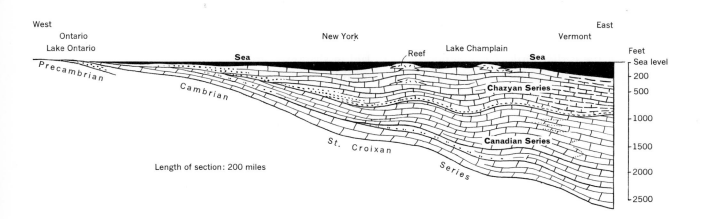

West   East
    Ontario   New York   Vermont
    Lake Ontario   Lake Champlain   Feet
                                      Sea   Reef   Sea   — Sea level
    Precambrian   — 200
                          Cambrian   Chazyan Series   — 500
                                                                                                        — 1000
                                St. Croixan   Canadian Series   — 1500
    Length of section: 200 miles   Series   — 2000
                                                                                                        — 2500

159

conglomerates are found along the margin of the Precambrian rocks on the coast of Labrador near Belle Isle, opposite Newfoundland, and in northwestern Vermont. And the general distribution of lower Ordovician facies around the continent is with argillitic belts farther from the craton than carbonate belts.

The fault in the St. Lawrence at Quebec (Fig. 9–7) carried the argillitic facies over the carbonate-rock facies which, though absent directly to the north, would lie to the south below the thrust plane, and would have been to the north of the boulder-bearing argillites of the thrust sheet before it was thrust. Mud flow and submarine sliding are processes that seem competent to enable gravity to move such large blocks on a gentle slope of soft mud into localities where the argillites were being laid in hundreds or a few thousand feet of water. The lower Cambrian limestones must have been raised, possibly by faulting, along the margin of the carbonate belt and perhaps the earthquakes dislodged them and started them sliding down the bottom slope.

Similar conglomerates of medial Ordovician age (Cow Head) along the west coast of Newfoundland have boulder beds alternating with graptolite shales, as we have seen (Fig. 8–9). Fossils in the limestone boulders are in the same order as in the undisturbed limestones in other areas. Probably the carbonate rocks formed shoals northwest of the deeper-laid graptolite-bearing muds and cherts. Shocks dislodged the limestones so that they slumped into deeper water. And the fossils in the boulders of limestone are only a little younger than the associated shales with graptolites, and they enable comparison in terms of time of the faunas of the two facies. In Quebec, however, some of the boulders are considerably older than the containing shale matrix. Thus for hundreds of miles along the St. Lawrence River and Gulf, there is evidence that the earlier Ordovician carbonate rocks laid in shallower sea floors, passed rather abruptly into more rapidly subsiding troughs in which muds settled with limestone blocks dislodged from time to time from the northwestern shelf.

CRATONAL MARGIN IN NEW YORK

The increasing instability of the cratonal margin is learned from study of Ordovician rocks from Quebec to the southern Appalachian Mountains. To recite the events is but to illustrate the way in which the structural changes are recognized, for the details are only of limited importance. Yet each is a part of the evidence of the increasing crustal stress that was to culminate in a revolution.

West of the Adirondack Mountains in northern New York, the Black River Group of the late Bolarian consists of about two hundred feet of

Fig. 9–5. Montmorency Falls, near the St. Lawrence River northeast of the city of Quebec (G. Hunter). The falls are over Precambrian granite of a fault-line scarp that has upper Ordovician (Cincinnatian) shale to the south, in the foreground. Just above the falls, middle Ordovician Trentonian limestone unconformably overlies the Precambrian, rock that would lie far below the shales in the downthrown side of the fault.

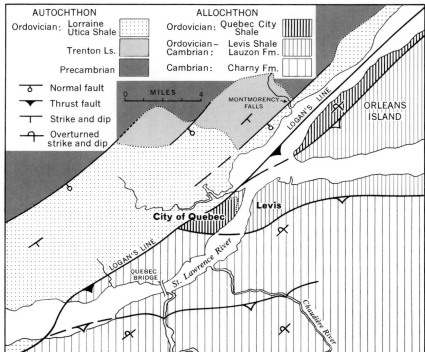

Fig. 9–6. Geologic map of the vicinity of Quebec and Levis in the province of Quebec (after F. F. Osborne). The rocks in the area south of the St. Lawrence River have been thrust northwestward, the trace of the fault, Logan's Line, lying generally below the river. The thrust fault is clearly younger than late Ordovician, for it cuts rocks of that age; it is probably considerably younger, of an orogeny in the Devonian Period.

carbonate rocks (Fig. 8–18*A*) that was laid in seas extending northeast-ward into Quebec: the sediments thin to disappearance in the Adirondack area (Fig. 9–8), regaining their thin limestone development along Lake Champlain and eastward in western New England. If we continue with successively younger stages, the lower Trentonian carbonate rocks laid in the shallow waters that spread over the craton (Fig. 9–9) thin from as much as two hundred feet on the west to ten feet or less along a line that can be drawn from about Quebec City, through the Adirondacks and central Pennsylvania to eastern West Virginia, then thicken again to the east, much as did the Black River. Presumably all of these rocks change to argillites eastward in the volcanic geosynclinal belt. The margin of the craton and miogeosynclinal belt were blanketed by limestone in the Bolarian and early Trentonian, subsiding least along a trend that can be called the Adirondack line for future reference. Not only was there little epeirogeny over a large area, but highlands were not near enough on the east to have made the waters muddy and turbid. This marks a lull before a storm!

Fig. 9–7. *Structure and restored sections near Quebec City. A. Structure section across the St. Lawrence River just east of the city of Quebec, an interpretation of the present relationships. Logan's Line is a fault that has brought Cambrian and Ordovician sediments of an original deeply subsiding geosynclinal belt northward upon medial and late Ordovician sediments that were originally laid some distance away. Although the thrusts are generally called late Ordovician Taconian, they may well have been formed in the middle Paleozoic Acadian Revolution. The high angle fault below Montmorency Falls is in a different fault system.*

B. Restored section of the relationships of the rocks as they are interpreted for medial Ordovician time, showing the contrast between the calcareous and argillitic facies, and the slide of boulders of the former into the latter. The small sketch shows the relations between the original stratigraphic section and the planes along which the thrusts developed.

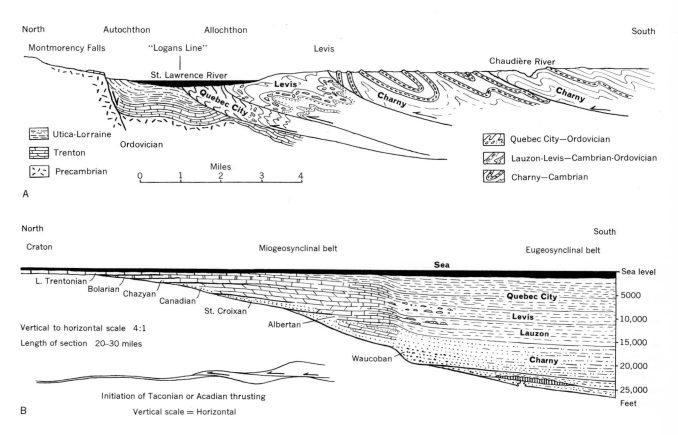

The immediately succeeding medial Trentonian was introduced by continuing widespread limestone deposition that overlapped older beds on the Adirondack line near Quebec as at Montmorency Falls (Fig. 9–5) and along the Mohawk Valley south of the Adirondacks (Fig. 9–10). But subsequently a great change came about to the east; we have discussed the changing facies from carbonate to argillite (Fig. 8–12). Not only does two hundred feet of limestone west of the Adirondack line grade into a thousand feet of shaly argillite eastward (Fig. 8–13), but that becomes a half-mile of quartz-silty and sandy shale (Fig. 9–11). A high source of terrigenous detritus arose to the east, so that streams from this new land in the western part of the eugeosynclinal belt carried the detritus to the eastern margin of a deeply subsiding trough in eastern New York. The quiet of the earlier Trentonian was interrupted by this phase of orogeny. For the first time, highlands to the east supplied terrigenous sediments to the miogeosynclinal belt, which previously had received well-sorted quartz sands from the craton or

Fig. 9–8. Restored sections at a succession of times across the Adirondack Mountains region of New York, showing the deformation that took place within the Ordovician Period.

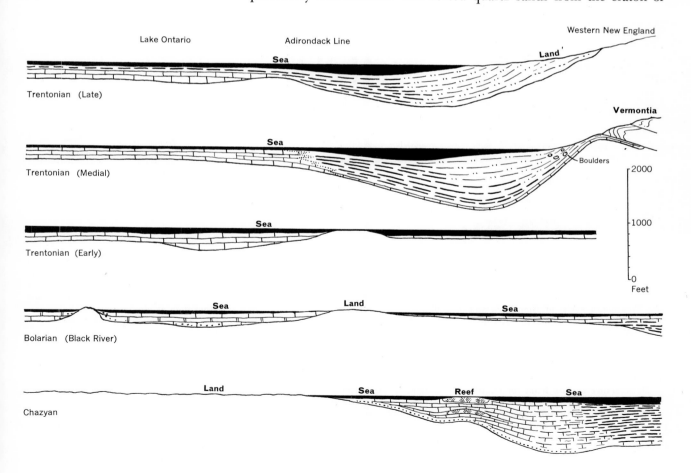

163

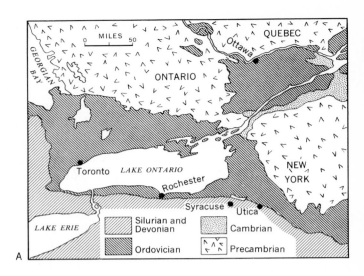

A

C

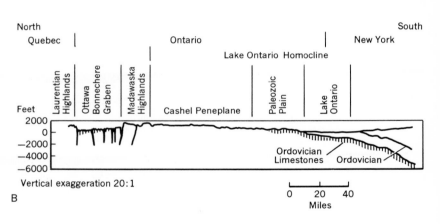

B

Fig. 9–9. *Primary structures and textural changes* that bear on paleogeography, on depths of water, and on directions of current flow in water and wind in the Trentonian rocks of New York and Ontario. *A.* Map of the belt of outcrop of the Ordovician System in southeastern Ontario and northwestern New York, north and east of Lake Ontario; rocks of the Trentonian Series are exposed in the middle part of the Ordovician belt.

B. Structure section from the Ottawa River to central New York, across the outcrop belt. Ordovician sedimentary rocks have been eroded to the north and east of the main outcrop except in faulted areas along the Ottawa River; and the rocks are

concealed beneath Lake Ontario and southward. Information is gained from surface exposures and from the very few wells that penetrate the series down the dip.

C. Exposure of limestones near Newport, New York, showing the beds of differing characters, representing the changing local depositional conditions. There are light-weathering calcilutites and calcisiltites (originally calcite-aragonite or "lime" muds), darker ledges of calcarenite and coquinite (lime sands and shell accumulations), and darker partings and bands of argillaceous shale. Careful records are made of these successions wherever they are ex-

posed, with notations on the lithology of each bed, its thickness and primary structures, the kinds of fossils, their abundance and orientation, and such other data as give promise of being useful. The time-stratigraphic zones are traced through recognition of distinctive lithic units, and the presence of fossils that have been found to have restricted ranges.

D. Thus the distinctive trilobite *Cryptolithus* persists in and is known only from a lithic zone recognized from southeastern Ontario to the Mohawk Valley south of the Adirondack Mountains (drawing after H. B. Whittington).

E. The changes in thickness in the

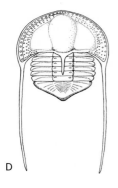

D

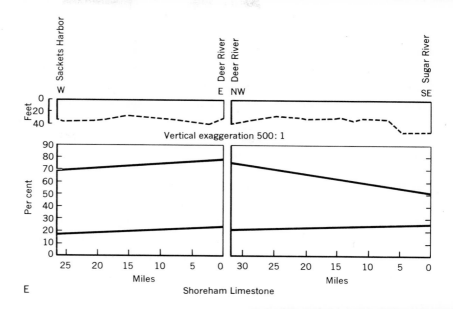

Sackets Harbor

W

Deer River
Deer River

E NW

Sugar River

SE

Feet
0
20
40

Vertical exaggeration 500:1

Per cent

90
80
70
60
50
40
30
20
10
0

25 20 15 10 5 0

Miles

30 25 20 15 10 5 0

Miles

E

Shoreham Limestone

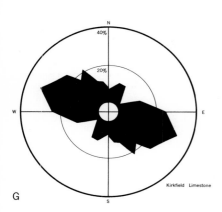

N

40%

20%

W E

S

Kirkfield Limestone

G

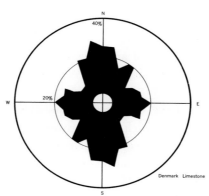

N

40%

20%

W E

S

Denmark Limestone

F

Shoreham Limestone, and of the proportions of calcite sandstone (calcarenite) to the finer textures, are plotted on diagrams from northwest to southeast (P. A. Chenoweth). Statistical tests show that the coarsening to the southeast is significant, and it is interpreted as related to shallowing of the sites of deposition. Though the sea may have shallowed, the data along a single line do not define the directions of the ancient shores. The relations of associated beds are shown in Fig. 8–12.

F. Ripple marks in the Kirkfield limestone that underlies the Shoreham, along Deer River, New York. Such structures may have formed as tidal currents drifted sands made of shell fragments; the ripples are more abundant in shallow environments, and have increasing wavelengths in deeper waters. As such structures are ephemeral on present shores, these must have been laid in some exceptional circumstances that permitted their preservation; perhaps they formed in an abnormal depth in a severe storm or unusually low tide. Each bed of rock has been laid in a short span of time, a matter of minutes or hours, but as thousands or millions of years are spanned in a few hundred feet, only a minute proportion of the time is represented by deposits that remain, the rest being interruptedly laid, removed, and relaid.

G. Orientation diagrams of ripples show preferred directions relating to current flow, or perhaps to the orientation of wind-directed waves; either would be affected by coastal shore forms. With the plotting of many local data, patterns of changing orientations lead to interpretation of paleogeographic setting. Similar orientations have been plotted of the directions of elongations of fossils, such as of straight cephalopods. Percentages pertain to readings within ten degrees of each direction.

165

carbonate rocks laid in clear waters. Later Trentonian carbonates continued to accumulate for a time west of the Adirondack line, but in latest Trentonian argillaceous muds drifted as far west as Ohio and Michigan. The Cincinnatian also had active deformation to which we will return presently.

RELATIONS ALONG THE TRENDS

The discussion has concerned the deformation across the trend of the margin of the craton and the geosynclinal belts in the latitude of the Adirondack mountains, without reference to the extent of the deformation along the belts. Through medial Ordovician time, the belt lying between the craton and the volcanic geosynclines was warped into elongate doubly plunging troughs complementing linear islands to the east from which coarser sediment came at times of uplift. The deformation proceeded quite actively in one part of the belt while another part remained rather stable, only to have the relations changed or reversed from time to time. The belt was one of interrupted turmoil, bending and weaving at times, and changing little or standing quietly at others. The medial Trentonian carbonates in the latitude of the *southern* Adirondacks passed into a half mile or more of argillites becoming coarser, more sandy, eastward and upward—indicative of a rising source land to the east.

Across the *northern* Adirondacks and Lake Champlain, equivalent middle Trentonian rocks, perhaps about as thick, pass upward not into coarser sandstones but rather into argillites with radiolarian chert, indicative of deepening water in which sedimentation did not keep pace with subsidence. There are slump blocks of early Trentonian limestones two thousand feet up in middle Trentonian shales along Lake Champlain, so folds of at least that magnitude must have developed by medial Trentonian from which rock masses slid westward. To the southeast, from southern Lake Champlain to beyond Albany, middle Trentonian shales have blocks of rocks such as are in the Cambrian of the overthrust sequences, the northern Taconic Range; perhaps the highlands rising to the east in the volcanic geosynclinal belt became unstable so that these highest rocks were transported westward as thrust sheets (Fig. 9–8), dislodging blocks that slumped into the deepening trough to the west. The recumbent folds (Fig. 9–12) in the limestones in western Vermont such as in the Vermont Marble so widely used in buildings are attributed to these movements. The red and green argillites of the Cambrian of the Taconic thrust sheet are the roofing slates of Vermont and eastern New York.

Far to the south in southern Virginia and Tennessee, the section of

Fig. 9–10. **Ordovician rocks in the Mohawk Valley, New York.** *A.* **Thin Trentonian limestone overlying Canadian dolomite and underlying Trentonian black shale in Canajoharie, New York; the shale is shown more fully in Fig. 8–13. The Bolarian Black River beds and the lower Trentonian limestone which are hundreds of feet in northwestern New York thin to disappear near Canajoharie, and then again separate the Canadian from the middle Trentonian shales eastward along the Mohawk Valley. At Canajoharie, the Canadian is succeeded by ten feet of basal middle Trentonian calcarenite, absent nearby, which is disconformably overlain by middle Trentonian black shales containing graptolites.**

B. **Ordovician shaly sandstone near Schenectady, New York. The sandstone is an eastern facies of the middle Trentonian, more than a thousand feet thick, which grades westward through black shale to limestones of two hundred feet or less exposed near Utica and westward; the thick deposit of sandy sediment resulted from the rise of lands in western New England and the subsidence of a trough along the eastern border of New York through Albany. The sandstones are marine-laid, for they have a few marine fossils, such as graptolites.**

Fig. 9–11. *Taconic rocks in eastern New York.* *A.* **Folded argillites and cherts, probably Cambrian of the Taconic thrust sheets, along the Taconic Parkway east of Poughkeepsie.**

B. **Middle Ordovician Trentonian shale along Highway 151 east of Albany, west of the Taconic thrust sheets or allochthon. Some of the boulders are of rocks like those in the allochthon; others, such as the largest, are of Trentonian sandstones, laid west of the rising Taconic thrust front. The blocks slid westward from the front of the advancing Taconic thrust sheet and from autochthon raised in front of it into the muds of late Trentonian seas to the west (J. M. Bird).**

A

B

A

B

167

A

B1

B2

Fig. 9–12. *Folds produced in Ordovician limestones in Vermont by the Taconian Revolution. A.* Southward along the axial plane of an anticline in upper Canadian argillaceous dolomite north of Middlebury, Vermont near the Dog Team Tavern.

B. Northward along another fold in a thrust sheet below the Taconic thrust fault west of Brandon, Vermont, showing the thickening of beds at the axis, and the cleavage in the limestones on the two limbs of the fold diverging from parallelism to the axial plane.

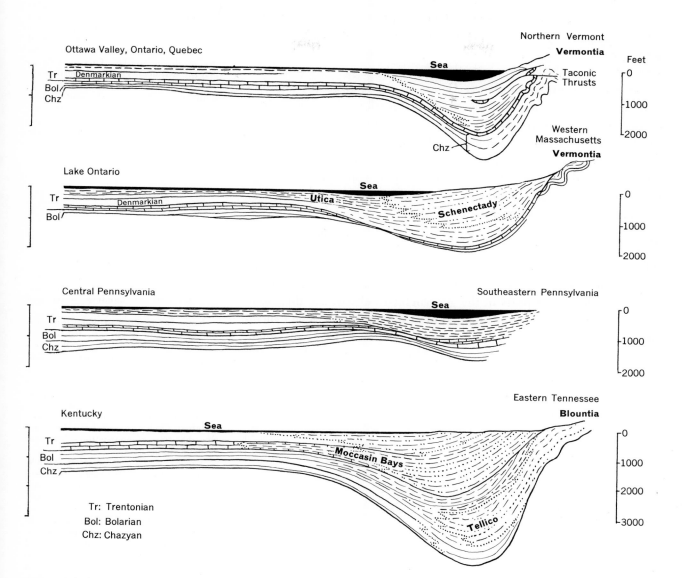

Fig. 9–13. Restored sections of middle Ordovician rocks across the flexure of the Adirondack Line southward from New York. The sections illustrate the variable conditions that developed from place to place along the belt; conclusions based on observations along one line of section apply only within geographic limits. Each line of section has changing conditions through time, as was illustrated in New York (Fig. 9–8).

the lower Bolarian is much like that of middle Trentonian south of the Adirondacks. After widespread distribution of Chazyan carbonate rocks, land rose to the southeast, contributing graywacke and conglomerate on the southeast, passing into silty argillite in the miogeosyncline, then grading into thin calcareous rocks across a flexure on the trend of the Adirondack line (Fig. 9–13). Unlike the conditions in New York, the early Trentonian in this Appalachian region is terrigenous within the arc centered in eastern Tennessee, the reddish silty argillites and sands grading northwestward in marginal marine argillaceous calcareous deposits and then into marine fossiliferous limestones in northern West Virginia, Kentucky, and central Tennessee (Fig. 9–14). The non-volcanic geosynclinal belt and the regions to the southeast had rising lands at times and associated warping and bending through parts of the belts that produced sites for thick sediments, the belt became relatively quiet through other stages.

SUBSIDING CRATONAL MARGIN

The Cincinnatian marks the introduction of new relations on the cratonal margin that were to continue through much of the Paleozoic in varying degrees. Uplifts in the geosynclinal belts spread their sediments far within the area of the craton of the Cambrian. The Cincinnatian rocks of the region of the Great Lakes southward have a maximum thickness of about three thousand feet in Pennsylvania (Fig. 9–15).

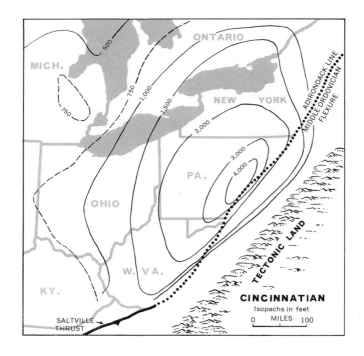

CINCINNATIAN

Isopachs in feet

0   MILES   100

Fig. 9–14 (right). Map of the lithofacies of the lower Trentonian sediments in southwestern Virginia and adjacent states. A lithofacies map is one showing the proportions of different lithologies through a thickness of rocks rather than the distribution of sediment types on the sea bottom at an instant of time; but if conditions did not change markedly, the two would be similar. This map shows that the sediments changed progressively from the coarser terrigenous ones near to the source lands on the southeast to dominant limestones in more distant sea bottoms. The changes through time are shown in a restored section along a single line.

Fig. 9–15 (left). Map showing the thickness of the Cincinnatian Series in Pennsylvania and surrounding states and provinces, an *isopach map*.

Isopachs are lines of equal thickness of rock; if the rocks are flat-lying, a drilled well will penetrate the thickness shown on the map. They may represent structural changes through a span of time, but only if the top surface of the measured sequence was originally of the same form as the bottom one—for instance, if each was an horizontal plane. Original thickness will have been reduced because of compaction subsequent to deposition.

The Cincinnatian shows greatest thickness in central Pennsylvania, diminishing southeastward, partly because upper beds were eroded prior to Silurian time. The convergence toward the west and north probably shows decreasing subsidence; but as the rocks in central Pennsylvania were laid on a subaerial delta plain, whereas those farthest northwestward were laid below the sea, there was some slope to the depositional surface that tends to make isopachs be somewhat greater than corresponding structural deformation. In regions where sediments were laid in shallow water, isopach maps give a good approximation of the regional deformation through a span of time; but where there were great differences in depths, they may indicate the amount of filling of a marine basin. Each must be considered critically with these factors under review.

170

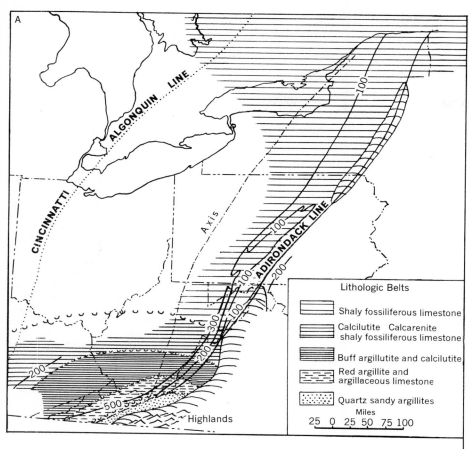

A

ALGONQUIN LINE

CINCINNATTI

Axis

ADIRONDACK LINE

100

100

200

200

300

200

500

Highlands

Lithologic Belts

Shaly fossiliferous limestone

Calcilutite Calcarenite
shaly fossiliferous limestone

Buff argillutite and calcilutite

Red argillite and
argillaceous limestone

Quartz sandy argillites

Miles
25 0 25 50 75 100

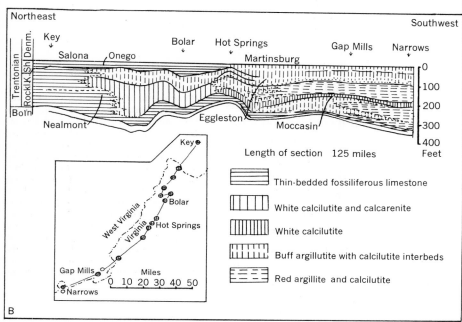

Northeast

Southwest

Key

Bolar

Hot Springs

Gap Mills

Narrows

Salona

Onego

Martinsburg

Trentonian
Rock I. K. |Sh| Denm.

Bo'ln

Nealmont

Eggleston

Moccasin

0

100

200

300

400
Feet

Length of section 125 miles

Thin-bedded fossiliferous limestone

White calcilutite and calcarenite

White calcilutite

Buff argillutite with calcilutite interbeds

Red argillite and calcilutite

Key

Bolar

West Virginia

Virginia

Hot Springs

Gap Mills

Narrows

Miles
0 10 20 30 40 50

B

171

Thicknesses decrease gently away from the maximum in directions from southwest through west to northeast, but more rapidly toward the southeast; the body of sediments is roughly a semi-lens in three dimensions, thicknesses of a thousand feet or more lying within an arcuate line passing from southern West Virginia through eastern Ohio and southwestern Ontario. But the stages within the Cincinnatian are differently distributed. The lower and middle stages continue in increasing thickness to the limit of preservation, but the youngest, the Richmondian stage, is bevelled by unconformity along the southeast so that the Silurian lies on older beds (Fig. 9–1). The Richmondian also disappears as it approaches the Adirondack Mountain region, as though that area ceased to subside. Thus a great basin-like surface subsided in eastern Pennsylvania and surrounding regions, but toward the close of the epoch and period, elevation on the southeast changed that margin from one of deposition to one of erosion.

The rocks of the Cincinnatian in the east are all terrigenous, mostly argillaceous, becoming finer textured westerly and northerly from Pennsylvania (Fig. 9–16). The rocks coarsen upward in each locality and coarsen outward through time. The coarser sediments, conglomerates with pebbles of quartz and quartzite, spread farther and farther, because streams flowed progressively faster from lands which rose faster than they were eroded; rivers were able to carry successively coarser particles to the sites of deposition. The youngest sediments in the center of the lens in Pennsylvania are red sandstones and conglomerates (Juniata) that grade westward and northward into reddish argillaceous siltstones (Queenston). The fossils beneath the red beds show that their base becomes successively younger as one examines sections farther and farther away from the center. At first marine waters covered the whole of the area of present exposure and extended somewhat eastward; sediment came from the rising lands to the east. By Richmondian the eastern source land spread so that stream-laid deposits entered the sea in Pennsylvania. The delta was built gradually outward, until it had reached southwestern Ontario, Ohio, and central West Virginia by latest Richmondian time. Similarly, Richmondian red sandy shales overlie fossiliferous early Richmondian in southeastern Quebec (Fig. 9–17). The sediment from rising lands in the marginal geosynclinal belts had driven the sea out of the region south of the Great Lakes. This might be attributed to withdrawal of marine waters because of lowering of sea level—eustatic fall. But in that case Richmondian rocks should have limited distribution on the continent, whereas limestones of the Richmondian and of the late Trentonian are

Fig. 9–16. Restored section of the Cincinnatian series from southwestern Ontario to the Atlantic near the close of the Ordovician Period. The progressive coarsening of the textures on the southeast and the spread of the red sediments of the Juniata-Queenston are evidence for rising highlands and a retreat of the sea from the detritus-filled trough or exogeosyncline formed in front of the rising land. Southeastern Pennsylvania, the site of deposition of thick marine sediments in early Cincinnatian, had been raised to form part of the land from which the sediments were being eroded by late Cincinnatian time.

Fig. 9–17. Late Ordovician terrigenous sediments along the Nicolet River south of the St. Lawrence River east of Montreal, Quebec. The lower beds, on the right, are thin-bedded sandstones and shaly siltstones containing marine fossils. They are overlain in the upper left by red argillaceous siltstones of the Queenston formation that is widespread as the latest Ordovician marginal marine sediment in the Appalachian region and northeastward. The late Ordovician terrigenous rocks in southeastern Quebec are more than two thousand feet thick, and were laid in a trough that subsided northwest of the rising highlands of the Taconian Orogeny, independent of the trough centered in Pennsylvania.

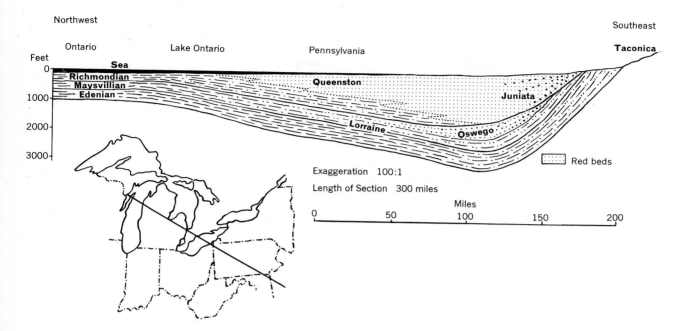

Northwest

Ontario   Lake Ontario   Pennsylvania   Southeast

Feet   **Sea**   **Taconica**
0

Richmondian
Maysvillian   **Queenston**
1000   **Edenian**   **Juniata**

2000   **Lorraine**   **Oswego**

3000   Red beds

Exaggeration 100:1

Length of Section 300 miles

Miles

0   50   100   150   200

very widespread, probably formed in the greatest inundation that the continent has experienced.

The rocks along the southeast were folded at least from Pennsylvania northeastward, for the Silurian lies unconformably on the Ordovician sediments. This is the basis for the Taconian Revolution. The region of the Taconic Mountains of western New England and New York has thrust faults below argillaceous Cambrian and Ordovician sequences. There has been controversy about their magnitude of displacement and their time of movement. As conglomerates in the middle Trentonian argillites below the thrusts of southwestern Vermont have boulders that seem like rocks in the thrust sheet, the displacements of the thrust sheets must have taken place within the Trentonian Epoch. The fact that rocks as young as Richmondian are folded beneath thrusts in southern Quebec (Fig. 9–17) does not give the thrusts there an upper age limit—they could be of any post-Ordovician age; those along Lake Champlain (Figs. 8–16, 9–18) and the St. Lawrence River may be of Devonian age. The sediments below the angular unconformities overlain by early Silurian in eastern Pennsylvania are as young as middle Cincinnatian, so there was late Ordovician folding and erosion in some regions. The Taconian had many phases and aspects.

LATE ORDOVICIAN INTRUSIONS

There were accompanying events. Far to the east in New Hampshire, Maine, and New Brunswick, granitic intrusives entered the Ordovician,

Fig. 9–18. Thrust fault, the Highgate Springs Fault, exposed along Lake Champlain south of St. Albans Bay, Vermont. Ordovician uppermost Canadian marble has been thrust over upper Trentonian shales, which define the earliest possible age. Regional structural studies lead to the interpretation that this fault and the Champlain Fault to the east (Figs. 8–15B, 8–16) are more likely of the Devonian Acadian Orogeny than of the late Ordovician Taconian Orogeny.

Silurian

Northwest

Strait of Belle Isle   Long Range   White Bay   Notre Dame Bay   Bonavista Bay   Southeast

Avalon Peninsula   St. John's

Thrust sheet   sea   sea   Ordovician   Sea level

Precambrian gneiss and granite   Silurian   Serpentine   Cambrian   10,000

Western Platform   Taconian   Silurian   Precambrian metasediments   20,000

Ordovician   Serpentine   intrusion ?   Ordovician   and volcanic rocks   Feet

Cambrian   argillite facies   lava   Avalon Platform   Vertical exaggeration 8:1

As in middle Ordovician   Cow Head   Volcanic Geosynclinal Belt   0   25   50   100

A   conglomerate   Miles

site of later thrust fault

B

Fig. 9–19. *Taconian Orogeny in the volcanic eugeosynclinal belt in Newfoundland. A. Restored section across Newfoundland in the Silurian Period (after M. Kay).* In Cambrian and Ordovician time, a central, deeply subsiding belt accumulated sediments and volcanic rocks, lavas, and fragmental tuffs and agglomerates, to thicknesses of thousands of feet; the oldest known exposed rocks are lowest Ordovician. On the east the Cambrian sediments of the Atlantic Province (Fig. 6–9) are rather constantly of argillites with some carbonate rocks of thickness of one or two thousand feet, lying on Precambrian metasedimentary and volcanic rocks that are little deformed; Ordovician is preserved only in limited areas, such as that of the iron ore-bearing sandy shales of Wabana, near St.

John's. On the west, the argillites of the central belt graded into carbonate rocks of a western miogeosynclinal belt, a platform relative to the central volcanic belt, on which a few thousand feet of carbonate rocks overlay sandstones; the basement rocks, now exposed in the Long Range, are gneisses and granitic intrusions. The Taconian Orogeny is evidenced in central Newfoundland by the introduction of slide conglomerates with large boulders in medial Ordovician, granitic-pebble conglomerates first appearing in the Trentonian graptolite-bearing argillites. Coarse conglomerates become widespread in the Silurian, some seeming to be derived from intrusive rocks that invaded Ordovician; the Silurian geosynclines may have been fault bounded, as has been proposed in Scotland (Fig.

8–6). Limestone boulder conglomerates and graptolitic argillites of western Newfoundland (Fig. 8–9) seem to be in thrust sheets that moved westward in late Ordovician, for they apparently lack Silurian sediments. Serpentine and other ultrabasic rocks have also been attributed to the Taconian Orogeny.

B. Conglomerates with boulders of intrusive granitic rocks and sedimentary rocks eroded from highlands formed in the Taconian Revolution of the late Ordovician and deposited in early Silurian deposits in the volcanic geosynclinal belt of northeastern Newfoundland; the exposure is near Lewisporte, Notre Dame Bay. Silurian conglomerates in the immediate area lie with little structural discordance on late middle Ordovician graptolite-bearing argillites.

175

and were in turn raised and eroded before being covered by Silurian. Late Ordovician and early Silurian conglomerates northeastward to Newfoundland have boulders of intrusive rocks associated with graywackes (Fig. 9–19). Records of the flow of currents show in flute-marks and in other sole-markings, impressions of the grooves made in bottom sediments by sliding rocks; thus there are indications of the direction of the slopes of the sea floors of the time (Fig. 9–20). Similar records show northwestward current flow in middle and upper Ordovician sediments in Pennsylvania. Scattered exposures of serpentine and ultrabasic intrusives from Newfoundland to Georgia are attributed to the revolution. Inasmuch as similar ultrabasic rocks are associated with modern volcanic archipelagoes such as the West Indies and East Indies, and the sediments and volcanic rocks of the Ordovician are like those now forming in such regions, the coastal parts of the present North America in early Paleozoic time are thought to have been in a stage comparable to the island arcs.

A region east of the Cambrian craton that had been one of great subsidence was in Ordovician time deformed further as increasing areas to the east were being raised into mountains. These lands produced sediments that were distributed by streams, first into limited belts along the edge of the craton, and in the late Ordovician into a great deltal plain spreading westward through the region adjoining and south

Fig. 9–20. *Evidence of current direction in submarine troughs in northeastern Newfoundland. A* (left). Flute-casts on the bottom of a bed of upper Ordovician or lowest Silurian sandstone on South Change Island, near Fogo. These sole-markings are on the base of a bed, casts of impressions of flow-scoured depressions in muds that were previously laid. The vertical bed faces southeast and flow was from the lower left as viewed, or from southeasterly direction; lineations in other beds in the bottom of the view are of the same orientation. Such direction of flow is local, and may not be the direction to the source of the sediment.

B (right). Diagram of the directions of sole-markings measured in Silurian beds at several localities on New World Island, showing their consistency; the percentages are of the number of readings within 10 degrees of each 5 degree directional ray, to smooth out the small errors in observation. A second sketch represents the source data by sectors, and a third gives the modal or most frequent direction, the angle of the arrowhead representing the sector within which about 70 per cent of the data are recorded.

Fig. 9-21. Summary of the terrigenous products of Ordovician uplifts in the Appalachian region. A succession of highlands along the southeast rose at several times within the Bolarian, Trentonian and Cincinnatian epochs of the Ordovician Period. The terrigenous sediments of the rising tectonic lands Blountia and Vermontia in the Bolarian and Trentonian were restricted to rather narrow subsiding troughs on the southeast of the Adirondack Line. Cincinnatian sediments, derived from the rising Taconica, were much more widely dispersed, finer silts and clays spreading over nearly all the northwestern part of the map area at the end of the period; see Fig. 9-15. The Bays-Moccasin sands and clays of the early Trentonian were red, as were the Juniata-Queenston of the late Cincinnatian. Sands of southeastern derivation also prevailed in the early Silurian, following the Taconian Orogeny.

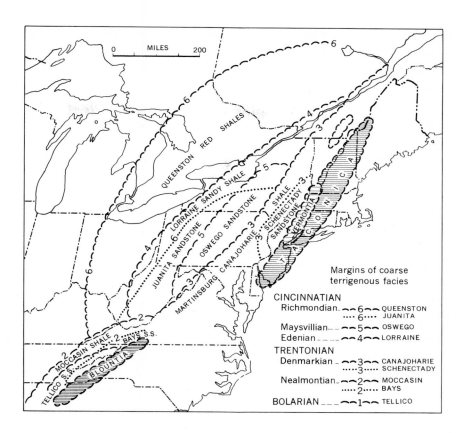

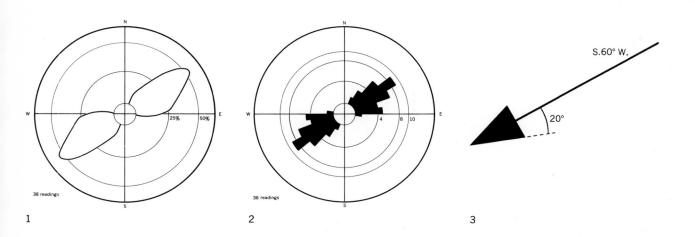

of lakes Erie and Ontario. The Ordovician history of the belts that were miogeosynclinal and eugeosynclinal in the beginning of the period is one of mobility, the interrupted bending and pushing of the earth's crust (Fig. 9–21). By the end of the period, Ordovician and older rocks had become folded and thrust westward in the Taconian Revolution, forming highlands that continued to produce sediments from their eroding surfaces.

With the relative cessation of deformation and elevation by early Silurian, the highlands remained. Hence it will not be surprising to learn that the early Silurian sediments were laid in deltas that continued to fill a subsiding region centered in Pennsylvania. Gradually, as the land was denuded and the elevations were reduced, streams flowed more sluggishly and carried finer and finer detritus into the sea, which as the basin continued to subside, gradually encroached upon the area of the earlier deltas. By the close of the Silurian, clear marine waters covered virtually all of the region that in the beginning of the period had received the delta-laid sediments from the recently formed Taconian mountains. This stability was to be short lived, however, for we will return to later orogeny in the same territory. The record that has been discussed is one of the gradual change of a part of the continent through a period of nearly one hundred million years, still a short time compared with the periods of the Precambrian but taking more than a quarter of the Paleozoic Era.

Fig. 9–22. Surface of a bed of limestone in the Cincinnatian shales exposed at Cincinnati, Ohio (The Smithsonian Institution). Most of the fossils are brachiopods, particularly exterior and interior surfaces of the semi-elliptical *Sowerbyella*, and less frequently the nearly circular, finely striated *Dalmanella*. The twig-like pitted fossils are bryozoans, and there are a few other kinds of brachiopods.

10
Silurian Reefs and Salt Basins

Silurian dolomite erosion remnants: the Flowerpots, Georgian Bay National Park, Lake Huron, Ontario (Geological Survey of Canada).

10 *Silurian Reefs and Salt Basins*

The Silurian System was named by Murchison from sections in Wales, and the name soon came into controversy with Sedgwick's Cambrian System, as has been stated. Similar rocks were known in New York at the time and were classified soon after the state geological survey was established in 1836. Their exposures produce the setting for the renowned falls of the Niagara River on the boundary with Ontario.

NIAGARA RIVER SECTION

The visitor to Niagara Falls (Fig. 10–1) sees the great river plunge more than one hundred and fifty feet over ledges forming the cliffed walls of the gorge. The cliff is of Silurian limestone (the Lockport), and calcite and dolomite rocks with some cherty zones; in places lenticular structures contain coralline masses, the adjoining bedded limestones lapping against the ancient reefs. About one hundred feet below the crest of the falls, and better exposed along the route of the old gorge railway, the base of the cliff-forming limestone lies on shaly and silty argillaceous rocks (Fig. 10–2) having a few limestone interbeds (the Clinton) (Fig. 10–3); they succeed two sandstones separated by a zone of argillitic shales and thin limestones (the Albion) exposed farther north along the gorge. The underlying red shaly siltstones (the Queenston) are Ordovician. Thus Ordovician Queenston red shale is succeeded by Albion sandstone and shale, Clinton shale, and Lockport limestone. Above the falls, dolomite rock (Lockport) continues in rapids through another fifty feet of thickness to the upper end of Goat Island; other exposures and wells show that the whole limestone unit is about two hundred feet thick. The strata from the top of the red shale to the top of the limestone constitute the standard section of the lower or Niagaran* Series of the Silurian System in North America; the Albion, Clinton, and Lockport are groups having divisions that are formations. In the following pages, rocks that are thought to be the time equivalents of the groups will be called Albionian, Clintonian, and Lockportian, as though they were stages.

---

*The name Niagaran originally was applied to the Clinton and Lockport beds and the older sandstone was classified with the underlying red beds in the "lower Silurian"—subsequently named Ordovician. When the sandstones were found to be Silurian, they commonly were made a separate series or added to the Niagaran; the latter procedure is followed here. The present system is essentially of beds called "Ontario group" in the first New York survey reports in 1842 and, subsequently, the Ontarian.

Fig. 10–1. Air view of Niagara Falls from the northwest, with New York on the left and Ontario on the right (Photographic Survey Corporation). The river flows over a falls with Silurian limestones at the crest; the gorge below has been cut in the less resistant underlying Silurian and Ordovician shales and sandstones. Goat Island on the left separates the American Falls from the Horseshoe Falls, each with falls of more than one hundred and fifty feet.

Fig. 10–2. Columnar section of the rocks exposed at Niagara Falls; the flow cuts away the less resistant shales below the Lockport limestone at the crest, causing retreat by the breaking away of the blocks at the crest of the falls. At times there are rock falls on the wall of the gorge, such as the one that destroyed a generating station on the American side in 1955.

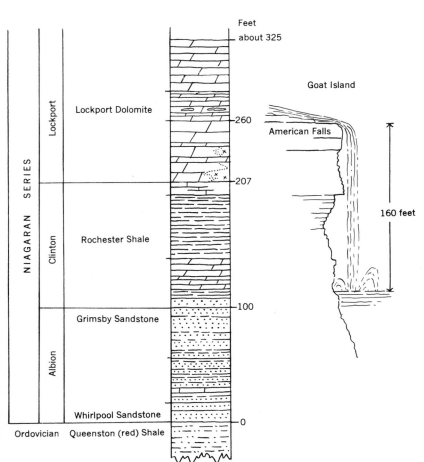

			Feet
			about 325

Goat Island

American Falls

160 feet

NIAGARAN SERIES

Niagaran Series column:
- Lockport — Lockport Dolomite — 260, 207
- Clinton — Rochester Shale
- Albion — Grimsby Sandstone — 100
- Albion — Whirlpool Sandstone — 0
- Ordovician — Queenston (red) Shale

Fig. 10–3. Silurian exposed along the Whirlpool Gorge of the Niagara River, western New York. The exposures are of upper Ordovician red sandy shales overlain by the lower Silurian sandstones and shales of the Albion and Clinton groups. These beds are less resistant to erosion than the succeeding Lockport limestones, so have been cut away beneath the limestones that form the crest of Niagara Falls, and the Niagara Escarpment.

183

The upper or Cayugan Series of the American Silurian is poorly exposed along the Niagara River, for the rocks are shales concealed by drift and soil. Wells drilled to the south near Buffalo penetrate one thousand feet of Cayugan shales passing up into the well-bedded limestones. For the present, discussion will concern only the Niagaran. The Cayugan Series of Michigan and nearby regions will be considered subsequently.

The Niagaran Series in New York has fossils like those in the British and Swedish Silurian limestones, such forms as the chain-coral *Halysites* and the brachiopod *Pentamerus* (Fig. 10–4). The Silurian of Wales, like the Ordovician, is most typically a sequence of argillites and gray-wackes containing many graptolites that have been assigned to many zones in three series. The graptolite-bearing rocks can be shown to grade into limestones having a rich coral and brachiopod fauna of shelly facies, particularly in western England; the Swedish Island of Gotland in the Baltic Sea is famed for its Silurian coral reefs (Fig. 10–5). Other sequences in North America with richly fossiliferous graptolite beds are widespread as in Nova Scotia, Oklahoma, central Nevada, and British Columbia; perhaps the best is the section of a mile or so of argillites exposed on Cornwallis Island in the Canadian Arctic at 75° North Latitude, 95° West Longitude, near the north magnetic pole. We will be concerned with North American rocks having fossils such as are found in the shelly facies of the Silurian of Europe; though discussion will pertain to a limited area in eastern North America, Silurian rocks are known in the West (Fig. 7–7), and are extensive along the Atlantic Coast (Fig. 9–19).

The stratigraphy of the Niagaran series of eastern North America is controversial, for the correlation of the rocks in time involves interpretations of the effects of the differing environments of deposition and

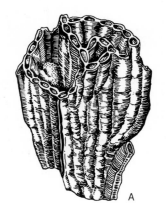

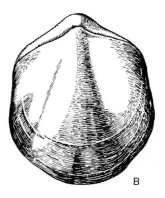

Fig. 10–4. *Fossils that are distinctive of the Silurian System. A.* The colonial coral *Halysites*, one in which the vertical tubular corallites are arranged in intersecting ribbons; though corals of this sort, chain corals, first appear in the middle Ordovician, they are particularly common in reef assemblages in the Silurian and unknown thereafter.

B. The brachiopod *Pentamerus* pentamerid brachiopods are known in rocks older and younger than Silurian, but large forms such as the typical genus are virtually limited to the Silurian and lower Devonian.

Fig. 10–5 (right). Silurian limestone on the island of Gotland, Sweden, in the Baltic Sea. The island is noted for the development of reefs of coralline limestone that pass laterally into bedded sediments, calcareous and argillaceous; the rocks are richly fossiliferous and were the source of some of the first-described Silurian fossils in the eighteenth century.

A

B

Fig. 10–6. *Niagaran exposures in Ontario. A.* Silurian Whirlpool sandstone lying disconformably on the upper Ordovician Cincinnatian Queenston red shale along the face of the Niagara Escarpment, Credit River west of Toronto, Ontario.

B. Silurian Lockport dolomite forming erosion remnants, the Flowerpots, in Georgian Bay National Park, Lake Huron, Ontario (*Geological Survey of Canada*).

of possible geographic separation of provinces on the fossil populations in rocks in scattered areas. The portrayal of the geography of successive times depends on these subjective judgments of lithic and faunal facts; alternate interpretations produce differing history. First, let us summarize the relevant data on the Niagaran Series westward from the falls.

LITHOLOGIES, FOSSILS AND TIME

Paleozoic rocks dip gently away from the Precambrian rocks exposed in the southern part of the shield of central Canada. Thus southward dipping Silurian rocks of Ontario (Fig. 10–6), northern Michigan, and eastern Wisconsin lie between Ordovician rocks to the north and younger sediments to the south (Fig. 10–7). The Lockport limestones that form the cliffs of the Niagara gorge are so much more resistant to erosion than the older Silurian and Ordovician shales, and the succeeding Cayugan shales, that they form a cuesta or escarpment that trends northwestward from Niagara Falls through Ontario to the tip of the Bruce Peninsula separating Georgian Bay from the main body of Lake Huron; the cuesta, emerging again in Manitoulin Island and islands to the west in northern Lake Huron, continues to the upper peninsula of Michigan north of Lake Michigan. Westward and southwestward, the limestones form the peninsulas restricting the entrance to Green Bay in Wisconsin, and become poorly exposed in the belt of outcrop along the western side of Lake Michigan to Chicago, Illinois. Throughout this extent, corals abound in reef masses within the limestones (Figs. 10–8, 10–9), but toward the top, distinctive (Guelph) gastropods, cephalopods, and brachiopods increase as the coralline elements disappear. This latter change has been attributed to modification of environment through time, perhaps through marine waters becoming more than normally salty.

Consider next the stratigraphy of the Silurian rocks lying below and to the northward of the limestones of the cuesta. The lowest Silurian, the Albion, can also be traced westward to Michigan, but only by careful field studies, for it has marked facies changes (Fig. 10–10). Whereas it is mostly sandstone in New York, it becomes argillaceous and shaly in southern Ontario, and farther northwestward the shales pass into limestones that prevail from northern Michigan to Illinois. The equivalence of the beds is established, because *tongues* of shale penetrate into the sandstone on the east and the limestone on the west; fossils differ in the several lithic facies. The correlation of the Clinton Group (Figs. 10–2 and 10–11) that lies between the Albion and the Lockport introduces problems. The shaly argillites in New York have abundant fossils in some of their formations. Iron ore, iron oxide

Fig. 10–7. Geologic map showing the distribution of the Silurian System in the region of the Great Lakes. The Niagara Escarpment is prominent from near Rochester, New York to near Milwaukee, Wisconsin; the Silurian limestones dip gently southward from the more easily eroded shales of the upper Ordovician. There are few exposures of the Silurian in the other belts of outcrop, such as the broad area in Indiana and Ohio. Silurian quartzites form prominent mountains in the Appalachian region in the southeast of the map.

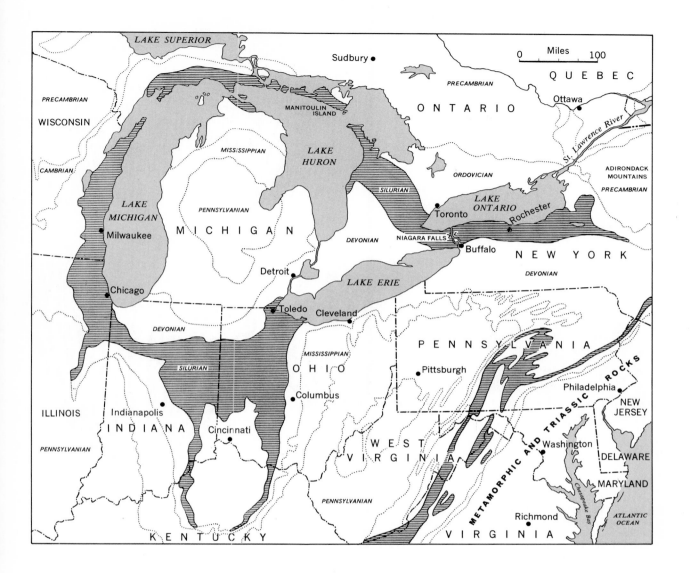

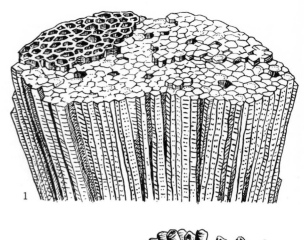

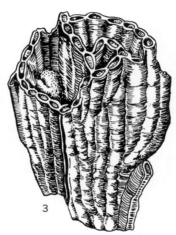

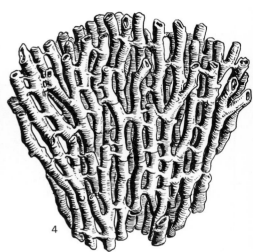

Fig. 10-8. *Corals from Silurian limestones.* Representative corals such as those that built the reefs in the Silurian: (1) and (3) tabulate corals *Favosites* and *Halysites*, (2) and (4) rugose colonial corals *Strombodes* and *Syringopora*. All are found in many places; those illustrated are from Michigan.

The coral reefs of the Silurian have calcareous colonial corals as principal or significant constituents, but as with all coral reefs much of the rock is made of skeletons of other organisms and of calcareous detritus. Coral is a comon name applied to several orders of organisms of the class Zoantharia of the phylum Coelenterata—animals having tentacles around the mouth of a many-celled sac-like body cavity with differentiated cells; some of the cells are for digestive functions. The geologist finds the resistant calcareous skeleton, commonly conical or horn shaped in simple corals, and more cylindrical in colonial forms; each has a conical depression or calyx in which

an individual polyp lived. Several larger groups of forms can be recognized. The Paleozoic corals are principally in the order Rugosa, having both single individuals and colonies of individuals living in calyces that seem radially symmetrical, but which have the partitions within each calyx arranged bilaterally. A second important order of Paleozoic corals is the Tabulata, with tubular corallites that have rather irregular septa and other internal structures. In the Silurian most reef corals are colonial rugose and tabulate corals; a most distictive tabulate is the halysitid or "chain" coral, in reference to the appearance of the corallites in cross-section; halysitids are present in the Ordovician too, but not after the Silurian. Corals are distinguished by their shapes and the arrangement of their septa and other internal structures.

Coral reefs of the Silurian attracted particular attention because they are common in Silurian rocks in the Great Lakes region as well as in Europe, as in

western England and on the Swedish island of Gotland. They are of particular interest to the geologist because of their bearing on paleoclimatology. They are sedentary, attached to the bottom, though their larvae float. Reef-building corals are restricted in the present to warm marine waters, below about 35 degrees latitude, to rather shallow depths of a few tens of feet, though solitary corals live to great depths and in quite cold waters. They prefer clear to turbid, muddy waters, and thrive in turbulent, agitated seas. The oldest coral reefs of the Ordovician Chazyan have very few corals, but they formed on shallow marine platforms that were subsiding less rapidly than their surroundings, a condition that also controlled the formation of some later reefs. But in many subsiding regions, the corals were able to grow quite as rapidly as the basement subsided and hence to accumulate thicknesses comparable to the depth of sinking.

Fig. 10–9. Reconstruction or diorama of a coral reef such as might have been seen in Silurian seas in Illinois (Chicago Natural History Museum). The view shows several types of colonial corals, such as the chain coral, *Halysites*, that were important reef builders. Trilobites, bivalved brachiopods, coiled cephalopods and stalked, flower-like animals, crinoids, are also shown.

replacing calcareous granules or oolites and fossil fragments, is a distinctive rock mined from thin beds at Clinton, New York, in the eastern part of the exposure. The Clinton shales gradually thin westward to disappear north of the western end of Lake Ontario, the Lockport coming to lie on the older Silurian Albion Group.

On Manitoulin Island, north of Lake Huron, and in northern Michigan and southwestward through Wisconsin to Illinois, limestones intervene between the beds traced as the base of the Lockport and those at the top of the Albion. These limestone rocks have the stratigraphic position of the Clinton of New York. But they are not shales and iron ores such as prevail in New York, nor do they have many of the fossils that are found there. They are more like the Lockport of New York and contain some of the kinds of corals that in New York are restricted to the Lockport. Albionian beds in Ontario are shaly like those in the typical Clinton, and they have many fossils much like those in the New York Clinton. So the fossils in the Clinton shales of New York have greatest similarity to those in similar *rock* in the older Albionian beds in Ontario; and the fossils in probably Clintonian limestones in Michigan are most like those in the younger Niagaran limestones in New York. The fossil assemblages in similar rocks representing the same sort of environments are more alike in successive stages than they are to rocks formed in other environments of the same time. The statement that fossils are similar is one that can be depended on only after the most critical study of collections by paleontologists who have become expert on the particular kinds of organisms found.

Lockportian coralline limestones are along this belt of outcrop and southward and are recognized with more limited success in wells. They prevail in an area north of a sinuous line from southern New York to northern Kentucky and western Illinois (Fig. 10–12). To the south, in Tennessee and southeastern Missouri, Lockportian limestones are not recognized. Study of the rocks removed from the Chicago drainage canal that extends southwest from Lake Michigan revealed that coralline reef limestones have interbeds of reddish argillaceous shale. Wells in Illinois show that argillaceous reddish and gray shales like the interbeds were laid among the coral reefs and become dominant to the south (Fig. 10–13). Fossils in the shales exposed in Missouri and Tennessee are so unlike those in the New York Lockport limestones and similar to those in shales of the type Clinton of New York that they were thought to be of Clintonian age until the discovery of the relations at Chicago. Thus Lockportian rocks of argillaceous facies have fossils resembling those in the argillaceous Clinton which was of a more similar environment than the typical coralline Lockport. The

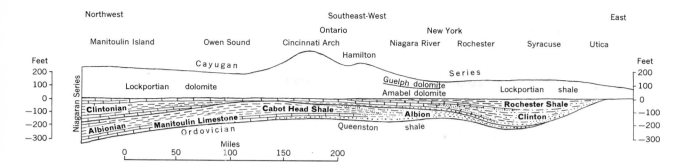

Fig. 10–10 (above). Diagram showing the relations of the lower Silurian Niagaran Series from Manitoulin Island and northern Lake Huron along the Niagara Escarpment to central New York. The line of section along the outcrop is south-easterly to the western end of Lake Ontario at Hamilton, then easterly into New York. The Albion Group changes from terrigenous detrital rocks on the east to carbonates on the west that continue into Michigan; the Clinton Group in New York is argillaceous, whereas the Clintonian of the Michigan Basin margin is limestone. The carbonate rocks of the succeeding Lockport Group are shown only in outline, their base being drawn as an horizontal line. Fossils found in the several units are affected not only by their age, but by the lithologies that relate to original conditions of deposition.

A

Fig. 10–11. *Silurian rocks exposed along the lower gorge of the Genesee River in Rochester, New York. A.* Sandstone in the bluff by the road is the basal Silurian Albionian Grimsby reddish sandstone, only partly exposed, overlain by a three-foot band of white Thorold Sandstone in the base of the Clintonian Stage that extends to the top of the bluff. There are some eighty feet of shales, with interbedded colitic iron ore in the dark cliff recess, and a few limestones (Ward's Natural Science Establishment).

B. Detail showing the bedded Grimsby below the Thorold Sandstone and succeeding Clinton shales and sandstones (R. G. Sutton).

B

191

science of paleoecology concerns such interrelations between ancient organisms and their physical and biologic environments.

LATERAL CHANGES WESTWARD FROM NEW YORK

Lithologies and fossils can be examined and compared but determinations of time are evasive. Continuous belts of exposures or closely spaced wells generally have tongues of thin units of distinctive lithology, often with characteristic fossils representing effects of extensive changes in conditions. Otherwise, fossils that change progressively may be useful guides to time; but there will be differing opinions on which forms are time controlled rather than more distinctive of environment. When rocks are called Clintonian or Lockportian, there is the assumption of assurance that time-determining facts are known. The judging of the evidence is the scientific base of stratigraphy. Lockportian rocks can be placed in lithologic belts. North of a line from New York to Illinois (Fig. 10–12) in a belt of a few score miles in width, the coralline beds are considerably of calcite limestone, with 10 to 15 per cent of argillaceous and siliceous matter that drifted in among the reefs. Northward as far as Lockportian is exposed, even to the Arctic, the rocks are predominantly reefy and of dolomite (calcium-magnesium

Fig. 10–12. Early Silurian paleogeography of eastern North America. Silurian rocks are preserved today only in a limited part of the area of the map, so much of the geography is interpreted. The initial deposits are thought to have been laid in embayments in the south central United States (Alexandrian) and maritime Canada (Anticostian). The distribution of seas along the Atlantic Coast is very hypothetical, for few Silurian rocks are recognized—perhaps because most rocks in this belt are metamorphosed and hence do not retain preserved fossils. Highlands in eastern United States produced coarse terrigenous sediments, as shown by the distribution of these rocks; the facies changes are shown also in Fig. 10–10. In late Alexandrian-Anticostian, or in early Clintonian, seas spread widely over the region.

The fossils are different in the separated earliest Silurian embayments, because perhaps of their geographic isolation; fossils are also different in the terrigenous rocks than in the calcareous ones. The calcareous rocks in the northeast contained a brachiopod, *Virginia* (Fig. 10–18), whose possible descendants spread into the northern United States with spread of the seas in early Clintonian. On the other hand, the ostracodes (Fig. 10–17) associated with the northeastern shales seem to be the ancestors of somewhat similar forms found in the Clintonian shales in New York. Thus geographic controls, and local conditions of deposition as evidenced in rock types, determined the kinds of fossils present in the beds in different regions as well as the differentiations that developed through time.

Clintonian rocks from eastern New York, southward in the Appalachian Mountains, have beds of iron ore, frequently oölitic—with small concentrically zoned particles—and fossiliferous. The ore is the source for the steel mills near Birmingham, Alabama.

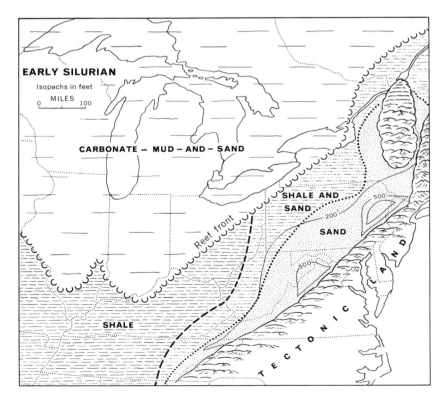

EARLY SILURIAN

Isopachs in feet

0   MILES   100

CARBONATE — MUD — AND — SAND

Reef front

SHALE AND SAND

500

200

SAND

500

SHALE

TECTONIC LAND

carbonate) of high purity; these rocks were originally calcite and aragonite (calcium carbonate) for they have the same organic constituents as the calcite rocks to the south. Alteration to dolomite took place soon after deposition, the shallow marine water in seas of restricted circulation and in a climate of high evaporation passing into the previously laid calcite sands and silts below the sea floor, adding magnesium while removing calcium.

Some of the variations in thickness can be attributed to warping during deposition. But in Illinois, thickness decreases rapidly in passing from the reefy belt into the argillaceous belt to the south (Fig. 10–13), where water was deeper. The whole area subsided, but it acquired greater thickness where reefs built up to the level of the sea than where deeper water with the limited agitation permitted the settling out of fine argillaceous muds; the finest terrigenous detrital clays were carried in suspension from the streams issuing from the highlands formed in the Taconian Revolution along the Atlantic Coast. The original thickness of the argillaceous clays had been further reduced by compaction, whereas the coralline limestones compacted little.

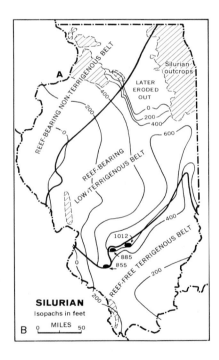

SILURIAN
Isopachs in feet
0   MILES   50
B

Fig. 10–13. *Stratigraphy of the Silurian System in Illinois (after H. A. Lowenstam). A. Restored section of the Silurian late in the period. The thick section of reef-bearing limestones, principally of calcite, formed in shallow water facing deeper water to the south and having a broad shoal to the north. Terrigenous muds settled in the subsiding basin to the south; debris from the reef front spread into the margin of the shales. Thickness of the muds will have been reduced by compaction more than that of the carbonate rocks to the north, so the restoration may exaggerate the water depth. To the north of the reef-front, calcareous sediments and organ-* isms laid in shallow waters became the present dolomite soon after their deposition. The lowest Silurian is bedded limestone throughout the area. Because the coralline rocks are similar to those in the Lockport of New York and Ontario, they have been called Niagaran, but they may extend into the Cayugan Series.

B. Isopach and lithofacies of the Silurian of Illinois. From the section we can see that the isopachs cannot be converted into a pattern of subsidence, because the sediments had differing depths of water. The principal belts of rock proportions or lithofacies are affected principally by the distribution of the deeper water and more southerly terrigenous facies.

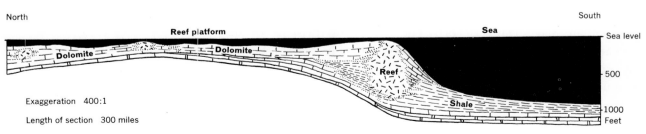

North   South

Exaggeration 400:1

Length of section 300 miles

A

ENVIRONMENTS AND FAUNAS

Environment was a very important factor in determining which organisms should appear in the sediments of the Niagaran. To say that faunas are alike or unlike is somewhat subjective, for it depends on the judgment of the observer; but when the differences relate to the kinds of animals, and even to the genera rather than the species, they are such as to make contrasts easily apparent. It has been stated that fossils in the argillaceous Albionian in Ontario resemble those of the typical Clinton and also those of the Lockportian of Tennessee. Faunas like the latter are present also in Oklahoma in argillaceous Lockportian facies. They are succeeded by other argillaceous rocks that have a great many of the same genera, forms so similar that some are scarcely, if at all, distinguishable. Yet there are a few missing forms and a few additional ones that show that these upper argillites are much younger and were deposited in the Devonian Period.

The carbonate facies of the Lockportian has an abundance of corals. In New York, where the environment is represented only in the Lockport, these might seem diagnostic of Lockportian time. But in Michigan, where there are carbonates in the Clintonian, similar corals appear. In fact, some of the same genera are present in the extensive limestones of the late Trentonian and Cincinnatian that are widespread over the northern and western continental interior; these ancestors lived in the same favorable environment, though there are distinctive associated forms. Thus to utilize fossils in classification of beds requires extended experience and knowledge, for though similarities in some instances are quite conclusive evidence of synchroneity, in others they are only indicative of similar environment; experience has shown that many forms, ancestors and descendants, endured—little changed—through millions of years.

GEOGRAPHIC PROVINCES

Geographic factors that isolate seas in parts of the continent also influence the faunas. In southeastern Missouri, the lowest of Silurian is a sequence of more than one hundred feet of limestones, of the Alexandrian Stage (Fig. 10–14). The successive divisions of the Alexandrian are progressively more widespread (Fig. 10–15). Seas of the earliest Silurian advanced northward in the Mississippi Valley region, and eastward toward Ohio. Farther east in Pennsylvania, the oldest beds in the thick lens of coarse detritus derived from the Taconian Highlands are also Alexandrian; so a second seaway advanced northward toward New York, where it deposited the Albion, with a

Fig. 10–14. *Lower Silurian Alexandrian limestones along the Des Plaines River near Joliet, southwest of Chicago, Illinois (Illinois Geological Survey).* A. The upper part, Kankakee Limestone showing a small organic structure or reef. B. Underlying this is the Edgewood Dolomite.

Fig. 10–15. Early Silurian marine invasions. The map illustrates the conventional view that at the beginning of the Silurian Period, the Anticostian Sea advanced from the northeast and the Alexandrian Sea advanced from the south, their waterways meeting by Clintonian time. The waterways in the New England-Maritime region are drawn to show the associated tectonic and volcanic lands in the belt; but their positions are rather hypothetical. In Clintonian time much of the region, except the highlands, had become covered by the seas. The control on which such a map is based is so limited that it gives little more than the general impression of the geography of the time. The region southeast of Logan's Line cannot be shown properly with respect to present geography because of the deformation that affected it in later orogenies, such as the thrusting and folding of the later Paleozoic disturbances.

A

B

peninsula extending southwestward through central Ohio, Kentucky, and Tennessee.

Anticosti Island, about one hundred and thirty miles long and the size of Delaware in the northern part of the Gulf of St. Lawrence, is a cuesta capped by Lockportian limestones (Fig. 10–16); at least the fossils resemble those in typical Lockport. Beneath are several hundred feet of argillaceous and calcareous sediments with abundant fossils, particularly brachiopods, bryozoans, and ostracods, minute bivalved crustaceans that look like miniature clams (Fig. 10–17). This fauna is very like that of the Clinton of New York, as one might now anticipate, inasmuch as it is in argillaceous shales. The ostracods of the early Silurian were changing rather rapidly, for the same kinds change perceptively in successive zones of shale rock laid in similar conditions. Study of the ostracods by those specially qualified indicates that the uppermost shaly beds in Anticosti seem time-equivalent to the upper Clinton of New York, the ostracods in the lower beds in Anticosti being peculiar to that section, unlike those in the Alexandrian and earlier Clintonian beds of the interior. There are a few other fossils, such as the brachiopod *Virgiana* (Fig. 10–18), that is widespread in lowest Silurian of the Arctic region and is found in limestone in northern Michigan in the late Alexandrian (Albionian) or earliest Clintonian.

It has been believed conventionally that the seas that occupied the Anticosti region, and a much larger area from which the records have been eroded away, advanced into the continent, so that these waters and those from the southern or Alexandrian province mixed in late Clintonian. Thus the Clinton of New York has organisms whose ancestors lived in both provinces, whereas the Alexandrian Stage of Missouri and the synchronous Anticostian Stage of Anticosti have few fossils in common even in shales laid in similar environments, because the

B

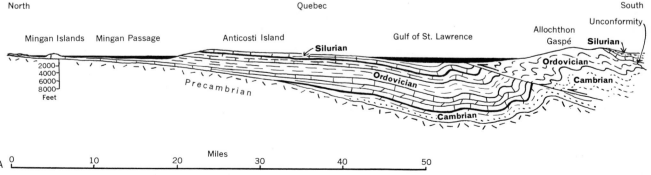

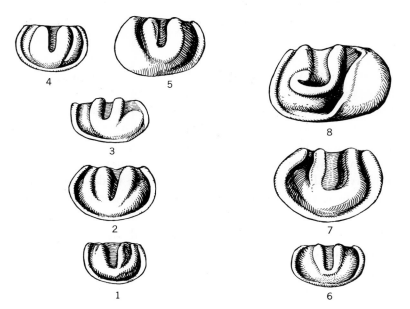

4   5

3

2

1

8

7

6

Fig. 10–16 (left). *The St. Lawrence River and Anticosti Island. A.* Structure section across the St. Lawrence River from the north shore of the St. Lawrence River at the Mingan Islands through Anticosti Island to the peninsula of Gaspé, Quebec. Anticosti Island has an excellent section of fossiliferous late Ordovician and early Silurian fossiliferous sediments in a seemingly conformable and continuous succession. Lower Ordovician limestones lie on the Precambrian rocks off the coast of Labrador; the section assumes that older Paleozoic rocks are preserved under Anticosti Island. To the south of the St. Lawrence River on Gaspé Peninsula, the Cambrian and Ordovician rocks are like those at Levis near Quebec, graptolitebearing argillites; to the south Silurian sediments lie unconformably on the folded Ordovician, evidencing Taconian structure. Because of the proximity of the undeformed section of calcareous rocks on Anticosti to the Logan's Line thrust fault that must lie under the Gulf of St. Lawrence, it seems improbable that the thrust is late Ordovician and more probable that it is of the late Devonian Acadian Orogeny.

B. Jupiter River, Anticosti Island, Gulf of St. Lawrence, Quebec, showing the exposures of the lower Silurian Becscie and Gun River shale formations. The faunas from these rocks are considered representative of the lowest Silurian of the northeastern faunal province (J. F. Caley, Geological Survey of Canada).

Fig. 10–17. *The Ostracoda.* Ostracoda from the lower Silurian of Maryland and Virginia, and of Anticosti Island, in the Gulf of St. Lawrence, Quebec (after Maryland Geological Survey); the species appear from oldest at the bottom to youngest at the top: (1) *Zygobolba anticostiensis*; (2) *Zygobolba decora*; (3) *Zygobolbina emaciata*; (4) *Mastigobolbina lata*; (5) *Zygosella postica*; (6) *Bonnemaia rudis*; and (7) *Mastigobolbina typus.* The valves are mixed, some right, others left; some male, others female with brood pouch; thus (4) has a left valve, male and a right valve, female.

Ostracoda are a subclass of Crustacea in the phylum Arthropoda. Trilobites, another sort of arthropod, are of such size that they can be seen without enlargement. Ostracods are minute bivalved organisms, usually about the size of the head of a pin, rarely as large as the nail of a little finger; hence they must be studied under a microscope. The shells, calcareous or chitinous, are concentrated by washing away clays in which they are preserved, the fossils being in the residue; or they may be seen on the surface of broken rock. In some fortunate instances they are silicified and can be obtained by leaching away the surrounding limestone by means of acid.

The minute size of the shells preserves most of them from destruction in the drilling of wells, so they are retained in the cuttings. Micropaleontology concerns the study of such minute organisms as ostracods and foraminifera, the conodonts and colecodonts that are minute parts of larger organisms, and bryozoans, large animals that can be studied most advantageously by cutting sections for microscopic study. These types of microfossils are described elsewhere in the text.

Ostracods are identified and classified principally by the patterns of surface ridges, pits, and other sculpture. They change appreciably through time, and in some sequences, forms within a genus or family develop rapidly enough to permit close differentiation of the containing beds. Ostracods are frequent and useful in each system from the Ordovician to the Tertiary.

The Silurian ostracods are representatives of species of several closely related genera that have been found in successive horizons in earlier Silurian argillaceous rocks. The ostracods change gradually through time and are generally present, so are useful guides to chronology in many systems.

marine connections were so distant and circuitous that few organisms could travel through the forbidding environments of intervening waters.

FOSSILS IN CHRONOLOGY

The use of fossils in interpreting chronology and paleogeography demands a great fund of knowledge based on experience, as the organisms of a time differ from place to place because of the conditions of their environments and because of their being isolated from one another by geographic factors. The history of the early Silurian has been interpreted as one of seas advancing into the interior. The faunas were influenced indirectly by the Taconian Revolution of the late Ordovician, in that the detritus from the highlands produced turbid waters and muddy bottoms in contrast to the clearer waters and harder bottoms of areas more distant from the source lands. Differential subsidence affected water depths and directed the course of advancing seas.

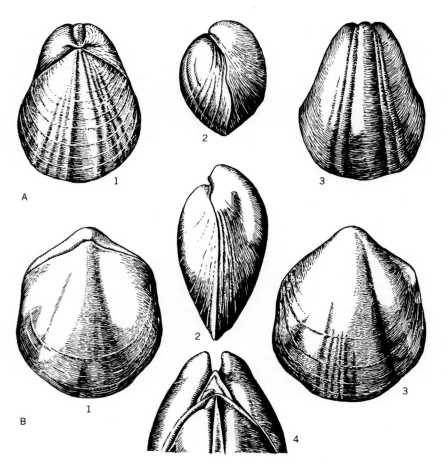

Fig. 10–18 (left). Silurian Brachiopoda, genera of pentamerids. *A. Virgiana* from the lowest Silurian of Anticosti Island in the St. Lawrence River, Quebec, and *B, Pentamerus*, from somewhat younger beds in New York; (1) brachial, (2) lateral, and (3) pedicle views of each; (*B4*) is an internal mold, showing the impression of plates that project on the interior of the valves and are distinctive or pentamerids. Note difference on groove on brachial valve of *Virgiana* compared to broad fold on *Pentamerus*. *Virgiana* is a representative fossil of a northern province of invasion but is distributed widely in early Silurian rocks in west Texas, Nevada, and Manitoba, for instance.

Fig. 10–19 (right). Restored section of the Silurian System from northwestern Michigan to eastern Pennsylvania (after L. I. Briggs and H. L. Alling). The section crosses two principal areas of subsidence, the isolated basin within southern Michigan in which thousands of feet of salt-bearing sediments accumulated in the Cayugan Epoch, and the sinking region in the Appalachian area of Pennsylvania that received terrigenous sediments from lands to the southeast formed in the Taconian Orogeny in the Ordovician, lands that were eroded to lowlands during the Silurian Period. The section is drawn in two parts.

SUBSIDENCE AND ISOSTASY

The Silurian presents striking evidence of the manner of deformation of the interior of the continent. The Taconian Revolution culminated late in the Ordovician, so that by the end of the period, the Cincinnatian Series of Pennsylvania and adjoining states formed a lenticular deposit more than a half-mile thick. There is a theory, the *theory of isostasy,* that the earth's crust is so in equilibrium that areas that become loaded will sink because the addition of matter to the area of loading disturbs the equilibrium. Material in the fluid zones of the inner earth must move away from the subsiding areas. The average density of the surficial sediments is less than that of the rocks in the outer part of the earth that must be replaced. Hence, if sediments are added to a part of the area of the earth, the load will not depress to as great a depth as the thickness of the sediments. The surface of deposition should gradually rise as the base sinks, for though the load would cause subsidence, the subsidence would be less than the thickness of the load.

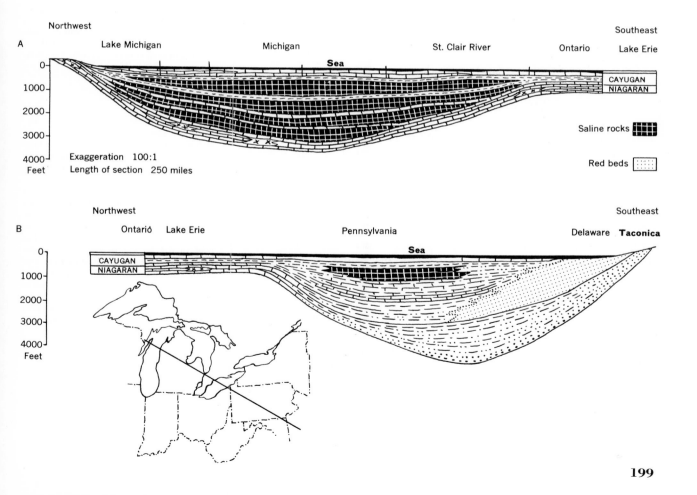

199

Normally, then, if there is no change of sea level, eustatic movement, the surface of deposition in an area will rise, and the sediments become shallower marine and ultimately non-marine unless some additional cause of subsidence is operating. The late Ordovician sediments are a load deposited in complement to the rising Taconian highlands; the subsidence on the margin of the craton is complementary to sediment-producing uplift outside the craton—hence this sort of geosyncline has been called an exogeosyncline, "from outside" geosyncline. The sediments pass upward from marine into non-marine at a time when there seems to have been eustatic rise, as can be judged by the great extent of late Cincinnatian seas (Fig. 9–15). The succeeding Silurian lenticular geosyncline-filling deposit begins with coarse conglomerate (Fig. 10–19), believed to have been stream-laid; the deposit has widespread sandstones (Fig. 10–20) and terminates with widespread marine limestones that even spread over the eroded surface of the land that formed the flank of the Taconic Mountains. Thus in spite of the load of sediments added to the area that subsided, the surface of deposition came to be lower than when the base of the deposit was laid. Subsidence must be initiated by some cause, and it must be due at least partly to causes other than adjustment due to loading. Evidence of an independent cause of subsidence is even more impressive in a case in which there is no complementing highland to produce terrigenous detritus.

EPEIROGENY IN THE MICHIGAN BASIN

The changing pattern of crustal warping or epeirogeny is shown in the Silurian of Michigan (Fig. 10–21). The Niagaran rocks are carbonates, less than 200 feet thick in the central part of lower Michigan, increasing to nearly 1000 feet along the northern shore of Lake Michigan and more than 500 feet at the western end of Lake Erie. As these rocks were shallow laid—they are Lockportian coralline limestones in the upper part—warping in Michigan during the early Silurian took the form of differential subsidence which was least in the central part of lower Michigan.

The deformation in later Silurian, Cayugan, is very different. From more than 4000 feet in the center of lower Michigan, thickness decreases gradually outward to 1000 feet or less along lakes Michigan and Huron. Put in another way, if the top of the Cayugan were a horizontal plane at the close of the epoch, the bottom would have formed a basin having depth of more than 4000 feet in the center and less than 1000 feet at the lakes. The series contains hundreds of feet of rock salt or halite which, when refined from rock quarried in the mines and from brines pumped from wells in southeastern Michigan, is a princi-

Fig. 10–20. *Silurian sandstones in the Appalachian Mountains (H. P. Woodward).* A. Seneca Rocks, along the upper Potomac River in eastern West Virginia. The quartzite ledges stand vertically, with older beds toward the southeast; the flat surface facing on the left in the view is the bottom of a bed. The Seneca Rocks are in the westernmost anticline of the Appalachian Mountains, the Silurian rocks descending and then flattening beneath the Allegheny Plateau to the west.

B. Beds of Silurian Clintonian sandstone along the James River north of Eagle Rock, Roanoake County, Virginia.

Fig. 10–21. *Maps of the Cayugan Series in the Great Lakes region (after L. I. Briggs and H. L. Alling).* A. Isopach map showing the thickness of Cayugan sediments, in the subsiding basins in Michigan and in Pennsylvania.

B. Lithofacies map, one showing the proportions of different constituents making up the successions in the region.

C. Paleogeographic map, presenting the interpretation of the geography that lead to the formation of the saline deposits. The topographic and bathymetric basins in an area of high evaporation concentrated salts from marine waters that entered them through restricted access channels. Slight deformation affected the geography so as to alter the rate of evaporation from time to time.

A

B

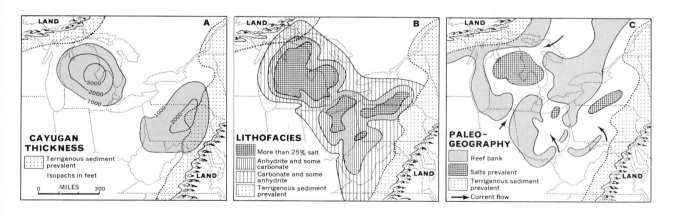

LAND   A

CAYUGAN
THICKNESS

3000
2000
1000
1000
2000

LAND

···· Terrigenous sediment
     prevalent

—— Isopachs in feet

0   MILES   300

LAND   B

LITHOFACIES

▦ More than 25% salt
▥ Anhydrite and some
     carbonate
▯ Carbonate and some
     anhydrite
···· Terrigenous sediment
     prevalent

LAND

LAND   C

PALEO-
GEOGRAPHY

░ Reef bank
▦ Salts prevalent
···· Terrigenous sediment
     prevalent
→ Current flow

LAND

201

pal source of table and industrial salt. These salt beds and the limestone beds that separate them also converge gradually toward the margins of the basin (Fig. 10-19), so that the series is made of a succession of lenses of salt and limestone—the effect of differential sinking that progressed gradually as deposition continued. The deformation of Michigan in late Silurian has similarities to the formation of the geosynclines in the Cambrian; in each situation relatively rapid subsidence accompanied deposition of sediments in the sinking areas. But the Cambrian miogeosynclines were extended belts lying on the margin of the relatively stable craton; whereas the Michigan basin-forming subsidence was local and within a part of the craton that was sinking by itself. Such structures have been called intracratonal basins that are independent of source lands, or autogeosynclines, but they are "geosynclinal" only in the sense that they had subsidence progressing through a long time. They are distinctive quite aside from what they are to be called.

EVAPORATION AND SALTS

The subsiding area served as a great evaporator into which marine waters came from outside and were concentrated until salt precipitated (Fig. 10-22). Marine water yields only a few feet of salt for every hundred feet evaporated. In the Silurian, this salt cannot have come from marine water isolated in a deep basin (Fig. 10-23), for such a

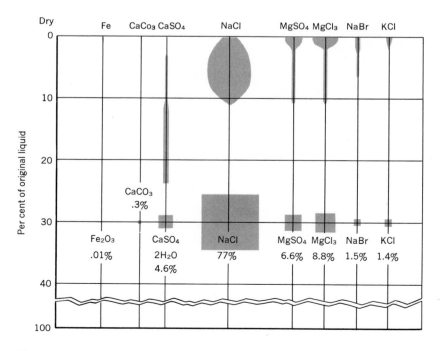

Fig. 10-22. Order of precipitation of salts from normal sea water. The numbers at the left are the percentage of the original sea water remaining after evaporation. The vertical lines and their expansions show the relative order and quantities of precipitates of several salts; these constitute about 3.5 per cent by weight of the original sea water, and are quantitatively proportionate to the areas of the squares. The diagram represents the salts as though they precipitate in the simple compounds; actually the salts to the right of sodium chloride (halite) precipitate as complex hydrous salts, too complex to be shown here. The essential features shown by the diagram are that calcium sulphate (gypsum) begins to precipitate when 25 per cent of the original water remains; gypsum continues to be deposited until 10 per cent of the liquid remains; sodium chloride (halite) is precipitated in quantity as the last 10 per cent is evaporated and other salts when only "last bittern" or residual concentrated brine remains.

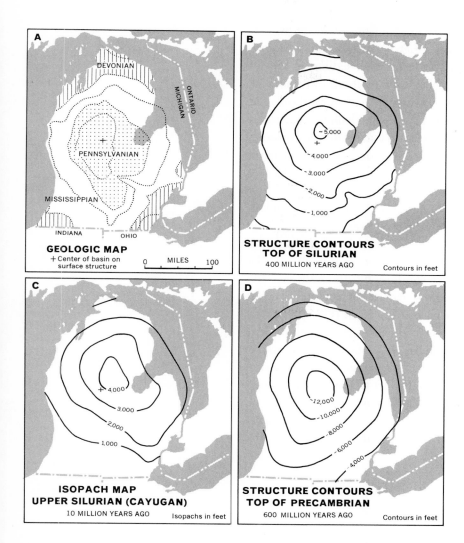

A	B
DEVONIAN	
ONTARIO	
MICHIGAN	
PENNSYLVANIAN	-5,000
	-4,000
	-3,000
MISSISSIPPIAN	-2,000
INDIANA OHIO	-1,000
GEOLOGIC MAP	**STRUCTURE CONTOURS**
+Center of basin on surface structure	**TOP OF SILURIAN**
0 MILES 100	400 MILLION YEARS AGO Contours in feet

C	D
+4,000	-12,000
3,000	-10,000
2,000	-8,000
1,000	-6,000
	-4,000
ISOPACH MAP	**STRUCTURE CONTOURS**
UPPER SILURIAN (CAYUGAN)	**TOP OF PRECAMBRIAN**
10 MILLION YEARS AGO Isopachs in feet	600 MILLION YEARS AGO Contours in feet

Fig. 10–23. *Maps showing the deformation of the lower part of the state of Michigan (after K. K. Landes and G. V. Cohee). A. Geologic map showing the present distribution of systems at the surface;* as the Pennsylvanian System is only a few hundred feet thick, and the state has rather a flat surface, the surface of the base of the Pennsylvanian has subsided less than one thousand feet in the 300 million years since the rocks were laid.

B. Structure contours, lines showing the present elevation of the surface of the top of the Silurian System, representing the deformation in the 400 million years since it was laid.

C. Isopach map, showing the thickness of the sediments of the Cayugan Series, the upper Silurian rocks that were laid in fifteen million years or so; the base of the series had subsided several thousand feet further in the middle of the basin than on the edges during this relatively short time span.

D. Structure contour map showing the present elevation of the surface of the Precambrian rocks below the Michigan Basin, a surface dating from somewhat more than 500 million years ago. The maps show that the basin form developed rapidly through the later Silurian, continued in diminishing rate until the Pennsylvanian, and virtually ceased by the close of the Paleozoic Era.

basin would have had to be not a few thousand feet but tens of miles deep for evaporation of normal marine water to yield the quantity of salt in the Upper Silurian! Evaporation lowered the sea level, so that sea water flowed in from outside this rather isolated basin. The subsidence cannot be attributed to loading, for the sediments are all chemical and organic precipitates which formed there only because there was a subsiding area. Moreover, the area had not subsided in this way during the early Silurian. It is one of the challenging puzzles of geology that a region subsiding less than its surroundings through early Silurian should reverse this trend and subside as though material below flowed out in all directions to create a hole in the surface of the crust.

The Cayugan rocks consist of carbonate rocks, of calcite (calcium carbonate) and dolomite (calcium-magnesian carbonate), of such salts as gypsum (calcium sulphate), and halite (sodium chloride), and terrigenous detritus. As the salts were precipitated from marine brines, the distribution of the salts areally and stratigraphically indicates the manner of ingress of waters. The highest calcium-magnesian ratio in the carbonates is in the northeastern part of the basin (Fig. 10–21), so the marine water seems to have entered from the direction of northern Lake Huron; from other aspects of analysis, it seems that the evaporation in the Michigan Basin was separated from that in the Ontario–New York area by land and shoals in a southwest trending belt through western Lake Erie, and that this saline basin was separated from the terrigenous sediments in the Pennsylvania region by a shoal of uncertain nature trending eastward south of Lake Erie through Pennsylvania and northeastern Ohio.

RATES OF DEFORMATION

There is a striking coincidence between this behavior of Michigan and that in later times (Fig. 10–21). The top of the Silurian is no longer a plane or a shallow basin as it must have been in late Silurian. The top of the Silurian now has very nearly the same basin form that the bottom had at the end of Silurian time—that is, the top of the Silurian is now about a mile below sea level in the center of the state, but a thousand feet or so along the margin of the lakes. Thus the deformation into basin form during the ten or fifteen million years of the Cayugan was almost as great as that in the four hundred million years or so that have elapsed since. Most of the post-Silurian deformation developed during the later Paleozoic, for the Pennsylvanian rocks, deposited about three hundred million years ago, are depressed only a few hundred feet within the central half of the Michigan Basin. But if this relatively rapid rate of deformation in the late Silurian seems like a cataclysm, the statement in another form—that it subsided on the average at a rate of about one foot in two thousand years where thickest —makes it apparent that the warping of the earth takes place very slowly.

The very gentle dip toward the center of the state of the younger Paleozoic beds gives to the geologic map the impression that the rocks are like bowls that have been stacked one within another, their edges beveled down to a plane. The "bowls" are not of the same convexity; it is as though one had a set of bowls in which each has a concave upper surface and a somewhat more sharply curved lower one, successively lower (and older) contacts being more strongly concave. It was not that parallel sheets of sediment covered all of lower Michigan until bowed down into the basin form; the bowing took place progressively through a long time, most rapidly of all in the late Silurian and not at all in the early Silurian. This is one example of the way in which the stratified rocks indicate the manner of development of the interior of the continent.

Fig. 10–24.   A restoration of a Silurian sea bottom in western New York (Chicago Natural History Museum). The large arthropod is an eurypterid, *Eurypterus.*

11
The Devonian Period and the Evolving Vertebrates

Upper Devonian sandstone, Ithaca, New York.

11 *The Devonian Period and the Evolving Vertebrates*

When Murchison in 1831 began his work in south Wales, the "Old Red Sandstone" formed the base of the established succession of stratified rocks, the Secondary of Britain, older rocks being classified as Transition or Primary. The Old Red is not really a sandstone but is a sequence of thousands of feet of shales, sandstones, and conglomerates, frequently red, with abundant lava flows in some regions; it is widely distributed from the Orkney Islands off northeast Scotland (Fig. 11–1) to the south of Wales and west in northern Ireland. Quite similar rocks are known in northeastern Greenland and Spitzbergen. Fossil fishes recovered from the beds, particularly by quarrymen in Scotland, had attracted popular interest thanks especially to the efforts and the unusual abilities of one Scot by the name of Hugh Miller (1802–1856). Miller grew up in the village of Cromarty, on the Scottish east coast, where there were extensive outcrops of the Old Red Sandstone. When he was a young man he worked in the quarries, and soon developed an overwhelming interest in the fossil fishes that constantly were being found during the course of quarrying operations. He educated himself, studied the fossils of the Old Red, corresponded with the leading geologists of the early nineteenth century, and in time became a respected authority on the Devonian prefishes and fishes of Britain. He was a skilled writer, and in 1841 he published *The Old Red Sandstone,* one of the great classics of geology, a book that had gone through nineteen editions by 1874 and had been read by countless thousands of people.

Fig. 11–1. Upper Old Red Sandstone, Devonian non-marine sedimentary rocks, in the Hoy Cliffs, Orkney Islands, Scotland (Her Majesty's Geological Survey).

CALEDONIAN OROGENY

The Old Red sediments were laid from streams and lakes in subsiding troughs adjoining the mountains raised in the late Silurian Caledonian Orogeny; they contain some of the earliest known land plants (Fig. 11–2). There is commonly unconformity at the base (Fig. 11–3), and unconformities are frequent within the Old Red, showing continuing movements through a long time. On the other hand, in some areas (as for example along the Welsh border) deposition was essentially continuous from the Silurian into the Devonian. As we have learned, the graywackes below the Old Red, with their abundant marine faunas, became the Silurian System of Murchison in 1835. The Old Red was considered to be a younger system.

In Devonshire, south of Wales, extensively exposed graywackes

(Fig. 11–4) were studied by Murchison and Sedgwick together in 1836. In spite of their judgment that the rocks looked very like the Cambrian rocks that Sedgwick had but recently described in North Wales, the fossils that they collected and accumulated did not seem like those they had seen in Wales in either Cambrian or Silurian, nor were they like faunas in the Carboniferous System that succeeded the Old Red and had been known for a good many years. They submitted the collections to William Lonsdale (1794–1871), a paleontologist, who judged that they were intermediate in age between forms in the Silurian and those in the Carboniferous. The two men were dismayed, for the rocks did not look like the Old Red that held the same stratigraphic position to the north. But when they found fossils resembling somewhat those of the Silurian in the lower part of the Devonshire graywacke, and some suggesting Carboniferous in the upper, they concluded that the strata were intermediate, and applied the name Devonian System to the sequence in 1839. Thus the Devonian was distinguished by the insight of a student of fossils, rather than by the usual stratigraphic methods applied in the field.

Fig. 11–2. *Early land plants in the Old Red Sandstone. A.* Restoration of the middle Devonian primitive plant, the psilophyte *Asteroxylon*, from the Old Red Group at Rhynie, Aberdeenshire, Scotland (after H. N. Andrews, Jr.).

B. A shoot of the plant *Psilophyton* from lower Devonian sediments in Bear Butte, east of Yellowstone Park, Wyoming. Similar non-marine limy shales are known from localities southwestward to the Grand Canyon, where the Temple Butte Formation is interruptedly present; westward they pass into the marine limestones of the miogeosynclinal belt in Utah and Idaho (E. Dorf).

The Old Red Sandstone of Scotland was known as a source of fossil fishes long ago, for the quarrymen recognized them readily. In the early twentieth century, a geologist saw plant remains in chert in a wall in northern Scotland, which he traced to its source in beds of Rhynie chert and sandstone of the middle Old Red Sandstone of early Devonian age. The plants were remarkably well preserved spore-bearing shoots that grew from rhizomes spreading beneath the

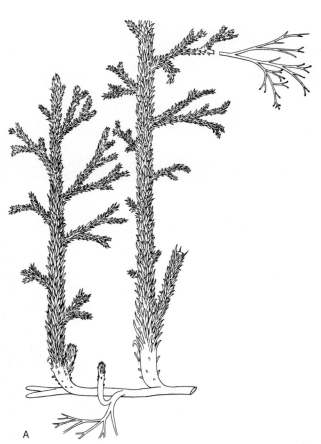

A

B

ground surface; they formed a peat-like mat, repeated in several beds with intervening sands. These *Psilopsida*, of the phylum Thallophyta, are the first well-known land plants.

The simplest and among the oldest of the known vascular plants belong in the psilophytes. *Psilophyton* was described more than a century ago from lower Devonian sandstones of Gaspé, Quebec, but its characters, though reasonably well portrayed, were only confidently accepted with the discovery of the Scottish specimens. Similar plants have been found in Wyoming east of Yellowstone Park, in southwestern Newfoundland, and in Maine; others are associated with late Silurian graptolites in Australia. The Rhynie plants are small, less than a foot high, have branching shoots with spiny or leaf-like structures rising from rootless underground stems—rhizomes. Spores were produced on slightly enlarged stem tips. The simplicity of these plants lies in their having only food- and water-conducting tissues surrounded by woody tissue. Psilopsida are principally of biologic interest.

The rocks of Devonshire are deformed and not richly fossiliferous. The marine beds are the equivalents of the non-marine Old Red Sandstone, but the interrelations are very poorly shown, and the faunas of the two facies of such different environment and origin naturally have little in common. Soon after the system had been created, its authors learned of excellently exposed strata rich in similar fossils along the Rhine (Fig. 11–5); and the west German section of Devonian rocks has ever since been recognized as the classic for study and comparison. Only in recent years after more than a century of interrupted study have most of the fossil zones of the German section been found in Devonshire.

NEW YORK SECTION

While these discoveries of Devonian formations and faunas were progressing in Europe, the New York geological survey, organized in 1836, found similar richly fossiliferous rocks widely distributed through the plateaus south of the Erie Canal and west of the Hudson. But whereas the Old Red Sandstones are coarse detrital products of the

Fig. 11–3. Devonian Old Red Sandstone unconformably overlying the Silurian on the coast of Berwickshire, southeastern Scotland, evidencing the Caledonian Orogeny (Her Majesty's Geological Survey). The locality was studied by James Hutton (1726–1797), who recognized that the relationships required that older sedimentary rocks be deformed and reduced by erosion before the deposition of the younger, which contains pebbles of the older type.

erosion of Silurian and older sediments during the pre-Devonian Caledonian Revolution, the earliest Devonian rocks in New York, like the youngest Silurian, are limestones (Fig. 11–6). The two regions on the opposite sides of the Atlantic differ fundamentally, in that the British sections are in a volcanic eugeosynclinal belt, while those in New York lie on the margin of the central stable craton. But the higher part of the New York Devonian succession does contain abundant debris from the erosion of highlands—a product not of the Caledonian Revolution but of similar orogeny that progressed in northeastern North America in the later Devonian. Orogenic movements are not universal and synchronous even in the more mobile belts of the earth.

The North American Devonian can be classified into several stages that need not concern us at the moment. It will suffice to consider that the lower* Devonian rocks are predominantly calcite limestones with some quartz arenite in New York becoming more argillaceous and siliceous southeastward into New Jersey and eastern Pennsylvania as they thicken. The middle and upper Devonian, on the other hand, present an exceptionally fine record of the sedimentation resulting from rising lands to the east. As the Devonian in New England and the Maritime Provinces of Canada presents the complementing record of mountain building in the volcanic geosynclinal belt, the integration of the records in the two regions will be of interest. The lower Devonian will then be reconsidered in showing that the later Devonian orogeny was not even persistent within the Atlantic coastal belt. Finally, the record of the Pacific marginal belts will be reviewed to show the contrasting persistence of pattern of deformation in that region.

LATER DEVONIAN DELTAL SEDIMENTS

The middle and upper Devonian rocks of New York and Pennsylvania and westward above the Onesquethawan were almost entirely terrigenous —argillaceous quartz sands, silts, and clays (Figs. 11–7, 11–8)—the principal limestone (Tully) being one of a hundred feet or less at the top of the middle Devonian (Fig. 11–9). The form of the whole deposit is semi-lenticular with the southeast side rather straight (Fig. 11–10); isopachs show the maximum thickness of more than two miles in a narrow belt in the Appalachian Mountains of Pennsylvania and the Virginias, gradual westward convergence reducing thickness to less than

* The term "Lower Devonian" is formally applied in Europe to rocks corresponding in age to the pre-Onondaga rocks in New York that were placed in the Helderbergian, Oriskanian and lower Ulsterian (Schoharie) Series when those terms were defined, the upper Ulsterian (Onondaga) being in the Couvinian Stage—lowest "Middle Devonian." Unfortunately the term "Ulsterian Series" has been applied also to include all three American series, the Helderbergian, Oriskanian ("Deerpark"), and Ulsterian ("Onesquethaw") "Stages," as well as to the combined Deerparkian and lower Onesquethawan.

Fig. 11–4. Lower Devonian shales and argillaceous sandstones near Lynton in Devonshire, southwestern England, in the type region of the Devonian System (Her Majesty's Geological Survey). The rocks are sparsely fossiliferous, and were not as useful as a basis for correlation as those in the Rhine region of Germany.

Fig. 11–5. Devonian limestones in the Rhenish Paleozoic Massif in the Eifel near Gerolstein, Western Germany (R. Brinkmann). The cliff-forming dolomites, in the upper part of the middle Devonian, contain the brachiopod *Stringocephalus*, which is widely distributed in western North America—see Fig. 11–19. The slopes below are crinoidal marls in the lower middle Devonian.

a thousand feet along a curved isopach running from extreme southern West Virginia through eastern Ohio into Lake Erie. The original deposits have been eroded away eastward from the Catskill Mountains of New York and the Appalachian Mountains to the southwest since Devonian time, but as the easternmost preserved rocks are dominantly non-marine; the source land that yielded them was not far to the east. The form of the deposit in three dimensions was similar to that of the late Ordovician in the same region, but the thickness is several times greater, the westward convergence being at a higher rate. As the medial and late Devonian probably had a duration of two or three times that of the late Ordovician, the rate of thinning and of deposition may not have differed appreciably.

Although the Devonian of the Appalachian region was the thick sequence that led to the concept of the geosyncline, it is now known not to have the long trough form that is often thought distinctive of a geosyncline. The deposits were laid in a structure sometimes called a foredeep, a depression that received sediments coming from the erosion of tectonic lands formed by deformation outside the craton or central stable region.

FACIES FOSSILS AND GUIDES TO TIME

The Devonian sediments grade from coarser facies in the southeast toward finer in the equivalent beds to the northwest. With such changing facies, it is not surprising that the fossils also change, and that there are some difficulties in recognizing horizons of synchrony—beds laid at the same time. The rocks change upward, tending to be finer at the base and coarsest at the top in any locality. This progression is not wholly regular, and it is the deviations from gradual change that give time-datum planes of distinctive lithology. Through limited spans of time, conditions seem to have been rather uniform over large areas. In one instance coralline argillaceous sands unlike beds above or below can be traced for many miles. A thin conglomerate bed with distinctive pebbles can be followed from section to section. A sharp contact between silts and sands and an overlying black shaly argillite forms a distinctive plane for scores of miles. One consistent limestone and other thinner ones are significant in the midst of the terrigenous rocks. A zone of red sands and silts gradually diminishes westward, because stream plains that were the depositional sites graded into marine beds farthest west when the middle of the wedge of red beds formed; such relations in section are known as tongues—a tongue of one facies invades that of another. Time planes always lie within such tongues, never cross them. Thus beds within the same tongue can be correlated

Fig. 11–6. A few feet of uppermost Silurian (Rondout) and lower Devonian (Manlius) limestone lying disconformably on upper Ordovician shale, barely exposed, at Indian Ladder in Thacher State Park, near Albany, New York (New York State Museum). Westward, thickening Silurian rocks separate the two systems, as in the Niagara region; eastward, the Manlius lies unconformably on shales deformed in the Taconian Orogeny, shown in Fig. 9–1B.

Fig. 11–7. Upper Devonian sandstones and shales exposed for five hundred feet along the Genesee River in Letchworth State Park, south of Rochester, New York (New York State Department of Commerce). The facies is intermediate between the more sandy shales to the east and the black shales to the west.

215

in time—the principle of *intratongue correlation*. It must be recognized that drawn sections are but generalizations of innumerable local relations.

Fossils in rocks of similar facies in the oldest beds generally resemble those in the youngest—so much so that until the early years of this century it was assumed that the combined similarity of fossils and of lithologies was indicative of identical age; thus certain brachiopods such as *Cyrtospirifer* (Fig. 11–10) are associated with facies (Fig. 11–11). Though the fossils in the same facies in successive stages do resemble each other, the differences that can be recognized on critical study give assurance that the faunas changed with time. These are the facies fossils that are the remains of organisms tolerant of only limited environments. In a few cases, animals not only lived in the region for a limited time, but were tolerant of a variety of conditions so as to be found in several equivalent facies; thus the brachiopod *Hypothyridina* (Fig. 11–10), distinctive of the thickest limestone (Tully), continues eastward (Fig. 11–8) into silts and sands that would otherwise not be thought of the same age. These forms of greater tolerance are particularly useful guides to time correlation.

THE FISHES

Although the changes in invertebrate faunas in similar facies are slight, or subtly expressed, through the range of the Devonian sequence, the same is not true for the vertebrates. The fish faunas of the lower Devonian are markedly different from those of later stages of the period, and illustrate in a striking manner how much more primitive the vertebrates were at the beginning of the period.

RELATIONS OF FACIES

As the relations are of solid, three-dimensional forms, they cannot be represented fully on the planes of sheets of paper and, therefore, test the imagination. We can think of each rock representing a span of time as bounded on top and bottom by planes of identical age, the surfaces not quite parallel, but generally diverging toward areas of greater subsidence and more rapid deposition. Laterally, the textures or sizes of particles between the planes remain fairly constant along arcuate lines that roughly parallel the shores and equal water depths and run about at right angles to, or normal to the direction in which streams carried the sediment to sea. Such a band might be, for example, one of better sorted sands and silts deposited along the shore lines through a stage of time (Fig. 11–12). The lateral limits are ill-defined for the gradation is one of continuous progression rather than one of discrete interruptions. Hence, normal to the band, a sand zone passes gradually

A

Fig. 11–8. *Middle and upper Devonian terrigenous rocks in Ontario.* *A.* Middle Devonian argillaceous shales with limestone interbeds near Arkona, Ontario, south of Lake Huron. The rocks are filled with well-preserved fossil corals and brachiopods that wash out of the clays and are easily collected. The beds are stratigraphically a facies of the Devonian rocks of New York that pass through progressively more sandy beds eastward into the red, stream-laid sandstones and conglomerates of the eastern Catskill Mountains. The black shales of Kettle Point succeed the shales at Arkona.

B. Upper Devonian black shale with concretions at Kettle Point, east of Sarnia, Ontario along southern Lake Huron. During the late Devonian carbonaceous black shale facies extended in a belt from southern James Bay, south of Hudson Bay, through Lake Huron, eastern Ohio, and Kentucky to southwestern Virginia and eastern Tennessee. Few fossil plants and animals are seen in these beds, but there are many microfossil plant spores and conodonts, tooth-like organisms of uncertain origin, that permit the differentiation of the time-stratigraphic units within the shale. The concretions at Kettle Point are of calcite that seems to have replaced the shale, perhaps during its consolidation.

B

into a silt zone through every intergrade—the limits are more easily shown by lines on a diagram than by observations in the field! The belts of successive grades on a map, coarser toward the eastward and finer toward the westward, are the *facies of a stage*—bands on the sea floor having the same kind of sediment because of similar depths and distances from sources of detritus; and as the environment of a band is fairly constant, so are the fossil organisms. As seas advanced, the bands moved eastward and upward, as they retreated, westward and upward. Occasional changes were rapid, so that a facies of one stage jumped over to lie on quite a different facies of an older stage, giving a sharp change in the lithic sequence locally—an excellent guide to the correlation of sections. Generally the gradation in the New York Devonian was of coarser over finer—a gradational overlap.

RISING SOURCE LANDS

The general coarsening of the sediments upward required that at each locality the currents and streams were competent to carry somewhat larger particles to that point than previously—either because the source lands were rising to give higher gradients, or the shore approached more closely to give greater chance that marine currents might move materials from the deltas and beaches. Thus the lithologies in the

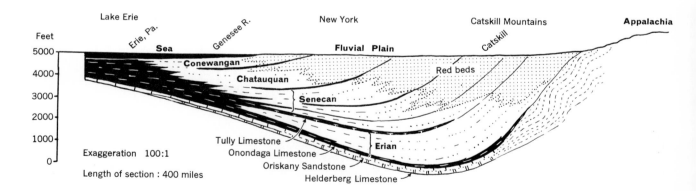

Fig. 11–9. Restored section of the Devonian System in New York State at the close of the period. The lower and early middle Devonian rocks are principally limestones. The advent of the Acadian (Shickshockian) Orogeny to the east is evidenced in the introduction of coarsening terrigenous sediments that generally spread westward, coarser, non-marine facies on the east passing through marginal marine sandy beds into offshore argillaceous black shales in the west. As the coarser rocks spread progressively though interruptedly westward, the highland source seems to have been rising as the foredeep to the west subsided; sedimentation exceeded subsidence, so that sediments at any place became gradually shallower marine-laid and then fluviatile or stream laid. At one stage, a limestone unit, the Tully Limestone, spread westward across the state from equivalent terrigenous beds in eastern New York; this formation is characterized in one part by the persistence of the brachiopod *Hypothyridina* (Fig. 11–11). The higher Devonian has an interesting distribution of the brachiopod *Cyrtospirifer* in the shallow-water sands and silty shales; the fossils of this genus persisted within this facies, representing an environment, through time (Fig. 11–12); certain lenses and tongues of other lithologies that penetrate the facies band show it to be an environment that moved westward interruptedly with the progress of time. Planes of the same age lie within the limits of tongues.

Devonian show that the seas retreated from Pennsylvania and New York as the lands rose to the east. Whether there was orogeny, in the sense of folding or faulting, at least there was rock raised to be eroded to form the terrigenous detritus.

ISOPACHS AND ISOLITHS

The thickness of sediment and the texture are somewhat independent variables. *Isopachs* are lines of equal thickness on a map, and *isoliths* are lines separating rocks of dissimilar character, whether of composition, color or texture (Fig. 11–12). In the case of the middle and upper Devonian, they have some similarities and some differences. As one approaches the area of greatest thickness from the westward, the rocks gradually coarsen, the arcs of the isoliths separating successively larger textural grades of one time approximating the trend of those for the isopachs for successively greater thicknesses of sediments of spans of time. But as one passes the area of maximum thickness, the progression to coarser grades continues, whereas the thickness decreases. Thickening and coarsening toward the source characterize the western part of the great lenticular deposit, while thinning accompanies coarsening as the highlands are approached. Moreover, as the finer-textured rocks represent offshore, deeper-water facies, the greatest depths of the water were not over the area of greatest thickness, but far to the west. The faunal belts correspond to those of the lithofacies, for each is the response to environments.

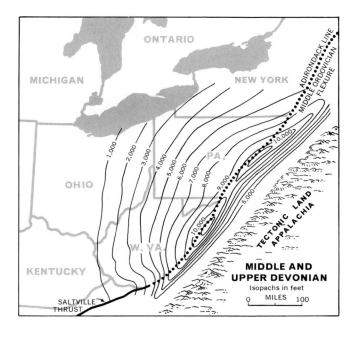

Fig. 11–10. Map showing the thickness of middle and upper Devonian sediments in eastern United States—an isopach map (after G. H. Ashley and H. P. Woodward in M. Kay). The rocks are almost wholly terrigenous detrital sediments derived from the tectonic land Appalachia that was raised to the southeast in successive movements in the Acadian Orogeny; they were laid in a progressively subsiding foredeep or exogeosyncline within the border of the early Paleozoic stable region or craton.

We will next convert these observations into the record of deformation, the tectonic history revealed by a semi-lens of some twelve thousand feet deposition; the surface of deposition was probably a nearly horizontal plane. The basal beds are dark shaly argillites that lie on the calcite limestones of the low middle Devonian (Onesquethawan); if this be a disconformity, surface of erosion, as some have believed, the relief was very small, and the elevation must have been practically at sea level. The top of the upper Devonian section was not a horizontal plane, for the rocks are non-marine, streamlaid gravels toward the land on the east, and were off-shore black marine muds in Ohio and southern Virginia. Hence the thickness does not give a true record of the deformation. If the base was a horizontal plane about at sea level at the beginning of deposition—and this can only be an approximation, the subsidence has been less than the thickness in the east, and has been more than the thickness in the west. And even if the sea level has risen or fallen through eustatic change, the same relative factor remains. Moreover, the thickness of some of the rocks

Fig. 11–11. *Devonian brachiopods from New York.* (1) *Hypothyridina,* is a rhynchonellid brachiopod that persists in the Tully Limestone and time-equivalent terrigenous rocks to the eastward; (a), (b), and (c) are posterior, anterior and side views.

(2) *Cyrtospirifer,* is found in late Devonian sandy shales and sandstones rising in age westward in New York, as shown in Figure 11–12; view from the brachial side.

(3) Bed of sandstone with impressions of the shells of *Cyrtospirifer,* external molds and internal molds; upper Devonian, Cooks Greek, Steuben County, New York (The Smithsonian Institution).

(4) Spires on the interior of a spiriferid brachiopod, *Mucrospirifer,* from middle Devonian exposed along the shore of Lake Huron near Alpena, Michigan; brachial view (The Smithsonian Institution).

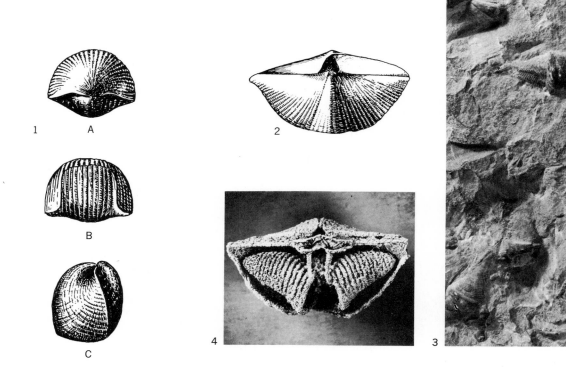

will have changed since Devonian time, because they have compacted under the weight of overlying strata and the pressure of later folding. So the axis of the geosyncline subsided greatly, but the exact amount of subsidence is dependent on interpretation of data rather than simple record of thickness.

The geosyncline, like that of the late Ordovician, is one that developed on the margin of the craton in complement to rising lands in the linear geosynclinal belts of the earlier Paleozoic to the east. The sediments came from outside the continent toward the interior. Such a geosyncline has been called an exogeosyncline, "from outside geosyncline" to emphasize its different character from the miogeosynclines and eugeosynclines that were discussed in the Cambrian and Ordovician. The name is not as important as an understanding of its distinction. We can examine the record of the area to the east, New England and the Maritime Provinces, to find the complementing evidence of the region of source.

Fig. 11–12. Diagram of the intertonguing facies of the upper Devonian in New York, and the association of the brachiopod *Cyrtospirifer* with the sandy shales (after H. R. Greiner). At one time, rocks containing the fossil were thought to be of one age. Stratigraphic studies showed that tongues of each facies penetrated others. Planes of synchrony, or isochronous surfaces, lie within such tongues. The lithologies and the associated faunas were found to generally rise in time westward, but in an irregular manner. The thickness is hundreds of feet.

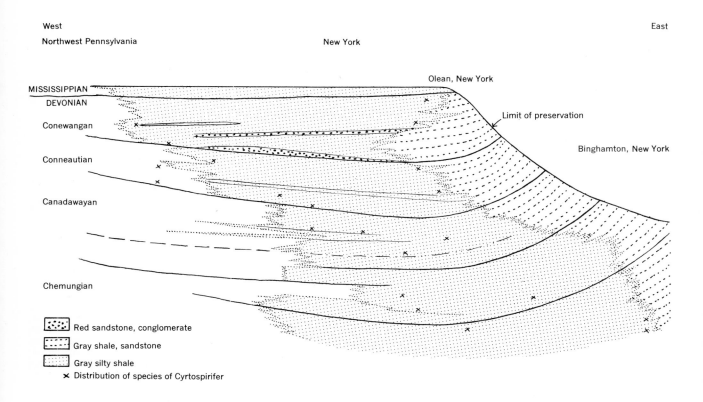

West   East

Northwest Pennsylvania   New York

Olean, New York

MISSISSIPPIAN

DEVONIAN

Conewangan

Limit of preservation

Conneautian

Binghamton, New York

Canadawayan

Chemungian

Red sandstone, conglomerate
Gray shale, sandstone
Gray silty shale
✕ Distribution of species of Cyrtospirifer

221

Mount Washington (6288 feet) in the White Mountains of New Hampshire (Figs. 11–14, 11–15), is made of schists, sillimanite schists, that have yielded severely altered lower Devonian brachiopods. Within the White Mountains and in their vicinity are extensive Devonian schists intruded by granitic magma that has so altered the originally argillaceous rocks that as one progresses toward such an intrusive the argillites become first biotite bearing, then change to successive belts of schists with garnet, staurolite, and sillimanite, minerals that represent reorganization of the chemical constituents of the quartz-argillaceous rocks under increasing temperatures as they approach the intrusive. The original Devonian sediments were several miles thick, predominantly of graywacke, but with thick sequences of lava flows. Until the finding of the brachiopod, the rocks were thought to be Precambrian, for they are quite like ancient rocks of similar composition and metamorphism. The early Devonian history of the area was quite as it was in the Ordovician, a deeply subsiding belt having rising linear lands that were being eroded, and volcanoes and other volcanic sources. Similar lower Devonian sequences are known from Newfoundland to Massachusetts.

In eastern Maine, near Eastport, granitic rocks like those that intrude lower Devonian are succeeded by another sequence, about a mile of conglomerate and sands with included basaltic lava flows (Fig. 11–15). The conglomerates have pebbles of granitic rocks like those unconformably under them (Fig. 11–16), and the sediments have plants resembling those of the latest Devonian. Along the south side of the Gaspé Peninsula in Quebec similar sediments contain placoderms, especially the interesting genus, *Bothriolepis* (Fig. 11–17), in life seemingly a bottom-living scavenger that perhaps imitated the habits of the armored ostracoderms. In these rocks, also, are the fossils of various bony fishes, including some very fine lungfishes and cros-

Fig. 11–13. Diagram representing the manner in which the facies interpenetrate, as in the Devonian of New York and Pennsylvania. Each lithic type of facies grades into others, generally coarser to the east and finer to the west, but with tongues of each entering others. Thus each lithic type or facies within a time span such as a stage is bounded by isochronous planes limiting the stage at the top and bottom, extends like a ribbon along a trend, and has serrate lateral limits of textural change, toward coarser on one side, toward finer on the other; the latter intertonguing surfaces may be indistinct, statistically selected surfaces representing points in a continuously grading pattern in successive beds; but often sudden changes produce abrupt lateral migrations in lithic belts from time to time, so that these times are well defined at the base of tongues.

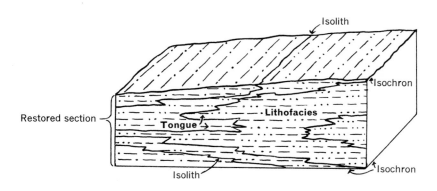

Fig. 11–14. Mount Washington, New Hampshire (6288 feet), highest point in New England. The mountain is composed of schists, deformed and altered by intrusions in the Acadian Orogeny in middle Devonian time (New Hampshire Development Commission).

Fig. 11–15 (below). Restored section in latest Devonian of the rocks of the Devonian System in New England. The early Devonian sedimentary and volcanic rocks were folded and intruded during the Acadian or Shickshockian Orogeny within the period, the mountains that were raised then being eroded to supply the coarse terrigenous sediments of the medial and late Devonian exogeosyncline to the west in New York and Pennsylvania (Fig. 11–9).

In preparing such a restored section, one must first sketch the surface profile as it is interpreted at the time of portrayal, in this case in latest Devonian. The successive events must be portrayed in order from last to earliest. Thus latest folds must further deform structures formed in earlier orogenic phases, with generalization that presents the sequence of events. Such a section is drawn to illustrate a concept of the events as they are interpreted.

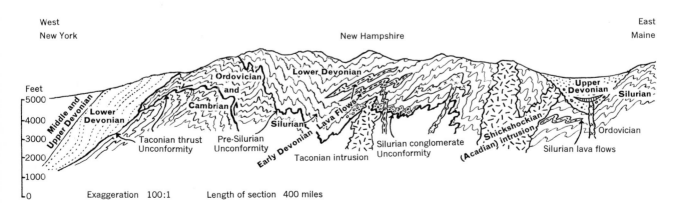

sopterygians. They seem to have lived in streams, for marine fossils do not accompany them; they are like placoderms found elsewhere in rocks determined as latest Devonian.

RADIOACTIVE ISOTOPES AND DATING

In Nova Scotia, sedimentary rocks containing fossils like those in the Onondaga Limestone of New York—basal middle Devonian, are intruded by granitic rocks that contain minerals with potassium-argon isotopes giving a calculated age of somewhat more than three hundred and fifty million years. This gives a rather definite time for an intrusion during the Devonian Period.

The record in New England and the Maritime Provinces is of deeply subsiding early Devonian geosynclines with intervening rising islands and volcanoes, then a time of orogeny and intrusion of plutonic rocks, the *Shickshockian* or *Acadian Disturbance,* named for the range of mountains in the center of the Gaspé. The name Acadian unfortunately is applied to both the middle series of the Cambrian System and to the orogeny in the Devonian. Although Shickshockian is a name assigned to the Devonian orogeny to avoid this confusion, Acadian is the term most widely used. These events in middle and late Devonian produced lands that furnished coarse latest Devonian sediments to areas of lower elevation within the region. They also supplied the coarsening detritus that swept west from the orogenic belt into the exogeosyncline of Pennsylvania and adjacent states; the geosyncline subsided as the highlands rose.

The Acadian Disturbance affected only the northeastern part of the Atlantic coastal eugeosynclinal belt. It may have produced the folds and some of the thrust faults along the Hudson Valley, Lake Champlain (Fig. 8–15C) and the St. Lawrence River, along Logan's Line. The isoliths of the terrigenous sediments have an arcuate pattern that shows the source to have been in Pennsylvania and northward. The lower Devonian carbonate and quartz arenite rocks of New York thicken and pass into more argillitic facies southeastward, and ultimately into the miles of Devonian in the volcanic geosyncline a few score miles farther east. But southward, the few hundred feet of lower Devonian passes into well-sorted beach sands southeastward. The axis of the depositional trough and of the shores trend across the miogeosynclinal marginal flexure of the Cambrian and the Adirondack line of lesser subsidence and flexure in the medial Ordovician, directed more westerly southward than the early Paleozoic trends. Such Devonian sediments as are present in the belt of thick non-volcanic earlier Paleozoic of the South are thin carbonates and well-sorted quartz sandstones that spread over

Fig. 11–16. *Upper Devonian stream-laid conglomerates and sandstones*, products of lands raised in New England and Maritime Canada in the Acadian Orogeny during the Devonian Period. *A.* Conglomerate near Perry in extreme eastern Maine; the boulders include pebbles of intrusive rocks such as intrude older Devonian fossiliferous rocks in the region.

B. Similar conglomerate and sandstone near Escuminac, along Chaleur Bay, on the south shore of the Gaspé Peninsula, Quebec; associated shales contain specimens of primitive fish-like animals, placoderms.

Fig. 11–17. Primitive fishes, placoderms, of the genus *Bothriolepis* from the upper Devonian Escuminac beds along the north shore of Chaleur Bay, on the south of the Gaspé Peninsula, eastern Quebec (American Museum of Natural History).

A

B

an unconformity bevelling the Silurian and Ordovician rocks; and they are succeeded by late Devonian or early Mississippian black shaly argillites, such as spread farthest from the Shickshockian highlands. The pattern of deformation of the continent is ever-changing. The stability of the Devonian is but a phase in that pattern, to be followed again by great mobility.

PACIFIC BELTS

The deeply subsiding volcanic belt on the Atlantic Coast had its counterpart on the Pacific Coast. But in the western United States and Prairie Provinces of Canada, subsidence with deposition of carbonate rocks continued much as in the earlier Paleozoic systems in a belt along the western margin of the craton. Thus a mile or so of carbonate rocks persists in Nevada and western Utah (Fig. 11–18), diminishing to disappearance eastward near the line of disappearance of the lower Cambrian. There were no highlands sufficiently near the miogeosynclinal belt to yield terrigenous detritus along this part of the belt except for some black shaly argillites of latest Devonian. But far to the north in northwestern Mackenzie Territory, west of the river of that name, the sequence in the Devonian is much the same as in New York and Pennsylvania, carbonate rocks of the earlier Devonian passing up into terrigenous and finally non-marine coarse detrital sediments, indicative that in this region, too, highlands had risen in the adjacent earlier subsiding belt. The widespread limestones contain somewhat different faunas than the more terrigenous equivalents in the East (Fig. 11–19). Limestone reefs are present in many localities, and are the reservoirs for the petroleum in important fields in Alberta and Mackenzie.

REEFS AND SALTS IN PRAIRIE PROVINCES

The Devonian rocks of Alberta and adjoining provinces and states have many features analogous to those in the Silurian of the Great Lakes region. In each there are extensive salt deposits and belts of reefs related to differential subsidence and restriction of circulation of marine waters on marginal parts of the craton. Recall that the rocks of the Cambrian System were much thicker in the Rocky Mountains along the British Columbia-Alberta border than eastward under the plains. The flexure on the margin of the miogeosyncline extended through southwestern Alberta to western Wyoming and central Utah. Thus much of Alberta and of the provinces to the east and Montana, was on the craton. In the succeeding periods, this marginal cratonal region remained rather stable so that Ordovician and Silurian rocks are thin or absent and Devonian sediments generally lie on Cambrian, and

Fig. 11–18. Lower Devonian limestone along Ikes Canyon in the Toquima Range, central Nevada. The limestone, dipping to the right, westward, lies disconformably on middle Ordovician limestone, having been laid on an arch that was along the western margin of the carbonate belt (see Fig. 8–14); chert-slate-volcanic rocks of the eugeosynclinal belt, in the smooth-surfaced peak on the extreme right skyline, have been thrust over the limestone. The limestone contains fossils previously known only in central Europe in rocks that have been assigned to either uppermost Silurian or lowermost Devonian.

Fig. 11–19. *Fossils distinctive of the middle Devonian (Erian) Stringocephalus fauna of western North America and Europe.* (1) *Stringocephalus*—an internal mold of the brachiopod is shown viewed from the brachial side and showing the median septum, a ridge on the interior of the brachial valve that caused an impression on the material that filled the inside of the valve after the death of the animal; the internal mold of the two valves has the form of the original soft parts—it is their cast. *Stringocephalus* is distributed through northern Europe and Asia, and in western and northwestern North America southeastward to Manitoba, in Givetian, later middle Devonian limestones.

(2) *Buechelia* and (3) *Omphalocirrus*, the associated snails or gastropods, are found as far southeastward as northern Michigan, in an enormous quarry at Rogers City that supplies limestone to the steel industry. Relationships there show that the beds are time equivalent to the Hamilton Group, or Erian Series of New York; the presence of different fossils in that region may relate to the rocks being generally terrigenous, argillaceous. *Stringocephalus* and its associates seem to have preferred to live in clearer limy seas.

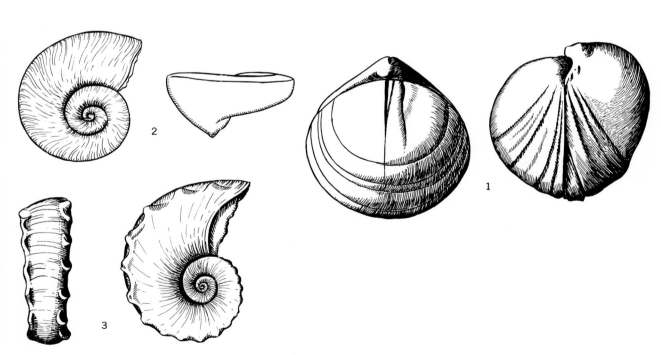

occasionally on Precambrian. In the early and medial Devonian (Fig. 11–20), the craton had relatively little subsidence in a belt trending from southern Alberta northwestward to the Peace River region in northwestern Alberta and northeastern British Columbia; in the latter area, late Devonian rocks lie on Precambrian that was exposed on a low island in earlier Devonian. Eastward of this marginal cratonal rather stable belt was a broad one of somewhat greater subsidence trending from northern and northwestern Alberta to east central Alberta, across southwestern Saskatchewan to northwestern North Dakota and southwesternmost Manitoba. Within this subsiding trough, 1500 feet or more of lower and middle Devonian rocks accumulated, about a half in the thicker sections being halite and particularly anhydrite; the belt is particularly notable for its reserves of potash

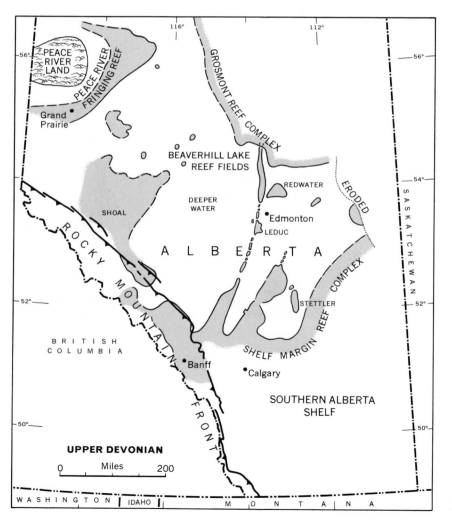

Fig. 11–20. Paleogeographic map of the Upper Devonian in Alberta, showing the distribution of the organic reefs bordering more open seas and the low island in the northwest (after H. Belyea and D. J. McLaren). The Devonian reef limestones form reservoirs that have produced petroleum—which was distilled in the long past from organic matter, probably the soft structures in marine organisms.

salts. The details of the distribution are not as well described as are those in Michigan Silurian, but the lands and shoals to the west along the miogeosynclinal margin must have restricted ingress of marine water, so that evaporation increased salinities to the point of precipitation.

In later Devonian, seas spread across the marginal arch and salines are infrequent. But reefs, principally of stromatoporoids with algae and corals (Fig. 11–21), rose on the shoals as subsidence progressed, while finer sediments some of them argillaceous and shaly, were laid in the deeper water. In extreme cases, the reef rocks to thicknesses approaching a thousand feet through upper Devonian time were flanked by finer sediments in the intervening depositional basins (Fig. 11–22). The rocks in the reefs were porous, so that petroleum accumulated in them, or migrated into them, forming the great oil fields that have made Alberta a principal petroleum producer.

A section from the shoals through the reefs into the basins is quite similar to that described in the Silurian of Illinois, for instance, but the dimensions are greater, for the Devonian rocks on the cratonal

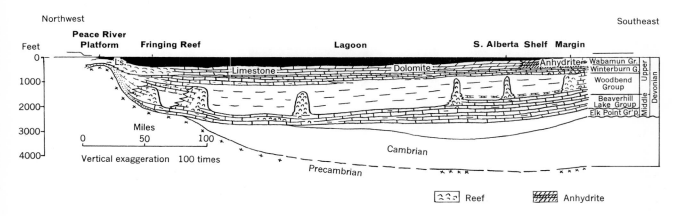

Fig. 11–21. *Restored section through limestone reefs in the upper Devonian of Alberta (after H. R. Belyea).* The province of Alberta produces nearly two-thirds of the oil from Canadian wells, and most of this comes from the porosity of the limestone reefs of the Devonian. The rock is of several constituents having differing proportions from place to place. Some of the rock is composed of the carbonate deposits of hydrocorallines called stromatoporoids, of corals and algae, the rest of fragments broken from these to form detritus, of the shells of other organisms that lived in the reefs, and perhaps of some inorganic calcium carbonate. The petroleum probably comes from the alteration of the soft parts of these organisms, and it is contained in the porosity of the rocks in which it formed, or into which it gradually flowed. Calcareous algae and corals built limestones of changing kinds from the Precambrian and Ordovician, respectively, and are still reef builders. The hydracoralline stromatoporoids, on the other hand, are distinctive of the Paleozoic through the Devonian.

Hydrozoa, hydracorallines, like Anthozoa, the corals, are a class of the phylum Coelenterata, relatively simple animals in which cell colonies are arranged to surround an interior sac, the individual cells having somewhat differentiated responsibilities but lacking arrangement into organs. Hydroids and jelly-fish are soft-bodied members of the hydrozoan order, but have little prospect of fossilization.

The order *Stromatoporoidea* is one of the types of hydrozoans that secreted calcareous skeletons, stromatoporoids are important rock builders in the Ordovician, Silurian, and Devonian systems. They formed layered structures having rods or pillars normal to the layers, growing to rather large hummocky masses of calcium carbonate rock. Although they can be distinguished if well preserved by careful study of polished surfaces and sections, their general appearance is much like that of the calcareous algae, and indeed some of the reefs have both organic types.

margin of Alberta reach as much as four thousand feet, several times the maximum in Illinois. In the Devonian example, the principal saline deposits were formed prior to the reef forming, because the region as a whole subsided with time, whereas the conditions became progressively more restricted in Michigan in the Silurian. But in the latter region, hundreds of feet of salt formed again in the medial Devonian, for once more the interior of the Michigan Basin subsided with restricted ingress of marine water; and the conditions were to recur still again in the Carboniferous.

WESTERN VOLCANIC BELT

Because the Pacific margin was a region of early Paleozoic subsidence, the earliest rocks became so deeply buried that they are exposed now only in the cores of great anticlinoria of later origin. The oldest fossils known in central and northern California are Silurian, found in strata in the northern Sierra Nevada. To the west, in the southern Klamath Mountains (Fig. 11–23) near Shasta Dam, Devonian carbonate rocks

Fig. 11–22. Devonian limestone cores from wells drilled in the Redwater oil field near Edmonton, Alberta (sections by E. Klovan). The Devonian reefs of Alberta produce hundreds of millions of barrels of oil each year. The petroleum comes from the porosity of reef limestones. The core at far left is composed of stromatoporoids (the large, laminated structures), corals (smaller with vertical sections of the corallites), and a matrix of broken fragments and organic detritus. Solution within the reef limestone is shown by the stylolites at the top of the section, a higher bed penetrating a lower one with solution along the dark surface that has some insoluble residue. In a second core, on the right, the dark patches are void of rock, and served as a reservoir for the petroleum that probably was derived from the soft parts of organisms buried in the sediments. The convex structures are sections of brachiopod shells.

230

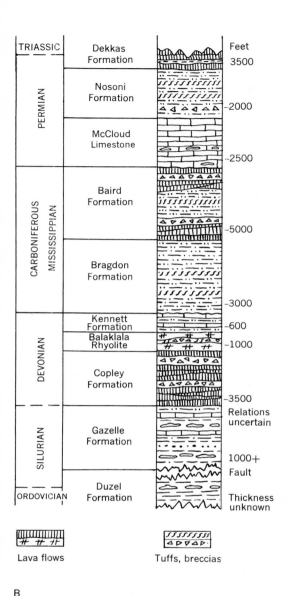

			Feet
TRIASSIC	Dekkas Formation		3500
PERMIAN	Nosoni Formation		2000
	McCloud Limestone		2500
CARBONIFEROUS — MISSISSIPPIAN	Baird Formation		5000
	Bragdon Formation		3000
DEVONIAN	Kennett Formation		600
	Balaklala Rhyolite		1000
	Copley Formation		3500
SILURIAN	Gazelle Formation		Relations uncertain
			1000+ Fault
ORDOVICIAN	Duzel Formation		Thickness unknown

Lava flows   Tuffs, breccias

B

A

Fig. 11–23. *Late Paleozoic rocks of the Klamath Mountains, northern California.* A. Aerial view eastward from the Marble Mountains, composed of late Paleozoic and early Mesozoic metamorphic rocks, across mountains with Paleozoic rocks in the middle distance to Mount Shasta, a volcanic peak in the Cascade Range (United States Geological Survey).

B. Columnar section of Paleozoic rocks exposed in the eastern Klamath Mountains, northern California. Late Ordovician or Silurian corals in this sequence are the oldest fossils known in the region; they are in beds thrust over rocks containing Silurian trilobites. Original relations are uncertain with younger volcanic rocks and the succeeding limestones with abundant and varied Devonian fossils, apparently deposited in shallow seas. Thousands of feet of Carboniferous sediments and volcanic fragmental rocks contain Mississippian fossils that are most like those known in Europe and northern Asia; Pennsylvanian seems absent, the Permian then represented by sedimentary and igneous rocks. The full section is four or five miles thick, and shows the deep but interrupted subsidence in the volcanic geosynclinal or eugeosynclinal belt; terrigenous sediments must have come from erosion of tectonic and volcanic lands (after J. P. Albers and J. F. Robertson, W. P. Irwin, and others).

to a thickness of as much as one thousand feet lie on the eroded surface of older volcanic rocks, the limestones containing brachiopods, corals and organisms such as show them to be shallow marine sediments. The underlying volcanic rocks in turn succeed a few miles of meta-sedimentary rocks including radiolarian cherts, and some volcanic rocks that are probably earlier Paleozoic, but not on direct evidence. The coralline Devonian accumulated on a surface that had subsided deeply in earlier periods, had been deformed and its sediments eroded, but then stood fairly stable near sea level for a time. Northwestward, the Devonian in extreme northwestern California and adjacent Oregon is a mile of argillite and lavas, and similar thickness in the northern Sierras suggests that the small area in which there is record of middle Paleozoic history had geosynclines separated by island lands and submarine platforms. Wherever Devonian rocks are found along the Pacific border, they have the great thicknesses that require subsidence and evidence complementary uplifts; lavas and fragmental volcanic rocks indicate immediate or nearby volcanism. These characters are so long continued and persistent in the belt that repetition of the records becomes prosaic.

The Devonian System has many interesting features. The correlation of the record of deposition in New York and southwestward with the evidence of disturbance in New England and Maritime Canada shows not only their interrelations, but also the dependence of the fossil record on geographically changing conditions of environment as well as the effects of time.

Fig. 11–24. The earliest fossil forest as exhibited in a diorama in the New York State Museum, Albany, New York. In the foreground are the rocks as they were exposed in Gilboa, New York, when a dam was built for the New York City water supply; in the background is a reconstruction of the forest, in which *Eospermatopteris* is prominent (Fig. 11–25).

A

B

Fig. 11–25. *The tree-ferns of the middle and upper Devonian. A. Eospermatopteris* restored, from the low upper Devonian sandstones at Gilboa, in the northern Catskill Mountains, New York (W. Goldring, New York State Museum).

B. Archaeopteris, fertile and sterile fragments of fronds (from H. N. Andrews, Jr.).

The land plants developed rapidly in the Devonian. The primitive lycopsids and psilopsids of the early Devonian formed matted foliage rising for a foot or so from moist soils. But by the middle of the period, there were forests of early tree-ferns.

The *Pterophyta* or ferns have a few large and usually branched leaves with spore-bearing organs on their under sides, or on separate spore-producing stalks. The earliest pterophytes are fern-like forms that resembled the earliest psilopsids. In the medial and late Devonian, the land flora included plants of several types that were of large size, including pterophytes. Nearly a century ago, large stump casts were discovered near Gilboa in the northern Catskill Mountains in New York, and in the early years of this century, many fine specimens were quarried in the building of dams for the water supply system of New York City. These tree-ferns, *Eospermatopteris*, have trunks tapering rapidly from a diameter of several feet at the base, and had crowns that probably rose to thirty or forty feet. They are known from several horizons and localities in the middle Devonian sediments where they are changing facies from the orig-

inal marine silts and sands on the west into stream-laid deposits on the east. The most widely distributed fern of the late Devonian, *Archaeopteris*, is well displayed in the shore cliffs on the south coast of the Gaspé, Quebec.

The *Lycopsida* include the common modern club-moss, sometimes used for green decorations. They are easily recognized by the presence of rhombic scars arranged in a steep spiral, leaves emerging from the summits of these scars. Perhaps the oldest land-living vascular plants are lycopsida found in the Silurian of Australia. These primitive lycopsids were but forebearers of some of the largest and most abundant plants of the Carboniferous when they were coal forming.

One of the particularly interesting features of the Devonian System is its contained record of fossil vertebrates, some of which have been briefly mentioned, a record that encompasses in its general aspects the complete radiation of the fishes. The Devonian is commonly called "The Age of Fishes," in recognition of the remarkable evolutionary radiation among these early back-boned animals. The period is regarded by the students of vertebrate evolution as a time of particular significance in the long history of the backboned animals. The period also was one of rapid development of land plants (Figs. 11–2, 11–24, and 11–25).

The vertebrates at the beginning of Devonian times were dominantly the jawless ostracoderms and the geologically short-lived placoderms. Some of the "higher" fishes were living in lower Devonian times; the sharks appeared, as did the bony fishes. These latter fishes proliferated during the later phases of Devonian history, so that by the end of the period they were represented by a respectable array of chondrosteans, the most primitive of the bony fishes, by early lungfishes and by advanced crossopterygians, these latter the direct ancestors of the amphibians. Indeed, the evolutionary step from fish to amphibian can be followed in the fossil record as preserved in upper Devonian rocks. Within these sediments are found crossopterygians so advanced that their fins show that pattern of bones characteristic of the legs of land-living vertebrates, and amphibians so primitive that they still retain the fin rays of the fish tail. This is one of the classic examples of evolution from one large group of animals to another.

One can say with justice that after the Devonian, evolution among the fishes was largely a matter of filling in the details, among the land-living vertebrates of building a complex evolutionary superstructure upon the basic tetrapod foundation established in the uppermost Devonian amphibians. Perhaps this consideration will help to emphasize the importance of the Devonian period in the history of vertebrate evolution.

FISHES AND AMPHIBIANS

The fossil record upon which these various conclusions are based is particularly well preserved in the region where the Devonian was first studied, in the Old Red of England, Wales, and Scotland. And in the northern European region the record is augmented by materials found in the Baltic states, in central Germany, and in Russia. To the far north, important Devonian fishes are found in Spitzbergen, while across the top of the world to the west, the eastern coast of Greenland is the region where epochal discoveries of upper Devonian amphibians, known

Fig. 11–26. The basic evolutionary radiation of the fishes during Devonian times (E. H. Colbert, 1955). The jawless fishes, represented in the Devonian sediments by various ostracoderms, were the most primitive known vertebrates, and persist today as the lampreys and hagfishes. The sharks and the bony fishes of the Devonian were ancestral to the two major groups of modern fishes. The placoderms, prominent in Devonian waters, failed to survive beyond Paleozoic times.

Fig. 11–27. Reconstruction of late Devonian sea bottom in western New York; a diorama (Chicago Natural History Museum). Reefs with simple and colonial corals have a few associated crinoids with long slender stems. On the sea floor are straight and coiled cephalopods, and a few brachiopods and trilobites, such as the bizarre creatures in the foreground. The small conical perforated structure nearby on the left reef is a sponge and there are snails, gastropods, below and to the right.

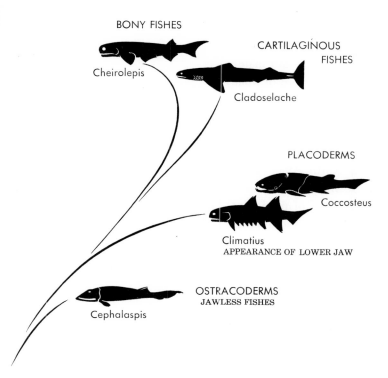

BONY FISHES

Cheirolepis

CARTILAGINOUS FISHES

Cladoselache

PLACODERMS

Coccosteus

Climatius
APPEARANCE OF LOWER JAW

OSTRACODERMS
JAWLESS FISHES

Cephalaspis

as ichthyostegids, have been made during recent decades by Scandinavian paleontologists.

The East Greenland ichthyostegids are known from abundant materials. In eastern North America is an all too fragmentary record of what may be perhaps an even more primitive amphibian, *Elpistostege,* from the upper Devonian of Escuminac or Scaumenac Bay, Quebec. Not far away, at Campbellton, New Brunswick, are lower Devonian sediments with fossil fishes.

The record is continued at other localities in North America, of which the lower Devonian of Beartooth Butte, Montana, and the upper Devonian Cleveland shales of northern Ohio are of particular importance. Elsewhere in the world, Devonian fishes are found in Asia, Australia, and Antarctica.

Almost all of the localities mentioned above represent continental deposits, laid down in rivers or lakes or perhaps in estuaries. Such were the environments of the Devonian vertebrates, which would seem to have been predominantly fresh-water dwellers. It may be that the vertebrates had their origins in the sea—this is a point that has been hotly debated in recent years—but certainly the fossil record would seem to indicate that by Devonian times the vertebrates were virtually limited to continental waters, while the oceans were the domain of the invertebrates.

After the close of the Devonian Period the development of backboned animals followed various trends. In late Paleozoic times the great continental platforms were the locales in which much evolutionary progress of the vertebrates took place, but during this same geologic span there was an invasion of the oceans, the beginning of a trend that in time established the most numerous and the most varied of the vertebrates, the marine fishes.

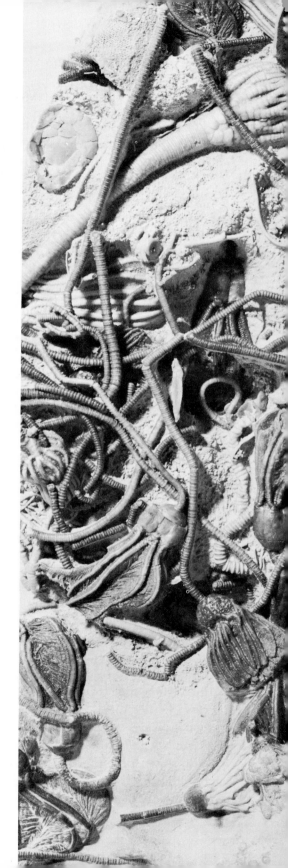

12
The Carboniferous—
Cyclic Movements
and Coal Swamps

Crinoids in limestone of the lower Carboniferous, Legrand, Iowa (Iowa Historical Archives).

12 *The Carboniferous—*
Cyclic Movements and Coal Swamps

The rocks that succeed the Old Red Sandstone of the Devonian System in the northwest of England (Fig. 12–1) were named Carboniferous in 1822, because they are coal bearing (Fig. 12–2). In the interior of North America, a great surface of erosion—a regional unconformity—separated this system into two parts or subsystems, often treated as separate systems. The lower, the Mississippian, gained its name from the river, along which these rocks are well exposed in western Illinois and adjoining southeastern Iowa and Missouri (Fig. 12–3). The caves near Hannibal, Missouri, made famous in Mark Twain's stories, are in limestones of the system, as are the great Indiana limestone quarries (Fig. 12–4) that yield the most widely used American building stone. The Mississippian rocks in the Mississippi Valley have long been known for their beautifully preserved echinoderms, particularly crinoids and blastoids (Fig. 12–5). The upper subsystem, the Pennsylvanian, was so named in the coal-bearing region of the western part of the state of Pennsylvania; it included the youngest sediments in the Appalachian Mountains and plateaus. The Carboniferous Period had a span of about 70 or 80 million years, the Pennsylvanian perhaps 50 million; the late Devonian (350 million years ago) and end of the Paleozoic Era (230 million years ago) seem rather well dated from studies of disintegration of radioactive minerals in intrusions of those ages. Apportionment of the spans of the Carboniferous and Permian periods has been made from the judgment of the record of their events.

THE MISSISSIPPIAN IN THE EAST

The Mississippian subsystem or "system" is generally divided into the Kinderhookian, Osagian, Meramecian, and Chesterian series, names that come from the sections in Illinois and Missouri. When thicknesses of the Kinderhookian and Osagian sediments are plotted, they are greatest in a belt in the Appalachian Mountains and increase also to the center of lower Michigan and into southern Illinois. The textures for any one time show a succession of arcuate belts, the coarsest in the east (Fig. 12–6). In Osagian, terrigenous clastic sediments extend to a line passing west of Michigan, through central Illinois, western Kentucky and into eastern Tennessee. Limestones accumulated in the clear waters beyond, rocks that are renowned in paleontology for their beautiful fossil echinoderms (Fig. 12–5). The silts and clays of Michigan and eastern Illinois passed into coarser sands and gravels, becoming

Fig. 12–1. Carboniferous shales and sandstones forming a hill near Ingleborough, western Yorkshire, England, with Carboniferous limestone in the foreground (Her Majesty's Geological Survey). The name Carboniferous was given to these rocks in the north of England, for they contain important coal deposits in the Pennine Hills region near Newcastle.

Fig. 12–2. *Carboniferous plants. A.* A block of coal from the Pennsylvanian of Illinois. The massive, bright, jet-like vitrain, V, has a larger proportion of the less resistant plant cell material than the dull fusain, F, made more largely of more woody matter; clarain, C, has streaked bands of intermediate character. Each is fully composed of altered plant matter (J. C. Frye, Illinois Geological Survey).
B. Bed of Pennsylvanian shale showing impressions of fossil fern-like plants from Olyphant, Pennsylvania (The Smithsonian Institution).

238

A

B

239

non-marine in their eastern facies. The sediments were winnowed at the shore, forming bands of porous sand on the geographies of successive times, reservoirs from which oil and gas fields yield petroleum products. The surface of sedimentation was a plain over which the currents carried the detritus, the coarsest being carried least far, because fewest currents were competent to move the particles. At the same time, gradual crustal warping took place so slowly that, though it affected thickness, the sediment textures were laid in quite independent patterns, related to distance from source and directions of current flow.

The medial Mississippian Meramecian Series is also widely distributed. In the Appalachian region, the rocks increase from small thickness in Ohio and northern Kentucky to more than one thousand feet in a geosyncline in eastern and southern West Virginia (Fig. 12–7), southwestern Virginia and eastern Tennessee. Wells have penetrated them throughout this region; the thicknesses, represented by isopachs on a map, have northeastward-trending lines, showing progressive increase toward the southeast. If the uppermost unit was laid in an essentially horizontal position, and the lowest as well, there was a flexure paralleling the trend, with relatively greater subsidence to the southeast—a miogeosynclinal pattern of deformation. When the rocks are further subdivided, the earlier units are found to disappear northwestward along this flexure, and all units to thin in that direction; land

Fig. 12–3. Lower Carboniferous (Mississippian) limestone in Mississippi River bluff at Alton, Illinois. The bedded limetones, of the Osagian Series, made largely of the shells and fragments of organisms, pass into facies of silty and sandy terrigenous shales in the states to the east of Illinois. The Mississippian Subsystem is named from the exposures along the river in Illinois, Missouri, and Iowa (Monkmeyer Photo Service).

northwest of the flexure in the earlier Meramecian was gradually covered as the sea advanced from a belt of more rapid subsidence. Meramecian seems to have been a time when the Acadian-formed highlands, Appalachia, had been denuded until they no longer furnished coarse detritus; and evidently they persisted about as long in the southern Atlantic coastal region as in the northern, where we have seen that the medial Mississippian is also dominantly non-terrigenous.

The carbonate deposition in the geosyncline was followed in the West Virginia Chesterian by formation of many thousands of feet of sediments. The earliest Chesterian is a calcite limestone, composed of fossil fragments. Then the lithologies progress irregularly through alternating terrigenous rocks and carbonates with marine fossils and fossiliferous marine sandy and silty rocks into coarser conglomeratic sediments, many of them red, containing occasional plants but lacking marine organisms. Evidently highlands rose again in the old eugeosynclinal belt to the southeast—another pulse in the long history of deformation.

INTERIOR WARPING

In view of the absence of Meramecian terrigenous rocks in the Appalachian region, and their overlap westward toward land that in the early part of the epoch extended from central Ontario to central Ken-

A

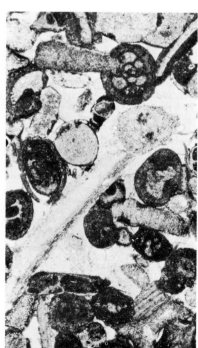

B

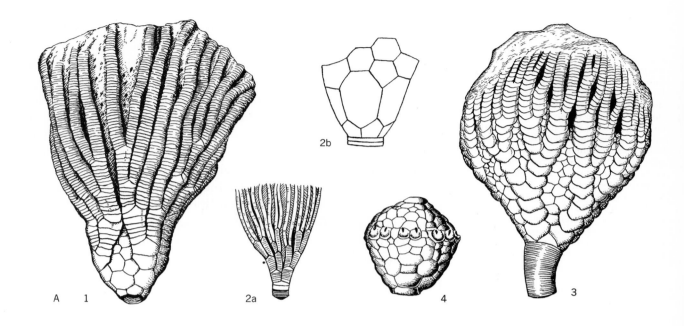

Fig. 12–5. *The echinoderms. A.* Crinoidea. (1) and (2) *Cercidocrinus* and *Blothrocrinus*, inadunates from Iowa; (2*a*) is a reduced illustration and (2*b*) shows the arrangements of plates at the base of the calyx. (3) *Forbesiocrinus*, a flexibiliate crinoid from Kentucky. (4) *Uperocrinus*, a camerate from Tennessee. All are from the Carboniferous (Mississippian) rocks.

B. Drawings of camerate crinoids of the genus *Batocrinus* from the lower Carboniferous limestones in Illinois; some of the arm plates remain in one specimen (from Wachsmuth and Springer's monograph of beautiful drawings published in 1897).

C. Blastoidea. Carboniferous (Mississippian) blastoid *Pentremites* from Kentucky; (1–3) side, summit, and basal views.

Phylum Echinodermata. The phylum contains marine organisms that have enclosing body shells made of regularly arranged closely joined small plates, generally with five arms or food grooves, as in the common starfish. Some echinoderms are free-swimming or free-moving, as with the starfishes and the echinoids or sand dollars seen along our present seashores. Others are fastened to the bottom by a long stalk, as are the crinoids or sea lilies that inhabit some present tropical seas, and the blastoids that lived with them in the Paleozoic.

Class Crinoidea. The most significant and prevalent echinoderms in the Paleozoic, the crinoids, are particularly well represented in the Carboniferous System. The typical crinoid has a long stalk made of perforated calcite disks, joining a short cup or calyx from which extend long branching arms made of small plates. At their death, the animals commonly disintegrate so that the plates become drifting fragments that accumulate as sand. Such crinoidal limestones are frequently the only evidence of the former presence of the animals. The plates in the calyx of a crinoid are remarkably uniform in their arrangement within each genus, successive rows of plates having remarkable geometric patterns. The classification of crinoids is based on the orderly pattern of the plates. They are grouped into those in which only the lowest plates in the calyx are rigid (subclass Inadunata), those having plates in the calyx that are not rigid (subclass Flexibilia), and those in which the calyx is like a rigid ball (subclass Camarata). The arms in all the types issue from the calyx as a feather-like crown. Crinoid fossils are common from the Ordovician through subsequent systems. They reach their greatest stratigraphic value in the lower Carboniferous, where they have been zoned in the central United States so as to enable close correlation of crinoid-bearing rocks.

Class Blastoidea. Another class of echinoderms, the blastoids, is likewise abundantly represented in some lower Carboniferous rocks. Similar to crinoids in that they are attached by a stem, blastoids are distinguished in their having their five ambulacra, or food channels, radiating from the top of the calyx, with a few intervening plates. Spreading, slender arms emerged along the groove, but these are rarely preserved. Blastoids are found principally in Carboniferous and Permian rocks, though rarely in beds as old as Ordovician.

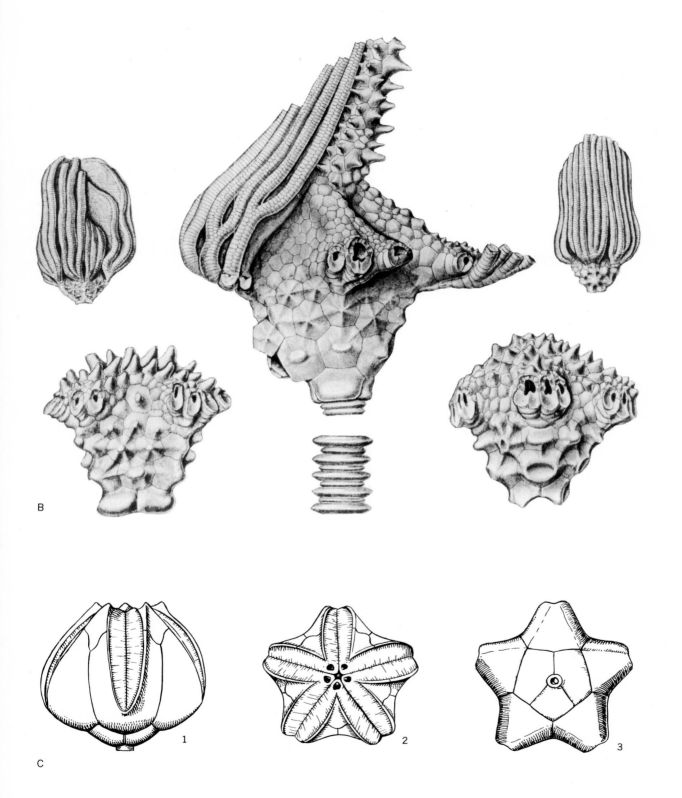

B

C

1

2

3

tucky, it should not be surprising that the Meramecian of the interior states is also predominantly nonterrigenous, of calcite limestone and saline rocks, and that these overlap eastward toward the same low arch along the Cincinnati line. In the gradually subsiding trough west of this line, from southern Michigan to southern Illinois, a few hundred feet of limestones show continuing regional warping like that of the Osagian. As an effect of the restricted ingress of water into southern Michigan during earlier Meramecian, when there was land along the Cincinnati line; high evaporation resulted again in precipitation of saline minerals, as it had in the two preceding periods. The salts form the supply for the chemical industries in central Michigan. As some of these salts are of potash and bromine such as precipitate only when sea water has been evaporated to about 1 per cent of its original water content, there must have been isolation at times (see Fig. 10–22). With the spread of seas over the Cincinnati Arch on the east in late Meramecian, normally fossiliferous limestones return to the sequence. Clear, shallow water then spread from the Mississippi Valley to beyond the preserved sections in the Appalachian Mountains.

The Chesterian Series in Illinois and adjoining states has a peculiar sequence of lithologies, essentially of calcite limestone and quartz sandstone alternating about ten times to a total of more than a thousand feet (Fig. 12–8); sands lie on the surfaces of the preceding limestone, and some have thin seams of plant material—poor grade coals. The thickness of the series as a whole increases toward a south plunging axis passing through the center of the state (Fig. 12–9). Each of the repeating successions of rock types or *cyclothems* likewise tends to thicken near this axis and thin toward the sides of the trough, but the lithologies are not simple sandstone-limestone alternating pairs. Warping took place as deposition proceeded.

PALEOGEOLOGY IN ILLINOIS

The warping did not cease with deposition of the youngest preserved Chesterian rocks, however. Pennsylvanian rocks lie with unconformity on the Chesterian in such a way that the base of the Pennsylvanian, the unconformity, bevels across successive older formations having an inverted U-shaped pattern on the map of the unconformity surface (Fig. 12–9); the formation outcrops diverge southward, surrounding the youngest beds in the center of their arcs. This form would be given in the beveling of a plunging syncline, so the rocks were bowed into a south plunging fold and then beveled by erosion after the youngest formation was laid. Part of the thinning of the series is through erosion of the older formations on the margins of the trough. A map showing

Fig. 12–6. *Paleogeographic maps of the early Carboniferous in eastern United States. A.* Distribution of types of sediments in the Osagian Series; the terrigenous sediments were largely derived from tectonic highlands remaining from the Acadian Revolution along the Atlantic Coast (after K. R. Walker).
B. A more detailed map of Ohio and adjacent states very early in the Carboniferous (in Kinderhookian time); though the highlands to the east were the principal source, studies of sedimentary petrology show that some detritus was carried from the north into a delta lying between the advancing deltal front on the east and the shoals along the little-subsiding Cincinnati Arch in western Ohio (after J. F. Pepper).

Fig. 12–7. Restored section of the lower Carboniferous (Mississippian Subsystem) from Ohio toward the Atlantic Coast in Virginia at the close of the Mississippian. The lands raised in phases of the Devonian Acadian Orogeny yielded coarse terrigenous sediments into a subsiding trough to their west at the beginning of the Mississippian, the Pocono Group being Kinderhookian and Osagian. The Greenbrier Limestone Group, Meramecian and lower Chesterian, was laid when source lands had been denuded and were low. With renewed rise of lands in the later Mississippian, the Mauch Chunk Group of coarser detrital sediments formed; they are overlain by the Pennsylvanian Subsystem. In contrast to the rising source lands on the southeast and the subsiding foredeep, or geosyncline, the northwestern margin in Ohio was relatively stable, gently subsiding; some of the lower detrital sediments were laid from streams flowing southward from the area of the Canadian Shield.

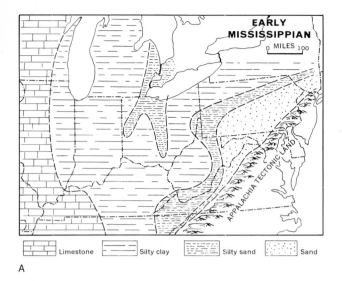

EARLY
MISSISSIPPIAN

0 MILES 100

APPALACHIA TECTONIC LAND

Limestone   Silty clay   Silty sand   Sand

A

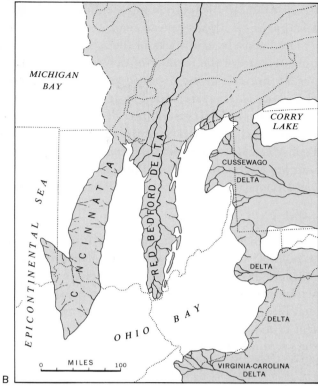

MICHIGAN
BAY

CORRY
LAKE

C I N C I N N A T I A

R E D   B E D F O R D   D E L T A

CUSSEWAGO
DELTA

E P I C O N T I N E N T A L   S E A

O H I O   B A Y

DELTA

DELTA

VIRGINIA-CAROLINA
DELTA

0   MILES   100

B

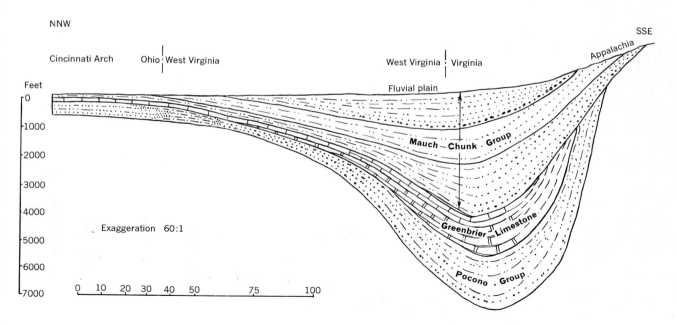

NNW   SSE

Cincinnati Arch   Ohio │ West Virginia   West Virginia │ Virginia   Appalachia

Feet
0   Fluvial plain

-1000   Mauch — Chunk   Group

-2000

-3000

-4000   Greenbrier   Limestone

-5000   Exaggeration 60:1

-6000   Pocono   Group

-7000
       0   10   20   30   40   50   75   100

the distribution of the rock units on an erosion surface concealed by a later cover is a *paleogeologic* map.

The Pennsylvanian rocks of the Appalachian region give an exceptional record of the manner of deformation of that region during the period. Because of the regional structure—the broad Allegheny Synclinorium, an elongate folded basin having lower Permian strata in its center—the older parts of the Pennsylvanian are more extensively exposed and preserved than the younger (Fig. 12–10). The original Pennsylvanian of the state of Pennsylvania had five groups of which four are Carboniferous, the Pottsvillian, Alleghenian, Conemaughan, and Monongahelan, series; the fifth (Dunkard) is now classed as Permian.

COAL CYCLOTHEMS

In southern West Virginia, the Pennsylvanian is about a mile thick, but it is 1000 feet or less in western Pennsylvania, eastern Ohio and northern West Virginia. The sections are preponderantly of detrital sediment eroded from lands to the southeast, for some units coarsen in that direction, and limestone and other marine rocks are more prevalent in the northwest. The sediments are not irregularly distributed in the section but tend to form repeating sequences, cyclothems, a suc-

Fig. 12–9. Paleogeologic and isopach maps of the Mississippian of Illinois, showing the tectonic development of the area. *A.* Isopach map of the Kinderhookian, Osagian, and Meramecian Series, the lower part of the Mississippian Subsystem. If the base was laid on a virtually horizontal surface, and the top was similar, the isopachs show that more than one thousand feet of differential warping took place during the epochs (after J. N. Payne).

B. Isopach map of the Chesterian Series, the upper Mississippian, showing the thickening from north, west, and east into the southern part of the state; part of the thickening is through addition of younger beds that were eroded away during the formation of the overlying pre-Pennsylvanian unconformity. In order to learn whether deformation had progressed during Chesterian, we must examine the pattern for single time-stratigraphic units, as shown in Figure 12–8; they show slight divergence toward the axis of the isopach trough (after L. E. Workman).

C. Paleogeologic map of the unconformity on the top of the Mississippian strata when they were covered by the succeeding Pennsylvanian Subsystem. The pattern of the outcrops, with the youngest beds in the south, shows that southern Illinois had been deformed into a southward gently plunging syncline that had been eroded, beveled, by the time of deposition of the Pennsylvanian (after L. E. Workman).

D. Map showing the age of Pennsylvanian sediments overlying the pre-Pennsylvanian unconformity; the sediments spread northward, overlapped, during the period (after H. R. Wanless).

Fig. 12–8. Mississippian sections in the southeastern part of the Illinois Basin in Indiana and Kentucky (after D. H. Swann). The limestones have been separately shown from a succession of wells; the associated rocks are quartz sandstones and silty shales. Although some limestones are persistent and rather constant along the line of sections, others are variable and lensing; there is repeti-

tion of lithologies and sequences suggesting alternation of conditions of deposition. Although changes in thickness are not apparent in nearby sections, there is a general thinning and convergence of units toward the eastern sections. The diagram cannot show details in lithologies and fossil content that support some of the correlations between the wells.

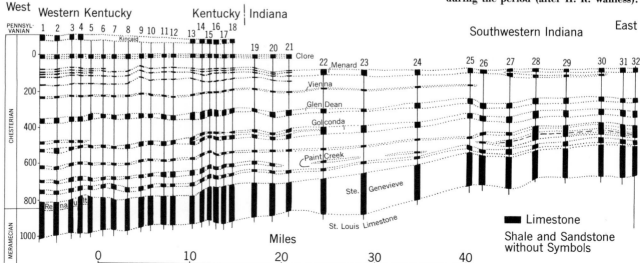

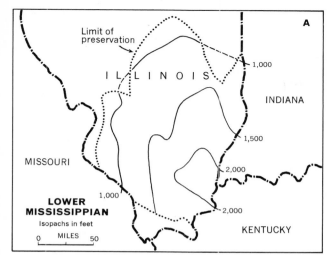

A

Limit of preservation

I L L I N O I S

INDIANA

MISSOURI

1,000

1,500

2,000

1,000

2,000

KENTUCKY

LOWER MISSISSIPPIAN

Isopachs in feet

0 MILES 50

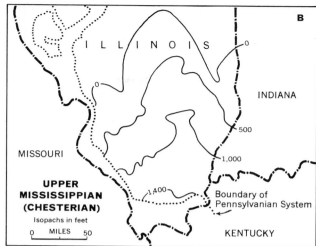

B

I L L I N O I S

0

INDIANA

MISSOURI

0

500

1,000

1,400

UPPER MISSISSIPPIAN (CHESTERIAN)

Isopachs in feet

0 MILES 50

Boundary of Pennsylvanian System

KENTUCKY

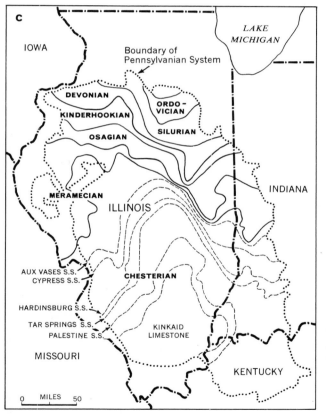

C

IOWA

LAKE MICHIGAN

Boundary of Pennsylvanian System

DEVONIAN

ORDO-VICIAN

KINDERHOOKIAN

SILURIAN

OSAGIAN

INDIANA

MERAMECIAN

ILLINOIS

AUX VASES S.S.
CYPRESS S.S.

CHESTERIAN

HARDINSBURG S.S.

TAR SPRINGS S.S.

PALESTINE S.S.

KINKAID LIMESTONE

MISSOURI

KENTUCKY

0 MILES 50

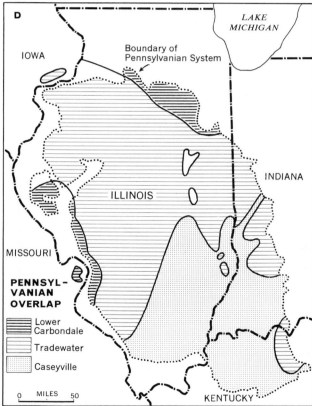

D

IOWA

LAKE MICHIGAN

Boundary of Pennsylvanian System

INDIANA

MISSOURI

ILLINOIS

PENNSYL-VANIAN OVERLAP

Lower Carbondale

Tradewater

Caseyville

0 MILES 50

KENTUCKY

247

cession from fluviatile sands, silts, and clays into coal and then marine clays with fossiliferous limestones. Erosional unconformities are sometimes recognized, and in some parts marine beds are lacking. There are about 90 of these cyclothems in southern West Virginia; thus if they were of equal duration, each represents about a half-million of the fifty million years or so in this period. And as the succession in each portrays a sequence of events taking place in about the same span of time, a single lithologic unit such as a coal is about synchronous with a unit of the same kind in another place if it lies in the same cyclothem. The continuity of the units is demonstrated best by careful surface mapping, though faunal and floral evidence is sometimes helpful; although generally quite persistent over considerable areas, many units are quite variable.

The simplicity of cyclothem sequence in the Carboniferous coal-bearing strata is not universal or perhaps even general. We have seen the splitting and lensing of beds in the upper Mississippian of Illinois (Fig. 12–8). Often the stratigraphic units change rapidly or are cut away by channels, relations that are prevalent in some of the thicker sections, as in central Pennsylvania. The variations are attributable to their deposition in an environment of shifting stream channels and distributaries, and intervening swamps, compounded by the effects of crustal warping, eustatic changes, and differential compaction. The relationships can be compared to those in modern deltas such as those of the Mississippi, to be described in Chapter 22.

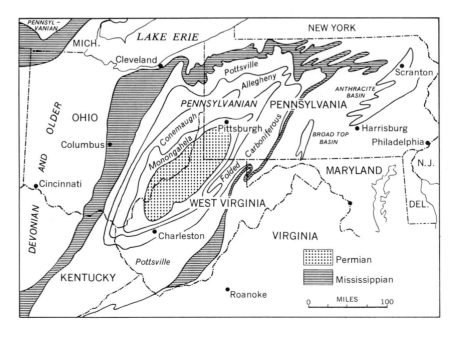

Fig. 12–10. Geologic map of the Carboniferous rocks of the Appalachian plateaus of West Virginia and Pennsylvania and surrounding region. The lower Pennsylvanian rocks in southern West Virginia are three hundred miles south of the exposures in western Pennsylvania and eastern Ohio, so the stratigraphy of the lowest series is interpreted from wells. But the youngest Pennsylvanian rocks are but 100 miles apart in the same section, so they have only a limited distance in which they are preserved and in which their lateral changes can be determined. In the Allegheny Synclinorium that lies below the plateau, the Pennsylvanian rocks thicken considerably southward; see Figs. 12–7 and 12–11. Far to the northeast are the separated Anthracite Coal Basin, a synclinorium, and the Broad Top Basin, also coal bearing; other smaller infolded bands are not shown.

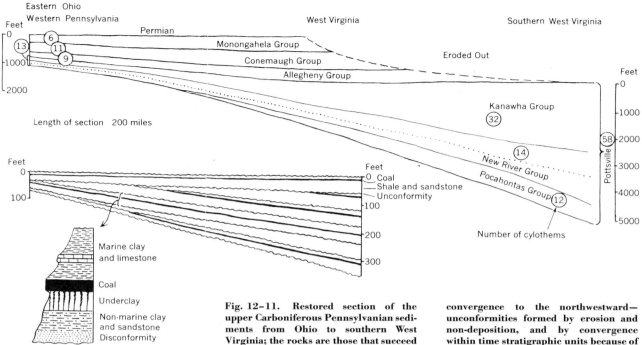

Fig. 12–11. Restored section of the upper Carboniferous Pennsylvanian sediments from Ohio to southern West Virginia; the rocks are those that succeed the Mississippian shown in Fig. 12–7. The lithologies are generalized, for the successions have many cyclothems, and the number in each group is noted. A representative cyclothem is shown in the lower left, and the factors producing the convergence to the northwestward—unconformities formed by erosion and non-deposition, and by convergence within time stratigraphic units because of greater relative subsidence on the southeast than on the northwest. The section originally continued southeastward to source lands along the Atlantic Coast, as in the Mississippian section.

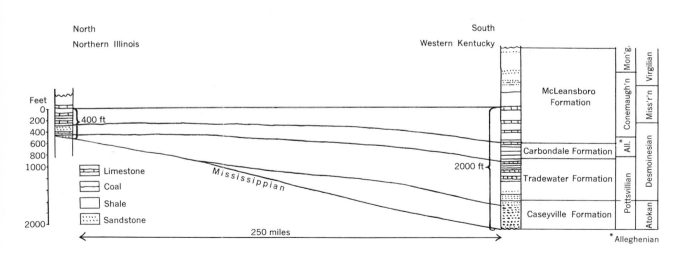

Fig. 12–12. Columnar sections of the upper Carboniferous Pennsylvanian sedimentary rocks at the northern and southern margins of the Illinois Basin showing the progressive overlap of the strata northward. The columnar sections are very generalized, for the rocks consist of a succession of cyclothems having several rock types in each.

249

The Pottsvillian Series, the oldest, probably represents most of Pennsylvanian time, the earlier and medial parts of the period. The thickness of that series is but 250 or 300 feet in western Pennsylvania, but nearly 4000 feet in southern West Virginia and may exceed two miles in Alabama (Fig. 12–11). The Pottsvillian Series formed a great elongate tongue of sediment that has been somewhat folded and eroded away in the subsequent periods. In southern West Virginia, of the 90 cyclothems about 60 are in the 4000 feet or so of Pottsvillian, the remaining 30, in the succeeding three series. But in eastern Ohio and western Pennsylvania, the 250 and 300 feet of Pottsvillian contains only about ten or a dozen cyclothems; a coal some 300 feet and 10 cyclothems from the base was continuous originally with one about 4000 feet and 60 cyclothems from the base in southern West Virginia. The average thickness of a cyclothem in Pennsylvania is about 30 feet but is more than twice that in the south. The lower Pottsvillian, almost a thousand feet in southern West Virginia with a dozen cyclothems, is not represented by deposits in the northern area; and very few of the more than a dozen in the similar thickness of middle Pottsvillian persist into Ohio. The Pennsylvania and Ohio sections have principally upper Pottsvillian, 8 or 10 of about 30 found in West Virginia. These data are indicative of the sort of history that prevailed.

Part of the diminution is through loss of the lower cyclothems by overlap on an unconformity. Some additional changes are through interruptions by erosion surfaces within the cyclothems sequence, and by convergence within single cyclothems. Throughout the sequence, records of the orientation of primary structures in the sediments show that transport was from southeast to northwest, hence that the surface of the land at any single time sloped northwestward, even though the base of the system sloped progressively more to the southeastward. Such an arrangement of strata could have come about through the progressive subsidence of a trough passing northeast-southwest through southern West Virginia, receiving sediment from tectonic lands to the southeast, and having northwest flowing streams reaching the sea on the northwest. Differential movements were such that the surface below the Pottsvillian subsided more than a half mile during the epoch in the southeast, while it sank only some 250 feet in the northwest relative to the surface of deposition at the end of Pottsvillian time. Through the earlier part of the epoch, any sediments that may have been laid in the northwest were not preserved, and that margin of the geosyncline subsided only a little in medial Pottsvillian; the differential continued into late Pottsvillian, but with sufficient subsidence on the geosynclinal margin to permit the preservation of some of the cyclothems through-

Fig. 12–13. *Pattern of sediment flow in the Pennsylvanian sandstones in eastern United States (after P. E. Potter and Raymond Siever). A.* Cross-bedded sandstone ledge in Illinois; the bedding surface is nearly horizontal, but the laminae within the bed dip toward the right, the current having flowed in that direction. Cross-bedding varies locally, but from many readings the general directions of transport can be determined (Illinois Geological Survey).

B. Map showing the average direction of dip of cross-bedding and its variation around the Illinois Basin and the western margin of the Allegheny Synclinorium.

C. Paleogeographic map showing the flow pattern of the currents that transported the sands in the Pennsylvanian Subsystem in the area.

A

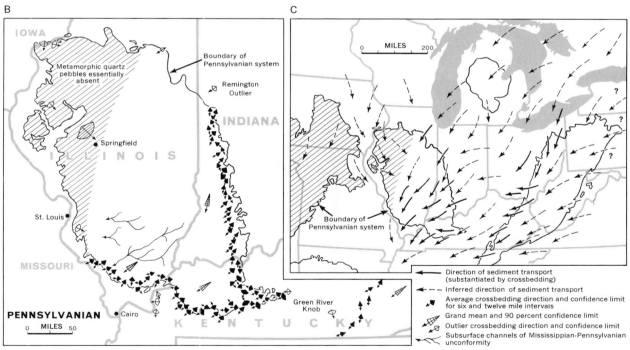

B

IOWA

Metamorphic quartz
pebbles essentially
absent

Boundary of
Pennsylvanian system

Remington
Outlier

INDIANA

• Springfield

I L L I N O I S

St. Louis •

MISSOURI

PENNSYLVANIAN

• Cairo

0   MILES   50

K E N T U C K Y

Green River
Knob

C

0   MILES   200

Boundary of
Pennsylvanian system

Direction of sediment transport
(substantiated by crossbedding)
Inferred direction of sediment transport
Average crossbedding direction and confidence limit
for six and twelve mile intervals
Grand mean and 90 percent confidence limit
Outlier crossbedding direction and confidence limit
Subsurface channels of Mississippian-Pennsylvanian
unconformity

251

out the region. Single cyclothems have been traced from thickness of a very few feet in northeastern Ohio to as much as 500 feet in eastern Tennessee. The progressive differential subsidence must have been superimposed on a pulsing factor that formed the cyclothems, which will be considered later.

MANNER OF WARPING

A very similar relationship pertains from south to north through Illinois. In the south some 2000 feet of Pennsylvanian sediments includes early Pennsylvanian strata that are lacking in the northern sections, the base lying with great regional unconformity on the beveled structures of the Mississippian and older systems (Fig. 12–12). The Pennsylvanian sediments were deposited on a plain that cut across Mississippian rocks that had been so gently folded that there is no apparent angular discordance in exposures. Thousands of wells pass through the Pennsylvanian into the older systems. Sections made by plotting data from these wells show that the full section of the Mississippian in southern Illinois is reduced northward progressively by the removal at the unconformity of the upper units. The *paleogeologic map* (Fig. 12–9C) showed what would appear if we were to strip the Pennsylvanian off the erosion surface beveling the Mississippian; the surface is the very prominent regional unconformity. The magnitude of the interval is now more apparent from our knowledge of the younger Mississippian and older Pennsylvanian sediments in West Virginia and their stratigraphy.

Fig. 12–14. *Plants that formed Carboniferous coals. A.* Pteridosperm, frond of *Lyginopteris* from Scotland (H. N. Andrews, Jr.).

B. Lycopods *Lepidophloios* and *C. Sigillaria* from the Carboniferous of Scotland (Her Majesty's Geological Survey; British Museum of Natural History).

D. Arthrophyte *Calamites* from Sweden (Riksmuseum); and foliage of the calamites, *Asterophyllites* (*E*) and *Annularia* (*F*) (H. N. Andrews, Jr.).

Several sorts of plants contributed their matter to the coals of the Carboniferous System. Although many of them look very much like ferns, few have yielded spores such as should identify their clasification as *Pterophyta*. On the other hand, some of the fern-like plants have been found to bear seeds on unmodified or slightly differentiated leaves, never in cones. Because of their seed bearing but fern-like appearance, they were called *Pteridospermophyta* or pteridosperms. From an economic standpoint, they are of greatest interest in their contributing to the formation of coal. But some of the basic information about such plants comes from the study of material that was silicified or enclosed in carbonate soon after burial to form the uncompressed coal-balls found in sediments, in contrast to the severely compressed remains found in the associated coals; the name is applied to the petrified plants even though many are not at all in the shape of balls. In some instances, woody fiber remains intact.

Pteridosperms are known first in the lower Carboniferous, though it is suspected that some Devonian plants may belong in the phylum. They continue through the Paleozoic into the early Mesozoic, where they are considered to be quite distinct.

Several other plant groups contributed to the Carboniferous coals. The *Lycopodophyta*, represented by the present-day small ground plants such as the club-moss, *Lycopodium*, are first known from surprisingly similar small Silurian plants from Australia. The lycopods were trees in the Carboniferous, *Lepidodendron*, with the prominent rhombic scars of leaf bases and *Sigillaria*, vertically ribbed with leaf scars alternate on adjoining ribs, are universal. Mature trees of *Lepidodrendon* are known to have reached heights of 100 feet. The cones or strobili, which produced spores, are elongate structures at the tips of slender twigs or lateral on larger ones. The leaves are small, needle-like to lanceolate, as long as two feet. The name *Stigmaria* is applied to the roots of these plants.

A

B

A third phylum of plants, the *Arthrophyta*, is represented today by the common "horsetail" *Equisetum*, with the characteristic jointed shoots, ranging to several feet high. The leaves are small, forming whorls at the joints. Some of the shoots have sporangia-bearing organs. In the Carboniferous, the *calamites* were abundant representatives of the arthrophytes, the name being applied to the cases of the stems. The leaves that formed whorls at each joint, standing out from the stem as in *Annularia*, are useful in distinguishing successive forms. And the cones, found in many coal-balls, are known in excellent preservation and great variety. Some of the Paleozoic arthrophytes have spore structures very like those of the modern *Equisetum*. Another type of arthrophytes, represented by *Sphenophyllum*, have distinctive wedge-shaped leaves; innumerable cones are known from these plants, too. As the arthrophytes are first recognized in the Devonian, the plants evolved rapidly within the later Devonian and earlier Carboniferous periods.

The fourth phylum of coal-forming plants in the Carboniferous is the *Coniferophyta*, which are represented principally by the class *Cordiatales*, the cordiates, though some *Coniferales* appear in the Permian. The cordaites formed trees up to 100 feet high, bearing slender leaves to a yard or more long, spirally arranged on the trunk. The distinction between the cordaites and true conifers is not simply defined.

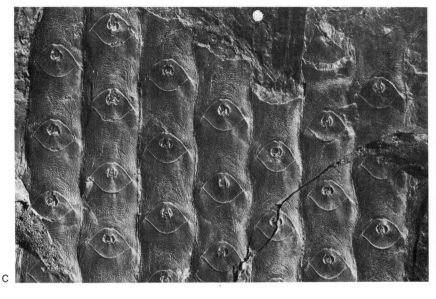

C

D

E

F

The sediments in the Pennsylvanian in Illinois are successively younger northward at the base, just as in the East, overlapping the unconformity (Fig. 12–9D). We might establish a sequence of fossils in southern, full sections, and find that the beds lying over the unconformity are progressively younger northward. But from the many wells, the relations can be recognized by tracing the distinctive lithologies, such as the coals; it can be found that they progressively approach the unconformity northward, and that older units terminate and become overlapped.

SANDSTONES AND THEIR SOURCES

The origin of the sandstones in the Pennsylvanian has been studied quite systematically. The sands came from the northward and northeastward; streams flowed southward and southwestward, and came from lands whose nature can be judged. A single exposure of the basal Pennsylvanian, such as a rock cliff along a stream, may have two or three sandstone beds, some with cross-bedding (Fig. 12–13). In modern streams, the secondary laminae in such beds dip in the direction of current flow. The directions of dip in three units such as might be seen in a single exposure are likely to be varied, just as the course of a southward flowing river will have many stretches that are not directly to the south. If we gather data from several rock exposures in a square mile or so, we can establish an average direction, a mean, and by means of mathematical statistics, determine how satisfactorily this mean indicates the probable true average direction that we would gain if we measured a very large number of beds. The range can be represented by an arc extending on each side of the mean direction, such as one representing the angle within which 70 per cent of all readings are likely to fall, a conventional method (Fig. 12–13B). Similar arrows and arcs can be compiled from other small areas in Illinois and in the other states, and from these we can prepare a map showing the trends of stream flow within the whole region (Fig. 12–13C).

The sands have minerals that show their sources in their compositions and properties. For example, in western Illinois the sands have quartz grains that the mineralogist can recognize as having come from igneous rock such as granite, with the resistant mineral tourmaline in well-rounded grains, and with little feldspar. Eastward, the sands have quartz of metamorphic rock origin, more angular tourmaline grains and more feldspar. The first sands seem to have come by erosion of earlier-laid sands such as might have formed the surface of the region to the north in Pennsylvanian time—perhaps areas of Cambrian sandstone-surfaced land. The second type of sands seem to have come from the

Fig. 12–15. Columnar sections of coal-bearing cyclothems, showing the ideal order of sequence of lithologies such as might be found in Illinois (after J. M. Weller).

Fig. 12–16. Exposure of some of the succession in a cyclothem in the upper Carboniferous at Fitchian, Illinois (Illinois Geological Survey). The beds near the stream are underclay succeeded by blocky coal; and limestone forms the overhang on the left but is cut out on the right; succeeding are black shale, gray shale, limestone, and at the top in the shadow, sandy shale; the coal in this instance is thin and not commercially useful at this place; the beds above it are marine.

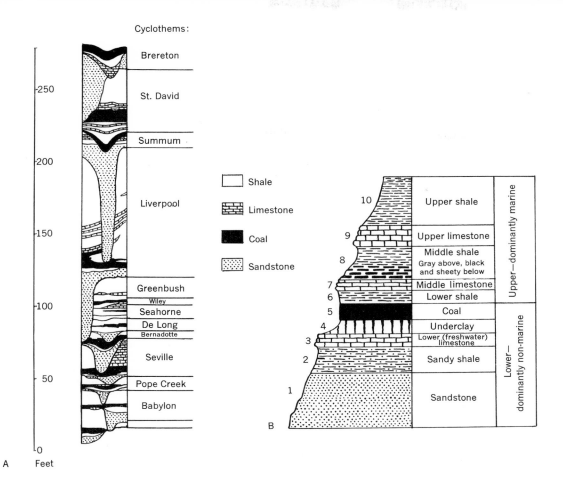

Cyclothems:

Brereton

St. David

Summum

Liverpool

Greenbush

Wiley

Seahorne

De Long

Bernadotte

Seville

Pope Creek

Babylon

A Feet

Shale

Limestone

Coal

Sandstone

10	Upper shale	Upper—dominantly marine
9	Upper limestone	
8	Middle shale Gray above, black and sheety below	
7	Middle limestone	
6	Lower shale	
5	Coal	Lower—dominantly non-marine
4	Underclay	
3	Lower (freshwater) limestone	
2	Sandy shale	
1	Sandstone	

B

erosion of metamorphic rocks such as may have been exposed in mountains in western New England, or possibly from exposed Precambrian rocks in southern Quebec or Ontario. Thus the source and direction of flow of Pennsylvanian sands can be reconstructed from the field and laboratory observations, with application of statistical methods to determine the probable validity of the conclusions. Such studies are the field of interest of the science of sedimentary petrology.

FORMATION OF COAL

The most important rock in the Pennsylvanian System in western Pennsylvania is coal (Fig. 12–2), formed through the accumulation and preservation of plants (Fig. 12–14). The coal is in quite constant beds that are associated with many other sorts of rocks, sandstones, shales, clays, and limestones, that are quite systematically repeated in the same or similar order, thus are in cyclothems. In western Pennsylvania, there are some 25 or 30 of these cyclothems in about 1000 feet of Pennsylvanian (Fig. 12–15A); thus the average thickness of each cyclothem is 35 or 40 feet. If marine beds are present, they invariably lie above the coal; clay is directly below it; and some distance below is sandstone, judged from its sedimentary structures to have been stream laid; in similar cyclothems in Illinois (Fig. 12–15B and 12–16), the sandstone lies on an unconformity that cuts deeply into underlying cyclothems.

The base of the latter cyclothem is an erosion surface. For some reason, streams that had been eroding at the locality of the section began to deposit sands and clays. The area became progressively less well-drained, until it was swampy, supporting a forest that has become converted to coal (Fig. 12–17). The swamp was gradually inundated by the sea, and as the sea advanced, the clearer offshore waters came to bear marine animals whose calcareous shells and fragments accumulated to form the limestones. But the sea withdrew, and erosion began again. If it had withdrawn to its former level, and there had been sufficient time, erosion should have quite fully removed the record of the preceding advance—there would not be any preserved cyclothems. There must have been alternate presence and absence of the sea, but in order that the sediments be preserved, there must have been additional progressive rise of sea level, or progressive subsidence of the basement or both, eustatic rise, epeirogenic sinking or both (Fig. 12–18). If it were eustatic rise alone, there should be the same thickness of rocks in cyclothems wherever those of the same time are found and if, for instance, the upper limestones were laid at the same depth, the thickness between the successive limestone beds should represent the

Fig. 12–17. Paleogeographic map showing the depositional basin in which coal was laid at one time in the Pennsylvanian (after H. R. Wanless). The deltas on the east extended into a low marshy plain that reached to the open marine conditions in Kansas and Nebraska on the extreme west. Forests thrived in the moist climate.

Fig. 12–18. *Factors controlling the formation of a succession of cyclothems.* In the upper drawing, point A retains a constant elevation relative to the center of the earth. The undulating line A—Bx shows the rise and fall of sea level relative to point A through time, the eustatic movement being 100 feet in 400,000 years. The straight line A—S shows the position of the original surface at A through time, under epeirogeny causing subsidence at a rate of 40 feet in 400,000 years.

From the lower left, enlarged from a section of that above, we can see that the original surface at A will be at X in 100,000 years, 60 feet below the level of the sea at B_1 but 10 feet below the level of A; in 250,000 years at Y, the surface will be 25 feet below A, but sea level will also have dropped to that level; in 500,000 years, the sea at B_3 will again reach a level 50 feet about A, and the original surface at A will have subsided to 50 feet below A and 100 feet below B_3.

In the lower right, the succession of events is to show the development of progressive cyclothems. With incremental deepening during the first 100,000 years, sediment is assumed to have filled the subsiding surface to a depth of 40 feet, leaving a water depth of 20 feet. Inasmuch as water level drops much more rapidly than the original surface of coal A subsides, the marine sediments and ultimately the coal and underlying sediments are eroded to a maximum depth at B_2 in 300,000 years. Then with the rise of sea level of 100 feet in the succeeding 200,000 years, a second cyclothem is laid, starting with stream-laid sediments and progressing to coal and marine sediments as in the first cyclothem. The process would then be repeated with successive events as long as the movements continued.

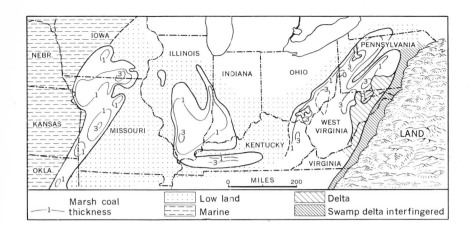

—1— Marsh coal thickness	Low land	Delta	
	Marine	Swamp delta interfingered	

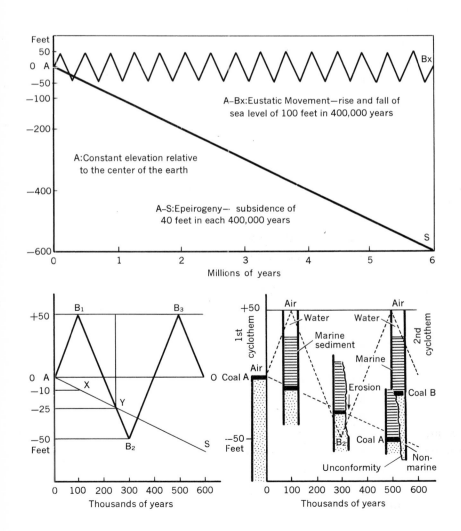

A–Bx: Eustatic Movement—rise and fall of sea level of 100 feet in 400,000 years

A: Constant elevation relative to the center of the earth

A–S: Epeirogeny— subsidence of 40 feet in each 400,000 years

Millions of years

Thousands of years

Thousands of years

eustatic rise in the interval of time intervening. But epeirogeny must have been an important factor for units differ in thickness and separation from place to place, depending on the amount of subsidence in each locality. The alternate withdrawal and return of the sea must have been caused by some factor additional to the subsidence that preserved sedimentary successions of variable thickness. Several have been considered.

EUSTATIC MOVEMENTS AND CYCLES

The repetitions could be brought about by some means of alternately increasing and reducing the volume of water in the oceans, or at least so changing the form of the ocean basins as to alter the sea levels. In the southern hemisphere, deposits of glacial till, tillite, and associated fluvio-glacial rocks, such as laminated or varved slates, are widely distributed in late Paleozoic. The oldest of these in Australia have been thought to be late Mississippian. If there were alternate glaciation and deglaciation on a large scale in Australia or in any other continent, water would successively be contained in glacial ice and added to the oceans, resulting in pulsing movements of sea level. The cyclothems would represent deglaciations in some part of the earth, and the separating erosion surfaces, the times of withdrawal of water during glaciation. Such eustatic movements should have been universal and it is true that cyclothems are known in many Carboniferous successions on several continents. Present methods of correlation do not enable us to determine that the cyclothems in the widely scattered successions are precisely synchronous though it has been computed that they have durations of a half-million years or so, too short a time for the geochemist to distinguish by isotopes and disintegration, or the paleontologist from organic evolution.

Repeated successions could be formed in other ways. The relative effort of epeirogenic pulsing rise and fall of land would be like that of eustatic fall and rise of sea level at a single place or in a limited area —the difference would be in the universal synchronous affect of the latter. Climatic effects of regular alternation and severe enough effect on erosion to cause the introduction of quantities of sediments at intervals in a generally subsiding depositional basin would produce alternating successions. And the distributions of streams entering such a basin might shift their deltas to a succession of areas along a coast, returning at another time to repeat the succession. Such factors are surely important in contributing to deviations of cyclothems from uniformity and lateral persistence whether or not eustatic control is pertinent. As we shall see, cyclothems also prevail in some rather open

Fig. 12 – 19 (above). Carboniferous
swamps were inhabited by small amphibians of primitive form (J. Augusta and
Z. Burian, 1956). Here we see a reconstruction of a late Carboniferous amphibian fauna, as based upon discoveries made
at Nýřany, Czechoslovakia. The genera
shown are "*Branchiosaurus*" on the rock,
Microbrachis at the bottom of the picture,
Urocordylus represented by the uppermost two animals, and *Dolichosoma*, the
limbless, eel-like form. This was a period
in earth history when the tetrapods, the
air-breathing vertebrates, were generally
small and rather limited in their adaptations to life on the land (Spring Books,
Paul Hamlyn Ltd.).

Fig. 12–20. Standing cast of the trunk
of a tree preserved in the upper Carboniferous sandstones at Joggins, Nova
Scotia, near the head of the Bay of Fundy
(G. W. Webb). For scale, note hammer
along side of the upper part of the cast.
The presence of the trunk evidences the
rapid deposition of the enclosing non-
marine sediments.

marine successions wherein climatic effects or shifting distributions seem not to be possible factors. So epeirogenic movements are surely necessary to allow preservation of cyclothems in subsiding areas, and to create their varying thicknesses and lithologies. Eustatic movements seem to be necessary to introduce the pulsing movements represented in some of the successions.

TETRAPOD DWELLERS IN THE COAL SWAMPS

It is to be expected that the vast coal deposits of Pennsylvanian age, derived from the luxuriant tropical forests which flourished in broad swamps some three hundred million years ago, should yield many clues as to the vegetation of that distant age. It is also to be expected that the fossil remains of swamp-dwelling animals should occur in association with these coal beds. They do, perhaps not as frequently as might be expected, but nonetheless in some profusion at certain localities throughout the world, to round out our knowledge of Pennsylvanian swamp ecology.

Among the most important and numerous inhabitants of the Pennsylvanian swamps were the early tetrapods, the four-footed amphibians and reptiles frequenting the streams and the quiet waters that spread widely across the swampland, climbing the primitive trees or seeking refuge beneath the ancient fern-like plants. It was in truth an age of amphibians, when these vertebrates, the descendants of late Devonian ancestors virtually had the land to themselves. These were the vertebrate pioneers on the continents, and during Carboniferous times they were exploiting the various ecological niches open to them. Most of these niches were in the swamps, perfect environments for amphibian exploitation and evolution, and it was here that the amphibians thrived, multiplied, and became diversified. They were almost unchallenged—almost but not quite. For even as early as Carboniferous times, the first reptiles had appeared to push their way into the territories where amphibians held sway. These first reptiles were small and not so very different from their amphibian cousins, and it seems evident that competition between Carboniferous amphibians and reptiles was fairly evenly divided. In short, the first reptiles had no great advantages over the contemporaneous amphibians, and certainly during the Carboniferous period the amphibians remained the dominant land animals.

The record of Pennsylvanian amphibians and reptiles is particularly well documented at three localities in North America, each of which is distinguished by a particular type of fossilization. At Linton, Ohio, in a cannel coal of middle Pennsylvanian age, are the skulls and skeletons of small amphibians and reptiles that once lived in the ancient

Fig. 12–21. Reconstruction of an early Carboniferous (Mississippian) sea as in northwestern Indiana (The Smithsonian Institution). The many crinoids are such as are preserved in the silty shales of the Osagian Series; starfish lie on the muddy bottom.

swamps covering this part of the continent. To the west, at Mazon Creek, Illinois, are similar tetrapods, preserved, as are the contemporaneous plants, in nodules within coal shales. But perhaps the most interesting occurrence is at Joggins, Nova Scotia (Fig. 12–19), where numerous trunks of Carboniferous trees are preserved in *upright* positions in the shales and coal beds. When these fossil tree trunks are broken open a small percentage of them contain the skeletons of amphibians. It would appear that these trunks are the fossil remains of dead trees that rotted, to become hollow while still standing. They were kept in a standing position by muds that washed in around their bases. Then, it would seem, amphibians occasionally crawled or fell into the hollow, upright stumps and were unable to get out. In these natural traps they died, were eventually buried, and their skeletons were preserved to give us today a fleeting but vivid glimpse of things that happened in the geologic past (Fig. 12–20).

13
The Carboniferous—
Tectonics in the Northeast,
West, and South

Canyon of the San Juan River in southeastern Utah, entrenched in limestones of the Carboniferous System.

13 *The Carboniferous—*
Tectonics in the Northeast, West, and South

The Paleozoic Era in North America is conventionally considered to close with phases of the Appalachian Revolution. Long before that, the Carboniferous Period had produced deformation of distinctive sorts: in the Maritime Provinces—New England region—great thicknesses of sediments accumulated in troughs formed after the Acadian Orogeny; in south central United States, a succession of events profoundly changed the paleogeography; in the central Rocky Mountain Region of Colorado and thereabouts, the usually stable craton developed differential movements of several miles amplitude; and in the Cordillera of Nevada and Idaho, there is evidence of strong orogeny in the western orthogeosynclinal belt.

ATLANTIC MARGIN AFTER THE ACADIAN OROGENY

The Carboniferous rocks from Massachusetts through the Maritime Provinces into western Newfoundland (Fig. 13–1) give excellent record of the events that changed a belt of deeply subsiding volcanic geosynclines into a rather stable addition to the continent. The late Devonian Acadian Disturbance folded the earlier sediments and volcanic rocks of this part of the Atlantic eugeosynclinal belt, and raised much of the region to highlands. Alternating troughs of subsidence and rising land welts, each a score of miles or more wide, developed in the orogenic zone, creating several northeast trending lowlands that received their non-marine sediments from the adjoining highlands of the older Paleozoic metamorphic rocks. The fact that the highlands were made of sediments shows that the new troughs had different disposition than the preceding eugeosynclines. In each of several lowland belts, the silts and sands of the early Mississippian entombed plants and non-marine animals; in their deeper parts, carbonaceous and sometimes petroliferous shaly argillites accumulated, the source of oil in small fields in New Brunswick.

Volcanism was not extensive, though hundreds of feet of lavas accumulated in a few districts, such as one near St. John in New Brunswick. Gradually the highlands were reduced, advancing seas formed long estuaries, and then spread widely in medial Mississippian extending beyond the non-marine sediments of the lowlands over some of the peneplaned surface of former highlands (Fig. 13–2); there was so little terrigenous detritus that the sediments became marine carbonate and saline rocks. Circulation was so restricted at times in the embay-

Fig. 13–1. Tectonic map of the Canadian Appalachian Region (after W. Poole and E. W. R. Neale). The Carboniferous System was laid in subsiding areas among rising highlands of earlier Paleozoic rocks that had been deformed and intruded in the Acadian Orogeny. The isopachs (after M. Kay and W. A. Bell) show an interpretation of the thicknesses of the Carboniferous rocks, which are dominantly terrigenous, though there are extensive limestones and saline rocks in the lower Carboniferous (after E. W. R. Neale, W. A. Nash, R. R. Potter, and W. H. Poole).

Fig. 13–2. Upper Carboniferous non-marine conglomerate lying with angular unconformity on Silurian sediments along the Jacquet River near Perce, eastern Gaspé Peninsula, Quebec (Geological Survey of Canada). The Acadian Orogeny that deformed the underlying rocks took place in the late Devonian Period.

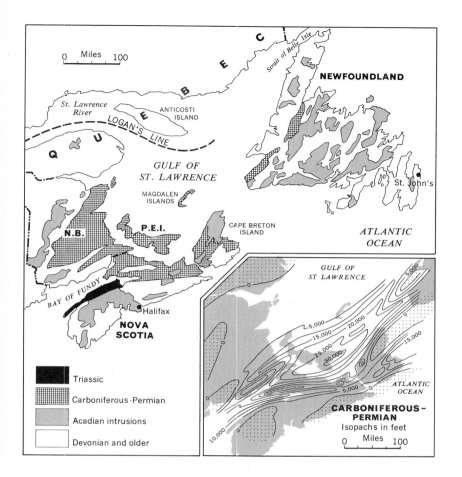

ments that hundreds of feet of gypsum, calcium sulphate (Fig. 13–3), precipitated from restricted marine waters under conditions of high evaporation in the arid climate that must have prevailed. The gypsum, an important source of building material is shipped along the Atlantic Coast southward as far as New York City. It may be coincidental, but the carbonate and saline rocks laid during a quiet interlude in the Carboniferous deformation in the Maritime region are about contemporaneous with the extensive limestones of the Meramecian of the Appalachians.

The region was not to remain quiet. In late Mississippian, highlands rose again in the areas of the preceding ranges, producing non-marine sands and muds that again accumulated in the lowland belts. The rejuvenation and reduction continued through most of the Pennsylvanian until as much as five or six miles of Carboniferous sediment (Fig. 13–4) now occupies synclinal troughs separated by the metamorphosed earlier Paleozoic rocks in the areas of the original highlands (Fig. 13–5). The deposits formed so rapidly in places as to enclose the trunks of standing trees (Fig. 12–20). In parts of the region, plants accumulated more slowly to thicknesses of scores of feet, principally in the Pennsylvanian; on consolidation they came to form coals mined in several districts in Cape Breton and the northern mainland of Nova Scotia. The small province of Prince Edward Island in the Gulf of St. Lawrence is underlain by little disturbed sediments, principally red sandstones and siltstones, that are late Carboniferous and Permian, lying in the center of the structural basin (Fig. 13–6).

THE END OF OROGENY

The Devonian Acadian Orogeny marked the last episode of severe folding, faulting and intrusion, of metamorphism of earlier sediments, in the Maritime Provinces and much of New England. The subsequent warping that permitted preservation of the miles of Carboniferous sediments virtually closed the long history of deep subsidence that had affected the region, though there was subsequent block faulting and warping in the Mesozoic Era. It is quite probable that some of the lowlands were bounded by faults rather than flexures and possible that the faults had lateral as well as vertical movements. A broad belt along the Atlantic Coast that had been subsiding deeply with volcanism and interrupted deformation and intrusion in orogenies through the early Paleozoic became a consolidated, more stable region that was added to the craton of the earlier Paleozoic. The original Cambrian craton grew by addition of the orogenically immobilized geosynclinal belts that flanked it on the southeast.

Fig. 13–3 (above).   Carboniferous gypsum exposed near Dingwall, Cape Breton Island, Nova Scotia (Nova Scotia Department of Mines). Gypsum is present widely in the Mississippian Windsor Group in eastern New Brunswick, Nova Scotia and southwestern Newfoundland. The proximity to the sea permits its shipment in large tonnage for use in construction industries in eastern Canada and the coastal cities in the United States.

Fig. 13–4.   Carboniferous (Pennsylvanian) sediments on the east shore of the Bay of Fundy at Joggins, Nova Scotia. View of the section of hundreds of feet of sandstones; interbeds of coal have been mined out of the cliff. Casts of standing stumps of trees (Fig. 12–20) show that sediments filled the external molds of decayed stumps before the hole had collapsed; the fillings have several kinds of non-marine fossils.

Deeply subsiding troughs with Carboniferous non-marine sediments continued southward to Rhode Island; thousands of feet of folded and metamorphosed sedimentary rocks, including some extremely "hard" carbonaceous coals, are exposed along Narragansett Bay (Fig. 13–7). If there were similar geosynclines farther southward along the Atlantic Coast, they have not been recognized; they may lie seaward from the presently exposed metamorphosed earlier Paleozoic rocks of the Piedmont.

Little has been said about disturbances along the west and south of the craton. The effects of orogenic movements become very evident in these regions in the Carboniferous, terrigenous deposits from rising tectonic lands spreading widely during the Mississippian. The thrust sheets in Nevada that place Ordovician and younger graptolite-bearing argillites, cherts, and volcanic rocks, eugeosynclinal rocks, on carbonate rocks and orthoquartzites laid in the miogeosynclinal belt were described in the interpretation of relations between such belts (Fig. 8–14). They will be used again in illustrating the problems in determining the time of displacement.

THE DATING OF THRUSTS IN THE WEST

The Devonian rocks of eastern Nevada and western Utah, thousands of feet of carbonate rocks and some quartzites, are succeeded by widely distributed argillites and quartz siltstones of a few hundred feet thickness (Fig. 13–8), bearing cephalopods and other fossils that sank into the carbonaceous muds on the sea floor, little damaged as would have been the case if they were tossed about in shallow, turbulent waters. Following widespread deposition of somewhat thinner lower Carboniferous limestone, the beds again become shaly siltstones and sandstones. As these terrigenous beds become coarser westward, containing chert-pebble conglomerates in central Nevada, they must have come from lands raised in latest Devonian time farther to the west.

Recall that the Acadian Orogeny was dated as within the Devonian Period, because late Devonian sediments lie unconformably on intrusions into early Devonian. Thrusts can usually be dated within the limits of the youngest beds that they cut through and the oldest beds that lie upon the thrust sheet and that were not transposed in the thrusting. The latter is a very difficult sort of evidence to obtain. The thrust sheets in Nevada contain the cherts, argillites, and volcanic rocks of the eugeosynclinal belt. Inasmuch as rocks as young as late Devonian limestones of the miogeosynclinal belt are beneath the thrust planes, the movements were of course younger than some time in the late Devonian. The earlier Paleozoic cherty rocks of the thrust sheets had

Fig. 13–5. Restored sections of Carboniferous sedimentary rocks in the Maritime Provinces of eastern Canada as interpreted for early Pennsylvanian time (after W. A. Bell and E. N. Belt). The upper section is an interpretation that the deposits formed in subsiding downfolded troughs separated by rising welts from which some of the sediment was derived. The prominent gypsum and limestone group is the Windsor, a record of marine invasion of the region; most sediments are non-marine, some of them coal-bearing. The lower section presents an interpretation that the subsidence of the troughs was accomplished along high angle faults, shown as though normal faults; possibly these are wrench or transcurrent faults, involving lateral movements along the plane of faulting. Probably the deposits accumulated in troughs in tilted blocks, partially fault bounded.

Fig. 13–6. *Carboniferous and Permian sedimentary rocks, Prince Edward Island, Gulf of Saint Lawrence (Geological Survey of Canada). A.* Cross-bedded Permian red conglomerate and sandstone by Point Prim Light, Hillsborough Bay. The cross-bedded units are separated by surfaces of erosion truncating the arcuate laminae.

B. Cross-bedded sandstone, probably upper Carboniferous, Kildare Capes, northwestern Prince Edward Island; the rocks are slightly folded.

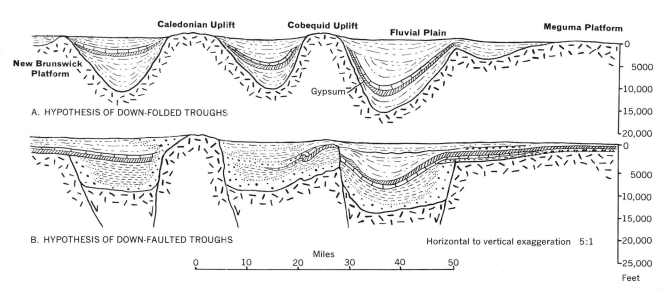

Northwest   New Brunswick

Bay of Fundy
Chegnecto Bay

Nova Scotia   Southeast

Bay of Fundy
Minas Basin   Windsor   Halifax

Caledonian Uplift   **Cobequid Uplift**   **Fluvial Plain**   **Meguma Platform**

**New Brunswick
Platform**

Gypsum

A. HYPOTHESIS OF DOWN-FOLDED TROUGHS

B. HYPOTHESIS OF DOWN-FAULTED TROUGHS

Horizontal to vertical exaggeration   5:1

Miles

0   10   20   30   40   50

0

5000

10,000

15,000

20,000

25,000

Feet

A

B

been folded and eroded before being covered unconformably by late Carboniferous (Fig. 13–9). There was deformation to the west in late Devonian and Carboniferous time; this is the Antlerian Orogeny. The Carboniferous has conglomerates with a dominance of chert pebbles, some of rather large size, to several inches diameter (Fig. 13–10). Although this shows that eugeosynclinal rocks were folded and elevated prior to the Carboniferous, or early in the period, it does not preclude the possibility that the unconformities were formed when the rocks were much farther west, and that the rocks both below and above the unconformity were then thrust eastward as a unit in an orogenic time later than the Carboniferous. The latter is suggested because there are also thrusts in central Nevada that cut through rocks of the Triassic and Jurassic, younger than the Carboniferous. But all faults need not be of one orogeny; there could have been a second time of thrust faulting, just as thrusts in western New England seem Taconian and Acadian.

The conglomerates of the Carboniferous in Nevada and Idaho have a few pebbles of limestones, but the fragments are almost wholly of cherts and quartzites such as are limited to the thrust sheets. They seem to have been laid near ranges raised along the eastern limit of the cherty rocks for some of the conglomerates are quite coarse. The Ordovician carbonates of the miogeosynclinal belt extend fifty miles west of the front of the overthrust cherty rocks (Fig. 8–14), so the Carboniferous conglomerates that lie on the miogeosynclinal rocks are

Fig. 13–7. Carboniferous conglomerate near Purgatory along Narragansett Bay, Rhode Island. The pebbles in the conglomerate have been deformed and elongated, as can be seen in comparing sections in three dimensions near the hammer; the pebbles are of quartzite in this locality. Thousands of feet of Pennsylvanian and Permian stream-laid conglomerates and sandstones were deposited in subsiding troughs within the earlier Paleozoic crystalline rocks; they were deformed, probably in the Permian Period. Some very metamorphic coals are so high in their proportion of carbon that they are difficult to burn.

more than fifty miles east of the original site of deposition of the cherty rocks, and probably as much as one hundred miles.

In the analysis, a reasonable manner of movement is by gliding of upper sediments from the crest of an axis raised to the west (Fig. 13–8*B*). There should then be an axis of old rocks to the west. As western Nevada has late Paleozoic and Mesozoic rocks of the surface, the hypothesis is that the thrust sheets moved from a belt of raised Cambrian basement that then subsided deeply during latest Paleozoic and early Mesozoic time. In this case the thrust sheets moved to their present positions during an orogeny (Antlerian) prior to the deposition of the early Carboniferous conglomerates. There were phases of deformation evidenced by coarse detritus of several ages along the margin of the Antlerian orogenic belt, shown also by unconformities within the Carboniferous (Fig. 13–11). Later orogeny formed thrusts that cut the Carboniferous.

OROGENY IN THE SOUTH CENTRAL STATES

The history was also complex in the geosynclinal region south of the Ozark Mountains in Arkansas and Oklahoma but is well-known from the presence of oil-bearing strata that have been studied in exposures and in wells. Osagian limestones in northern Arkansas and northeastern Oklahoma are overlain by argillaceous shales and sandstones that are believed Meramecian and Chesterian. The marine Pennsylvanian section of the south central United States has several divisions, the

Fig. 13–8. Restored section of Paleozoic rocks of central Nevada soon after their overthrust faulting, under two interpretations of the time of thrusting; see Fig. 8–14 for stratigraphic relations. *A*. Structure section interpreting the Carboniferous rocks as having been laid far to the west, then thrust to their present positions in a later orogeny. The interpretation either requires that the rocks be thrust upward as well as laterally, or that the originally eastward descending thrust was deeply depressed on the west prior to the time of the section; the latter hypothesis, that the original plane was eastward descending, would require that the top of the section have very great elevation.

** *B*. An alternative hypothesis that the structure formed prior to the Carboniferous, the section representing conditions just prior to Carboniferous sedimentation, the thrust plane being one on which the allochthon, the chert-slate-volcanic facies of the Ordovician on the west, has been carried by gravity gliding over the carbonate facies of the lower and middle Paleozoic on the east.**

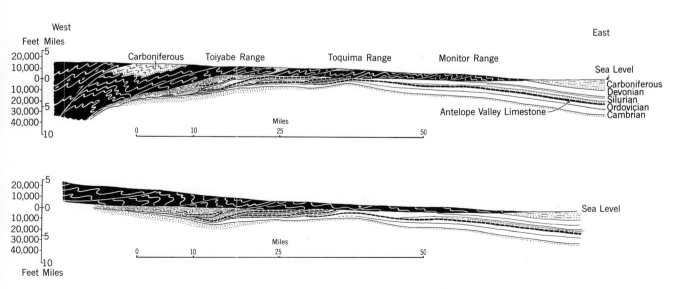

271

Springeran, Morrowan, Lampasan or Atokan, Desmoinesian, Missourian and Virgilian; the succeeding Wolfcampian Series, like the Dunkard, is Permian. The expression lower Pennsylvanian will be applied to the Springeran and Morrowan, and middle and upper Pennsylvanian to the succeeding pairs of series.

The late Mississippian (Fig. 13–12) and lower Pennsylvanian sediments are several miles of deposits derived from the erosion of islands to the south, exposed best in the Ouachita Mountains of Oklahoma and Arkansas and laid in a geosyncline subsiding most deeply in an arc passing through southern Arkansas, southeastern Oklahoma and beneath the coastal plain of eastern and south western Texas (Fig. 13–13). The slope of the sea floor and direction of currents are gained from studies of sole-markings (Fig. 13–14). The rocks overlapped northward as seas spread nearly to Missouri and southeastern Kansas; a rather narrow subsiding belt trended northwestward from eastern Oklahoma through southwestern Kansas—the Arbuckle-Ardmore geosyncline. The younger sediments in the Northern Ouachita Mountains of southeastern Oklahoma have great limestone boulders of the early Paleozoic cratonal facies. These have been attributed to submarine slumping. In their position along the margin of a deeply subsiding terrigenous-sediment bearing geosyncline they have similarity to the conglomerates described in the Ordovician of Levis, Quebec and western Newfoundland. The early Pennsylvanian then had a very deeply subsiding non-volcanic arcuate geosyncline, a miogeosyncline, to which marine waters were limited in the earliest epoch, and which had fluvial plains in its southern part gaining detritus from newly risen highlands in the earlier geosynclinal belt to the southward. The seas spread cratonward, and deformation along the cratonal margin of the geosyncline released boulders that slid into the argillaceous muds of the geosyncline; some were of rocks considerably older than the enclosing early Pennsylvanian.

RISE OF THE OUACHITA GEOSYNCLINAL BELT

In the medial Pennsylvanian (Lampasan and Desmoinesian), the area of the earlier geosyncline became covered by coarse detritus gained from mountains advancing northward into the geosyncline. The Lampasan sediments are restricted to the northern part of the early Pennsylvanian areas of deposition extending into embayments northward in southern Kansas and northwestward into Texas (Fig. 13–15). The coarser detritus contains cherty pebbles such as might have come from the Ordovician rocks of the graptolitic facies of the Ouachita Mountains, and there are some volcanic rocks. The rocks of the eugeosynclinal

Fig. 13–9. Carboniferous (late Pennsylvanian) sandy limestone on the right lying unconformably on Ordovician chert along Mill Canyon, Toquima Range, central Nevada (M. Kay and J. P. Crawford). The unconformity and the associated beds have been further deformed in later orogeny, so that the younger beds now dip steeply to the right or south. The unconformity is the direct evidence of the Antlerian Orogeny, which is shown here to be older than late Pennsylvanian; elsewhere, early (Mississippian) Carboniferous rocks are unconformable on pre-Carboniferous.

Fig. 13–10. Carboniferous conglomerate at Carlin Canyon, west of Elko, Nevada. The conglomerates, of late Mississippian or early Pennsylvanian age, contain rounded pebbles and cobbles derived from older Paleozoic rocks folded and elevated in an early phase of the Antlerian Orogeny that affected the margin of the volcanic geosyncline belt in several phases in the late Paleozoic. The conglomerates were laid from streams that reached the sea along a shifting shore extending northeastward through central Nevada into Idaho. The conglomerate bed in the view is one of those truncated by an unconformity of a later orogenic phase that is succeeded by latest Pennsylvanian and Permian limestones nearby, shown in Fig. 13–11. Similar conglomerates appear in several areas along the shore belt, the deposits ranging from late Mississippian into Permian; they resulted from a succession of uplifts of cherts, argillites, and volcanic rocks in the orogenic belt to the west.

belt to the south, now raised and deformed, were the source of the detritus. At about this time, normal faults broke the crust along bands extending in from the margin of the craton, forming a series of fault blocks and scarps that were somewhat eroded before they were buried beneath the deposits of a Desmoinesian sea that spread far inland—to the northern limit of preservation of the system in Iowa and Nebraska.

In the later Pennsylvanian, Missourian deposition was followed by strong folding in the geosynclinal branch that passed through the Arbuckle Mountain region of Oklahoma; later Pennsylvanian Virgilian non-marine sediments unconformably overlie the eroded and folded Missourian sediments in the Arbuckle Range (Fig. 13–16). By Virgilian, deposition had ceased in the region of the earliest Pennsylvanian geosyncline, which had become a source land; and the principal deposition was in areas wholly northwest of the site of early Pennsylvanian subsidence and deposition.

OUACHITA THRUSTS

At some undetermined stage the rocks of the southern part of the geosynclinal belt were driven northward scores of miles on great thrusts which crop out along the northern and western margins of the Ouachita Mountains and continue in the subsurface beneath Texas. The movement was later than Lampasan, the youngest Pennsylvanian in the thrust sheets, and was older than the Jurassic that overlies the eroded southern margin of the thrust sheets. As the fold system of the

Fig. 13–11. *Unconformity within the late Carboniferous in northern Nevada.* A. Exposure on United States Highway 40 at Carlin Canyon of the Humboldt River west of Elko, Nevada. Late Mississippian and possibly early Pennsylvanian conglomerates (Fig. 13–10) and shales were folded and eroded in a late phase of the Antlerian Orogeny before the deposition of the latest Pennsylvanian and Permian limestones and siltstones that overlie the unconformity.

B (right). Restored section of the relations in late Paleozoic (after R. H. Dott, Jr.), showing the deposition of Carboniferous sediments in troughs separated by source lands. Some of the thrust faults in the region are post-Carboniferous and perhaps much younger. Other thrusts may have formed in an early phase of the Antlerian Orogeny prior to deposition of the Carboniferous conglomerates, which seem to lie both on allochthon and autochthon, that is both on transposed and indigenous sequences (Fig. 13–8).

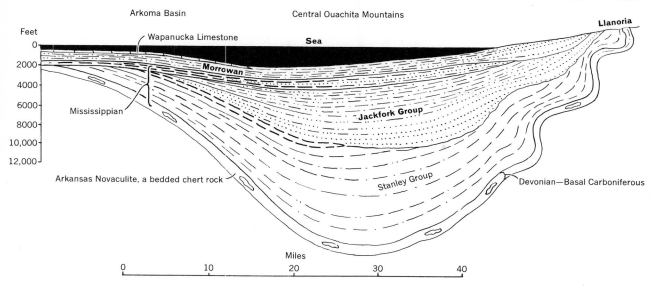

North   Arkansas-Oklahoma   Louisiana-Texas   South

Arkoma Basin   Central Ouachita Mountains

Wapanucka Limestone   Sea   Llanoria

Feet

Morrowan

Mississippian

Jackfork Group

Arkansas Novaculite, a bedded chert rock

Stanley Group   Devonian—Basal Carboniferous

Miles

Fig. 13–12. Restored section of the lower Carboniferous (Mississippian) sedimentary rocks in the Ouachita Mountains and Arkoma Basin in early Pennsyl- vanian time (L. M. Cline). The Ouachita Geosynoline received terrigenous clastic sediments from the source lands, tectonic lands raised along the southern margin of the geosynclinal belt; primary structures show that the trough sloped southward in the Arkoma basin and the axis westward in the Ouachita Mountains.

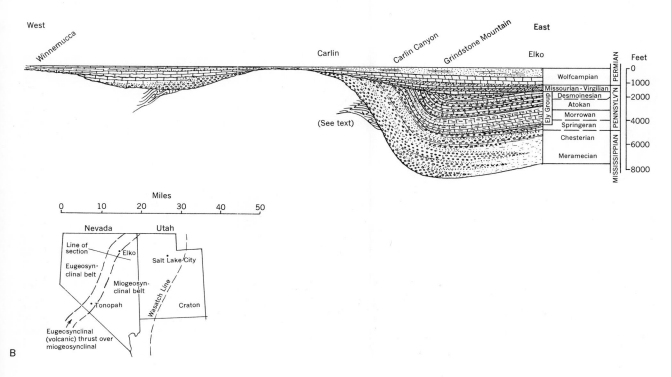

West   East

Winnemucca   Carlin   Carlin Canyon   Grindstone Mountain   Elko

		Feet
PERMIAN	Wolfcampian	0
	Missourian - Virgilian	1000
Ely Group / PENNSYLV'N	Desmoinesian	2000
	Atokan	3000
	Morrowan	4000
	Springeran	
MISSISSIPPIAN	Chesterian	6000
	Meramecian	8000

(See text)

Miles

Nevada   Utah

Line of section

Elko

Eugeosyn-clinal belt

Miogeosyn-clinal belt

Salt Lake City

Tonopah

Wasatch Line

Craton

Eugeosynclinal (volcanic) thrust over miogeosynclinal

B

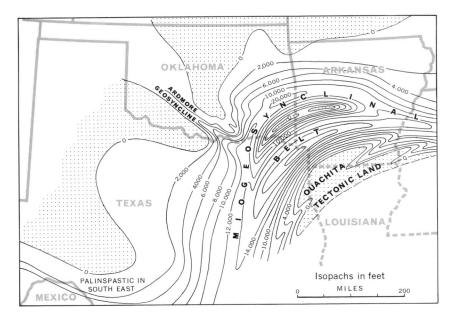

Fig. 13–13 (left). Isopach map of the lower Pennsylvanian (Springeran and Morrowan Series) in the south-central states (after C. W. Tomlinson in M. Kay). The map is quite hypothetical in the southeast, for Carboniferous rocks have been buried or eroded away in eastern Texas and southern Arkansas; moreover, the rocks in southeastern Oklahoma and adjacent Arkansas are in thrust sheets that have displaced northwardly in time later than that of the map, which is drawn palinspastically, showing their relations as interpreted prior to thrusting.

Fig. 13–14. *The flow of currents and bottom slopes in Carboniferous time in eastern Oklahoma shown in the sedimentary structures in the Arkoma Basin and the Ouachita Mountains. A.* Sole-markings in the bottom surfaces of sandstone blocks used in the wall of a building in Kosoma, Oklahoma. The block in the upper center, about a foot square, shows flute casts, the bottom of a sand bed having filled the depressions eroded by submarine currents in the top of an unconsolidated clay bed previously laid; the direction of flow, from upper left to lower right, is shown by the shallowing flare from the upstream deeper pit in the original clay. Other blocks show bounce-casts, the marks made by fragments skipping down slope, and linear grooves formed by sliding blocks. Such data when measured in the beds in natural exposures give information on original bottom slopes and current flow (L. M. Cline).

B. Map showing the flow in northern Arkoma Basin determined from the direction of slope of cross-bedding in the lower Pennsylvanian rocks—southward (after F. P. Agterberg and G. Briggs). Such current flow need not have been directly down slope as would be the case in the gravity-directed lineations produced by sliding or bouncing blocks.

C. Diagram showing the directions of flow and slope from sole-markings in upper Mississippian sediments in the Ouachita Mountains to the south of the Arkoma Basin, westward down the floor of the submarine trough. Tectonic lands that furnished the detrital sediments lay to the southeast. The data are for the beds that furnished the blocks in the wall in Fig. 13–14*A* (after O. B. Shelburne). Percentages pertain to readings within 10 degrees of each direction.

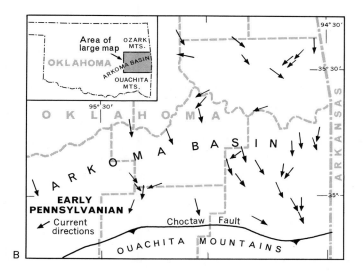

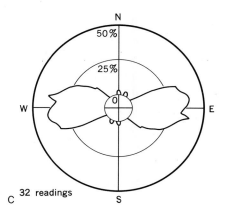

Arbuckle Mountains trends toward the thrusts and seem to deform them, they seem to be pre-Missourian, the time of deformation in the Arbuckle geosyncline. They may be later Paleozoic or early Mesozoic, however.

The Ouachita geosynclinal belt extends southwestward and westward into western Texas. There, later Virgilian sediments were folded, thrust-faulted and eroded prior to the deposition of overlying limestones of the Permian Wolfcampian. Thus, to the south of the craton, important orogenies in Oklahoma and west Texas can be dated by rocks associated with unconformities as post-Missourian and post-Virgilian respectively. Other movements are shown as in the incursion of detritus from lands to the south in early Desmoinesian. And some of the more profound, such as the Ouachita thrusting, cannot be placed as directly in time. The geosynclinal and orogenic belt passes northward to a quite stable region in the western interior states (Fig. 13–17); stratigraphy in this area will be discussed in Chapter 14.

The Pennsylvanian System records orogenies at several times within the period and in many regions. Each folding and rise of lands cannot have taken more than a part of the period. The span of a period is very long, for not only were rocks folded, but they were eroded to such a low plain that in several instances Carboniferous limestones lie unconformably on older Carboniferous folded sediments. The orogenies that have been discussed are restricted to belts that had subsided deeply along the continental borders, the geosynclines.

DEFORMATION WITHIN THE CRATON

The Pennsylvanian also presents evidence of profound deformation within the previously stable craton. In Colorado and New Mexico there is a peculiar combination of deeply subsiding troughs with rising lands of a different sort. Through the pre-Pennsylvanian Paleozoic, Colorado and northern New Mexico had been in the western part of a rather stable region, for the region has only hundreds of feet of marine sediments representing but fragments of the span of time. Thus in central Colorado, the section is of a few hundred feet each of Cambrian, Ordovician, Devonian, and Mississippian (Fig. 13–18). On the other hand, the Pennsylvanian has thickness of two miles or more in central Colorado. The earlier Pennsylvanian there is of carbonate rocks and some argillites laid to greater thickness than the sediments of the preceding systems, in subsiding basins within the craton. Mid-Pennsylvanian comes to contain an increasing proportion of coarse, commonly red detrital sediments, most of them stream laid. Highlands rose in a belt extending from northeastern Utah into northern New

Fig. 13–15. A succession of paleogeographic maps of the Pennsylvanian in southern Oklahoma, showing the profound changes in the geography through the later Carboniferous Period (after C. W. Tomlinson and W. McBee, Jr.).

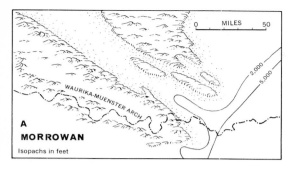

MILES
0   50

A
MORROWAN

Isopachs in feet

WAURIKA-MUENSTER ARCH

2,000
5,000

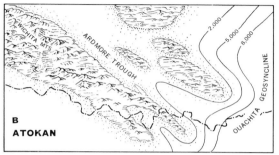

B
ATOKAN

WICHITA MTS
ARDMORE TROUGH
OUACHITA GEOSYNCLINE

2,000
5,000
8,000

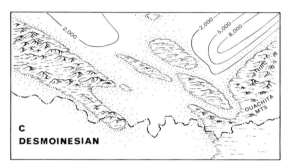

C
DESMOINESIAN

2,000
2,000
5,000
8,000

OUACHITA MTS

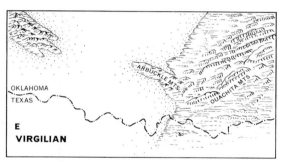

E
VIRGILIAN

OKLAHOMA
TEXAS

ARBUCKLE MTS
OUACHITA MTS

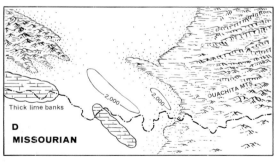

D
MISSOURIAN

Thick lime banks
2,000
2,000
OUACHITA MTS

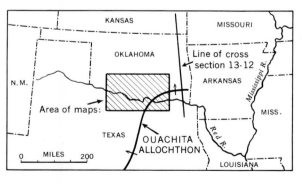

KANSAS   MISSOURI

OKLAHOMA   Line of cross
                         section 13-12

N.M.   ARKANSAS

Area of maps: MISS.

TEXAS   OUACHITA
             ALLOCHTHON

MILES   Red R.
0   200
                         LOUISIANA

279

Mexico, the Uncompaghra land, having deep subsiding troughs on each side (Fig. 13–19). Such intracratonal troughs, having complementing intracratonal source lands, have been called zeugogeosynclines, that is, yoked geosynclines in reference to their having subsiding areas associated with rising lands within the craton. That on the east was bordered by another north-trending highland on the east, stream plains separating it from Uncompaghra. To the east of Frontrangia, coarse sediments rapidly disappeared in the subsiding trough that encroached on the east on a little-subsiding Central Kansas Platform (Fig. 13–20).

On the west of Uncompaghra in Utah, sediments of older systems thickened rapidly along a flexure separating the craton from the miogeosynclinal belt. But Pennsylvanian sediments are quite thick in a trough within the craton in southwestern Colorado, passing northwestward into Utah—the Paradox trough that has a little-subsiding platform to the southwest, at times having reef structures and shoals. With high evaporation, this trough became another salt basin, similar to those we have described previously in the Michigan region. But in this case, the trough is complemented on the east by the Uncompaghra

A

Fig. 13–16. Air photographs of rocks folded in Carboniferous orogenies in Oklahoma. *A.* South flank of the Arbuckle Mountains ten miles north of Ardmore, Oklahoma; the area shown is about four miles wide. The beds are standing nearly vertically, and facing southward, so the oldest rocks, lower Ordovician (Arbuckle Group) are the light-colored bands in the north, principally limestones; middle Ordovician in a broad wooded dark band of limestones, sandstones and shales (Simpson Group) and a narrower light band of (Viola) limestone is succeeded southward by a narrow wooded band of upper Ordovician (Sylvan) shale and Silurian and Devonian (Hunton Group); Carboniferous rocks crop out in the extreme south. As the rocks stand vertically, thicknesses can be judged from the breadths of outcrops; the folding was late Pennsylvanian (Arbuckle Orogeny).

B. Western end of a syncline in upper Carboniferous (Jackfork) sandstones and shales in the Ouachita Mountains, southeastern Oklahoma, rocks folded in probably a medial Pennsylvanian Orogeny. The scale is like that in the preceding photograph.

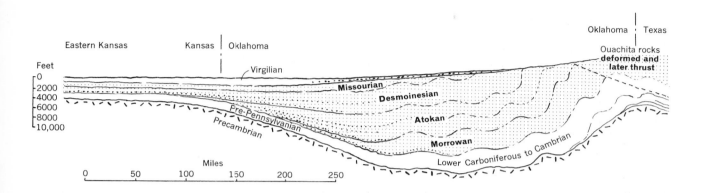

Eastern Kansas   Kansas | Oklahoma   Oklahoma | Texas

Ouachita rocks
deformed and
later thrust

Feet

Virgilian

Missourian

Desmoinesian

Atokan

Pre-Pennsylvanian

Precambrian

Morrowan

Lower Carboniferous to Cambrian

0
-2000
-4000
-6000
-8000
-10,000

Miles

0   50   100   150   200   250

B

Fig. 13–17 (above). Restored section of
the upper Carboniferous Pennsylvanian
Subsystem from Nebraska and Kansas to
Oklahoma and the Gulf Coast. The
Ouachita Geosyncline gained great thicknesses of Carboniferous sediments derived from tectonic lands to the south,
and deformed by orogenies within and
subsequent to the period. The older series of the Pennsylvanian are limited to
Oklahoma and southward; they are overlapped northward by the Desmoinesian
Series, which extends over much of the
continental interior. The Pennsylvanian
rocks of the western interior are relatively thin and have persisting lithic units
in cyclothems (see Fig. 13–22).

highland, so has coarse terrigenous sediments on that margin contrasting with the reefy limestones on the platform to the southwest. To the northwest in the region of Great Salt Lake is Pennsylvanian accumulated to a thickness of perhaps four miles (Fig. 13–21); although this is in the margin of the previously subsiding miogeosynclinal belt, the subsidence is not in the form of a belt as in earlier periods but is an elliptical depressed area, basin-like.

PROBLEMS OF CAUSES

These relationships emphasize the problems that remain in understanding the behavior of the continent. A region that had stood at rather constant level, with little sinking, through much of the Paleozoic, for some reason became an epeirogenically active one in the late Pennsylvanian. There is a record during the Silurian in Michigan of strong subsidence in previously rather inactive crust. In the Carboniferous of Colorado, thick sequences of coarse sediments accumulated in long elliptical troughs that subsided as much as two miles during the Carboniferous and Permian. But in this case, there must have been complementing rise on their margins of highlands of Precambrian rocks of the previously stable craton. These events contradict the general experience that great differential movements are restricted to the linear geosynclinal belts off the margins of the craton. The deformation in Colorado is epeirogenic, not orogenic—the rocks were differentially warped, not strongly folded and moved laterally. And when one draws a section to scale, the bending is not as great as one imagines when he sees the sections with the vertical scale greatly exaggerated.

Let us next consider some other relations that may possibly be dependent on the orogenies. We have discussed the cyclothems of the Pennsylvanian, and noted that in some regions, they are separated by erosion surfaces. These have been attributed to fall of sea level, perhaps related to continental glaciation in the southern hemisphere. But the Pennsylvanian of the interior lies with very great disconformity on

Fig. 13–18. Restored section of the Carboniferous sedimentary rocks from Colorado to Kansas. Source lands rose in central Colorado, supplying terrigenous sediments to adjoining subsiding troughs in the previously stable cratonal interior. Eastward, carbonate and saline rocks accumulated in the basin to the west of the stable to slightly subsiding Central Kansas Platform.

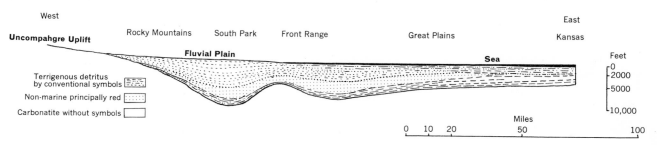

Fig. 13–19. Isopach map of the Carboniferous and Permian systems in Colorado and northern New Mexico (after C. B. Read and G. H. Wood, and M. Kay). The region, which had been quite stable through early Carboniferous time, developed great differential elevations between raised highlands and subsiding troughs, the sediments eroded from the former filling the latter so that much of the section is non-marine. In rate of subsidence and thickness of sediment, the troughs are comparable to some geosynclines; though called zeugogeosynclines, they have few of the characteristics of the orthogeosynclines along the continental margins.

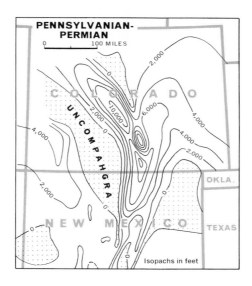

Fig. 13–20. *Upper Carboniferous, Pennsylvanian red beds in Colorado (J. H. Rathbone, from United States Geological Survey).* A. Pennsylvanian Fountain red arkosic sandstone lying at the extreme left on Precambrian crystalline rocks along the Denver and Rio Grande Western Railroad near Plainview, northwest of Denver. The prominent east-dipping ledge on the right is Lyons Sandstone of Permian age. The rocks were tilted in the Laramian Orogeny in late Mesozoic time.

B. Pennsylvanian red sandstones of the Maroon Formation in Maroon Peak (14,158 feet) southwest of Aspen, Pitkin County, central Colorado; the view is to the southwestward. The sediments were laid in a long narrow basin that gained detritus from rising lands on each side.

A

B

283

the Mississippian and older rocks; we have seen that northwestward from West Virginia, and northward from the Ouachitas, earlier Pennsylvanian sediments disappear and are overlapped by younger series. In the western interior, as in Kansas, the cyclothems (Fig. 13–22) have little evidence of erosion between them. But erosional unconformities of greater magnitude separate the Desmoinesian, Missourian, Virgilian, and lower Permian Wolfcampian series in several places. If these are due to marine withdrawal, the sea level must have fallen to considerably lower level at these longer time intervals, permitting erosion of valleys into preceding deposits. There must then be some additional cause of withdrawal. The hypothesis has been advanced that they took place at times of orogeny.

Perhaps the orogenies that have been recorded in two regions in pre-Pennsylvanian, pre-Virgilian and pre-Wolfcampian have something to do with this withdrawal. The evidence is compatible with the hypothesis. But it does not prove the hypothesis. And in view of the many structural events that must have taken place in other continents as well as our own, there will not be any difficulty in finding an event that is about synchronous with any withdrawal. It is hard to conceive a mechanism by which an orogeny would sufficiently enlarge an ocean basin as to reduce sea level over the whole world by hundreds of feet, then return the level to the original after the orogenic climax. But possibly there are orogenies in many regions at intervals of time. It is a subject that requires very close correlations of events in time all over the earth in order that its truth be demonstrated or seem convincing; such correlations are not now practical on a refined scale. The Carboniferous reveals much that is known about the deformation in North America and that many problems await means of solution. The correlation in time of the rocks of the Carboniferous and Permian systems is made principally by means of fossils found in the strata, particularly through the identification of minute one-celled organisms, the foraminifera.

Fig. 13–21. Carboniferous section in Mount Timpanogos, 12,008 feet in elevation; view eastward from the lowland about 4500 feet south of Great Salt Lake (L. Hintze, Brigham Young University). The high peak and the mountains with foreground expose thousands of feet of folded Pennsylvanian limestones forming part of a section of nearly five miles of Carboniferous and Permian sediments. A few miles to the northeast, the equivalent beds thin to a few thousand feet. Timpanogos Cave in the limestones is a National Monument.

MICROFOSSILS IN CORRELATIONS

The foraminifera secrete a chambered skeleton, commonly calcareous, that has a great range of forms—some are straight, others coiled in a variety of ways. There were many kinds in the Paleozoic, but in the Pennsylvanian there arose a type that is peculiarly useful in the classification of the rocks, the fusulines, which varied somewhat in size and external forms but commonly were only a few millimeters long, spindle shaped, and about like a grain of wheat or oats (Fig. 13–23).

The fusulines varied systematically with time, and they are univer-

Fig. 13–22. Divisions of a Pennsylvanian cyclothem exposed west of Lawrence, Kansas; the beds are marine limestones and shales of differing environments of deposition (Kansas Geological Survey). The lower formation is the Lawrence Shale; the upper formation, the Oread Limestone, has a thicker limestone member succeeded by two thinner limestones with intervening shales.

sally distributed. They are so minute that they can be recovered from the cuttings from wells, permitting classification of rocks far below the ground in the *subsurface* as well as those exposed to view and touch. They are so complex that differences can be observed readily. The forms are coiled around an axis, the coils having differences in rapidity and form of enlargement, and in the structures of the shell layers, of folds in the shell walls, and in many other characters that are constant in forms of the same time and place, but change in forms of other levels and localities. Experience has shown that the differences are progressive, that in a succession of specimens from a long sequence of rocks, forms change systematically with time, and the same changes take place in successions in other places and regions. The fusuline forms seem to evolve. Thus it is judged that the changes are useful in recognizing spans of time, that the forms of the same sort in the progression in one place are of the time of those in another. Fusuline foraminifera are known only from the Pennsylvanian and Permian systems. Other types of foraminifera, some of them somewhat similar, are known in earlier, and particularly in later systems, but the fusulines are distinctive of the late Paleozoic. The dating within these systems is thus accomplished by the identification of these microscopic fossils by measurements of the external form, and internal features observed under the microscope in extremely thin sections cut through them in various orientations. Such study is known as micropaleontology, the study of fossils under the microscope.

Though fusulines are found quite universally, they do not persist in all kinds of rocks within each region. Beds that contain fusulines may not contain brachiopods or clams or cephalopods; each species has a range of environments that it will tolerate, so that some will live in conditions that exclude the others. In some instances, successions of lithologies have associated successions of fossils, which when repeated form other examples of cyclothems that we mentioned in discussing the Mississippian rocks. In Kansas and Nebraska, marine limestone with fusulines is in the midst of a succession having other forms in succession above and below until we reach silty and sandy plant-bearing sediment, either stream laid or deposited along the margin of the sea. Successive forms represent deeper and deeper water environments; the fusulines are thought to have inhabited waters of a hundred foot depth or so, perhaps living in the limy muds of the quiet sea floor and drifting into piles like sand grains as they died. Other environments support cephalopods, which include goniatites that are particularly useful in classification. Commonly they are found in somewhat limy argillites—muds laid in offshore quiet bottoms into which

Fig. 13–23. *The Foraminifera.* Fusulinid foraminifera from the Carboniferous (after J. J. Galloway). (1) *Endothyra* from the Mississippian of Indiana. (2–4) Sections cut along the length of the specimens, which are from the Pennsylvanian: (2) *Fusulinella* from Oklahoma; (3) *Fusulina* from Kansas; and *Triticites* from Nebraska. Enlargement about ten times.

Phylum Protozoa. Organisms of the phylum Protozoa are distinctive in having but one cell, with its nucleus and associated protoplasm. Many of us have known of the living *Amoeba* that can be seen under a lens in collections from fresh-water ponds. Such soft animals are not preserved in rocks. But several classes of protozoans have representatives that form shells that are preserved. The most important are the Foraminifera and the Radiolaria. Some classify these simple organisms and the comparable Protophyta, commonly placed in the plant kingdom, in a separate kingdom of organisms, the Protista; this practice is followed in the *Treatise on Invertebrate Paleontology.*

Foraminifera secrete shells that are surrounded by their soft protoplasm; the shells are composed of one of several substances, in complex and distinct arrangements. One of the families of the Foraminifera is that of the Fusulinidae, which developed and were abundant in the Carboniferous and Permian; they are very useful in distinguishing stages within

Fig. 13–24. Reconstruction of a late Carboniferous (Pennsylvanian) sea as in north-central Texas, west of Dallas. (The Smithsonian Institution). The view has many forms of animals, such as sponges, cup corals, brachiopods, crinoids, and echinoids.

those periods. Fusulinids are foraminifera having calcareous shells of spherical-to spindle-shape, commonly of a fraction of an inch in length, such that they can be seen readily if one is searching carefully; they resemble grains of wheat. Although fusulinids differ in their external appearance, more profound differences become recognizable when they are sectioned and examined under the microscope. The shells are tightly coiled, with longitudinal partitions within the walls representing the progressive outer chamber walls, having perforations through which the cell material could pass; these structures are evident in sections cut at right angles to the axis of coiling. In longitudinal sections, the chambers have wavy structures that result from the fact that in many types the septal walls are ribbed or plicate. The fusulinids can be divided into several subtypes or subfamilies on the basis of the wall structures, and these in turn have progressive developments that permit the recognition of ages of beds through the stages of development of the fossils.

Most foraminifera are smaller than fusulinids; they abound in many marine rocks of the late Paleozoic and succeeding eras. Because they are so small, they are not all destroyed in wells drilled for oil and are recovered in the cuttings. Experience has shown them to be useful in recognizing stratigraphic units in subsurface penetrations.

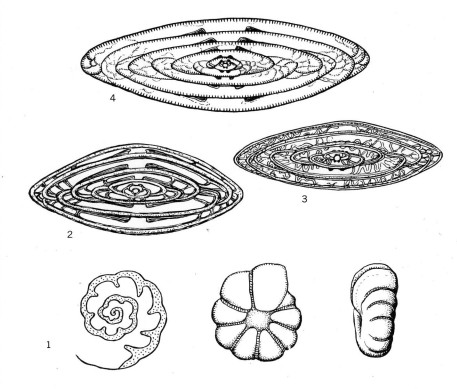

the fragile shells might drift and sink to be well preserved. The goniatites, like the fusulines, have complex structures that permit the recognition of significant differences in successive strata; they will be discussed more fully in Chapter 15.

The organisms found in the Pennsylvanian beds thus permit the classification of the rocks in widely scattered regions into several series and stages. In the fifty million years of duration of the period, it is possible to place strata with sufficient fossils into several divisions, each with a span of five or ten million years or so. Within more limited regions, persistent lithic units permit correlation of rocks in smaller spans of time, perhaps to limits of a hundred thousand years or less in some of the cyclothem sequences.

14
The Permian System and Continental Instability

Grand Canyon of the Colorado River (Union Pacific Railroad). Permian sedimentary rocks forming the upper wall of the Canyon in the distance are the lower Supai Shale, the cliff of Coconino Sandstone and the Toroweap and Kaibab limestones. The base is a disconformity at the broad platform, the Esplanade, on the lower Carboniferous Redwall Limestone. The view is from the east.

14 *The Permian System and Continental Instability*

Czar Nicholas I sponsored a geologic survey of European Russia in the late thirties of the nineteenth century. Roderick Murchison, who had gained a high reputation for his study of the Silurian, was employed with two distinguished colleagues to direct this undertaking. In the western flank of the Ural Mountains, he discovered a sequence of fossiliferous marine rocks overlying limestones containing Carboniferous fossils (Fig. 14–1). In 1841 he named them the Permian System, from the district having the city of Perm as its capital. The rocks of western Europe that seem of this age are non-marine, plant and vertebrate-bearing. The richly fossiliferous marine sequence in western Texas has many faunas similar to those in European Russia, and has come to be the standard of reference for North America. The Texas Permian has been divided into successive Wolfcampian, Leonardian, Guadalupian, and Ochoan series. Although the first is older than the base of the original Permian of Murchison, Wolfcampian and equivalent beds have come to be classified as Permian.

AMERICAN STANDARD SECTION

The Permian in the southwestern interior states shows the succession of events that led to the formation and development of carbonate rocks and of saline deposits perhaps better than any other on the continent because of the wealth of subsurface information available through wells that have been cored in this oil and gas producing area. In the early Permian, seas spread widely within the region (Fig. 14–2). To the south, lands that had risen with the Pennsylvanian orogenies in the Ouachita geosynclinal belt probably continued interruptedly from Arkansas through Oklahoma and eastern Texas. In the southern part of western Texas, rocks were folded, thrust-faulted, and raised to become eroded in a pre-Wolfcampian orogeny, so the early Permian lies on quite a complex paleogeology. On the northwest were the highlands elevated during the Pennsylvanian in about the position of the present Rocky Mountains in Colorado and northern New Mexico. Each of these lands had bordering coastal plains on which the meandering streams spread sediments eroded from the highlands, carrying the finer-texture particles to the interior sea. The waters spread northward into Kansas, Nebraska, and eastern Wyoming. Wolfcampian limestones bearing fusulinid foraminifera (Fig. 14–3), descendents of those in the Carboniferous, formed persisting marine deposits

Fig. 14–1. *Permian rocks in the Ural Mountains region, eastern Russia. A.* Shikan Tra-tau, an erosion remnant of reef limestone of the lower Permian near Sterlitamak, south of Ufa, Bashkiria, on the west slope of the southern Ural Mountains; the less resistant shales that flanked the original reef have been eroded away. Limestones accumulated on the eastern margin of a platform that approached a geosynclinal belt in the Ural region.

B. Lower Permian (Artinskian) sandy shales, with some conglomerates and distorted beds, along the railroad east of Kungur, near Perm on the west slope of the Urals. The terrigenous sediments of the geosyncline, derived from lands to the east in westernmost Siberia, are limited to the geosyncline, contrasting to the limestone facies of the platform to the west. In later Permian time, extensive salt deposits accumulated in the northern Urals, ultimately being buried beneath stream-laid sediments from which primitive reptiles have been recovered.

Fig. 14–2. Paleogeographic maps of the Permian in West Texas and southeastern New Mexico at four times within the period (after P. B. King). The maps are paleolithologic, showing the distribution of kinds of bottom sediments in the progressively restricted seas.

A

B

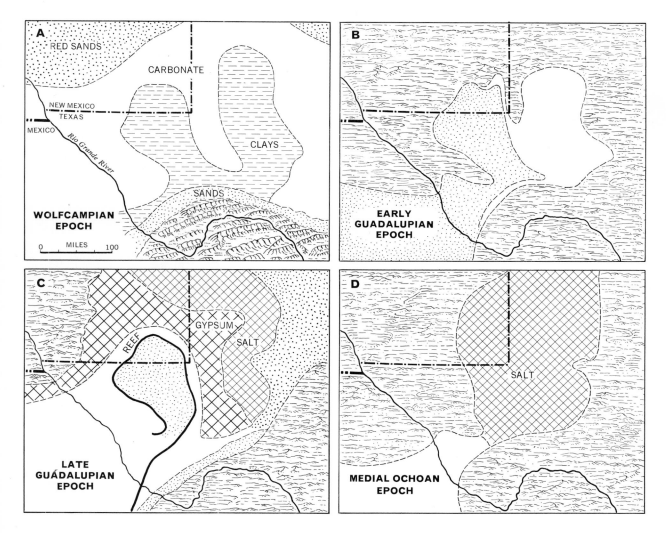

A — RED SANDS / CARBONATE / NEW MEXICO / TEXAS / MEXICO / Rio Grande River / CLAYS / SANDS / WOLFCAMPIAN EPOCH / MILES 0 100

B — EARLY GUADALUPIAN EPOCH

C — REEF / GYPSUM / SALT / LATE GUADALUPIAN EPOCH

D — SALT / MEDIAL OCHOAN EPOCH

that spread over much of the region, single lithic units of a few feet thickness continuing with little change along hundreds of miles of outcrop. The thin units comprise parts of cyclothems (Fig. 14–4) which ideally have lithologies recording alternate shoaling and deepening of the waters, with consequent changes in faunas. These pass into quartz sandstones and arkoses toward the lands. Where the subsidence was more rapid and water deepened more rapidly than deposits filled the subsiding basins, darker clays accumulated, and the present rocks are argillaceous limestones (Fig. 14–5).

Late Paleozoic sediments are not preserved eastward until one encounters the Wolfcampian Dunkard non-marine sediments in West Virginia, Ohio, and Pennsylvania, so the eastern limit of Wolfcampian seas is not directly recorded. The great inland sea was thus fairly surrounded by lands, though there must have been broad direct oceanic connections that permitted ingress of normal marine waters and their abundant organisms. A principal open seaway passed through western Texas and southeastern New Mexico into northern Mexico.

EFFECTS OF RESTRICTION OF MARINE INGRESS

Differential movements gradually restricted the connections of these interior seas. Arches rose across western Texas along the trend of the

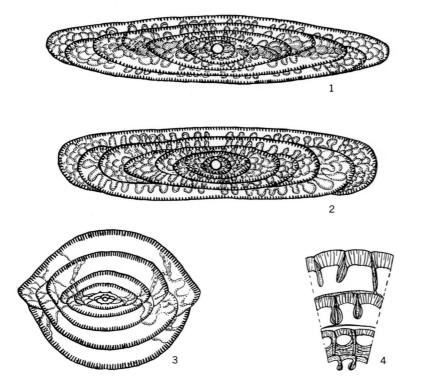

Fig. 14–3. *Fusulinid foraminifera from the Permian, to be compared with those in Fig. 13–23; enlarged about ten times* [(1), (2), (3), and (4) after J. J. Galloway]. (1) Diagram of types of shell walls to show the complexity in details that can differ significantly. (2) *Schwagerina*, characteristic of lower Permian. (3) *Parafusulina* of middle Permian. (4) *Polydiexodina*, of upper Permian; the specimens are from Texas. (5) Fusulinid foraminifera of the genus *Parafusulina* from the Permian Word Limestone northeast of Marathon, western Texas, about natural size (The Smithsonian Institution.)

present Arbuckle and Wichita Mountains of Oklahoma, so that the normal marine waters to the southwest were inhibited from flowing freely northward into the northern interior. High evaporation there in a region of arid climate reduced the level of the sea, causing general inflow. The commonest rocks are red and buff sands and silts that have interbeds of dolomite and of gypsum. If there had been abundant organic matter, such as plants on land or animals in the sea, the iron oxide would have been reduced to oxides of drab gray shades. Beds of rock salt reach such thickness in southern Kansas that they are mined as an important source (Fig. 14–6). Gypsum is of greatest economic

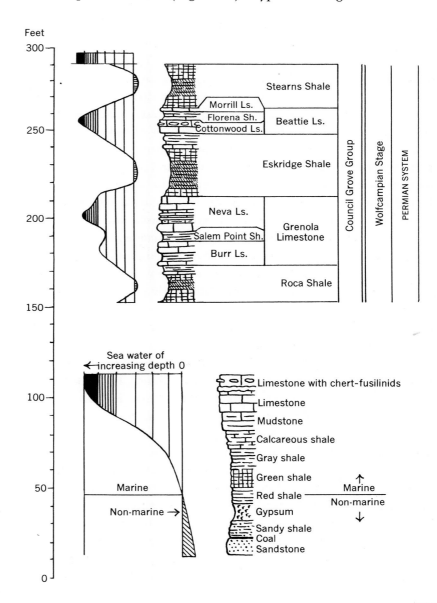

Fig. 14–4. A cyclothem in the lower Permian in Kansas, contrasting with those in the Pennsylvanian of eastern states in the dominance of marine beds. The differing strata are attributed to the effects of differing depths on the bottom sediments and the organisms that they supported.

importance in central Iowa, though widely distributed through the western interior. As has been stated in discussing older saline deposits (Fig. 10–22), sea water of normal composition must be evaporated to about one-tenth its original volume before salt and gypsum precipitate, so the saline rocks represent strongly concentrated marine brines, so salty that marine animals could not inhabit them. Thus it is not surprising that the oxides of iron were not reduced, and that the beds tend to be red.

REEFS AND FACIES

In the later Permian Guadalupian Epoch, further restrictions to the ingress of marine waters developed in southeastern Mexico and adjacent Texas, as intervening areas subsided rapidly (Fig. 14–7). Reefs grew on the lesser subsiding arches while deeper water sediments accumulated in the more rapidly sinking basins. The Permian reef of the Guadalupe Mountains on the southern border of southwestern New Mexico is spectacular (Fig. 14–8). The mountain front is of massive carbonate rock built of the skeletons of organisms. In front of the reefs to the south, water had a depth of nearly two thousand feet, occupying much the same position as the present lowland by the range. Waves striking the reef broke off masses and fragments that formed a slope of detritus passing into the deeper water muds at the base. And from time to time, muds were set loose, perhaps by earth tremors, to form a dense mixture of mud and rock fragments that flowed rapidly into the low basin and spread over the nearer part of its bottom. Behind the reef the water was shallow. Saline rocks, particularly gypsum and dolomite or dolomitized limestone accumulated in the evaporating waters, passing northwesterly into the plains laid by streams issuing from the mountains of the northern New Mexico and Colorado highlands. At times of low water levels, some of the terrigenous sediments swept through the reef front into the deeper water below. To the southwest, waters beyond the reef were open to the oceans, and accumulated richly fossiliferous calcareous muds and sands. Possibly a volcanic island belt lay to the south, for the sequence of a mile or two of Permian in southern Coahuila, Mexico, contains hundreds of feet of lava flows in tongues in marine sediments bearing typical Permian cephalopods.

Other reefs grew toward the northeast into western Texas on structurally higher areas that subsided less than their surroundings. Beyond, in Oklahoma, Kansas, and Nebraska, the restricted circulation and high evaporation discouraged the growth of organic structures and the strata pass into dolomite and gypsum-bearing red silts and sands.

The reefs so restricted entrance of marine waters from the southwest

Fig. 14–5. Restored section of a small time-stratigraphic unit, the Cottonwood Limestone, in the early Permian for hundreds of miles from northern Kansas to Oklahoma (after L. Laporte). The changing proportions of constituents in response to water depth and sources of terrigenous sediment are shown. The varying conditions affecting the life, the ecology, also had strong influence on the organisms; the study of ancient life associations is *paleoecology*. Note the great vertical exaggeration.

Fig. 14–6. Mine in Permian (Wellington) salt, Lyons, central Kansas (Kansas Geological Survey). The laminations are thin argillaceous shale partings, layers of clay that interrupted the precipitation of the sodium chloride. Such salt is the product of the evaporation to less than ten per cent of its original volume of normal marine sea water; see Fig. 10–22. Kansas is one of the leading producers of rock salt.

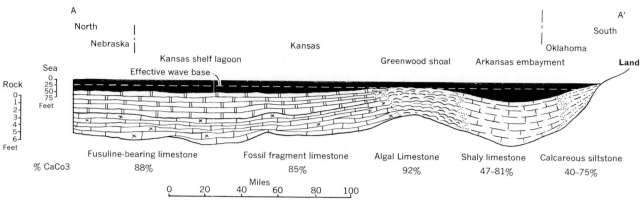

A North
Nebraska
Kansas
Greenwood shoal   Arkansas embayment
A'
South
Oklahoma
Land

Kansas shelf lagoon
Effective wave base

Sea
Rock
0
25
50
75
Feet

Rock
0
1
2
3
4
5
6
Feet

% CaCo3

Fusuline-bearing limestone
88%

Fossil fragment limestone
85%

Algal Limestone
92%

Shaly limestone
47–81%

Calcareous siltstone
40–75%

Miles
0   20   40   60   80   100

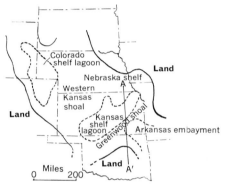

Colorado shelf lagoon
Nebraska shelf
Western Kansas shoal
Kansas shelf lagoon
Greenwood Shoal
Arkansas embayment
Land
Land
Land
A
A'

Miles
0   200

that in late Permian Ochoan time, thousands of feet of gypsum and rock salt accumulated in the basins that were periodically flooded and evaporated within the reefs (Fig. 14–2D). As normal sea water yields less than one-fiftieth the thickness of salts as the depth of the original fluid, the accumulation of thousands of feet demanded a basin gaining inflow of additional saline water as evaporation reduced its level, just as was the case with the Silurian in Michigan. Potash salts are present in limited areas. Marine water must be reduced to one per cent of the original fluid volume before such salts precipitate, and the quantity is but 2 per cent of that of all salts (Fig. 10–22). So the deposits are the last residue of the fluids that entered the basin, concentrated in the lowest part from the diminishing sea. They are succeeded by the stream and lake-laid red sands and silts of the latest Permian, these passing imperceptibly up into similar Triassic sedimentary rocks.

THE PERMIAN DELTA OF TEXAS

The inland sea of the early Permian was open to the south, as has been pointed out, by a seaway passing through what is now western Texas and eastern New Mexico. Low lands on either side of this seaway were inhabited by amphibians and reptiles in great numbers and of diverse aspect. Our knowledge of these early Permian land dwellers is based upon the many fossils that have been collected during the past century from the justly famous red beds of Texas and New Mexico, the sediments in Texas comprising the Wichita, Clear Fork, and Pease River groups of Wolfcampian and Leonardian age, the partially corresponding beds in New Mexico being those of the Abo and Cutler formations. The Texas sequence is particularly noteworthy, because it consists of a series of a dozen or so fossiliferous horizons, the faunas of which show various changes that may be correlated with changing environments, and the individual genera of which show the consequences of evolutionary development through time. The lower Permian continental sediments of Texas provide us with a remarkable case history of faunal development and animal evolution in one locality.

The Texas red beds were evidently deposited in a low delta, bordered on the west and the south by the seaway, mentioned above, and contiguous to a large land mass on the north and east. The rivers and streams that flowed across the land dropped their loads of sands and clays on the low delta, where there was such an abundance of life.

The climate of this part of the continent in early Permian time must have been moderately subtropical to tropical, with never throughout the year any severely cold temperatures. Such a climate was necessary for the continuation through so considerable a time span of great

Fig. 14–7. Restored section of the reef limestones and associated sedimentary rocks in the Permian of the Guadalupe Mountains, West Texas (after P. B. King). The beds had initial inclination from the shallow waters of the reef front to the deeper water of the bathymetric basin in front; the slope was sufficiently steep that some reef detritus slumped into the deeper botton. Fusulinid foraminifers and ammonite cephalopods tend to be dominantly in the reef and basin facies, respectively.

Fig. 14–8. Permian reef limestone in El Capitan (8058 feet), at the southern end of the Guadalupe Mountains, western Texas (J. Muench). The rocks in the foreground are the sandstones and shales that formed in the Delaware Basin that lay in front of the massive reefs, and that ultimately filled with the saline precipitates of later Permian.

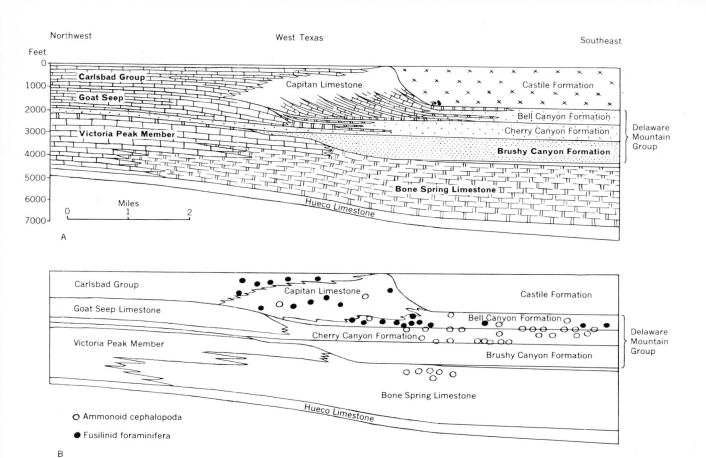

Northwest   West Texas   Southeast

Feet

Carlsbad Group

Capitan Limestone

Castile Formation

Goat Seep

Bell Canyon Formation

Victoria Peak Member

Cherry Canyon Formation

Brushy Canyon Formation

Delaware Mountain Group

Bone Spring Limestone

Hueco Limestone

Miles

0   1   2

A

Carlsbad Group

Capitan Limestone

Castile Formation

Goat Seep Limestone

Bell Canyon Formation

Victoria Peak Member

Cherry Canyon Formation

Brushy Canyon Formation

Delaware Mountain Group

Bone Spring Limestone

Hueco Limestone

○ Ammonoid cephalopoda

● Fusilinid foraminifera

B

297

numbers of large amphibians and reptiles—animals notoriously sensitive to adverse temperatures. The rainfall was sufficient to support a rather luxuriant cover of ferns, seed ferns, equisitales and conifers, under the benign cover of which many of the land dwellers sought protection from the direct rays of a hot sun. And the rain was sufficient to keep replenished the many streams and ponds in which lived numerous fish and hordes of amphibians (and even reptiles) that preyed upon these fish. This was an early faunal association of various continental vertebrates—fishes, amphibians, and reptiles—living within moderate variety of ecological environments.

ANIMALS AND THEIR HABITATS

The preceding sentence has been written with good reason, because it has been possible to distinguish the several environments on this delta from the nature of the sediments and of the fossils contained within these sediments. In essence, there would seem to be about four ecological habitats indicated in the Texas red beds, these being the uplands, the pond margins, the ponds and the stream. Of course, it must be remembered that these several habitats were all a part of the low delta, and that they graded into each other. The uplands were certainly not very high; rather they were the dry land between the streams and the ponds. And there was a gradual transition from upland to pond margin to pond or to stream. Moreover, it must be remembered that vertebrates are very mobile animals, and within the limitations of their own abilities they wander from one environment to another. Thus many of the amphibians and reptiles on the Permian delta frequently crossed from the upland to the margins of the water and into the water and back again, while the fishes, although in general confined to the water, were able in some cases to survive in the absence of water; these specifically were the lungfishes that burrowed in the mud during dry seasons, and their burrows have been preserved in some of the Texas red beds.

It may be recalled that the Carboniferous was a period in earth history when the amphibians and the reptiles were competing more or less on an equal footing, with perhaps the amphibians the dominant land animals. By Permian time, when the Texas red beds were being deposited, the balance had shifted very much in favor of the comparatively efficient reptiles even though some of the amphibians of this age were large and aggressive enough to have been direct competitors with some of the reptiles for the fishes and the other animals upon which all of them fed. This period was in fact the beginning of the Age of Reptiles.

Fig. 14–9. View from the South Rim of the Grand Canyon, showing the Permian section above the Carboniferous Redwall Limestone cliff in the distance (Union Pacific Railroad). The conspicuous angular unconformity is below the Cambrian Tapeats Sandstone, on the Precambrian Grand Canyon Group. A diagrammatic section is shown below. See also the chapter opening.

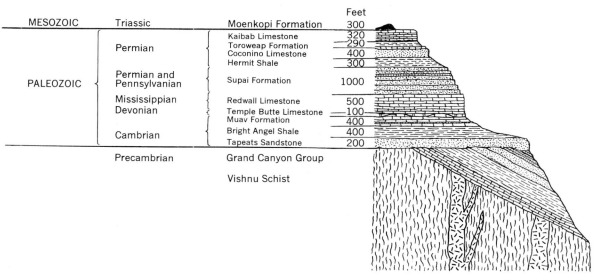

			Feet	
MESOZOIC	Triassic	Moenkopi Formation	300	
PALEOZOIC	Permian	Kaibab Limestone	320	
		Toroweap Formation	290	
		Coconino Limestone	400	
		Hermit Shale	300	
	Permian and Pennsylvanian	Supai Formation	1000	
	Mississippian	Redwall Limestone	500	
	Devonian	Temple Butte Limestone	100	
		Muav Formation	400	
	Cambrian	Bright Angel Shale	400	
		Tapeats Sandstone	200	
	Precambrian	Grand Canyon Group		
		Vishnu Schist		

On the uplands of the Permian delta were large reptiles that evidently fed upon plants, the *first order* consumers that obtained their energy directly from the green vegetation, which in turn had converted the kinetic energy of sunlight through photosynthesis into the potential chemical energy of organic molecules. These reptiles were the diadectids, and in the upper levels of the Texas sequence, the cotylorhynchids, the former as bulky as large swine (although of course with heavy, sprawling, clumsy legs), the latter equal in weight to small cattle. Then there were large carnivorous reptiles, the *second order* consumers, fin-backed dimetrodonts that preyed upon the large plant eaters, as well as upon other animals of the uplands. There were also smaller upland reptiles, especially the captorhinids that probably fed upon insects and other small game.

The pond margins were occupied by a mixture of amphibians and reptiles. The reptiles, particularly the grotesque, fin-backed edaphosaurs, fed in part upon plants and perhaps upon molluscs, while the long-jawed ophiacodonts were evidently fish eaters. But the principal predators upon fishes were the amphibians, the large eryopsids, five feet and more in length, and their small labyrinthodont relatives, many of which were only a foot or so long.

The water was peopled by various amphibians and by heavy-scaled bony fishes, by lungfishes, and by some fresh-water sharks, which lived abundantly in Permian streams and ponds.

Such was vertebrate life on the land during early Permian time. It was life of a varied sort, successfully lived by fishes, by numerous amphibians and by reptiles, which had by now become the dominant animals of dry lands where amphibians were ill-adapted for survival. The picture is nicely displayed by the successive faunas of the Texas sequence, and in looking at this particular faunal succession we get a reasonably full view of life on the land, all over the continent. For the elements that compose the faunas of the Permian delta in Texas are found elsewhere—in New Mexico, in Pennsylvania, and far to the north on Prince Edward Island. Indeed, closely related faunas are found in the continental Permian sediments of France—an indication that these early Permian amphibians and reptiles were widely distributed across the world in that distant age.

THE GRAND CANYON

The Grand Canyon of the Colorado River in Arizona (Fig. 14–9) has cliff-forming limestone at the rim, limestone that lies below the surface of the extensive Kaibab Plateau to the north. The rocks for two thousand feet below the plateau are Permian (Fig. 14–10). A steep cliff

Fig. 14–10. Canyon of the Little Colorado River, Arizona, just above its junction with the Colorado River (T. Nichols). The rocks in the bottom of the canyon are the Precambrian Grand Canyon Group, succeeded by a bench of Cambrian quartzite. The Permian extends from the light band of the Coconino sandstone to the top of the Kaibab Plateau on the skyline, composed of Permian limestones.

Fig. 14–11. Coconino Sandstone, a weathered surface of a bed, showing the footprints of a vertebrate animal (American Museum of Natural History). As the tracks generally are directed up the sloping cross-beds in the formation, and the cross-bedding is generally south dipping, the view should be one with the light from morning sun.

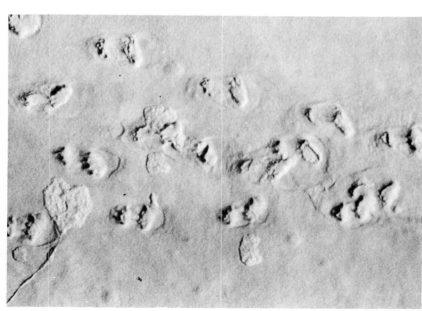

of Mississippian Redwall Limestone lies far below, the color being derived from the overlying broad terrace or "Esplanade" of Supai red sandstone and siltstones with an upper Hermit red shaly siltstone member; the Hermit has yielded plants, occasional insect wings, and footprints of vertebrate animals. The most conspicuous white cliff is of the Coconino sandstone, just below the terrace and cliff of the Toroweap and Kaibab formations that have marine Permian fossils somewhat like those in the Leonardian and Guadalupian rocks of western Texas.

FOOTPRINTS IN SANDSTONE

Trails of animals on the sloping surfaces of the laminae in the Coconino cross-bedded sandstone aroused curiosity about the animal that formed them, the conditions that might permit such fine preservation of the prints, and the origin of the sand (Fig. 14–11). The skeletons of the animals have not been found, but the tracks seem to be those of a primitive reptile, judged by analogy with known vertebrates in other areas. Similar trails have been seen in similar Permian sandstone as far away as the front of the Rocky Mountains north of Denver, Colorado. The Coconino is a well-sorted quartz sand such as might be found along a beach or in dunes along the shore near the sea. The rounded and frosted surfaces are like those shaped in air, though such grains might have been carried to the sea. As dunes are ephemeral, the forms moving as the sands drift, only the very basal layers such as form the lower part of the lee side of the dune will be preserved. The Coconino has footprints that always ascend the dipping cross-laminae. If we place footprints on the surface of a dune, they do not remain, because the dry sand slumps and rolls into the depressed surfaces, as all who have walked on loose dry sand can testify. Prints made on sands in a shallow sea are not sharply defined, since the grains do not retain the steep-sloping sides of the impressions, and so these are destroyed by the currents and waves. If a dune is wet, as after a shower, the imprint is retained, and may be buried by dust carried by the next wind.

The cross-laminae in the Coconino have raindrop impressions occasionally, the tracks of worms, and small ripples that frequently run down the slope of the laminae, as they may in modern dunes. So the Coconino seems to have been formed in dunes. Sand like the Coconino persists over a great area in Arizona and Utah. The cross-laminae have been recorded, and are found to dip preponderantly toward the south, as though that were the direction of the prevailing winds (Fig. 14–12). As wind directions are systematically arranged with respect to the earth's equator and poles, the data bear on the permanence

Fig. 14–12. Map of Arizona and states to the north. The directions of dune movements determined from orientation of cross-bedding in aeolian sandstones are shown; those directions in northeastern Arizona are from the Permian Coconino Sandstone (after N. Opdyke). The line extending northeastward from the Grand Canyon is the 5° N. Lat. meridian for late Pennsylvanian time as interpreted from studies of the paleomagnetism of rocks, a subject to be discussed in Chapter 19.

Fig. 14–14. Restored section of the Permian System late in the period from the miogeosyncline in Idaho and Utah across the western part of the craton. The Wyoming region formed a rather more stable part of the continent than the region of the Missouri River in Nebraska, where quite a full section of the Permian is present, the lower part of normal marine cyclothemic lithologies, the upper with dominance of restricted marine or non-marine sediments. Early Mesozoic erosion removed the upper part of the section toward the east.

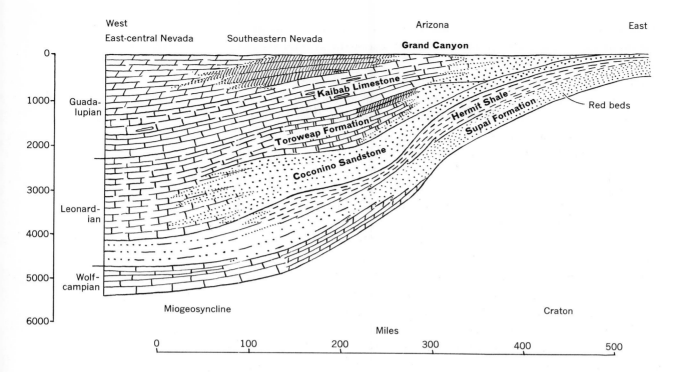

Fig. 14–13. Restored section of Permian sediments at the close of the period from Nevada to eastern Arizona (after E. D. McKee). Non-marine sediments on the cratonal margin on the southeast grade into marine sediments in the geosyncline to the west. Note that the Grand Canyon is in a belt of changing facies and westward thickening sections.

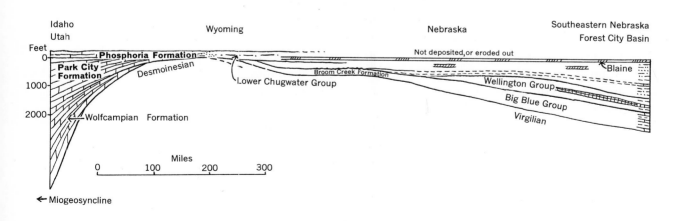

of those geographic references. Thus a sand formation such as the Coconino yields a great store of information about the conditions in the time of its deposition.

The beds below and above the Coconino are not the same throughout the region. Studies of the overlying Toroweap and Kaibab formations show them to change gradually from place to place (Fig. 14–13). As the lithologies change, there are associated changes in the contained organisms. Alternating sequences of lithologies, cyclothems, have red beds and gypsum in the deposits of lowest sea level, and fossiliferous limestones in those of highest water level. Studies of the faunas show that in the Kaibab, the earliest beds have fossils most like those of West Texas, suggesting direct marine connection, whereas the upper faunas are most like those of Idaho and Wyoming. The Toroweap passes into shore sands eastward within Arizona. Below the Coconino, the Supai and Hermit do not continue far to the west; evidently they too change, but into marine limestones, in a manner probably like changes in similar beds that can be demonstrated more readily in Idaho and Wyoming. Thus seemingly trivial details of the character of formations such as those in the Permian of the Grand Canyon lead to an understanding of many aspects of the conditions under which they were laid.

THE RED BEDS

The change from fossiliferous marine beds to red beds is better shown farther north (Fig. 14–14). The extreme western part of Wyoming and Idaho subsided more rapidly than the cratonal region to the east in earlier periods; this was the margin of the miogeosynclinal belt. The contrasts continued in the Permian, with marked change in conditions of sedimentations on the cratonal border. The Wolfcampian forms the upper part of a limestone sequence extending into eastern Wyoming from the interior states. Equivalent rocks appear sparingly in western Wyoming, becoming much thicker in the states to the west and southwest. A broad regional unconformity separated these Wolfcampian rocks from succeeding later Permian (Guadalupian) sediments, the latter lying on Pennsylvanian throughout central Wyoming. Later Permian sediments have been shown to grade from fossiliferous marine limestones on the west (Phosphoria) into preponderant red sandstones and shaly siltstones (lower Chugwater) in the east. Although the latter are red, they are probably marine laid, for they are generally plane bedded, single beds extending for long distances. There are lensing gypsiferous beds such as might have precipitated from restricted marine waters, and occasional heaps of clams and cephalopods of marine habitat, perhaps drifted from more favorable environments. Yet the environment must have been so unfavorable for normal marine life that the oxidized sediments supplied from adjoining lands, such as those in Colorado, were not reduced in the sea.

The conclusion that the limestones in the west correlate with the red beds in the east has been reached by plotting of carefully recorded sections as a succession of localities in mountain ranges across Wyoming. Some lithic units, such as quartz sandstones and limestones, persist from preponderantly limy sections in the west into the sections dominated by red sands and silts in the east. The rocks change gradually, limestone tongues, sections of geometric wedge-shaped bodies, gradually thinning eastward and losing marine fossils, some finally disappearing, while intervening red shale tongues thicken in complement. The correlation is essentially by means of sequences of lithologies in which new key beds are introduced in sequence as others disappear, the method much as we first discussed in the Precambrian of Minnesota.

PACIFIC VOLCANIC BELT

The stratigraphy of Paleozoic systems along the Pacific Coast has been referred to infrequently because they are very sparsely exposed. The Permian is the oldest system that is widely distributed, represented by many thousand feet of sediments and volcanic rocks in localities from western Alaska to northern California and Nevada (Fig. 14–15); four or five miles of submarine lava flows form the section in ranges in western Nevada. In central British Columbia, north of Prince George, thousands of feet of Permian contains a preponderance of fusuline-bearing limestones with interbedded lavas; conglomerates with granitic boulders are found in other parts of the province. The great surface extent of the Permian sediments is a matter of the structure of the Pacific border as well as the original prevalence of Permian rocks over a great area. The coastal belt generally has been one of great subsidence, older Paleozoic rocks being rarely exposed, because they have subsided to such depth that they seldom have been raised to the present surface. Moreover, as the deepest rocks, they have been the first destroyed by widespread intrusives in later periods. The Permian sediments and associated volcanic rocks are very similar to those in older systems that have less extensive outcrops; and similar rocks continue to prevail in the earlier Mesozoic Triassic and Jurassic systems, which were laid in deeply subsiding belts. From time to time, disturbances raised areas to erosion.

CLOSE OF AN ERA

The Permian Period closes the Paleozoic Era. The separation of the Paleozoic and Mesozoic was made by Sedgwick when he gave the "zoic" endings. The eras were based on the rocks that previously had been called "Transition" and "Secondary," originally lithic divisions; the later Paleozoic was included in the Transition. When the Paleozoic was defined, the top was placed at the top of the Carboniferous, which included what is now Permian, and the base of the Mesozoic was that of the Triassic.

APPALACHIAN REVOLUTION

In North America, it has been customary to consider the end of the Paleozoic to be marked by a revolution, the Appalachian Revolution. Although the Appalachian Mountains gained their structure at about that time, evidence shows only that the folding and faulting proceeded sometime later than the beginning of the Permian, that the folded rocks had been eroded to a peneplane that was faulted and covered by late Triassic rocks, and that there were intrusions that came within about the same span—as determined somewhat inadequately by radioactive mineral disintegration methods.

The term Appalachian Mountains is applied invariably to the ranges of folded and faulted rocks extending southwestward from Pennsylvania to Alabama and generally to the Allegheny Plateau to the west. The term is also used to refer to ranges of somewhat similar structure trending northeastward through western New England and Quebec, particularly to the mountains of Gaspé, southeast of the St. Lawrence. These latter mountains have been described and their structures attributed to Ordovician (Taconian) and Devonian (Acadian) orogenies.

The Appalachian Mountains from Pennsylvania southward have thrust faults on the southeast, particularly southward from southern Virginia, that cut rocks as young as middle Pennsylvanian (late Pottsvillian) in the South (Fig. 14–16). The folds within this region and in Pennsylvania extend northwest to a rather definite line, the Appalachian Structural Front, margining the Allegheny Plateau in Pennsylvania and West Virginia (Fig. 14–17). This line in the south follows the trend of the western margin of the Pennsylvanian miogeosyncline and linear Mississippian. In Pennsylvania the folds swing off northward in an arc that includes the thick late Ordovician and middle and late Devonian terrigenous sediments that were laid in the semi-lenticular bodies, exogeosynclines, within the cratonal margin; earlier and later sediments are relatively thin. Thus the folded Appalachians (Fig. 14–18) include

Fig. 14–15. Distribution of volcanic rocks in late Paleozoic and early Mesozoic sections in western North America (after M. Kay). The spots are representative of localities having volcanic rocks, though many others are known. In contrast to their abundance in the coastal belt, volcanic rocks are unknown to the eastward of the line on the east. Comparison of this map with that of the Ordovician shows that volcanism has been distinctive of a Pacific coastal belt, the volcanic belt contrasting with a nonvolcanic belt to the east of it throughout the Paleozoic and early Mesozoic eras.

Fig. 14–16. Structure section of the present through the Appalachian Mountains to the Blue Ridge and Piedmont to the southeast (after P. B. King). The structures in the sedimentary rocks are attributed to the Appalachian Orogeny; those in the crystalline rocks farther east pertain to events within and even prior to the Paleozoic Era.

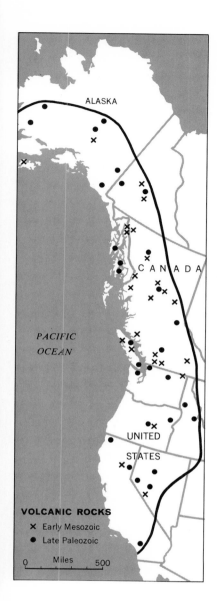

VOLCANIC ROCKS

× Early Mesozoic
● Late Paleozoic

0   Miles   500

the region where Paleozoic sediments were thickest, but the thick rocks were laid in different parts at different times.

The folds are gentle west of the Structural Front, extending into the area of Permian in the center of the Allegheny Synclinorium; lateral compression was greater southeast of the front (Fig. 14–19). Throughout this region, the best that we can do with direct dating is to follow the folds and faults southward and find where they pass unconformably below the Cretaceous rocks of the Alabama coastal plain, or follow the folds eastward in Maryland to the point where they are overlain by late Triassic sediments. Folds and faults affect Carboniferous rocks as far northward as western Newfoundland, but distinctive Permian is not known, though suspected in Prince Edward Island in the Gulf of St. Lawrence (Fig. 13–6). Granitic rocks in eastern Massachusetts intrude plant-bearing Carboniferous metamorphosed sediments. On direct evidence, then, the folds and faults in the Appalachian Mountains of the southeastern United States postdated early Permian and preceded late Triassic. There was sufficient time for the folds to be denuded to a deeply eroded surface, perhaps a peneplane, before the upper Triassic was laid unconformably across them.

The Ordovician, Silurian, and Devonian areas of subsidence were filled with sediment eroded from rising lands along coastal New Jersey to Virginia. The Appalachian folds to the west were formed late in the Paleozoic. They may have been produced by the gliding of the Paleozoic sediments northwestward from rising welts in the same locus to the southeast in a manner analogous to that discussed in the Carboniferous history of Nevada. Evidently the folds spread into the salient of middle Paleozoic sediments in Pennsylvania and into the more linear-trending geosyncline of Carboniferous to the south from Virginia to Alabama. The folds in the latter area break to give the linear thrust

Nashville Basin

Cumberland Plateau

Sequatchie Valley

Valley & Ridge Province

Blue Ridge

0   Miles   50

Blue   Ridge   Piedmont   Province

Great Smoky Mountains

Carolina Slate Belt

Deep River Basin

Atlantic Coastal Plain

0   Miles   50   100

faults that prevail in southwestern Virginia and Tennessee; as the thrusts increase in their displacement southwestward, the trends of the initial breaks became more southerly than the present fault traces. Although the intrusions in the Atlantic provinces, New England, and in the Piedmont of the southeast have been attributed in the past to the late Paleozoic, geochemical studies of isotope ratios show that many are Devonian or older. Thus the Appalachian orogeny is neither as extensive as we once assumed nor are its effects closely dated as immediately post-Paleozoic.

RESTRICTION OF LATE PERMIAN SEAS

In the western interior, the boundary between the Paleozoic and Mesozoic systems lies within a sequence of red sediments. A slight regional unconformity underlies the eastwardly onlapping lower Triassic in western Wyoming. In the region of the Permian in Russia, the sediments similarly pass from Paleozoic to Mesozoic within stream-laid sediments; vertebrate fossils permit separation. The Permian in England is non-marine or at least quite non-fossiliferous. Possibly the fact that there are so few places in the world where late Permian sediments contain marine fossils, and that these were so long unknown, led in part to the recognition of a marked distinction between the life of the Paleozoic and Mesozoic; the comparison was being substantially made between faunas in the Carboniferous, actually Pennsylvanian, and those in the Triassic. As discoveries have revealed sequences having more nearly continuous sedimentation from the Pennsylvanian to the Triassic, the differences in the faunas have lost their apparent catastrophic aspect. As anticipated, the forms of the earliest Triassic are so like those of the latest Permian as to be distinguished with difficulty, if at all. However, the very fact that such records are so scarce supports the placement of a division point in classification in such a position.

STABILITY AND MOBILITY

The generalization has been made that North America throughout the Paleozoic Era had a central more stable region margined by relatively unstable mobile belts. Stability is a quality of firmness, of resistance to change, a tendency toward return to a favored position. Mobility is fluidity, the property of being readily changed in position. Measures of mobility include those relating to change in vertical position and those of lateral displacement. We must consider whether the changes are of short duration or are long continued, whether they are rapid or slow, constant or interrupted, and whether they produce significant changes in the relative positions of parts of the earth's crust.

One measure of instability could be the displacements of the crustal surface from an average level. The level of the sea is an excellent datum of reference for any one time. If the level of the sea with reference to the center of the earth is changing through time, these eustatic changes will need consideration. If sea level remained static, the elevation of sediments formed at sea level at one time compared with sea level at another time would give the vertical crustal displacement in the intervening time span. If all Paleozoic sediments had been laid at sea level and if sea level remained constant, the thickness of Paleozoic rocks would be a measure of subsidence. These assumptions are of course not fully valid, but most preserved sediments probably were laid in depths within a few hundred feet of sea level. Moreover, sea level on the average remained rather constant, for shallow-laid sediments are extensively preserved on the continent today. So the comparisons of

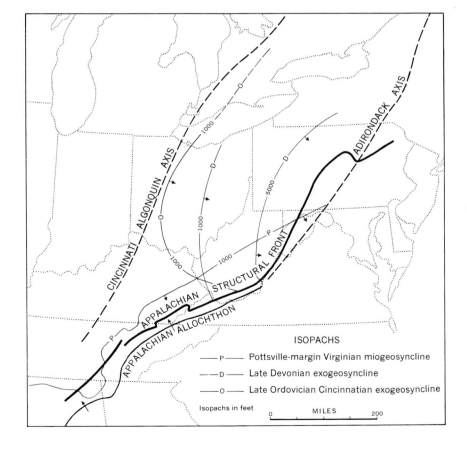

Fig. 14–17. Map of the southeastern United States showing the relationship between the trends of folds in the Appalachian Mountains, represented by the Appalachian Structural Front that is generally the westernmost principal steep limb of an anticline, to the pattern of deformation in subsiding basins in the area of the Allegheny Synclinorium in the Paleozoic, suggested by selected isopachs for each of several sediment masses (after M. Kay). The trends of Appalachian structures are similar to those of late Paleozoic depositional basins but somewhat transverse to earlier trends. As the folds have moved westward on thrust faults that lie above the basement (Fig. 14–19B), the position of the Appalachian Structural Front at the surface is displaced from the position that the rocks had when they were attached to the basement.

ISOPACHS

—— P —— Pottsville-margin Virginian miogeosyncline

— — D — — Late Devonian exogeosyncline

—— O —— Late Ordovician Cincinnatian exogeosyncline

Isopachs in feet   MILES

0   200

thicknesses give a method to use in the analysis of vertical changes of crustal surface level.

The greatest thickness of Paleozoic rocks in the continental interior, within the craton as it was defined for the Cambrian, is about three miles, and two miles is exceeded in several areas, most extensively along the southeastern margin of the craton. If we allow a span of 300 million to 400 million years for the Paleozoic Era, subsidence of three miles is at an average rate of about 50 feet in a million years, a maximum for the craton. Subsidence was not at a uniform rate through time; in the section of upper Cambrian to Pennsylvanian in central Michigan, nearly a mile of the three miles was laid in some 15 million years of the late Silurian, at a rate of perhaps 300 feet in a million years. Our knowledge of the lengths of the periods and their parts is not sufficient to warrant more than very general judgments.

The miogeosynclinal belts along the margins of the craton were defined in the Cambrian as having relatively greater subsidence. The Paleozoic rocks in eastern Nevada in the western miogeosyncline are about 30,000 feet thick at the maximum; thicknesses of 15,000 feet or more are quite general in this belt. The thickness of Paleozoic sediments in the Appalachian region in southeastern United States seems to approach or exceed 30,000 feet. These maxima give the subsidence of the crust for the era approaching 100 feet for each million years. The thicknesses of Paleozoic rocks in the miogeosynclines are of an average that is about that of the maxima within the craton.

Parts of the craton accumulated little or no Paleozoic rock. The average thickness over the whole of the interior has been estimated as less than a half mile, largely of Paleozoic rocks. The thickness on the average of 2000 feet for some 400 million years is only about 5 feet for each million years, but this indicates that the craton as a whole subsided relative to sea level, either because it sank while sea level was constant or because sea level rose on an average through time. But in the miogeosynclinal belts, the average of Paleozoic rocks of some three miles gives a rate of about 50 feet in a million years, some ten times that of the craton. Fifty feet in a million years is but a foot in 20,000 years, so instability is not very impressive in terms of the span of human history.

Measurement of the full thickness of Paleozoic sediments in the most peripheral, eugeosynclinal belts is hampered by the very fact that the belts were mobile, for the rocks are generally deformed, metamorphosed, and intruded by igneous rocks that have made much of the record obscure. Estimates of the thicknesses of each of several systems in many places in the eugeosynclinal belts exceed any known thickness

Fig. 14–18. *Sedimentary rocks deformed in the Appalachian Orogeny. A.* View southward along the axis of an anticlinal fold, Bolar Valley, in the Appalachian Mountains, in northwestern Virginia. The ridges that close in the distance are of Silurian quartzite in a southwest-plunging anticline having a core of relatively less-resistant Ordovician shale and underlying limestone. The dip is gentle to the left and steep to the right, so the fold is asymmetrical with its axial plane inclined toward the left (the east). Such structures are prevalent in the folded Appalachian Mountains; the rocks were deformed in the Appalachian Revolution in latest Paleozoic, probably in the Permian.

B. Susquehanna Watergap at Harrisburg, Pennsylvania, in Ordovician and Silurian rocks folded in the Appalachian Revolution (G. Heilman). The prominent north-dipping ridge is of Silurian Tuscarora Sandstone.

A

B

for the whole Paleozoic section in the miogeosynclinal belts and the craton. The rate of subsidence in eugeosynclinal belts indicates the propriety of calling them the true "mobile belts" on this criterion alone. The dominant rocks in these great thicknesses are terrigenous sediments and volcanic rocks, the former derived by the erosion of elevated nearby lands within the belts, and the latter from the displacement of lavas that came from the crust below the belts. So there were not only great downward vertical movements but also complementing movements of elevation, which produced the lands that furnished the sediments; and there was an outpouring of surficial lavas that had been expelled from beneath the subsiding troughs or had speeded depression by their very emergence. So just as subsidence is a measure of instability, so also is elevation.

RATES OF SUBSIDENCE

One important aspect of stability is that of constancy. Rates of subsidence changed in limited areas through time, for the thicknesses of systems of shallow-laid sediments are not proportional locally to the spans of time that they represent. Some regions that had subsided hardly at all for a long time became, for a time, the sites of relatively great subsidence; the thickness of the Paleozoic through the lower Carboniferous in central Colorado is only a thousand feet or so, all

Fig. 14-19. *Maps showing the lateral displacement of the Paleozoic rocks in West Virginia produced by the folding and faulting of the Appalachian Revolution (J. M. Dennison and H. P. Woodward). A. Map showing the estimated amount of compression in miles and in per cent of the present distance to that prior to orogeny.*

B. Diagrammatic structure section across the structural trends; note that the folds are considered to affect only the upper sedimentary layers, which are believed to have glided on a basement of Cambrian rocks.

C. A palinspastic map of West Virginia and surroundings to be compared with the present map; the palinspastic map shows the interpreted original extent of the rocks at the state boundaries; the open areas are those in which the rocks have moved over the original, now unexposed basement on thrust faults. Though the latitudes and longitudes seem distorted, one might say that their original trends have become distorted on the present geography.

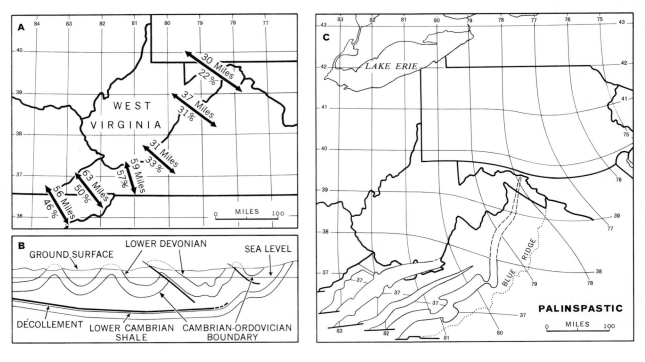

shallowly laid along coasts of low elevation. In the later Carboniferous, the area developed local subsidence of two miles or more, sufficient to accommodate thousands of feet of sediments, and produced comparable elevations to yield the terrigenous sediments. The causes of such marked contrasts in crustal behavior from time to time, from average subsidence of a few feet to that of more than 100 feet in a million years, is one of the mysteries that scientists have not solved.

Elevations produced in the distant past were not preserved; only the products of their erosion laid in areas of subsidence will be recognized in the sedimentary record. The volume and texture of the terrigenous rocks evidence the characters of the elevations. The only area within the continental interior that produced significant volumes of coarse terrigenous sediments during the Paleozoic was the Colorado-New Mexico region in the late Paleozoic, mentioned in the preceding paragraph. The volume of detritus has not been accurately measured, but must have been of the order of 20,000 or 30,000 cubic miles, representing erosion of a similar elevated mass. This is relatively small as compared to the amount that came to the eastern margin of the craton from elevations outside the craton. The volume of terrigenous clastic sediments in the late Ordovician and early Silurian in the exogeosyncline of Pennsylvania and surroundings has been estimated as about 200,000 cubic miles and there are large volumes of eastward derived terrigenous sediments in each of the succeeding Paleozoic Systems in the Appalachian region. The elevations to the east in the late Ordovician alone produced greater volumes of sediments than any within the interior. On the west, terrigenous sediments from peripheral uplifts did not encroach on the craton, nor in the miogeosynclinal belt in United States until the latter part of the Paleozoic Era.

In terms of subsidence and elevation, then, the interior of the continent was far less deformed than the miogeosynclinal belts, and these were less deformed than the more peripheral eugeosynclinal belts.

LATERAL CHANGES

Mobility is based also on changes in the lateral positions of parts of continents as expressed in the folding and thrust faulting of the crustal rocks. The interior was extremely stable in this respect, for such movements are virtually absent. Deformation within the Paleozoic is recorded in a few places along the margin of the craton by the presence of Paleozoic sediments unconformably overlying the eroded folds of earlier Paleozoic rocks, but the examples are so few as to make the exceptions negligible. Thus, in the Arbuckle Mountains of Oklahoma, latest Pennsylvanian sediments lie unconformably on folded older

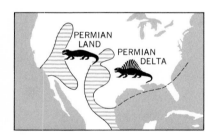

Fig. 14–20. In early Permian times there were well-developed and related but distinct tetrapod faunas, living on each side of the west Texas Permian sea. The amphibians and reptiles living in what is now central Texas and eastern New Mexico had evolved from common ancestors, but had been isolated from each other by the marine embayment for a sufficient time to develop distinct characters in their separated environments. This fact is epitomized by the presence of the fin-backed pelycosaurian reptile, *Dimetrodon*, in the Wichita and Clear Fork beds of Texas and the closely related genus, *Sphenacodon*, without the sail on the back, in the correlative Abo beds of New Mexico.

313

Pennsylvanian sediments. In the eastern miogeosynclinal belt, Silurian and Devonian rocks lie unconformably on folded Ordovician sediments in New York and eastern Pennsylvania. But Paleozoic rocks commonly are strongly folded in the more peripheral regions on east and west. In many places in the New England states and the Maritime Provinces of Canada, deformation resulted from strong orogeny with accompanying igneous intrusion within the Paleozoic; angular unconformities and radioactive isotope age determinations of intrusions and metamorphisms distinguish the times of orogeny. Some stresses seem to have produced transcurrent faults, steep fault planes having lateral slip between the opposing walls. In the west, similar dated orogenies are known within the Paleozoic.

Much greater lateral displacement seems evidenced in far-traveled thrust sheets. The determination of the amount of movement in a thrust fault depends on interpretation of the original relations of rocks in the overriding thrust sheets and in the underlying unmoved rocks. Displacements of fifty miles and more seem necessary in some instances in which rocks of the same age in the two sequences require that great separation on the original geography. Not only the displacement but the dating of such thrust sheets is rarely precise; they are clearly younger than the rocks they affect, but commonly the orogeny produced elevation and erosion, so that the surfaces were not covered until long afterward. There are also questions with respect to the significance of the displacements. They might be taken to mean that a part of the crust rose and overrode another part for a distance of fifty miles. But

Fig. 14–21. An interpretation of the habitats occupied by the Lower Permian tetrapods of Texas. Here on an ancient delta we see, from left to right: the pelycosaurian reptile, *Dimetrodon*, and the cotylosaurian reptiles, *Diadectes* and *Captorhinus*, living on the low uplands; the labyrinthodont amphibians, *Cacops* and *Eryops*, and the pelycosaur, *Edaphosaurus*, inhabiting the pond and stream margins; and the amphibians, *Trimerorhachis* and *Diplocaulus*, spending most of their time in the water.

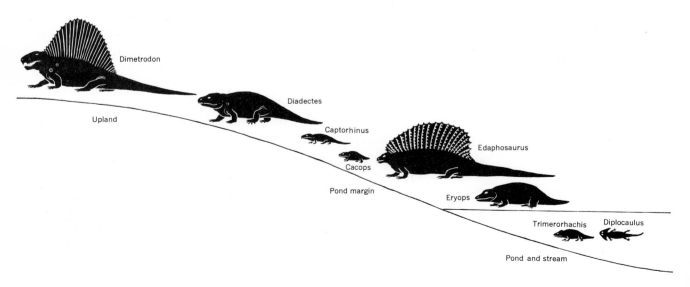

the overriding rocks are rather surficial parts of the crust. It has been suggested that they represent parts of the surficial crust that rose in the region from which they came, and spread out through gravity, sliding downward and outward without great displacement at depth. The surficial rock may glide a great distance on relatively immobile basement. These lateral movements are commonly associated with alteration of rocks under high pressures and temperatures, with production of metamorphic rocks and granitic rocks through flow. Whatever the forces that produced the orogenic structure, they were widely effective in the peripheral parts of the continent and practically unknown within the interior. Mobility is prevalent in the miogeosynclinal belts, and volcanism is inherent in the definition of the more peripheral eugeosynclinal belts. Stability is a relative term, but one that can be applied with little reservation to the continental interior.

Another very important aspect of lateral displacement on a much larger scale is that of the movements among continents relative to each other and to the rotational poles of the earth. For example, there is an hypothesis that during the Permian, the continents of Australia, southern Asia, Africa, and South America were contiguous, forming Gondwanaland. The subject of continental drift will be explored more fully on later pages.

The Paleozoic Era presents abundant evidence of the changing aspects of the earth's crust and the deeper forces that must cause the deformation.

Fig. 14--22. Restoration of the sea bottom in early Permian time in West Texas (The Smithsonian Institution). The pipelike and beaded structures in the upper right and the compact ones in the lower left are sponges. A spined nautiloid is in the lower foreground, and ammonoids are on the left. Many kinds of brachiopods are on the shelf—long hinged spirifers and large spined productids, as well as colonial cup-shaped productids. There are a few simple conical corals with crowns of tentacles. In the foreground are peculiar brachiopods, leptotids having a slotted brachial valve. The whole is a neat assemblage.

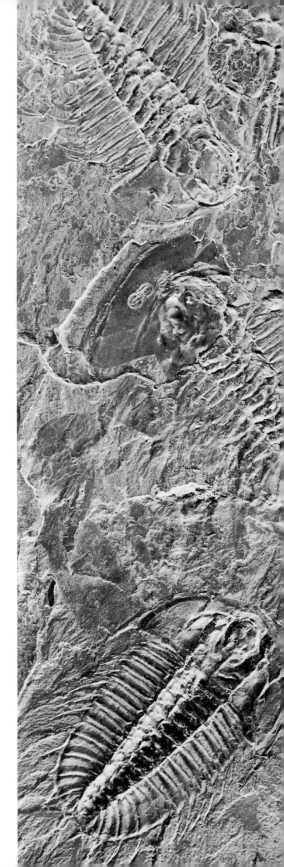

15
Paleozoic Life

Trilobites on a bed of Cambrian shale from Mount Stephen, southeastern British Columbia (The Smithsonian Institution).

15 *Paleozoic Life*

Some of the organisms of the several Paleozoic systems have been briefly shown to be significant and useful in many ways. Few objects in nature have such fascination to collectors as fossils. They have inherent interest quite aside from scientific applications. Some are attractive because they are so like familiar living shells. The collecting of clams and snails is a popular pastime on present shores as well as in the rocks. Other fossils are cherished because of their intrinsic beauty or because of their being so grotesque. The nodes or pits in a trilobite and the suture patterns of cephalopods have beauty in their orderly and intricate geometric patterns. And there can be feelings of awe and admiration for a specimen of an armored fish. Fossils were collected as curiosities many centuries before they were known to have scientific significance. To some, supernatural properties were attributed to them as amulets and idols. It is somewhat surprising that their true significance was not realized until late in history of human thought.

CLASSIFICATION

The collecting and classifying of fossils has been a beginning in science for many who have gained distinction. The boy who lives near a locality where fossils abound will soon sharpen his skill in distinguishing the several kinds of these objects and their characters, one of the first requisites of a scientist. This interest in classification may then develop into thought as to the reasons why, as well as the less sophisticated recognition of differentiating characters. But we will now concern ourselves with the more specific application that can be made of fossils in solving geological problems.

SOME APPLICATIONS

How are the fossils of the Paleozoic and later eras interpreted and used? They can determine stratigraphic and structural sequences, and are the most dependable means of classifying rocks in time. To take another point of view, the procession of changing life forms through the successive stages of the Paleozoic is a part of the story of the earth. Their associations with each other and with the including rocks, permit reconstruction of the geographies and the environments of the past. Distributions of fossils indicate patterns of migration and the geographic barriers that through time limited organisms to geographic regions. Distributions also bear on the directions of marine currents, as do the physical arrangement of the fossils. Organisms have changed through time as is clearly demonstrated by successions of fossils, and

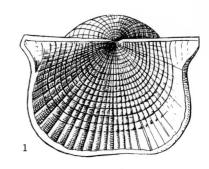

Fig. 15-1. *Representative brachiopods from the Carboniferous and the Permian.* (The Smithsonian Institution). *A.* Brachiopods from the upper Carboniferous Pennsylvanian subsystem. (1–8) Productids: (1, 2) *Dictyoclostus*; (3, 4) *Echinoconchus*; (5–7) *Juresania*; (8) *Chonetes*. (9, 10) Rhynchonellid *Welleria*; Strophomenids: (11, 12) *Meekella*; (13, 14) *Derbya*; (15, 16) Spiriferid *Neospirifer*, (16) is an interior, showing spires of brachial valve.

Permian brachiopods from limestone, the Glass Mountains, West Texas. The original calcareous shells were altered to silica; thus when acid etched away the limestone matrix, the siliceous fossils remained, preserving even the finest structures. The brachiopods are mostly productids, differing from some others illustrated from Carboniferous because of better preservation. Some of the forms, particularly *Prorichtofenia*, are aberrant—that is, they are forms differing grotesquely from the normal appearance of the organic group; this genus has a conical pedicle valve, attached or sessile, and grows in association with other specimens in a colonial manner; *Heteralosia* also is shown in a cluster of individuals. The genera illustrated on page 321 are productids.

A few words and illustrations cannot adequately convey the differences in the invertebrates of the Paleozoic Era. The species of the phylum Brachiopoda are the most widely distributed and generally prevalent of the fossils in the Paleozoic, though other animals that failed to be preserved must have existed in greater numbers. The fossils from stage to stage have so much in common that experts can have great difficulty in distinguishing rocks within these short spans of time. Even the average faunas of adjoining systems have many genera and a few species that are alike. But when one compares a large population of invertebrates from the extremes of the Paleozoic, the dominant forms are clearly contrasting. However, in illustrating the differences,

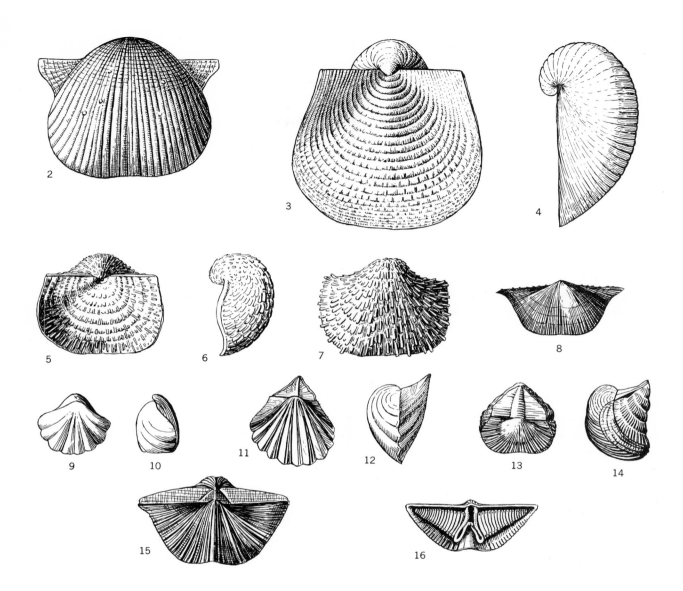

limitations of space intervene. Some 1000 species of brachiopods of 160 genera are described on more than 250 plates in a single book (G. A. Cooper: *Chazyan and Related Brachiopoda*) on North American brachiopods of the early medial Ordovician. Each species has variations that might take many illustrations. Some of the species are extremely rare, whereas others could be collected by the thousands and millions. Only a few were shown as illustrations of the variety of forms in the phylum in Chapter 8 (Fig. 8–8).

The Carboniferous has similar abundance of forms of brachiopods, and some are moderately similar. In the Carboniferous, the dominant groups are the spiriferids and productids. The former have broad slender shapes, with ribs in the fold of the brachial valve and in the sinus of the pedicle valve; but, most significantly, they have the coiled internal structures that distinguish the Spiriferacea, a group barely represented in the Ordovician. The productids have shells that are concave in the brachial and convex in the pedicle valve; prominent sur-

face spines emerge in some genera. These were unrepresented in the Ordovician, though other families of the pseudopunctate brachiopods were similar, and are represented also in the Pennsylvanian. In the Permian both have such exotic, misshaped forms that they are hardly recognizable as brachiopods (Fig. 14–22).

If we plot the distribution in time of the orders and families of brachiopods through the Paleozoic, we see that there are progressive changes.

these developments are among the most forceful evidences in support of the trends and processes of organic evolution. These are topics that deserve discussion. Furthermore, fossils are of great economic importance. Our principal sources of energy are in coal, oil, and gas, all of organic origin. These are the fossil fuels. And fossils are the principal constituents of most limestones, which include many important building stones.

DIFFERENCES THROUGH TIME

Paleozoic rocks contain fossils of great variety. Some strata include many different kinds of fossil organisms, others great numbers of similar forms. Collections, whether from a single stratum exposed at widely scattered places, or from horizons of differing ages, are never quite alike. Since recent animals and plants differ from place to place, a reflection of their varying habitats even when the conditions under which they live are not recognizably different, it is not surprising that collections of past organisms differ, particularly where the associated rocks are dissimilar. But it would not seem predictable from modern analogies, or even expectable, that the organisms would differ from time to time in strata that seem of the same character, thus presumably having been formed under similar environments.

Experience with Paleozoic rocks shows, however, that successions of differing fossils in separated regions have similar orderly and constant progressions. Thus fossils are useful in the determination of the relative ages of strata. After having learned the order of fossil succession, it becomes possible to use this order in judging the relative ages of rocks of unknown relative position. It is often possible also to tell which side of a bed is the top by the orientation of a fossil; thus a shell of a pelecypod will be most stable with its convex surface upward. This one fact will often permit a direct determination of the relative age of adjoining strata. Our experience shows the order of succession, and the organisms enable the classifying of the rocks against a standard time classification.

Organisms evolve through time, and thus show changes even when we cannot distinguish any differences in environment as reflected in the associated lithologies. This can be seen best by comparing collections of fossils from varied rocks laid in each successive large span of time, such as the periods. It takes little experience for a layman to distinguish an Ordovician assemblage (Fig. 8–8) from a Carboniferous one (Fig. 15–1), but differences between collections from successive systems, such as Silurian and Devonian, are not so obvious as this.

Fig. 15–1B. (1–2) *Echinauris*, pedicle and side views; (3) *Echinosteges*, brachial interior view and pedicle view showing attachment spines; (4–5) *Paucispinifera*; (6–7) *Muirwoodia*, pedicle and side views; (8) *Heteralosia*, a cluster of individuals; (9–10) *Prorichtofenia*, side and posterior views, pedicle valve, showing anchor spines and interior spines of brachial valve (The Smithsonian Institution).

The contrasts can better be seen in the organisms from similar rocks, such as seem most likely to have been formed in comparable habitats in the successive periods.

The nature of the differences among fossils of successive ages bear scrutinizing. For example, trilobites are commonly said to have been abundant in the Cambrian. They are certainly the most frequently seen of Cambrian fossils, though Cambrian rocks are not usually very fossiliferous, and very few types of organisms are likely to be found in such rocks. A collector in Cambrian rocks might normally expect to find a few brachiopods, and although several other phyla are represented by fossils in the system, the finding of specimens of each is rather fortuitous. The most remarkable and virtually unique fossils are the impressions of soft-bodied organisms found at Mount Field, British Columbia (Fig. 15–2). Thus to say that trilobites are dominant in Cambrian rocks is only relative, for they are present in greater variety and numbers in succeeding systems. But there are so many other fossils of such a variety of classes in these later rocks that trilobites are *relatively* not as abundant as in the Cambrian sediments. Their frequency

Fig. 15–2. *A Cambrian fauna from British Columbia. A.* Quarry in the middle Cambrian Burgess Shale near Field, British Columbia, from which remarkably well-preserved fossils were recovered by Charles D. Walcott (1850–1927), the man shown in the center of the picture (The Smithsonian Institution).

B. Remarkably preserved fossils from the Middle Cambrian of British Columbia. The illustrated organisms are (1) a jellyfish or medusa, a representative of the phylum Coelenterata; (2) a soft-skinned echinoderm—a holothurian, and (3) an annelid worm, and (4) types of crustacea, including a trilobite having preserved appendages. These are but a few of hundreds of specimens of scores of kinds of animals that were found in the shale. The illustrations are from somewhat retouched photographs of the specimens. (Fig. 7–19 is a restoration of the original conditions; see Chapter 15 opening for a natural assemblage.)

In the early days of the century, a paleontologist, Dr. Charles D. Walcott, Director of the United States Geological Survey, while climbing near Field, British Columbia, came upon beds of shale with distinctive Middle Cambrian trilobites. But this Burgess Shale Member of the Stephen Formation had in addition a great array of remarkably well-preserved organic remains—carbonaceous films and impressions of the structures of soft organisms. The living animals had been buried in soft muds, then compressed as the rock was indurated. They remained preserved until found a half-billion years later. The fossils show that the animals were much like their descendants of today, yet but for their fortunate preservation and discovery, there would be little or no record of the existence of many of these sorts of animals throughout the intervening time. The material impresses one with how limited is our knowledge of ancient life, because of the factors of preservation. It gives insight into the advanced development of some of the Cambrian organisms. And it tells us of how little some invertebrates have changed in the eras to the present.

A

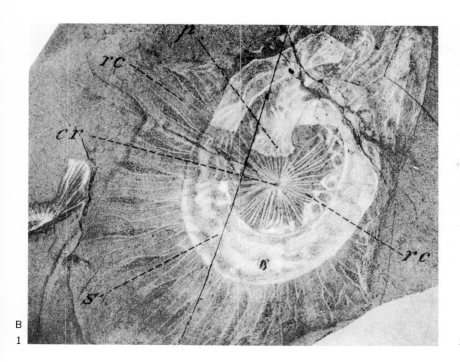

B
1

2

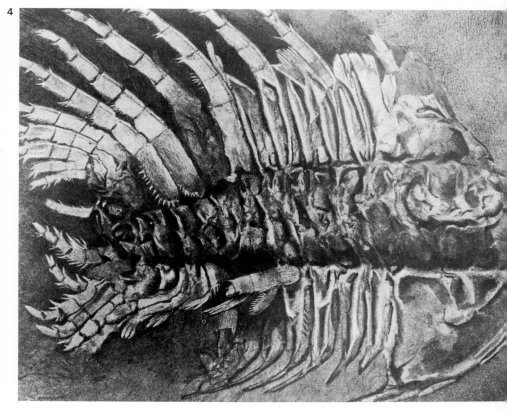

3

4

in variety and number decreases in the later Paleozoic; they are rare in the latest Paleozoic, and are absent in the Mesozoic rocks. So the dominance of certain forms is dependent not only on their own frequency, but also on that of other living forms of their time. Trilobites are limited to the Paleozoic Era but are more frequent in the earlier than in the later Paleozoic.

The frequency of trilobites, or of any other organisms, is dependent on the frequency of each of many different genera and species within the class. That is, just as the trilobites can be described as having been increasingly abundant in the earliest Paleozoic, and decreasingly so in the later, so also the successively smaller categories within the class—the families, genera, and species—have ranges. The distribution in time of these successively smaller categories is successively shorter. The frequency of the class as a whole is the cumulation of the frequencies of the smaller categories. And so it is with all the other organisms. For example, the pelecypods which include the clams, and

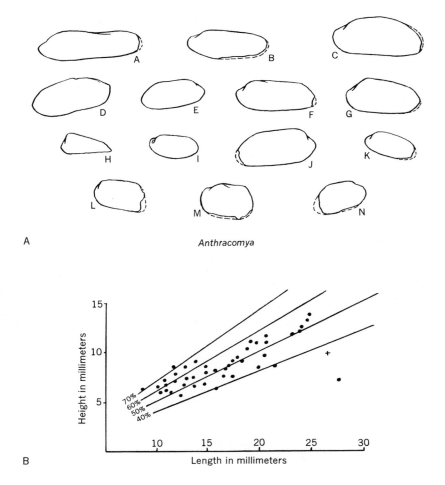

A   *Anthracomya*

B

Fig. 15–3. *Pelecypods.* Variation and evolutionary trend in the non-marine Carboniferous pelecypod *Anthracomya* from the coal-bearing rocks in Wales (J. Weir). *A.* Outlines of shapes of specimens from a single bedding plane.

B. A graphic plot of the distribution of the specimens in terms of lengths and heights of valves.

C. Comparison of the average shapes and sizes in a study of similar valves from three levels at another place.

Although some specimens from each stratigraphic zone or bed are similar in proportions, the general average is toward specimens having larger size and less elongate shape, as is shown graphically in the diagram.

The analysis of variation of pelecypods. The Phylum Mollusca has several classes, three of which are of particular importance in stratigraphy, the classes *Pelecypoda*, *Gastropoda*, and *Cephalopoda*. The first two are the clams and snails of the present beaches, the last a class rarely seen by the visitor to the shore. The cephalopods are quite as common as the other molluscs in the Paleozoic and Mesozoic, however.

The pelecypods or clams are bivalved molluscs, each valve commonly a reversed likeness of the other, the two shells being essentially symmetrical with respect to the plane of junction. In contrast, the two valves of brachiopods are quite dissimilar, and each is symmetrical with reference to a plane normal to their junction. The clams live not only in the sea, but in fresh-water lakes and streams.

the amphibians among the vertebrates, become increasingly abundant through Paleozoic time from their recorded advents, the former in the early Ordovician, the latter at the end of the Devonian Period. Their distinguishing differences through time are so universal and consistent that these fossils are useful in determining the ages of stratigraphic units. This is a product of our collecting experience, not the verification of a preconceived theory relating the causes of change or their effects.

EVOLUTION IN ORGANISMS

Perhaps the best way to determine the evolution of Paleozoic life is by comparing the organisms that occupied quite the same kind of habitat through the Paleozoic sequences. It is not possible to determine identical environmental conditions from lithologic records, so the organisms themselves must be used for judging the similarities of

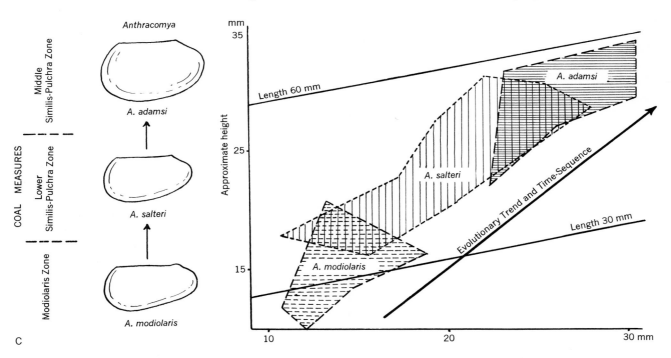

C

Because few other animals are preserved in non-marine strata, clams have been critically studied in the Carboniferous sediments associated with the coals of Britain. The assemblage of clams in one bed at one place, a population, has very great variation (Fig. 15–3A). The variability in each bed is distinctive, differs measurably from that in another bed, so the variations are stratigraphically useful.

In contrast to these population studies are the comparisons of growth stages in individuals in successive marine sediments in the Permian System in Kansas (Fig. 15–3B). They show progressive change in the shapes of adult clams. And there is similarity between earlier growth stages in the stratigraphically younger clams and the adult shapes of stratigraphically older ones. These populations are not as variable, and as the changes are progressive, they are also stratigraphically useful. Perhaps the marine clams were descendants of a generally interbreeding stock, whereas the clams in non-marine sediments living in separated streams represent descendants of different stocks.

Marine sediments have many pelecypods in Ordovician and later systems, and non-marine clams contain them from at least the Devonian. They are of greatest value in their defining facies bands, being most common in shallow marine water, as in beach sands and off-shore muds. They are the dominant invertebrates in the non-marine sediments of the Paleozoic.

325

environments. Thus study of the succession of most similar organisms, changing through time, should be instructive. There are coral reefs in the Ordovician and in each succeeding period; but the coral masses, though similar, are not of the same sorts of corals. Since periods are continuous in time and evolutionary lines are likewise continuous, it is inevitable that the genera at the beginning of one period are like their ancestors at the end of the preceding period. Consequently, corals in the late Ordovician are so like those in the early Silurian as to be of the same general kind, mostly of the same genera. Devonian corals show close resemblances to many in the Silurian; but the latter system has a few distinctive forms such as the chain coral, *Halysites* (Fig. 10–4), not known in the Devonian but present in the Ordovician, or some very like it. If we compare strata of widely separated ages, Permian reef coral with Ordovician reef coral, the resemblances are only in the broadest realms of classification, usually indicative of the same largest category, the suborders, very rarely of the same family, never of the same genus and species.

The fossils in Paleozoic rocks change through time, even when the fossil assemblage and the rock characters lead to the belief that the environments in the selected successive deposits were so similar that they should have supported similar life. Life as a whole was changing, but some of the changes proceeded at a slow pace, so that descendants of certain organisms survived with little apparent change for varying durations of time. Some corals found in reefs in the lower Silurian are so similar to corals in the middle Devonian that they are scarcely if at all distinguishable even by the expert; brachiopods collected from the middle Silurian limestones of the typical section in Wales include some that are virtually identical with specimens found in the lower Devonian in Tennessee or Oklahoma. But other associated forms in each horizon are unrepresented or unknown in the other. By way of contrast, the descendants of many organisms living at one time are enough changed at a later time so that the differences are clearly recognizable. This is clearly seen in the succession of amphibians and reptiles found in the Wichita-Clear Fork sequence, of the lower Permian of Texas. It should be added, in this connection, that we are increasing our skill in recognizing significant differences, and our methods of analysis, some of them statistical, are becoming increasingly refined (Figs. 15–3, 15–4).

CAUSES OF CHANGES

Why should there have been changes with time—how did they come about? To test the theory that organisms evolved, we may study in

Fig. 15–4. *Representative species of the marine pelecypod genus Myalina in the Carboniferous and Permian of Kansas (N. D. Newell).* A. The illustration on the right indicates the shapes of the successive growth stages of one species of *Myalina* in order to show how these stages in the ontogeny resemble the succession of forms that are found in the underlying beds.
B. A statistical analysis of species of *Myalina* from the Carboniferous and Permian in central United States. The upper and lower graphs show the ratios of lengths to widths, and the maximum lengths of specimens; with decreasing geologic age, the shells become relatively shorter and higher and increase in absolute size.

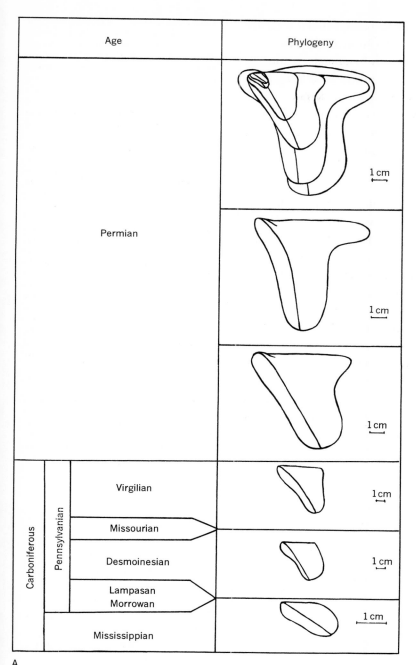

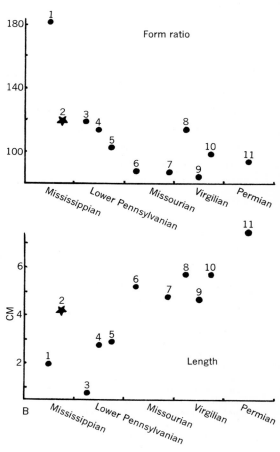

327

successive strata the organisms that are most nearly alike, with the assumption that the later forms are descended from the earlier ones. In many kinds of invertebrate animals the adult animals retain the skeleton of their youthful stages. Thus a coral grows by adding to its earliest calyx, a clam and snail add successive growth to their initial shells. Cephalopods are admirably suitable for careful study because they preserve the juvenile shell as the chambers are added to increase the size of the shells, and the partition separating the chambers, the septum, is peculiarly fluted in the later kinds, giving a distinctive basis for structural comparisons.

Although organisms seem to have changed through time, it has not been possible to determine time precisely by means of fossils. Differences of opinion with regard to the classification of the sequence within systems, and the correlation of strata from place to place is because judgments are subjective. The experience of one paleontologist with the range of organisms leads him to conclusions that differ from those of another who had different experience; one may put faith in organisms that give different conclusions, than the other, or one may attribute to environment or province the differences that the other attributes to time, even though they may agree completely in the identification of the fossil specimens. The judgments are not subject to rigid proof. Methods in the analysis of the differences and in the correlation of the variables are being improved; this research is one of the most interesting and rewarding prospects in stratigraphic paleontology.

Fig. 15-5 (right). *Cephalopods.* The Ordovician and Silurian nautiloid cephalopods shown illustrate the differing shapes (after A. F. Foerste). (1) and (2) *Ormoceras,* and (3) *Spyroceras,* orthoceracones from the Ordovician of Quebec; (4) *Richardsonoceras* and (5) *Centrocyrtoceras,* cyrtoceracones, curved forms from the Ordovician of Ontario and Quebec; (6) *Plectoceras,* a coiled nautiloid from the Ordovician of Wisconsin. (7) *Phragmoceras* and (8-10) *Hexameroceras* are Silurian forms from Ohio that have constricted openings in the body chamber. All have rather straight sutures; the internal siphuncle and septa (septum) are shown in (2).

Cephalopods are a class of molluscs that has flourished in changing forms from the early Paleozoic to the present. Of the two principal classes, the tetrabranchiata with external shells and dibranchiata with internal shells, only the Pearly Nautilus is a living representative of the former. The dibranchs are better known today, perhaps because the octopus, squid, and cuttlefish have such bizarre and terrifying appearances. The tetrabranchs are among the most useful fossils in the Paleozoic and Mesozoic rocks.

The tetrabranch cephalopods of the Devonian and earlier Paleozoic are fossils with calcareous shells, commonly a long conical tube, straight or curved, or in the form of a disk coiled in a plane. The external surfaces are varied, some having ridges and grooves; apertures are

Fig. 15-6. Ontogenic development. Trilobites, such as *Crassifimbria* (a genus found in the lower Cambrian in eastern Nevada), develop and discard a succession of external shells or carapaces as they grow, just as do some modern insects (after A. Palmer). The successive moults are sometimes preserved, and in the case of the illustrated examples, they were in limestone, from which they could be extracted because they were insoluble in acid that dissolved the rock. Study shows that there are as many as fifty such moults derived from the growth of one individual from its first form to the adult form. A few of the stages are shown; note that those on the left, the earliest in the growth series, are considerably enlarged relative to those on the right. In the instance of this rock and locality, the limestone yielded myriads of these trilobite moults, enabling the scientist to gain understanding of the manner of growth of the living animal.

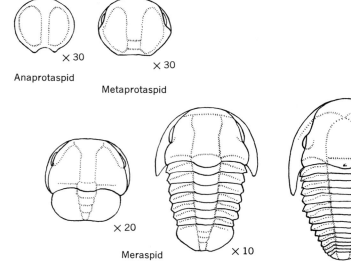

× 30
Anaprotaspid

× 30
Metaprotaspid

× 20
Meraspid

× 10

× 5

open or varyingly constricted. The distinctive characters, however, are the succession of transverse partitions or septa that divide the shell into compartments or chambers of increasing size with age, and the rod-like structure passing from chamber to chamber, the siphuncle. Tetrabranchs have two divisions, the nautiloids in which the septa are gently concave, and the ammonoids in which they are wavy or even pleated as they approach the chamber wall, producing an irregular pattern of contact or suture with the wall. When the chambers have been filled with sediment or mineral deposit and the outer shell of the cephalopod is removed, the fossil shows the trace of the septum against the chamber wall as a line of the thickness of the septum, permitting accurate comparison among cephalopods. The siphuncles are also distinctive, both in their position within the chamber, as whether near the outside of the coil, the venter, or nearly central, or even near the inside or dorsum, and in their form and internal structure.

Prior to the Devonian, the nautiloid cephalopods were found almost exclusively, whereas the ammonoids become sufficiently common in the Devonian to be useful stratigraphically. They preserve in the successive septa a record of the changing form of growth; the early septa of some of the late Devonian forms have sutures similar to the mature sutures of some of the earlier Devonian forms of similar shape and presumably of the same lineage. There are many lines of ammonoids in the Devonian, so the recognition of lines of descent or phylogeny is based on collections from many localities and stratigraphic levels. The distinctive cephalopods of the upper Devonian of western Europe, the clymenids, have only distant relationship to those shown in the illustration. Ammonoids are prevalent forms in the late Paleozoic (Fig. 15–8), but the nautiloids, first common and significant in the Ordovician, outlived them; the ammonoids disappeared at the end of the Mesozoic.

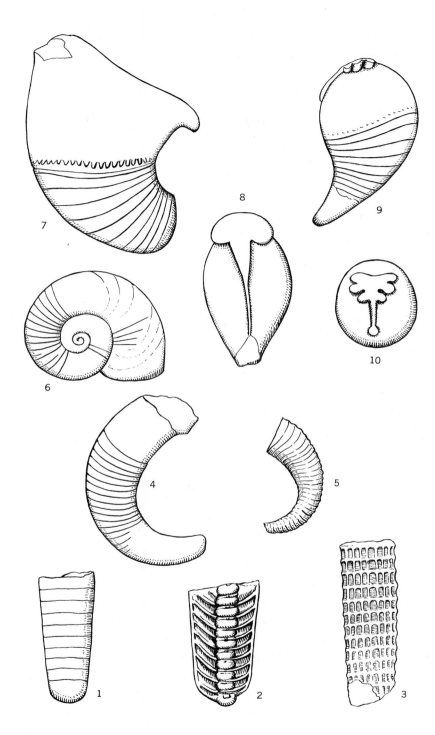

We must first select forms that seem most likely to be lineal descendants of one another. Obviously it is futile to compare the corals of one time with the clams or brachiopods of another. If we were to gather all the information about the cephalopods of the Devonian, for example, we would be able to arrange them according to their resemblances and differences; this is the basis of taxonomy. We might designate those that seem virtually identical as belonging to the same species, others of lesser likeness as of the same genus, and those showing still less likenesses as in the same family of cephalopods (Fig. 15–5). If, conversely, we then chose to examine carefully all those having such characters that we thought them of a family, such as the manticocerids, we might find among them forms that differed in such degree as to be placed in several genera.

ONTOGENY AND PHYLOGENY

In some other organisms, growth is gained by the production of a succession of shells or carapaces that are discarded, as in arthropods such as trilobites (Fig. 15–6) and ostracods. Thus there are many fossil skeletons that represent the successive stages in the development of each then living organism. Yet other organisms have a skeleton that expands without preserving the relics of its growth stages, as in the case of vertebrate animals like ourselves. The successive stages of growth of the individual is known as its *ontogeny*, its life history, in comparison with the adult forms of descendant organisms through time, known as the *phylogeny*. The cephalopods are one of the sort of organic groups in which the adult shell preserves the record of its earlier growth.

Let us examine the structures of one of these genera of cephalopods. If the outer shell is removed (Fig. 15–7), we see a surface showing the pattern of attachment to the outer shell of the septa that divided the chambered shell. The patterns of these junctures, known as sutures, can be traced. Within one individual, those of the later chambers differ somewhat from those of the earlier chambers. The different genera can be distinguished on the nature of the sutures of the later chambers. If we examine forms of the manticocerids from several stratigraphic levels in the Devonian, we find that some have their last sutures different from others, and also that some adult or at least large forms have last sutures that are more like the juvenile suture patterns of others than like other adults (Fig. 15–7C). It is as though the adult of one time resembled the youth of a later time, and that this youth on reaching adulthood developed a somewhat different pattern of suture. So we can suppose that the forms with the more complex sutures are the descendants of those with the simpler sutures, and that they retained

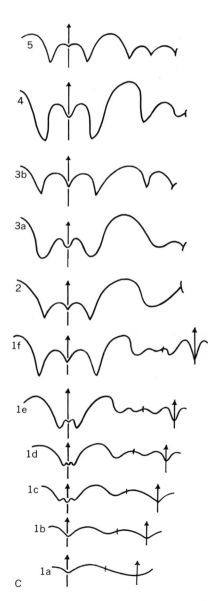

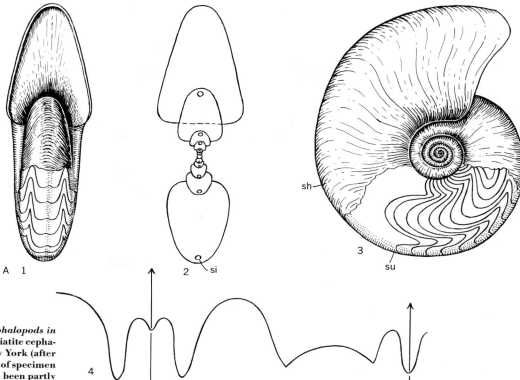

Fig. 15–7. *Ammonoid cephalopods in the Devonian System. A.* Goniatite cephalopod *Manticoceras* from New York (after A. K. Miller); (1) lateral view of specimen from which the shell (sh) has been partly removed to expose the internal mold, with sutures (su) distinctive of a goniatite; (2) cross-section through the specimen showing siphuncle (si); (3) ventral view (the ventral being the outer side of the shell, the dorsal the inner) with the body chamber above; (4) suture drawn with arrows toward the body chamber, the lobes being the curves away from the chamber, and the saddles being those directed toward it.

B. Surface of a bed of upper Devonian limestone from the Helderberg Mountains south of Utica, New York, exposing many specimens of goniatites, the straight *Bactrites* and the coiled *Manticoceras* (New York State Museum).

C. Ontogeny and phylogeny in Devonian goniatite cephalopods. The lower figures (1a–f) are the drawings of a succession of representative sutures in a specimen of the genus *Koenenites* from the Devonian of Michigan, showing the growth of complexity with age. The upper figures are sutures of several genera of goniatites that are thought to be possible linear or adult specimens of phylogenetic descendants of (2) *Ponticeras*, (3a) and (b) being *Manticoceras*, (4) *Koenenites*, and (5) *Timanites* (after A. K. Miller).

the simpler sutures in their earlier, or youthful, stages of development.

If all cephalopods were in a single line of descent, and the successive progressive forms completely replaced their ancestors, all would be simple. We would find that among the cephalopods, all of one age are more primitive than those of another later age. Unfortunately, it is not as simple as that. Some of the less complex sutured cephalopods had descendants that did not become so complex, these being simpler genetic cousins of the more complexly sutured descendants. As a whole, however, we can say that cephalopods of one evolutionary line tend to change in an orderly way with time (Fig. 15–8).

Thus we see that the character and distribution of various cephalopods in time seem to be related to the evolution of one form into another. At least the specimens in the younger beds are similar to those in the older except that the later chambers of the younger forms are separated by more convoluted septa than the comparable chamber of the older forms, while the earlier chambers of the younger forms have septa similar to those of the adults of the older forms. There are very few organisms in which the relationships of descent are as nicely shown as in the cephalopods. So we are to consider that if organisms changed through time, as shown by the stratigraphic record, and if a sequence of development can be shown in some of the organisms, it is reasonable to assume that development has taken place in other organisms, but in a manner often so complex, or often sufficiently obscure that its demonstration is not tenable.

Fig. 15–8. *Permian ammonoid goniatite cephalopods. A.* Complex suture patterns in two closely related genera, (1) *Medlicottia,* and (2) *Artinskia,* from western Texas.

B. Ontogeny and phylogeny in some Permian ammonoid goniatite cephalopods. The lower figures, (1) *Parapronorites* and (2) *Propinacoceras* are from Permian rocks in Texas and Italy; ammonoids of this sort are thought to have been ancestral to forms such as *Artinskia* (3); representative sutures drawn to show the ontogeny in specimens from the western Ural Mountains, Russia. Note how complex the suture patterns in these late Paleozoic genera as compared to those in the Devonian forms (after A. K. Miller and W. Furnish).

PROVINCES AND MIGRATIONS

Paleozoic fossils are a means of learning about the migration routes of the faunas of the past. The distribution of the fossils of a time is related to the character of the containing rocks. Sandstones formed from beach sands contain organisms that preferred the shore habitat, and deep

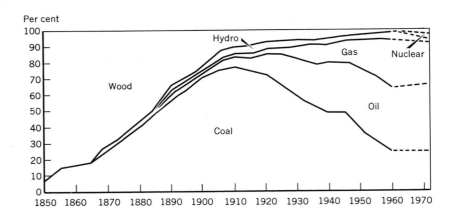

Fig. 15–9. Table showing the sources of energy used in the United States in the past century. All but hydropower and nuclear sources were gained through the consumption of organic material that accumulated in plants and animals; in the past, the source was in present wood, but as demands have risen, coal and then oil and gas have become principal sources.

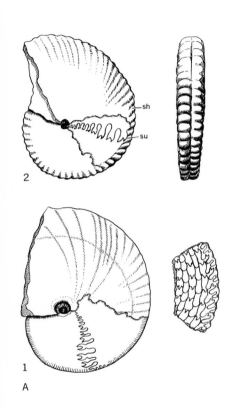

sh
su

2

1

A

3

2

1

B

water muds have shells of animals adapted to that environment, or that settled or drifted to the bottom. Some organisms seem to have been controlled by other factors, for they have distinct geographic distribution in addition to lithic factors. Geographic limitations may be the effects of specific physical conditions in the environment that restricted areas of habitation and have not been recognized. On the other hand, limits to free migration of the forms that developed in a particular region, will have caused them to remain in a geographic province. Among invertebrates it has been shown that Cambrian fossils in the several "provinces" have little in common. The differences long were attributed to distribution of lands that isolated the developing organisms. Another interpretation has been made more recently, that the geographic factors were dependent on habitats, and fossils represent environment-controlled communities, populations isolated by environmental barriers rather than inpenetrable land barriers. Any alternative explanations will still require routes of migration, so the distribution of fossils of a particular time is related to the routes available for migrations. These factors become more apparent when one is dealing with vertebrates that live on the lands.

ANIMALS ON THE LANDS

The distributions of land-living animals have thus been of crucial significance, since they bear upon the interpretation of past continental relationships. Modern amphibians are notoriously intolerant of salt water, and we can only assume that the same was true for amphibians of past ages. Even though amphibians are aquatic animals, their movements from one place to another were and are only by way of fresh-water streams and lakes, and short intervening stretches of moist ground or shaded avenues protected from the desiccating glare of the sun. Consequently, it seems axiomatic that when a single genus or very closely related genera of these vertebrates are found in sediments of widely separated continental regions, as is the case with regard to certain late Paleozoic deposits and fossils, a land connection of some sort must have existed between the areas concerned. Carboniferous and lower Permian amphibians in Europe and in North America show quite definitely that during the time they lived there were routes for the migrations of these animals between the continents.

Land-living reptiles, likewise, are restricted in their movements from one place to another by the distributions of the oceans. Therefore, the occurrence of middle and upper Permian mammal-like reptiles in South Africa and northern Russia is an indication of the presence of

Fig. 15–10. *The mining of coal in Illinois. A.* **The cutting machine on the left is digging beneath the seam of coal; subsequently the coal will be drilled and broken by explosives, preparatory to its being removed.**

B. **Face of a bed of coal in Pennsylvanian rock and an open pit near Leamington, Illinois; the augers bore into the face to bring out the coal of the seam (Illinois Geological Survey).**

A

B

continental connections between these regions during late Paleozoic time.

So it is that the correlation of the evidence from marine and continental fossils, together with other evidence, affords an interpretation of the geography of past ages.

DIRECTIONS OF CURRENT FLOW

Fossils have been of use in determining directions of flow in marine bottom currents. For example, graptolites in Ordovician shales in New York have definite trends in some instances; pointed gastropods in some of the limestones have their apices consistently directed, and straight orthoceracone cephalopods have been rolled into parallel arrangements. Growth in reef corals may be enhanced on the lee side of the colony. Such observations may be quite instructive when systematically investigated.

ORGANIC FUELS

The combustion of fossil fuels, coal, oil, and gas, is the source of more than 95 per cent of the energy produced in the United States. The fuels result from the accumulation, burial, and alteration of past organisms.

Fig. 15–11. Early drilling for petroleum. The Drake Well, as reconstructed near Titusville, northwestern Pennsylvania; the first commercial oil well in the United States, it was drilled in 1859 to a depth of 150 feet in Carboniferous sandstone. Petroleum was first produced in the preceding year in Canada from a dug well in Devonian rocks at Oil Springs, east of Sarnia. Both fields have continued to produce from small but very long-lived wells (Pennsylvania Historical Commission).

Although the use of coal in steam locomotives has ended and that in homes has considerably decreased, the consumption of coal in industry as in steam power-generating plants has increased, so the total use of coal is nearly the same as it was in the past. But the total energy use from all sources has doubled in the past twenty years, and coal now yields only about one-fourth of the energy produced (Fig. 15–10). Oil has come to yield about two-fifths and natural gas about one-third of the energy produced in the United States (Fig. 15–9).

COAL AND LIMESTONE IN ECONOMY

Paleozoic rocks are the predominant source of American coal (Fig. 15–10), about 95 per cent of that produced in the United States coming from Pennsylvanian rocks. Carboniferous rocks, aptly named, are also the main sources on most other continents. The origin of coal has been discussed. Of the oil produced and known in the United States, nearly half comes from Paleozoic rocks, about one-sixth from the Mesozoic, and one-third from the Cenozoic. In the world, however, Tertiary rocks contain about 60 per cent of the oil, and the Paleozoic about 15 per cent. Oil is thought to be produced from the distillation of the soft substances in marine organisms, principally invertebrate

Fig. 15–12. A modern drilling well in Grand County, eastern Utah. The objective is production from Carboniferous sedimentary rocks thousands of feet below the surface in the Paradox Basin. The location of such a well is preceded by extensive field studies of the exposed rocks and geophysical surveys evidencing the nature of the subsurface rock structures (Sinclair Oil Company).

animals, but probably in varying proportions from algae and other marine plants. The origin and occurrence of petroleum will be considered more fully in a later chapter.

Paleozoic organisms form the principal constituents of the most widely used American building stone, the Indiana Limestone, a rock produced from extensive quarries in southern Indiana. The limestone is made of minute organisms, such as foraminifera and gastropod shells, fragments of recognizable fossils such as bryozoans, and grains of fragments of shells of many kinds, all loosely cemented (Fig. 15–13). It is a calcarenite, a sandstone of calcareous composition. The thick beds have cross-stratification; geologists, on looking at building walls, are concerned that builders do not lay the blocks consistently with the bedding right side up. The Indiana Limestone is but one of many that are used in building. The Ordovician reddish Tennessee "marble" shows cross sections of many brachiopods in calcarenite. The Devonian "Ste. Genevieve golden marble" of southeastern Missouri is a calcarenite showing sections of coral masses. The Ordovician marble in Vermont (Fig. 15–14) is the product of metamorphism of limestone that may have had organic origin but has lost most of the fossil evidence. Building stones make most interesting sites for searching for fossils, for they are of many kinds, transported far from their sources. In the city of New York, for instance, there are many buildings which have interior and exterior stone brought by sea from Europe.

Fig. 15–13. Quarry in the lower Carboniferous Indiana Limestone near Oolitic, Lawrence County, Indiana (Indiana Geological Survey). This most widely used of American building stone is porous, essentially a sand made of fossils and fossil-fragments, and yet quite resistant to weathering (see Fig. 12–4).

HISTORY OF PALEOZOIC LIFE

Since we have reviewed some of the paleontological principles that may be learned from Paleozoic fossils, and some of the uses to which these fossils can be put, it may be useful now to summarize in a general way the story they tell. It has been pointed out that the fossils found in Paleozoic rocks show us evidences of evolution through long geologic ages, of the progress and the change in life through time as indicated by changes in the fossils through successive rock systems, of the migrations of organisms from one region to another and of the bearings that the past distributions of animals and plants, as indicated by the occurrence of their fossil remains, may have on the interpretation of ancient climates and environments. But these are lessons that can be learned from fossils of any geologic age—the same principles apply throughout that part of the geologic column in which fossils are preserved. To be more specific, what light is thrown by the evidence of Paleozoic fossils on the nature and progression of life during that long time span beginning with the advent of Cambrian history and

Fig. 15-14. Quarry in Ordovician marble at West Rutland, Vermont (Vermont Marble Company). The rock is very steeply dipping, as can be seen in the bedding in the pillar. The original limestone was composed of detritus of calcareous organisms; the sediment has been so severely deformed that the organic nature of the original is not discernable; the rock now shows a mosaic of calcite crystals in thin-section studied under the microscope. In some instances the crystals have been so arranged as to be parallel, in which orientation they transmit light readily along one axis; such marble is thus translucent.

continuing to the end of the Permian? It is a story unique in its own right.

BEGINNINGS IN THE PRECAMBRIAN

As a prelude to this discussion, the point should be made that life began in the ancient seas, long before the advent of Cambrian history. We can only speculate, at the present time, as to the manner in which life originated, perhaps several billion years antecedent to the advent of the Cambrian, but we can speculate with a certain realistic approach to the subject, nevertheless. This is owing to the fact that certain recent laboratory experiments seem to indicate the manner in which life may have arisen. In these experiments a mixture of water vapor, methane, ammonia, and hydrogen was subjected for several days to an electric spark, to produce a number of amino acids, among which were glycine and an alanine. These are prominent constituents of protein and can combine to form long molecules. From this experiment, we may suppose that in an ancient Precambrian "soup" of similar mixture in the Precambrian seas, frequent discharges of lightning might have initiated the formation of proteins that eventually lead to the origin of the first simple organisms.

CHANGING FORMS THROUGH THE PERIODS

However that may be, it is obvious that life did arise at a very early date, and went through a long history of evolutionary development for which we have virtually no fossil record. Finally, about six hundred million years ago, organisms had become sufficiently complex so that they contained hard parts, and this was the beginning of the fossil record. Cambrian life comes into view suddenly, and with a full panoply of diverse, living things. Life was lived in the sea, and it consisted to a great degree of various protozoans, sponges, coelenterates, brachiopods, some molluscs and arthropods. Such a listing encompasses in its large aspects much of the life of the seas during the long periods of Paleozoic history, yet if we go beyond this simple consideration of the animal phyla it becomes immediately evident that Cambrian life was a mere foreshadowing of what was to come later. The animals of the Cambrian were, on the whole, of primitive aspect, and a glimpse of a Cambrian sea bottom would reveal abundant life, but life showing much less variety than the life of subsequent ages. In brief, the key to the history of marine invertebrate life during Paleozoic time was ever greater diversification through the long procession of years. There was a progressive enrichment of life through time.

Fig. 15–15. In the continental Upper Permian sediments of northern Russia and South Africa are found closely related reptiles, indicative of overland migration routes between these two regions during the later phases of Permian history. Many genera in Russia and Africa indicate the accessibility of each area to the other for land-living animals. This is illustrated by the gigantic, carnivorous, saber-toothed gorgonopsian reptiles, *Inostrancevia* in Russia, and *Rubidgea* in South Africa.

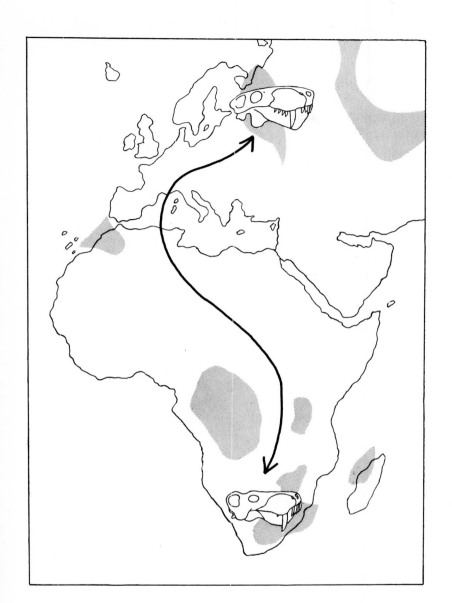

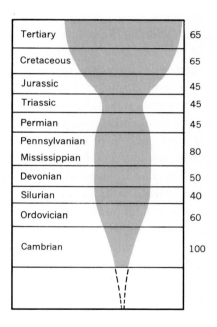

Tertiary	65
Cretaceous	65
Jurassic	45
Triassic	45
Permian	45
Pennsylvanian Mississippian	80
Devonian	50
Silurian	40
Ordovician	60
Cambrian	100

Fig. 15–16. The expansion of life through time (after G. G. Simpson, C. S. Pittendrigh, and L. H. Tiffany, 1957). The width of the shaded pattern is proportional in each geologic period to the known diversity of life in that period. The vertical dimension of the bars representing the periods of geologic time are proportional to the duration of each period.

In one respect, however, the life of the Cambrian may be set apart from that of later periods, and this is in the great development of the trilobites. These complex arthropods, discussed in Chapter 7, reached the zenith of their evolutionary development during Cambrian time. Their history after the Cambrian was one of steady decline and final extinction as the Paleozoic era drew to a close. Theirs was an early success in the history of life.

At first glance, it might seem surprising that some of the most anatomically advanced of the invertebrates should reach their peak so early in the record of life history. But this rise of the trilobites to their supreme eminence during Cambrian times is not truly surprising, if carefully considered. It is for one thing an indication of the fact that these animals must have had a long Precambrian history, as did their contemporaries, during which they largely attained the high grade of anatomical organization recorded by their Cambrian fossils. For another thing, the development of the trilobites in the Cambrian is an indication of the fact that there was a place in the ecology of that time for relatively active and in many cases relatively large animals, animals that could range across considerable distances in search of food, animals that could, in a word, "dominate" the scene wherever they might be. So it has been through time. In the years of the Cambrian, trilobites played the role of dominance. In the years after the Cambrian, they gave way to other dominant types, as we shall see.

This development of new dominant animals is adumbrated in certain Ordovician sediments of North America, namely the Harding Sandstone of Colorado, which contains numerous scales of ostracoderms, primitive jawless vertebrates. But there are only these scattered scales to hint at a long previous history of evolution during which the ancient vertebrates were almost certainly completely naked and, therefore, not subject to fossilization. During that time, these early members of the chordate phylum were probably very much under the domination of the well-established invertebrates, the trilobites, for example.

Then in the Silurian Period, new dominant animals became very prominent in the faunas of that distant age. The giants of the Silurian were the eurypterids, great marine "scorpions" which not infrequently attained lengths of eight or ten feet. At that time, the eurypterids were unchallenged; the ostracoderms, although well along on the several paths of their evolutionary development were, in comparison with the great eurypterids, very small and inoffensive animals. The seas still belonged to the invertebrates. Perhaps the domination of Silurian waters by the eurypterids may be the key to the rise of heavily armored fishes during this stage of earth history, for the armored ostracoderms, indicated only by scales in the Ordovician Harding Sandstone, became

Fig. 15–17. A reconstruction of a Silurian sea bottom, showing the giant eurypterid, *Eurypterus*, an animal attaining a length of ten feet (The Smithsonian Institution). The presence of gigantic predatory invertebrates such as this may account for the fact that the early jawless fishes, the ostracoderms, were for the most part heavily armored. The time of vertebrate dominance was still in the future.

Fig. 15–18. Armored fishes of early Paleozoic times, after restorations by Lois Darling (E. H. Colbert, 1955). Below are shown four ostracoderms, representative of the four orders of these jawless vertebrates that lived during the late Silurian and the early Devonian. Above are two acanthodian fishes, considered by many students as forming one large division of the placoderms. The one at the top is *Parexus*, of early Devonian age, and the one below it the well-known genus, *Climatius*, found in upper Silurian and Lower Devonian sediments.

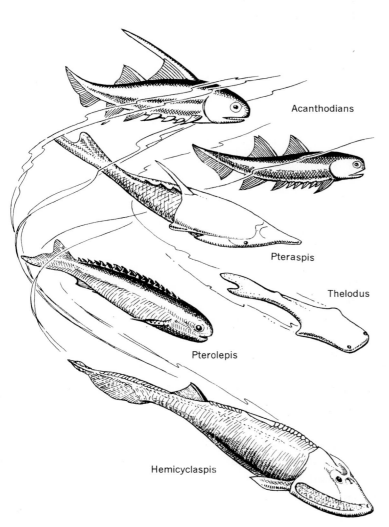

Acanthodians

Pteraspis

Thelodus

Pterolepis

Hemicyclaspis

numerous in the shallow waters of late Silurian seas and estuaries. They were armored for protection against the eurypterids, and perhaps against other invertebrates as well. They lived in a hostile world.

With them were other armored fishes, the early placoderms and the acanthodians, or "spiny sharks," the first of the jawed fishes. These primitive jawed fishes were rather small and, like their jawless relatives, were heavily protected against the aggressive eurypterids that inhabited the waters in which they lived.

By Devonian time life in the sea had become much more diversified than had been the case during the preceding Silurian period. Sea bottoms were inhabited by immense numbers of brachiopods and corals, echinoderms, sponges, and molluscs. Very probably a sea floor of that age would have appeared almost as richly and colorfully occupied by a kaleidoscopic array of attached, stalked, and freely swimming invertebrates as is a modern tropical coral lagoon. All of this is strikingly attested by some of the fossiliferous sediments of New York State, where along the middle reaches of the Hudson River and to the west there is a burial ground for Devonian life hardly equaled elsewhere in the world.

Adding to the diversity of Devonian waters was the array of fishes swimming in and out among the animals without backbones, or resting on the sea floor. Jawless ostracoderms, armored in various ways, were now much more diverse than had been their Silurian ancestors; indeed, the Devonian Period marked the culmination of ostracoderm evolution. It marked the culmination of placoderm and acanthodian evolution as well, and by late Devonian times some of the armored placoderms had become veritable giants, twenty feet and more in length. Such is *Dinichthys,* the great placoderm of the upper Devonian Cleveland shales in Ohio, a fish that without doubt dominated its environment even more completely than had the giant eurypterids of the Silurian seas.

In addition to the jawless ostracoderms and the jawed placoderms and acanthodians, the other major groups of fishes had become established in Devonian seas, rivers, and lakes. These, as has been mentioned, were the early sharks and the early bony fishes, the latter with well-formed skeletons and heavy, shining scales. As has been remarked in Chapter 11, the Devonian is rightly regarded by students of evolution as a very crucial and significant time in the history of the vertebrates, particularly because of development of certain bony fishes, the crossopterygians, in the direction of land-living tetrapods. In the crossopterygians the structure of the skull, the vertebral column, and the paired fins were of such form that from them the skull, the vertebral column, and the limbs of the ancestral amphibians (which appeared at the end of Devonian times, as we shall see), may be derived. Moreover, it is evident that the crossopterygian fishes had lungs that enabled them to breathe air. The same was true for the numerous Devonian lung fishes or dipnoans, an evolutionary line closely related and parallel to the crossopterygians, but one which never led in later ages to anything other than more dipnoans. And while these two groups of bony fishes were evolving through Devonian history in ways that have been of so much interest to modern students of evolution, the great majority of Devonian bony fishes were developing along lines ancestral to the bony fishes of late Paleozoic time, these in turn being the forms from which the great hosts of Mesozoic and subsequently of Cenozoic bony fishes were descended.

PLANTS IN THE DEVONIAN

The Devonian is important, however, not only for its record of life in the sea, but also for its remains of land-living organisms. There seemingly had been little land life previous to the Devonian Period; indeed the evidence of the rocks would seem to indicate that the lands were barren through all of the Precambrian and well into the beginning of Paleozoic history.

The first indication of life on the land comes from rocks of upper Silurian age containing the remains of plants. In the Silurian sediments, too, have been found a fossil scorpion—our first inkling as to the invasion of the land by animals. The record is scanty, as may be inferred from these brief remarks, but it seems likely that by late Silurian times

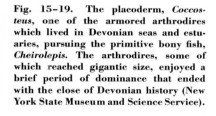

Fig. 15–19. The placoderm, *Coccosteus*, one of the armored arthrodires which lived in Devonian seas and estuaries, pursuing the primitive bony fish, *Cheirolepis*. The arthrodires, some of which reached gigantic size, enjoyed a brief period of dominance that ended with the close of Devonian history (New York State Museum and Science Service).

there was a considerable variety of life on the land in the form of primitive land plants and a limited array of invertebrates, probably arthropods for the most part.

But when we come to the Devonian fossil record, we find assemblages of plant remains attesting the presence of green marshes and subsequently of well-established, albeit very primitive, forests on the continents of the world. The early Devonian land plants covered low, coastal marshes and other wet habitats with mantles of low, spore-bearing thallophytes, some of which have been briefly described in Chapter 11. Once having gained a foothold on the land, plant life evolved rapidly, so that by middle and late Devonian time there were forests in which there grew lycopods, scouring rushes and primitive ferns. The evidence for such an ancient forest was strikingly revealed many years ago at Gilboa, New York, during excavations for the construction of a large dam (Fig. 11–24). The stumps of many primitive "trees" were revealed, plants that in life may have stood twenty feet or more in height. By the end of the Devonian Period such forests were probably extensive, making the land a habitat quite conducive to the support of animal life.

ANIMALS INVADE THE LANDS

And it was at this stage in earth history that the backboned animals migrated out of streams and ponds and into the marshy forests, where there were wide opportunities for a new kind of life. These first

Fig. 15–20. A comparison of the ancestral amphibian, *Ichthyostega*, from the upper Devonian sediments of East Greenland, with the advanced crossopterygian fish, *Eusthenopteron*, from the upper Devonian of Quebec. Restorations are shown above, the skeletons upon which these restorations are based, below. Note the close similarity in the patterns of the skull bones, the arrangements of the bones of the limbs, and the form of the vertebral column, all of which point to the direct derivation of the first amphibians from crossoptyergian ancestors. Of particular interest are the limb-like paired fins of *Eusthenopteron*, and the fish-like tail of *Ichthyostega*.

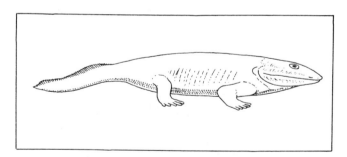

Ichthyostega

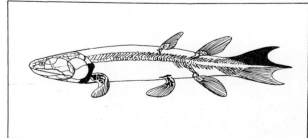

Eusthenopteron

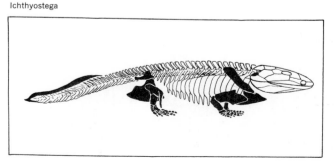

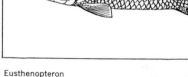

amphibians, the ichthyostegids, the remains of which are known from Greenland and from Quebec, have already been mentioned in the discussion of Devonian life. In spite of their immediate derivation from fish ancestors, the ichthyostegids were well adapted for life on the land; they had strong limbs for walking. Yet they retained certain vestiges of their fish ancestry, among them a rather fish-like tail, marking them as truly transitional types, the connecting links between two major vertebrate classes. From them the numerous Carboniferous amphibians arose.

The amphibians that lived during Carboniferous times prospered because there was available to them a favorable world-wide environment, an environment of tropical jungles which clothed the continents from the equator to high latitudes, north and south. It was an environment of low swamps, in which were luxuriant growths of ferns and other vegetation, all beneath a canopy of giant lycopods, tree ferns, calamites, lepidodendrons, sigillarians, and other trees. We have seen in Chapter 13 how the amphibians became established as the ruling land animals during Carboniferous time, a role they enjoyed briefly before the reptiles arose and became sufficiently advanced to displace the amphibians as rulers of the land, this being an event of early Permian time.

But even though the amphibians were dominant in the lush jungles of the Carboniferous, they shared the land with some extraordinary arthropods—giant dragonflies and cockroaches, as well as other insects, many of them remarkable for their large size. This stage in earth history was, in a sense, a golden age in the evolution of the insects. They were the only flying animals, they were free in the air, with no enemies there to prey upon them. Food supplies and temperatures were conducive to growth; consequently many of the insects grew to maximum sizes that could be attained by such animals. Just as later, during the middle and late Mesozoic, the dinosaurs reached the greatest sizes possible for land animals with an internal skeleton, so in the Carboniferous the insect giants reached the greatest sizes possible for land animals with an external skeleton and a tracheal system of respiration.

There were in addition to the insects various other land-living invertebrates in the Carboniferous forests, centipedes and scorpions, the ancestors of the spiders, and air-breathing gastropods. Undoubtedly many of the forest-dwelling invertebrates of that age were preyed upon by the amphibians which lurked in the undergrowth and frequented the edges of streams and ponds and marshes. All in all we are afforded from the fossil record a fairly detailed picture of Carboniferous land life.

Coincident with the abundant Carboniferous fossil record of life on the land, that of life in the sea continues with ever increasing richness. The early part of Carboniferous history, the Mississippian, was in many parts of the world a time of clear seas in which there were great coral reefs, and especially echinoderms. This was the age of blastoids and crinoids, when sea floors were abundantly populated by flower-like echinoderms, swaying back and forth in the ocean currents on their long, jointed stalks. Their fossils are well known in the central part of North America, in the Mississippi Valley area from which the subsystem is named. In fact, there was a considerable change in the composition of marine life with the transition from Devonian to Mississippian time, marked by the evolutionary expansion of the crinoids and the blastoids. Many of the sea dwellers that had been so abundant in Devonian times continued, however, into and through the Mississippian, these being the brachiopods and bryozoans, the sponges, and the molluscs, these latter animals evolving and continually increasing in diversity. The trilobites, in contrast, were declining. On the whole this was a time of great evolutionary expansion in the seas of the world.

Marine life continued in abundance through the latter part of the Carboniferous, the Pennsylvanian, the age of the great coal forests on the land. This was still an age of great diversity among the brachiopods, of continued evolution among the molluscs, and of coral reefs. Among the common and widely distributed Pennsylvanian fossils are the fusilinid protozoans, discussed in Chapter 13. They lived abundantly in many Carboniferous seas, so abundantly that almost alone they formed vast deposits of limestone.

With the advent of Carboniferous history the ostracoderms became completely extinct, as did the placoderms; the acanthodians persisted through the Carboniferous and into the beginning of Permian times. Consequently, there was a marked change in the composition of fish faunas during the transition from the Devonian to the Carboniferous, just as there was in the constitution of many invertebrate assemblages. The fish faunas of the Carboniferous were, in a word, more modern in appearance than those of Devonian times. The acanthodians were not particularly numerous; sharks and bony fishes ruled the waters. In the Carboniferous seas were many sharks of varied forms of habits, but few bony fishes, these latter being largely confined to the fresh waters, seemingly the habitat in which they had originated. Here the ancient palaeoniscoids, the ancestors of the advanced bony fishes that were to evolve so widely during Mesozoic and Cenozoic times, swarmed in

Fig. 15–21. A restoration of a Carboniferous forest, with a giant dragonfly, having a wingspread of some two feet (New York Museum and Science Service).

Fig. 15–22. A late Permian South African scene, as restored by John Germann (American Museum of Natural History). Here a mammal-like reptile, the carnivorous gorgonopsian, *Lycaenops* (about the size of a small collie dog), from the lower Beaufort beds of the Karroo Series, is shown stalking a small anomodont, *Dicynodon*, one of the presumably herbivorous mammal-like reptiles that evidently lived in vast numbers during Karroo times.

great numbers, to constitute the great bulk of fresh water fishes. The crossopterygian fishes and the lungfishes, so characteristic of Devonian fresh waters, were reduced to relics of their former state. Interestingly, fresh-water sharks were numerous inhabitants of the streams and ponds of Carboniferous time.

DIVERSITY OF LIFE ON LAND

Thus life developed through the later phases of the Paleozoic era, to reach a crisis of sorts in the Permian Period, the final chapter of Paleozoic history. For with the advent and the progression of Permian time there was a gradual shift in environmental conditions that was to have a considerable effect on the plants and animals of the world, this being brought about largely by the emergence of land areas, leading in turn to a wide-spread restriction of the low, tropical forests which had so characterized the late Carboniferous scene. Lands would seem to have become progressively drier and more barren of vegetation, and in some areas there were deserts. For example, the Permian Coconino Sandstone of the Grand Canyon region, mentioned in Chapter 14, gives graphic evidence of a dune sand, across which primitive reptiles wandered, to leave their trackways as evidence of their comings and goings.

But it must not be thought that deserts, even though prevalent in some regions, were the dominant environments. In the last chapter we saw that in Texas there was a low delta bordering the sea, and a profuse host of amphibians and reptiles lived on it. This delta was certainly a well-watered area, and the same must have been true of the higher lands to the northeast, where amphibians and reptiles quite like those of the Texas delta lived. Yet although the habitats of the early Permian tetrapods of North America were to a large degree well-watered, they probably were not the locales of luxuriant plant growth, as had been the vast lands covered by Carboniferous forests. Climates were more variable than in the preceding Paleozoic periods, and plant life had been altered in accordance with the new climatic conditions.

What life was like on the land in North America during late Permian times we do not know, for the record is missing. In South Africa lands were high, almost prairie-like, and were inhabited by great numbers of diverse reptiles, among which the progressive mammal-like reptiles were overwhelmingly numerous. Perhaps the same was true in North America, but on this point we probably will never have a definite answer.

PERMIAN MARINE ORGANISMS

As to life in the Permian seas, the North American record is much more complete, for it continues from the beginning to the end of the period. In the Permian limestones of western North America is the evidence of continued life on the sea bottoms, of great reefs, such as the massive limestone reefs that form the Guadalupe Mountains in western Texas, on which were clustered many invertebrates in varied array. Sponges and brachiopods were numerous and molluscs were prominent. Many of the lines of evolution that had originated in early Paleozoic times were being continued as the Permian Period drew toward its termination. But with the close of Permian time there were wide-spread extinctions of many invertebrate groups, extinctions so profound that they are frequently regarded as being parts of a broad wave of extinction, a "Permo-Triassic crisis" that marked the end of Paleozoic history. Some mention of this has already been made. It may be added here that the brachiopods, so numerous and characteristic of Paleozoic times, and even of much of the Permian, suffered a great decline at the end of Permian history, a decline so extreme that many lines of brachiopod evolution disappeared. Trilobites became extinct during the Permian Period. The same is true for various groups of echinoderms and of corals. Certain lines of molluscs that had been eminently successful during Paleozoic time died out with the close of Permian history, to be replaced, during the advent of the Mesozoic, by related groups. And this wave of extinction would seem to have affected the fishes, too. The Paleozoic marine fishes largely disappeared at the close of Permian time. Specifically, the reign of the paleoniscoid fishes, so widely divergent during the Carboniferous and Permian periods, had come virtually to an end. New groups of bony fishes were to replace them in post-Paleozoic waters.

Fig. 15–23. The effect of the transitions from Permian into Triassic time, and from Triassic into Jurassic time, on the development of the reptiles. The Permo-Triassic transition is generally regarded as a time of crisis in the history of life on the earth, and it will be seen that there was a great reduction in the number of genera between late Permian and early Triassic history. Most of this reduction was, however, within the mammal-like reptiles, and actually there was an increase in the number of reptilian orders, as may be seen. At the end of Triassic history, there was an even more marked decrease among the reptiles that affected the numbers of both orders and genera. In brief, the great revolution in tetrapod life (for the amphibians were also affected) was at the close of Triassic time.

	ORDERS	GENERA
Lower Jurassic	11	27
Upper Triassic	13	122
Middle Triassic	8	47
Lower Triassic	7	56
Upper Permian	4	199

Yet even though the end of the Paleozoic Era and the beginning of the Mesozoic was a time of great changes among many groups of organisms, it was not universally so characterized. Some groups of animals and plants lived through from the one great era into the next with very little change—with very little if any break in the continuity of their evolutionary development. This is particularly true of some of the land-living animals, particularly the tetrapods now found in South Africa and in Russia. In these areas there is a continuous record of the development of tetrapod life during the transition from Permian to Triassic time, and although there were some dramatic amphibian and reptilian extinctions at the close of the Permian, we can nonetheless see various evolutionary lines continuing into the Triassic without any marked effects on their development. The record of life on the land emphasizes the fact that time is continuous and life is continuous, in spite of the vicissitudes suffered by many animals and plants. So life continued even though an era had come to its end and, in coming to this end, had brought about the extinction of numerous evolutionary lines which had flourished through past ages.

16

The Triassic Period, When the Lands Were Dominant

Tracks of a dinosaur in the red sandstones of the Triassic Newark Group at Turners Falls, on the Connecticut River in Massachusetts (G. W. Bain). The specimens are the casts of tracks formed by the filling in the overlying layer. The illustration is somewhat reduced from the original size.

16 *The Triassic Period, When the Lands Were Dominant*

The Mesozoic Era contains three periods, the Triassic, Jurassic, and Cretaceous, differentiated by their faunas, which were among the first that were critically studied and recognized in Europe. One may wonder whether there is really a sound basis for placing a major stratigraphic boundary between the Permian and Triassic. Perhaps another way to consider the matter is one of whether the record as known today would have resulted in the same classification as was established on limited record in northwestern Europe a century and a half ago.

BASE OF THE SYSTEM

There are very few places in the world where late Permian rocks seem to be preserved, as in the island of Timor in the East Indies. There, where they are succeeded directly by lower Triassic rocks, the differentiation is made with greatest difficulty. Nevertheless, it is probably true that the differences in the fossils of the Triassic and Permian is as great or greater than that between faunas of any pair of the Paleozoic periods. And on the other hand, the similarities between fossils in the two systems is so great in some groups as to make the separation difficult; perhaps this is as it should be, inasmuch as it is assumed that the later are descendants of the earlier. The very fact that late Permian marine sediments are so rarely present is interpreted as indicating that lands were most extensive in late Permian, as is indeed the case in the South African Karro Series, a physical condition that in itself set this time off as being peculiarly significant.

Locally in North America, there are localities having record of marine withdrawal and erosion, some with intervening orogenic movement, but we have seen that such events were present recurringly within the Paleozoic. Thus along the Atlantic seaboard, late Triassic sediments in many places lie with unconformity on deformed Paleozoic rocks, rarely if ever on sedimentary rocks of Permian age, however, though perhaps on intrusives of that age. In the Grand Canyon region, lowest Triassic lies with disconformity, erosional unconformity, on middle Permian Kiabab Limestone. But in southeastern Idaho, the earliest Triassic marine siltstones and shales lie on similar late Permian sediments; they are distinguished by their faunas. The separation of the Permian and Triassic red beds in Wyoming is based on tracing the distinctive lithologic units in each system from Idaho.

Fig. 16–1. The Palisades of the Hudson River, rising about two hundred feet above the water on the New Jersey shore north of the George Washington Bridge to New York City; the view is from the east near Hastings, New York (A. Devaney).

The basalt-like igneous rocks are about 1000 feet thick, forming an intrusive sill that altered the underlying upper Triassic shale exposed nearer the water's edge. The columnar jointing, which gave the name to the Palisades because they seemed like a wall of vertical logs, is at about right angles to the surface of cooling; as the sill dips away to the westward, the columns tilt a little in the cliff. A zone rich in the mineral olivine persists a few tens of feet above the base of the sill, well-shown along a road below the cliff face north of the bridge. The intrusion probably entered in late Triassic. The top of the cliff is beveled by a peneplane that seems to pass below the Cretaceous rocks nearer the Atlantic Ocean.

Fig. 16–2. Map of the Atlantic Coast of the United States, showing the areas underlain by the sediments and lavas of the Newark Group of late Triassic age. Additional Newark rocks are exposed along the Bay of Fundy in Nova Scotia and New Brunswick (see Fig. 13–1).

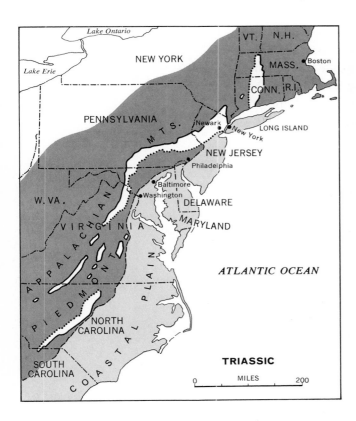

THE THREE-PART ORIGIN

Triassic, like Carboniferous, is a name that was given to the rocks of a time from their lithology rather than a designated place where they might be seen. Trias was applied in central Europe in 1834 to a three-division rock sequence that contrasted to a two-fold one, the Dyas, below. The latter term was dropped in favor of the place-named Permian, but Triassic has been retained, one of the few relics of the early classification wherein the sequence in central Europe was thought to be of universally distributed constant lithologic layers.

THE NEWARK GROUP ON THE ATLANTIC COAST

The most distinctive scenic natural feature in the metropolitan region of New York is the Palisades, a dark cliff face of basaltic lava, called trap by quarrymen, that extends for miles along the west side of the Hudson River opposite Manhattan Island and northward. The name was given because the nearly vertical columns in the face of the ridge resemble standing logs of the wall of a palisade, such as might have surrounded a frontier fort (Fig. 16–1). The Palisades is of an intrusive rock; there are several other ridge-forming lavas of Triassic age in northern New Jersey. They stand in conspicuous linear hills, for the basaltic rocks are more resistant to erosion than the extensive associated sediments in which there are fossils that enable us to date the sequence. Similar lavas can be seen in many other places along the Atlantic Coast (Fig. 16–2). They form the long ridge and peninsula separating the Annapolis Valley and St. Mary's Bay from the Nova Scotia shore of the Bay of Fundy. The high cliff of West Rock by New Haven, Connecticut, and the crests of the Hanging Hills of Meriden (Fig. 16–3) to the north are supported by Triassic lava flows. The later phases of the Battle of Gettysburg in Pennsylvania involved fighting on Big and Little Roundtop, of similar lavas. The sediments, generally red beds, extend from the northern Bay of Fundy to southern North Carolina in a series of elongate areas along the coast. The general provincial name given to these Triassic rocks is the Newark Group, from the city in New Jersey near many quarries that in the nineteenth century supplied the "brownstone" for residences in New York City. The area of Triassic west of the Hudson River in northern New Jersey, extending northward into New York, presents many of the features that characterize the group in all the areas (Figs. 16–4 and 16–5).

PALISADES OF THE HUDSON RIVER

The Hudson River follows a somewhat curved course from The Highlands of the Hudson River at West Point, where it passes through a

Fig. 16–3. The Hanging Hills of Meriden, in Connecticut, showing eroded lava flows within the Triassic sediments (J. P. Strang).

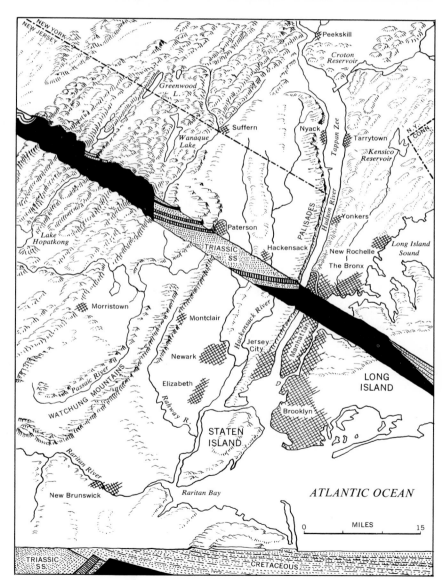

Fig. 16–4. Triassic Newark Group in New Jersey. Physiographic diagram of the northern end of the Triassic Lowland in New Jersey and New York, with a structure section showing the relations of the Mesozoic and Tertiary rocks to the crystalline rocks, which are Precambrian in the highlands on the left, perhaps lower Paleozoic under much of New York City (after Erwin Raisz). The Palisades along the west shore of the Hudson River opposite Manhattan are of an intrusive basalt or diabase sill, but the mountains farther west are of lava flows, each now tilted and more resistant than the associated sedimentary rocks. The Hudson River has eroded a subsequent valley in the lowest Triassic sediments between the Palisades sill and the similarly resistant crystalline schists of Manhattan. Cretaceous sediments underlie the coastal plain on a peneplane that also bevels the older rocks to the northwest.

gorge in Precambrian granitic rocks and schists, southward and south-westward to separate the island of Manhattan—underlain by gneisses, marbles, and schists that are Precambrian-like but are probably earliest Paleozoic—from the Palisades cliff and the gray and red sandstone below it. The river follows the outcrop of the relatively easily eroded basal Triassic sediments which lie unconformably on the more resistant crystalline rocks of the city (Fig. 16–6). At the base of the Palisades cliff below the west end of the George Washington Bridge, we can see a fine-textured dark trap-rock overlying bedded sediments; the glassy texture at the base of the trap resulted from hot magma being rapidly cooled at the contact with the cold sediments. We can see similar chilling of the upper contact of the trap on the gently west-dipping top surface of the Palisades, and fragments of overlying sediment floated in the lava. The Palisades are the face of a section of part of a thousand-foot thick plate-like intrusive mass of basic igneous rock of basaltic nature, a sill (Fig. 16–7). When studied in greater detail, the cliff shows progressive change in mineral and chemical composition upward, the minerals overlying the basal-chilled contact being those that crys-tallized first from the cooling magma, successively higher zones being accumulations of successively later settling crystals formed at progres-sively lower temperatures. There seems to have been some intrusion of the fluid residue of late stages into the already solidified early stages of the magma. The Palisades thus form a magnificent exhibit of the phenomena of crystallization from molten rock, a phase of physical chemistry.

If we go up the Hudson, The Highlands have an abruptly rising face of Precambrian rocks, a mountain front standing higher than the Triassic ridges. This mountain face continues in a straight course southwestward into New Jersey as the front of the Ramapo Mountains.

Fig. 16–5. Geologic map of the area of Triassic rocks along the Delaware River in New Jersey and eastern Pennsylvania (after D. B. McLaughlin). The map shows the conglomeratic facies of the Triassic along the bounding fault scarp grading into the finer textured sandstones and shales. A principal fault trends into the crystalline rocks of the northern high-land, offsetting the Triassic in the ad-joining lowland; such faults have been interpreted as transcurrent, the northern block moving laterally to the right rela-tive to the southern one.

Fig. 16–6. Structure section from New York City (Manhattan) westward to the Ramapo Mountains Precambrian rocks, showing the westward dipping Triassic sedimentary rocks (Newark Group) with the intrusive sills, such as the Palisades, and the lava flows of the Watchung Moun-tains. The Triassic rocks lie unconform-ably on the questionably Paleozoic schists of New York City, and are faulted down against the crystalline rocks on the west; the distribution of conglomerates along the west indicates that they were derived from the rocks of the fault scarp, though at a time when it preserved rocks that have since been eroded.

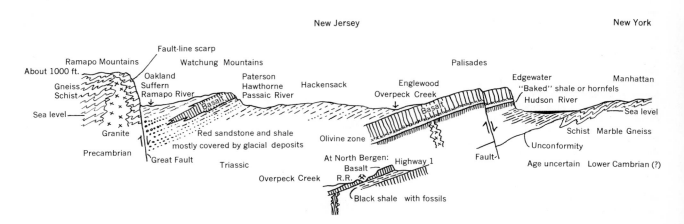

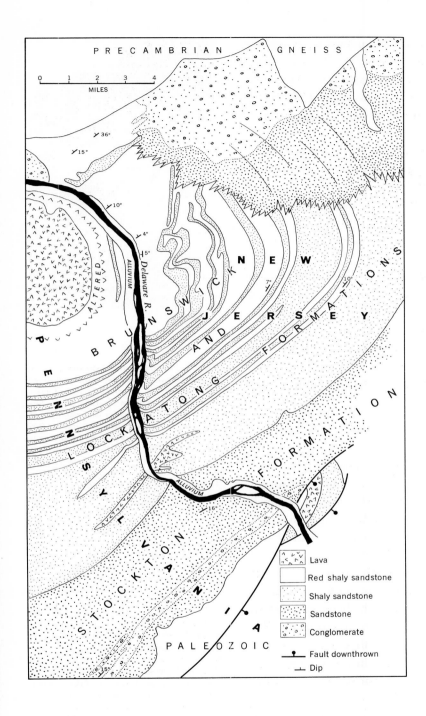

PRECAMBRIAN GNEISS

0 1 2 3 4
MILES

36°
15°
10°
4°
5°

Delaware R.

ALTERED

ALLUVIUM

NEW

BRUNSWICK AND

JERSEY

LOCKATONG FORMATION

7°

10°

PENNSYLVANIA

STOCKTON FORMATION

ALLUVIUM

15°

PALEOZOIC

Lava

Red shaly sandstone

Shaly sandstone

Sandstone

Conglomerate

Fault downthrown

Dip
15°

359

Triassic rocks occupy the area between the mountain front and the Hudson except for a small area where the base of the Triassic swings westward directly south of The Highlands, and another at Hoboken, across from Manhattan, where crystalline rocks like those of Manhattan lie below exposed Triassic sediments. If we traverse northwest from George Washington Bridge over the Hudson, we see very few rocks other than the traps, but there are a few exposures of red sandstones and shales, and of dark argillites, dipping gently northwestward toward the Ramapo scarp. As we go northward, the dips become westerly toward the scarp, the Palisades cliff curving in association with the beds it intrudes. So the structure at the surface is as though one cut a spoon longitudinally, then placed the Ramapo scarp in the place of the left half of the spoon.

Many of the sediments are red sandstones and shaly siltstones that seem quite barren of fossils. In the course of years, several skeletons of land-living reptiles have been found that indicate the age to be late Triassic. The dark argillites in places have myriads of minute bivalved crustaceans, *Estheria*, like forms now living in fresh-water lakes and brackish lagoons. The same beds have well-preserved skeletons of coelacanth fishes and, rarely, remains of gliding reptiles, some of the earliest known aerial vertebrates.

CYCLIC SEDIMENTS

The rocks have a cyclic succession (Fig. 16–8) cyclothems that seem particularly significant because being in non-marine rocks, they suggest that climatic control systematically varied at the site of sediment production. Plants occasionally are associated. Long ago, in similar rocks west of Richmond, Virginia, there were mines in coal of such poor grade that they were abandoned with the discovery of the rich deposits in the Carboniferous in states to the north and west. Raindrop impressions, salt-crystal impressions and dinosaur tracks indicate the non-marine origin of the red sediments. Some of the beds seem to have been laid in stream plains, others in lakes, lacustrine sediments. The cycles in lacustrine sections seem of about twenty thousand years span, with larger groups of about a half-million years duration, comparable to those in the Carboniferous. As they are not marine, they cannot be directly eustatic controlled, unless climatic fluctuations are correlated with rise and fall of seas. Such cyclic phenomena will be explained again when we come to discuss the rather recent glaciations of the Pleistocene Period.

Conglomerates (Fig. 16–9) have an interesting distribution. They are along those borders of the Triassic areas that adjoin fault-line scarps

Fig. 16–7. **Exposure of the olivine zone of the Palisades diabase intrusive along the highway at Edgewater, New Jersey, across from Grant's Tomb, New York City. The Palisades diabase, a basaltic rock, forms an intrusive about 1000 feet thick in the lower part of the sediments of the late Triassic Newark Group. The intrusive is sill-like, about parallel to the sediments, along the face of the Palisades, but is cross-cutting to its northward extension in New York (see Fig. 16–4). The olivine-rich zone is thought to be a layer of crystals that settled within the magma as it cooled, the rocks below and above having crystallized earlier and later. An alternative interpretation is that the olivine-rich zone was introduced as a sill of the last residue of the molten magma after most of the lava had solidified.**

The sketch at the right shows the changing constituents of the magma that form the sill, with the concentration of the mineral olivine near the base; this has been attributed by some to the settling of these early crystallizing, heavier particles at an early stage of cooling of the magma; some have believed that the olivine does represent a fractionation of the magma, but that it was intruded as a sill within the larger sill.

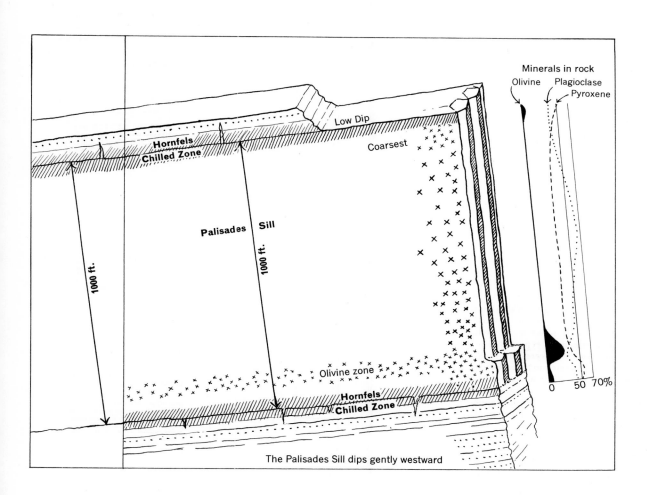

Low Dip

Hornfels
Chilled Zone

Coarsest

Palisades   Sill

1000 ft.

1000 ft.

Olivine zone

Hornfels
Chilled Zone

The Palisades Sill dips gently westward

Minerals in rock
Olivine   Plagioclase
              Pyroxene

0   50 70%

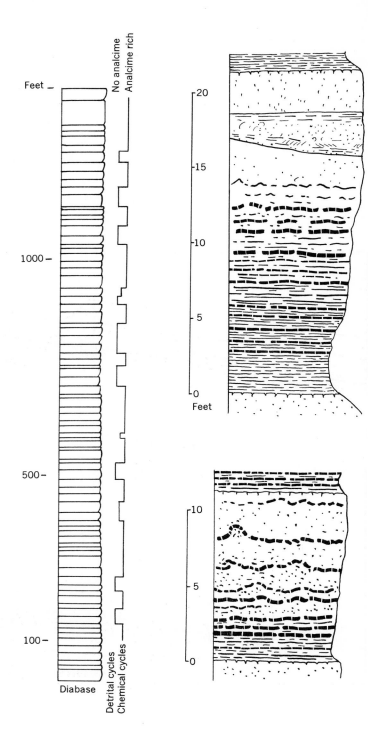

Fig. 16–8.   Cycles of sedimentation represented in the upper Triassic Newark series in New Jersey (after F. H. Van Houten). Such sequences, repeated hundreds of times in the succession, seem related to some phenomenon that affected deposition in a repeating manner in a continental environment, in sediments laid in streams and lakes. The upper illustration is of beds having more sand than in the lower. The source of the sediments was deeply weathered. Each cycle is believed to represent deposition through a span of one to two thousand years, and these are in turn assembled in large, cyclic units of the order of a half-million years duration.

Fig. 16-9. Conglomerate in the red Newark sandstone of the Triassic along the New York State Thruway west of the Tappan Zee Bridge. The exposure is about three miles from the fault-line scarp that forms the western wall of the Triassic Lowland, and that is thought to be along the fault that produced the escarpment adjoining the stream plains where the Triassic sediments were laid. The pebbles are mostly of gneisses and granites such as are in the Precambrian of the escarpment but some are pebbles of Paleozoic sediments such as originally were near, but were eroded away and now are only found farther west. A few fossil reptiles have come from the Newark sediments in nearby northern New Jersey.

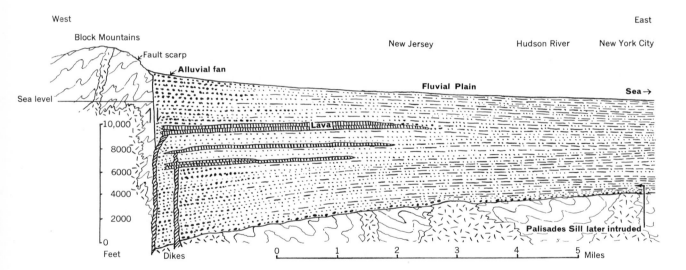

Fig. 16-10. Restored section interpreting the relations in New Jersey in the late Triassic Period. A rising fault block on the west produced detritus that was distributed by streams in the subsiding block on the east, the coarse sediments near the fault grading into finer ones at a distance. Lava flows are interpreted as issuing from fractures along the western side of the basin, and sills were intruded into the sedimentary succession. The thickness accumulated in the basin has been determined as more than twenty thousand feet.

(Fig. 16–10). The pebbles are like those in the scarp, though in northern New Jersey and adjacent New York, there are pebbles also of younger Paleozoic rocks that no longer remain preserved in the scarp. The scarp marks the position of a normal fault; the Triassic beds toward the north swing against the scarp and are cut off; they dip toward the base of the scarp. The conglomerates having rocks such as are in the scarps, must have come in part from erosion of the fault scarps. They are principally in a band of a mile or so right along the fault because they were deposited by streams flowing from a fault scarp in the same position. As the conglomerates are in all horizons from the oldest to the youngest, faulting must have been proceeding interruptedly throughout the time of deposition of the late Triassic sediments. The scarp in New Jersey continues southwest to the Pennsylvania, though it is not a simple, linear fault line.

HISTORY OF THE FAULT-BOUNDED TROUGHS

The region had been deeply eroded after the Appalachian orogeny; although the erosion surface was nearly a plane in parts of the region, it may have had more relief in others. In middle Triassic time, there was a very gentle seaward slope through the region. A rift or fault formed, with the southeastern side subsiding while the northwest rose. Streams carried rock fragments from the fault scarp into the adjoining basin, depositing them in shallow lakes in the lowlands, or flowing across the basin on a gradually sloping plain to the sea. Further faulting repeated the process, until a great wedge-shaped mass of Triassic sediments and lavas of three miles or more thickness accumulated in the fault-bordered trough. From time to time lavas flowed out, perhaps from fissures along the fault, or from openings penetrating the sediments to form the Palisades sill, which originally extended farther to the east, the present cliff having the position to which erosion has cut back the sill. The Triassic rocks were peneplaned subsequently in the Jurassic and covered by later Mesozoic; it is this erosion surface that gives such flatness to the top of the Palisades. Erosion has stripped the cover off, and now the ridges stand where the harder traps are at the surface, and the Hudson River, other valleys, meadows, and salt marshes are in the bands of the less resistant sandstones and shales. The stripped slope of the pre-Triassic peneplane has been recognized on the east shore of the Hudson opposite the Palisades.

Other areas along the Atlantic Coast are similar. In Connecticut and Massachusetts, the main Triassic-developing fault is along the east (Fig. 16–11); the depression contains four or five miles of sediments and flows (Fig. 16–11B). These have been eroded away on the west as

Fig. 16–11. Triassic rocks of the Connecticut Valley (G. W. Bain). A. Fault-line scarp on the east wall of the Triassic lowland northward from the Connecticut-Massachusetts state line; the hills on the right are of metamorphosed Paleozoic sedimentary rocks, the lowlands are underlain by the Triassic sedimentary rocks. B. Columnar section of the upper Triassic Newark Group in the Connecticut River Valley of Connecticut and Massachusetts (after P. Krynine). The conglomerates are a facies that persists along the eastern margin of the outcrop, for it was deposited from streams flowing from the fault-raised highlands that adjoined the subsiding valley block.

A

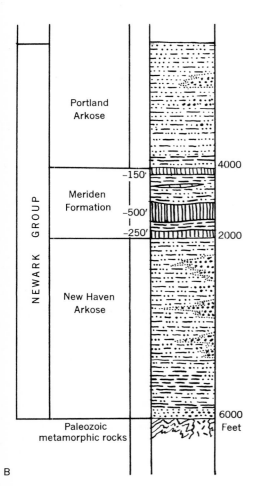

B

the New Jersey sequence has on the east, and Triassic is also preserved in an intervening small down-dropped belt in western Connecticut. The Triassic depositional plain may have extended from the west of the New Jersey belt to the east of that in Connecticut. The red sandstones of the Connecticut Valley in Massachusetts have yielded exceptionally fine footprints of dinosaurs (Chapter 16 opening photo).

BLOCKS FROM NOVA SCOTIA TO NORTH CAROLINA

The Bay of Fundy lies in a Triassic lowland, fault-bounded on the west, the Annapolis Valley of Nova Scotia lying, like the Hudson, in the lowland of the basal Triassic. The sediments in this area seem to have been laid in northwest flowing streams, so the basin must have been drained by a main stream near the scarp. Several miles of sediments and lavas in southeastern Pennsylvania dip toward the boundary fault on the northwest. In Virginia there are several belts, the principal northwestern one having the main fault on the northwest. In the largest area in North Carolina, near Durham, the sediments dip toward a fault on the southeast. These great rifts have a general trend like that of the structures of the Appalachian Revolution. There is a suggestion that they are faults that had considerable lateral displacement southward, on the oceanward side; rather than vertical movements, they involve slippage of blocks past each other. Faults of this sort are better authenticated and will be discussed in the later history of the Pacific Coast. Conventionally it is stated that the compressional forces of the late Paleozoic Appalachian Orogeny were followed by tension that produced normal faults, and Triassic sediments accumulated in structural depressions on the downthrown sides. The deformation characterized by development of tilt blocks, horsts, and grabens is taphrogeny, and subsiding areas developing by such a process are taphrogeosynclines. It has been shown that block movements may have been initiated in the Acadian–New England area within the Carboniferous Period (Fig. 13–5), or perhaps even in the Silurian (Fig. 9–19*A*). The region had changed greatly from the belt of subsiding troughs with volcanic and tectonic islands of the early Paleozoic.

PETRIFIED FOREST AND PAINTED DESERT

One of the popular national monuments in the west, the Petrified Forest, and the nearby Painted Desert in Arizona exhibit Triassic rocks. The term "forest" is a little misleading, for the deposit is of logs of coniferous trees that were laid among stream gravels after being rafted, rather than of the trunks of standing trees (Fig. 16–13). They have been altered by replacement of the woody fibres by silica. The upper

Fig. 16–12. Triassic rocks in North Carolina and Virginia. *A.* Fault on southeast side of Triassic basin along Atlantic Coast Line Railroad, Lee County, North Carolina (M. R. Campbell, United States Geological Survey.) A vertical fault in the center of the view has horizontal red sandstones on the left and schist on the right. The deep weathering in the region conceals bed rock.
  B. Triassic conglomerate near Orange, Virginia (C. K. Wentworth, United States Geological Survey.) The pebbles, somewhat rounded, have many kinds of rocks.

Fig. 16–13. Petrified Forest, Arizona (J. Muench). The stream-laid conglomerates of the Triassic contain large logs of the conifer *Araucarioxylon* that seem to have been carried in currents from other regions; thus the Petrified Forest is really not the site of the forest, but of the stranding of logs carried there by streams. The logs are as much as 120 feet long and 7 feet across. The wood has been replaced by beautifully colored silica, opal, but the silicification has destroyed details of the structure of the original plants. Much that is known of the flora of the times comes from plant impressions in associated beds.

Triassic sequence has in some parts of the southwest a persistent conglomerate, in Arizona known as the Shinarump, laid in the channels of streams that meandered over a pediment surface beveling the red sands and shaly silts of the middle and upper Moenkopi Formation (Fig. 16–15). The Moenkopi thins to disappearance in northeastern Arizona. It contains fossil vertebrates. The logs in the Shinarump and the Chinle beds are of coniferous trees that were very much like those occasionally found in the Newark Triassic in the East. They floated down from the forests where they grew to become lodged in the stream gravels and altered to their petrified state. The Shinarump conglomerate, not a separate formation but a basal phase of Chinle sedimentation, is normally up to about one hundred feet thick, and there are other very similar younger conglomerates in the region. They have been subject to exhaustive study, for the gravels of the ancient stream channels were the sites of deposition of carnotite, the yellow uranium-bearing oxide that is mined in scattered localities, principally in southwestern Colorado. The conglomerate grades up into the vari-colored purple, maroon to gray shaly siltstones of the Chinle Formation, well-displayed in the Painted Desert, with similar gravels such as those that contain the logs in the Petrified Forest and occasional beds of light-colored volcanic ash. The Chinle (Fig. 16–14), as much as five hundred feet thick, is for a continental deposit, very fossiliferous. A single lens within the Chinle, perhaps a deposit of flood-plain mud flat, yielded dozens of skeletons of primitive dinosaurs in a small quarry in northwestern New Mexico (Fig. 16–16).

EXTENSIVE RED BEDS

Rocks of the same sort, red siltstones and sandstones with interbeds of gypsum, representing the Triassic throughout an area of hundreds of thousands of square miles, extend north to Wyoming, where they form the Red Peak and Popo Agie (po-po-zhee) beds (Fig. 16–17), and east to Texas, where the Dockum beds are exposed. Areas in central western Colorado lack Triassic; the source of the sediments seems to have been in the highlands raised there during the late Paleozoic. Streams laid hundreds of feet of the red sands and silts in earlier Triassic. Then, non-deposition prevailed, or slight erosion, streams carrying all but the coarsest gravels to sites of deposition beyond the region. Sediments again accumulated on fluviated plains until the close of the period.

It has been possible to follow lithologic members of the Moenkopi westward and northward (Fig. 16–18). Gypsum beds and dolomite enter the red sand and silt sequence. Northwestward, in the northwestern

Fig. 16–14. *Restored sections of Triassic rocks on the western margin of the craton.* A. Section late in Triassic time from southeastern Idaho to central Wyoming, from the geosynclinal belt on the west to the cratonal platform. Much of the latter section is red sandstone and siltstone, with some interbeds of gypsum; the deposits were laid as non-marine or marginal marine sediments in an environment unfavorable for organisms whose remains might have reduced the oxides (after B. Kummel).

B. Section to the same scale of the preserved Triassic on the cratonal platform in Arizona (after E. D. McKee, and others). The Moenkopi beds thin eastward by overlap and convergence, the marine deposits becoming non-marine. The Petrified Forest and the Painted Desert are in the Chinle vari-colored volcanic ash-bearing shaly siltstones.

Fig. 16–15. Late Triassic Moenkopi red beds east of St. George in southern Utah. The interbedded shaly siltstones and sandstones, deposited on stream plains and in lakes, are a source of uranium in some of the mines of the Colorado Plateau. And they have yielded occasional remains of dinosaurs.

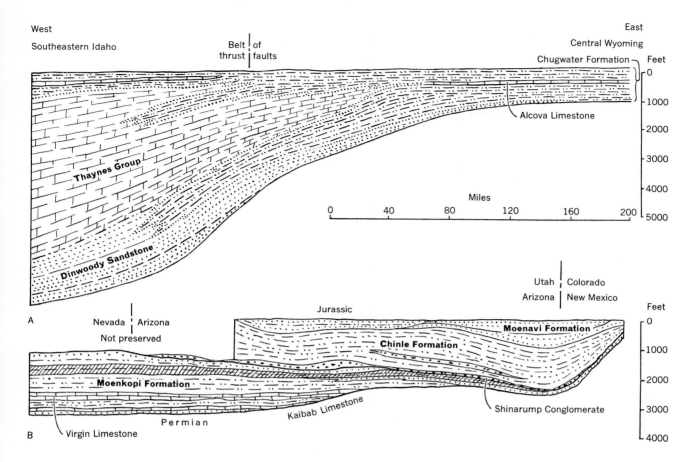

West
Southeastern Idaho

Belt of
thrust faults

East
Central Wyoming

Chugwater Formation — Feet

0

Alcova Limestone

1000

Thaynes Group

2000

3000

Miles

4000

Dinwoody Sandstone

0   40   80   120   160   200   5000

A

Utah | Colorado

Arizona | New Mexico

Nevada | Arizona

Jurassic

Feet

Not preserved

Moenavi Formation

0

Chinle Formation

1000

Moenkopi Formation

2000

Kaibab Limestone

Shinarump Conglomerate

3000

Permian

B   Virgin Limestone

4000

half of Utah, marine limestones and sandstones with characteristic lower Triassic pelecypods and cephalopods (Fig. 16–14B) grade into the red bed sequence in virtually the same area that subsided more rapidly in Cambrian and later Paleozoic periods; and the thickness increases to approach a mile. Interbedded red terrigenous sediments and fossiliferous carbonates in eastern Nevada are separated by more than one hundred miles from the next exposures of Triassic rocks to the west in central Nevada (Fig. 16–19). The marine Triassic rocks extend northward along the west side of the craton through Idaho into western Alberta and northeastern British Columbia, disappearing eastward—passing into red beds in Wyoming and into an unconformity in Western Canada.

GEOSYNCLINES AND LANDS IN THE NORTHWEST

Westward to the Pacific, Triassic rocks are known in many scattered areas from Alaska to southern Nevada and California and northwestern Mexico. Volcanic rocks are frequent, just as in the older systems in the same belt, and among the sediments are some having conglomerates with boulders and pebbles of older sediments and a variety of intrusives and extrusive volcanic flows, agglomerates, and tuffs (Fig. 14–15).

The subsidence in the Pacific belt was very great in some places. In southwestern Nevada, the Triassic has marine fossils ranging from ones similar to those in the oldest stages in Europe to ones like those in the youngest stages. The sequence is three or four miles thick, and is largely of lavas and volcanic sediments, particularly siliceous rocks (Fig. 16–19). Yet, a hundred miles to the northwest, a few thousand feet of lower and middle Triassic marine limestones succeed the section of several miles of Permian volcanic and sedimentary rocks.

VOLCANISM ON THE PACIFIC COAST

The Pacific belt was one having volcanism, and as we have recognized in several Paleozoic systems, the conditions differed from place to place, and volcanism was of differing intensity from time to time. The volcanic rocks of Triassic age on Vancouver Island and adjoining mainland in British Columbia are many thousands of feet thick, associated with argillites, carbonate sediments, and conglomerates with granitic boulders up to two feet in diameter. Similar lavas and sediments are widespread in northern British Columbia, the archipelago of southeastern Alaska and southern Yukon, and there are fossiliferous limestones with few basalt flows in central Yukon. The sequences are essentially like those found in the same belt in rocks as old as the Ordovician. The region has been interpreted as having similar belts of volcanic islands, and of

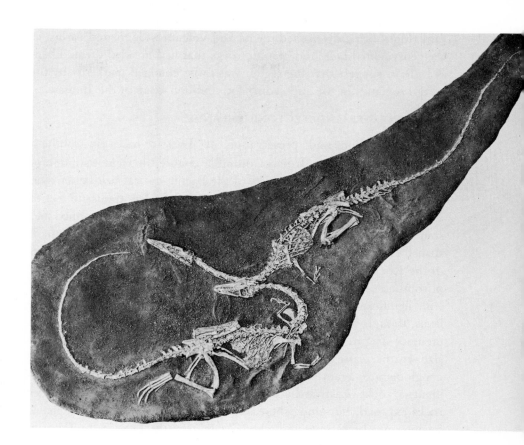

Fig. 16–16. Two nearly complete skeletons of the small, primitive dinosaur *Coelophysis* from the red-beds of the Chinle Formation of the upper Triassic in the quarry at Ghost Ranch, near Abiquiu, northwestern New Mexico (American Museum of Natural History).

Fig. 16–17. Triassic Chugwater Red beds, sandstones and shales, northwest of Casper in the southern Wind River Mountains, western central Wyoming. The long slope above the valley is composed of the bright-colored Red Peak Sandstone, in which are found reptilian footprints. This sequence is capped by the thin Alcova Limestone, in which is found a marine nothosaur. Above, the second slope is formed by the Popo Agie beds, containing a characteristic upper Triassic amphibian and reptile fauna. The Jurassic Nugget Sandstone lies at the crest.

The red beds grade westward into gray marine shales and sandstones in the western ranges of Wyoming and in southeastern Idaho. Similar red sediments are prevalent in the Triassic southward to Texas, New Mexico, and Arizona.

fold ridges bringing up sediments and volcanics deposited earlier in the geosynclinal belt, and granitic rocks that had invaded them. There is little to suggest that the region was to be changed markedly by the great orogenies in the succeeding forty million years of the Jurassic.

GEOSYNCLINES AND TECTONIC ISLANDS

The distribution and preservation of Triassic rocks in southern Yukon and adjoining British Columbia enable the demonstration of the mobile and changing aspects of the Pacific coastal belt better than in the preceding systems. Several sequences of a few thousand feet of Triassic rocks (Fig. 16–20) have been determined in the deformed and somewhat metamorphosed rocks in southern Yukon and northern British Columbia, and additional sedimentary sections have been found in the Rocky Mountain foothills to the east. The western succession in southeastern Yukon is dominantly of sandy argillaceous sediments, with significant conglomerates having boulders dominantly of extrusive lavas, though with some plutonic granitic pebbles and others of sedimentary origin. The boulders decrease eastward in quantity and in size. Farther east, near the center of the southern part of the territory, conglomerates again enter the sequence, this time principally of sedimentary rocks. Eastward a long interval has only Paleozoic and older rocks exposed, but along the east of the Rockies are sands and shales having some limestones and many Triassic fossils. Thus in the Triassic, a subsiding trough or geosyncline occupied most of the breadth of southern Yukon—lying between islands raised to the west along the

Fig. 16–18. Illustrations from Triassic Paleotectonic Atlas. The distribution and characters of Triassic rocks in the United States are summarized in an atlas prepared by the United States Geological Survey, of which these figures are representative. The originals are in colors that make the portrayal more vivid. The map (*A*) and section (*B*) are of the Colorado Plateau region in southeastern Utah and northeastern Arizona, see also Fig. 16–14*B*.

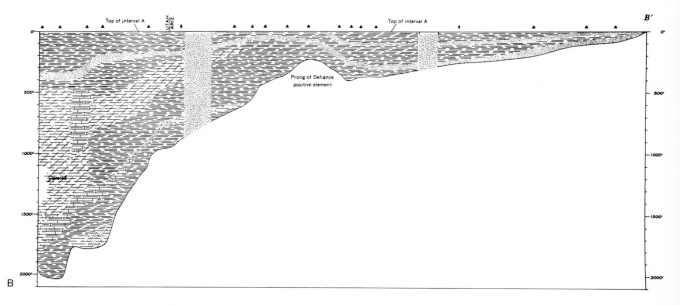

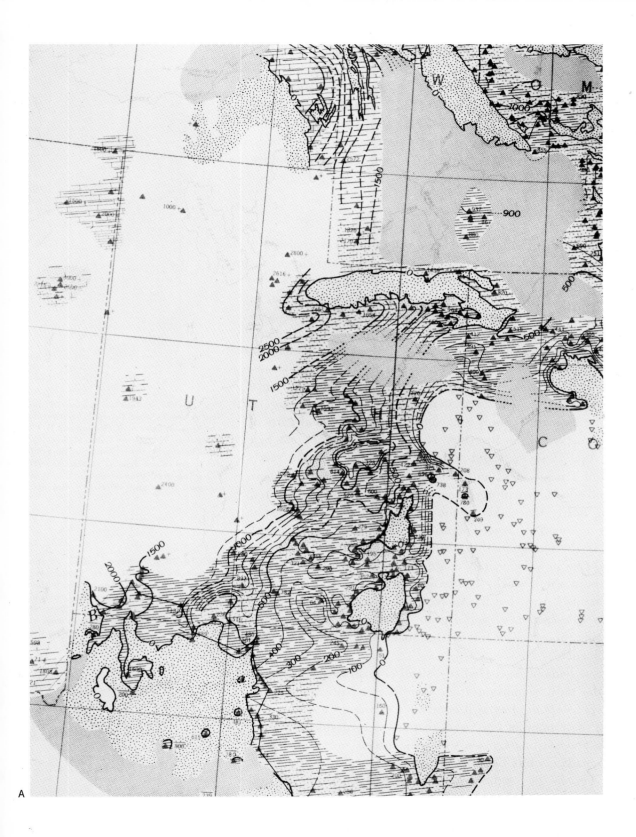

A

373

Alaskan border (Fig. 16–21) and having eroded volcanic rocks and other lands on the east toward the Mackenzie border with uplifted folds of Paleozoic and Precambrian sediments; to the east of these lay another subsiding belt that extended eastward to the cratonal margin, where younger Jurassic spread over the Triassic on to the eroded Paleozoic surface. These Triassic sediments did not pass eastward into red beds, as they did to the south of Montana and Wyoming, but thinned to disappear on the border of the craton.

ARCTIC RECORD

In discussing the distribution of facies in the Ordovician System, the Arctic Archipelago of Canada was shown to have geosynclinal belts along the northern margin of the craton. Later Paleozoic rocks are widely distributed in this great area. Exploration has proceeded at an increasing pace in recent years because of the accessibility through the use of planes and the interest in the economic resources, particularly in petroleum reserves. The Triassic and later Mesozoic rocks show in their changing facies that the islands were along the margin of the subsiding Arctic Ocean basin. Thus the Triassic, distributed from Prince Patrick Island on the west, to western Ellesmereland, has marginal marine and non-marine sands and silts lying to the southward of argillaceous marine sediments (Fig. 16–22). The relations are best known in Axel Heiberg Island and southern Ellesmereland. Nearly three miles of Triassic rocks on the northwest are dominantly of shaly argillites and siltstones with representative marine fossils, ammonites, representing nearly the whole of the Triassic, only the upper most beds being non-marine. Traced southeastward toward Ellesmereland (Fig. 16–23), the sediments thin, develop some unconformity in the middle, and pass into dominantly non-marine, plant-bearing sands and siltstones about a mile thick. The farther southeastward constituents have been eroded away, but the source lands must have been nearby in southern Ellesemereland and Devon Island. The latter is aptly named, having thousands of feet of Devonian vertebrate-bearing non-marine sediments.

The marine beds of Axel Heiberg Island contain fossils in several horizons, the most significant from their use in correlation being ammonite cephalopods. Just as it is possible to arrange a sequence of trilobites in the Cambrian into zones that are recognizable in Britain, Scandinavia, and maritime Canada, so also the succession of cephalopods in Arctic Canada is quite like that in northwestern British Columbia, or central Nevada, or southern Mexico, as well as in the classical

Fig. 16–19. Columnar section of the Triassic rocks near Tonopah, southwestern Nevada (from S. W. Muller and H. G. Ferguson). The section is notable because it has marine fossils representing many of the stages of the Triassic system, as well as having great thickness—much of it of volcanic rocks and siliceous sediments.

Fig. 16–20. Restored section of the Triassic from southeastern Alaska into southwestern Yukon, showing the Whitehorse geosynclinal trough between the volcanic and tectonic land on the west and a tectonic land on the east, each of which contributed coarse detritus to the intervening sea (after J. Wheeler).

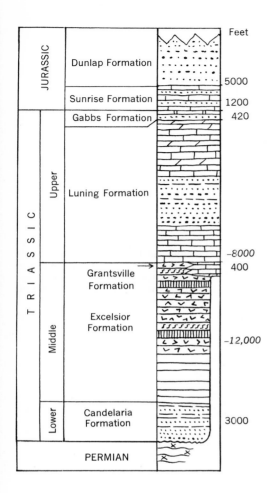

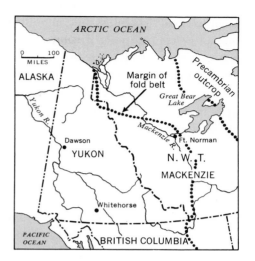

Fig. 16–21. Outline map of the territory of Yukon, northwestern Canada, and the bordering state of Alaska, the district of Mackenzie in Northwest Territories, and the province of British Columbia. Precambrian rocks are at the surface in the northeast; they are overlain by Paleozoic and younger rocks southwest of a line near the Mackenzie River. Triassic rocks are exposed in many widely distributed areas in Yukon and Mackenzie.

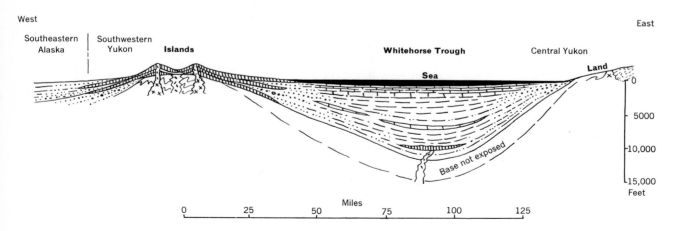

375

sections in Europe, particularly those in the Italian Alps. Most distinctive of the Triassic ammonites are the ceratites (Fig. 16–24), forms having serrate sutures on the lobes, the part of the suture directed away from the body chamber, but smooth sutures on the saddles. Although ammonite cephalopods are conventionally thought to be distinctive of the Mesozoic, we have seen that some developed in the later periods of the Paleozoic. Those of the Jurassic and Cretaceous are quite distinct from the Triassic forms, for the ammonites very nearly disappeared at the end of the Triassic, the few survivors leading to thriving and diverse descendants (Fig. 16–25).

UPPER TRIASSIC PALEOECOLOGY ON LAND

The distribution and the expression of the upper Triassic sediments in North America have been discussed at some length in the preceding pages of this chapter. What can be said, against this background, about the fossils that occur in these sediments? What do the remains of

Fig. 16–22. Map of the District of Franklin and the Queen Elizabeth Islands of northern Canada, with Greenland in the notheast, showing the principal structural provinces (after R. Thorsteinsson and T. Tozer).

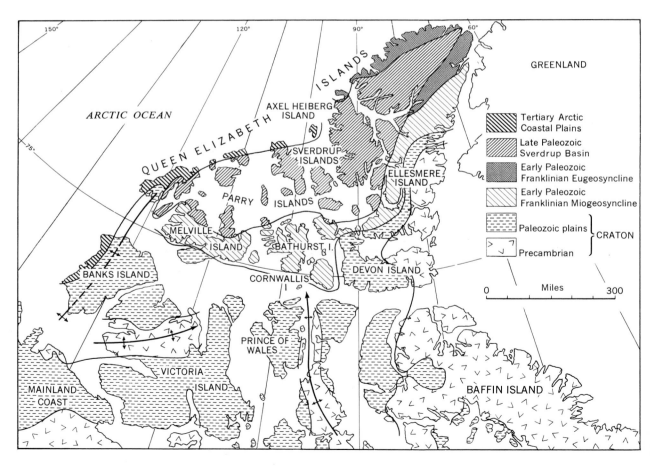

extinct animals and plants tell us about life on the North American continent during the later phases of Triassic history?

This question is one prompted not by idle curiosity but rather by its significance, because the upper Triassic continental sediments have yielded many facts about the nature of life on the land during late Triassic time. And the picture of land life obtained from the upper Triassic fossils of North America is additionally significant because it applies not only to this continent, but also to large areas in other parts of the world as well. Moreover, Triassic fossils from North America are significant, because they demonstrate to a considerable degree the new evolutionary trends that were to be so characteristic among the dominant land-living organisms, especially the backboned animals, at this stage of earth history.

Perhaps the most spectacular upper Triassic fossils found in North America are the great agatized tree trunks that make up the Petrified Forest of Arizona. These petrified logs, already briefly described, are

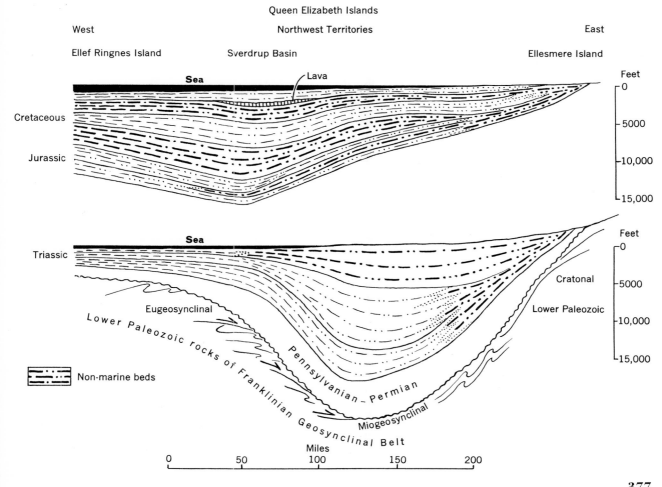

Fig. 16–23. Restored sections of the Mesozoic rocks of the Sverdrup Basin, northeastern Queen Elizabeth Islands, Arctic Canada (after R. Thorsteinsson and T. Tozer). The basin, subsiding deeply through the era, gained terrigenous detritus from source land on the continental side. The late Paleozoic and Mesozoic rocks lie with unconformity on the lower Paleozoic rocks laid in the volcanic and non-volcanic belts of the Franklinian geosyncline.

characteristic of the Chinle Formation and are to be found in many regions where this upper Triassic horizon is exposed. It may be assumed that extensive forests, of which these fossil logs are the visible evidence, once flourished widely across North America, in areas where tropical lands rose above the seas. Such areas existed not only in the southwest, where the Chinle and the closely related Dockum Series of Texas are exposed, but to the north, as is now indicated by the Popo Agie beds of Wyoming, and up and down the eastern seaboard where the Newark sediments are widely distributed. That these extended lands were tropical is indicated by the nature of the fossils.

The trees were predominantly conifers, many of them closely related to *Auracaria,* which now extends across parts of the southern hemisphere. Beneath these trees there grew a profusion of ferns and other low plants, to form in many places heavy masses of undergrowth on the forest floor. Although the Triassic scene was not one of rich jungle growth, as had been true of the Carboniferous and as was to be true of the later Mesozoic periods, there must have been a considerable amount and variety of plant life available as food and cover for many reptiles. The upper Triassic plant-eating reptiles, the first-order consumers, that roamed across the North American landscape, were primarily armored pseudosuchians, and perhaps procolophonids, small, lizard-like reptiles. These latter were holdovers from the Permian, but the pseudosuchians represent an important innovation in tetrapod evolution. These were among the first of the *archosaurians,* and the archosaurians—the pseudosuchians and their relatives the phytosaurs, the crocodilians, the two great orders of dinosaurs and the flying reptiles—were to be the overwhelmingly dominant land animals of Mesozoic times.

Fig. 16–24. *Triassic ammonite cephalopoda.* Ceratites, having serrate lobes and smooth saddles on the sutures: (1) *Otoceras;* (2) *Meekoceras;* (3) *Paraceratites.* Ammonites, having serrate lobes and saddles: (4) *Trachyceras;* (5) *Juvavites;* and (6) *Choristoceras.* The genera are distinctive of successive stages in the Triassic and thus are excellent time-stratigraphic guide fossils.

The ammonites were cephalopods having complexly serrate sutures, the junctions of the chamber-separating septa and the external shell; sutures are seen on the internal mold when the external shell is removed. Ammonites appeared in some frequency in the Devonian, though some may have lived as early as the Ordovician. Reference has been made to the late Paleozoic ammonoids in which the progression of forms in time suggests the succession of growth stages in individual specimens.

Ammonites are particularly useful in the classification of rocks in the Mesozoic systems. The animals had such complexity of suture patterns that those of close relationship can be recognized confidently. There were many lines of descendants through most of the era; however, ammonites very nearly disappeared at the beginning and again at the close of the Triassic (Fig. 16–24), only limited lines of descent surviving each crisis to again diverge. The related lines have similar shapes, and the members of a single line or family can in turn be distinguished by the complexities of their sutures. The animals were changing rather continuously and rapidly—geologic periods are of tens of millions of years, so the scores of divisions recognizable through ammonites in systems such as the Jurassic still represent very long spans in terms of human history. Cephalopods were free-swimming organisms, some of which lived on other life in the open sea; so they migrated from one shore to another, readily spreading around the earth in areas of favorable environments (Perhaps only the early Paleozoic graptolites had as world-wide a distribution). These qualities—complexity of structure, multiplicity of lines of descent, progressive changes in time, and facility in migration—permit their use as fossils distinctive of successive stages of time.

The ammonites that expanded rapidly in the Triassic, only to nearly disappear

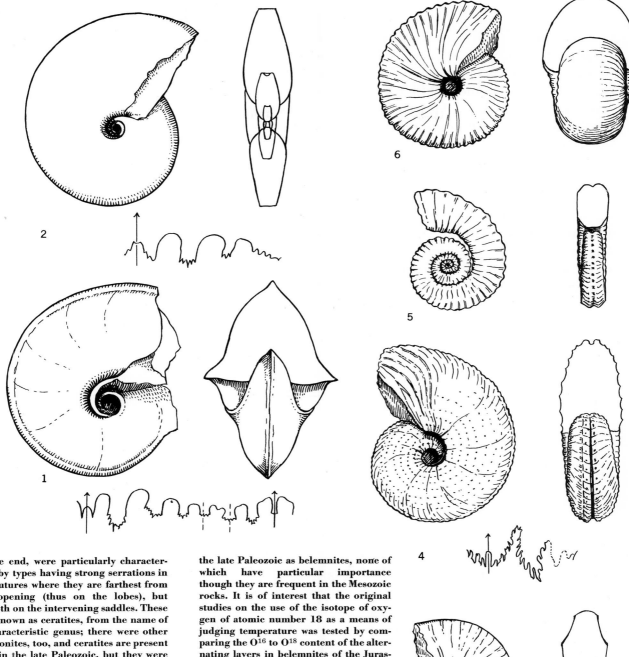

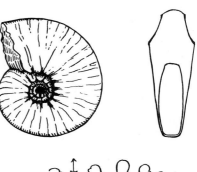

at the end, were particularly characterized by types having strong serrations in the sutures where they are farthest from the opening (thus on the lobes), but smooth on the intervening saddles. These are known as ceratites, from the name of a characteristic genus; there were other ammonites, too, and ceratites are present also in the late Paleozoic, but they were most dominant in Triassic.

The cephalopods that most of us know first were the Dibranchiata, the squid, cuttlefish, and octopus that live today and appear in popular fiction; they have internal shells. In contrast, the tetrabranch cephalopods, having external shells, comprise the nautiloids, goniatites, ceratites, and ammonites; they are represented in present seas only by the Pearly Nautilus. The dibranchs appear in

the late Paleozoic as belemnites, none of which have particular importance though they are frequent in the Mesozoic rocks. It is of interest that the original studies on the use of the isotope of oxygen of atomic number 18 as a means of judging temperature was tested by comparing the O^{16} to O^{18} content of the alternating layers in belemnites of the Jurassic (Fig. 17–5); as the layers were alternately varied, it was judged that the layers represented cooler and warmer seasonal waters in which the dibranchs were growing. Modern shells at higher latitudes growing in colder water have larger proportions of O^{18} than those growing in warmer waters. It is, of course, rather incidental that dibranch belemnites from the Jurassic of England were the only ones used here.

Desmatosuchus and *Typothorax* are characteristic upper Triassic pseudosuchians in North America that lived on high ground or perhaps frequented the edges of streams and lakes, where they could browse on the plants around them. With these large, well-protected reptiles was a huge beaked dicynodont, *Placerias,* also a plant eater, the last representative of a reptilian group that had flourished widely throughout the world during the Permo-Triassic history.

Probably the most aggressive of the second order consumers that fed upon other animals, the predators that probably hunted, among other things, the pseudosuchians and the dicynodonts, were the phytosaurs, crocodile-like reptiles especially characteristic of upper Triassic continental sediments in North America and northern Europe. These long-jawed reptiles, which reached lengths of ten to twenty feet and more, lived in and along streams and lakes where like modern crocodilians they lurked and hunted. They must have been the archpredators of their time.

These reptiles have been compared to crocodiles, but they were not crocodiles, nor were they directly ancestral to crocodiles. Here is

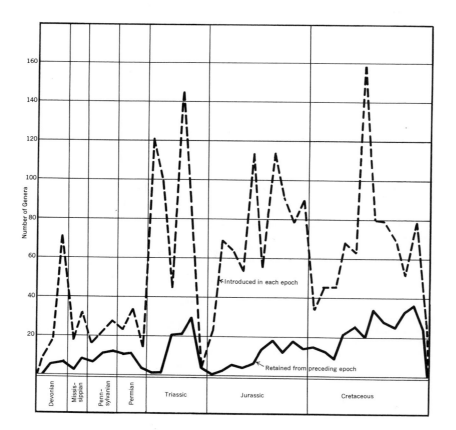

Fig. 16–25. A diagram to represent the relative abundance of genera of ammonoid cephalopods from the middle of the Paleozoic to the close of the Mesozoic Era (after B. Kummel). The lower solid line represents the number of genera in each epoch that were represented in preceding epochs—the genera that lived on from the past. The upper, broken line shows the number of genera that first appeared in each epoch. It can be seen that very few genera lived from the Permian into the Triassic (one only) and from the Triassic into the Jurassic period (none, though one left a descendant). The chart also shows the great number of genera in the late Devonian, and particularly in several epochs of the Mesozoic periods. As the Paleozoic periods are drawn as though with only two or three divisions, whereas the Mesozoic periods are divided by stages of unequal and shorter span, the data are not strictly comparable statistics; if the Mesozoic time spans were plotted on the same basis as the Paleozoic, their peaks would tend to be relatively much higher.

Fig. 16–26. A late Triassic scene in North America as restored by John Germann (American Museum of Natural History). The large reptiles are phytosaurs, thecodonts which preceded and anticipated, but were not ancestral to the crocodiles. Their abundant remains in North America and Eurasia indicates close land connections between these two regions during Triassic history. Along the edge of the swamp are ferns and giant horsetails, and in the background are auracarian trees, the fossilized logs of which are so strikingly abundant in the Petrified Forest of Arizona.

Fig. 16–27. The thecodont genus *Desmatosuchus*, above, and the large mammal-like anomodont, *Placerias*, below, these reptiles known from the upper Triassic Chinle and Dockum beds of Arizona, New Mexico, and Texas (*Earth Song*, C. L. Camp, 1952). *Desmatosuchus* is a large, armored pseudosuchian, ten feet or more in length, of presumed herbivorous habits. *Placerias*, likewise a plant-eater is related to large anomodonts of Asia and Africa as well as South America. These reptiles wandered widely during Triassic times.

a nice example of parallelism in evolution through time. The phytosaurs became adapted to a certain mode of life during late Triassic time and flourished profusely in certain parts of the world. But at the end of the Triassic they became extinct, and the crocodiles, which had just appeared during the final days of phytosaurian dominance, imitated them to an uncanny degree and so successfully that they have been on the earth from the beginning of the Jurassic to the present day.

It is probable that the mainstay of the phytosaurian diet was the fishes living in rivers and streams. These fishes were predominantly armored types, with heavy, shining scales, the fishes often designated as "ganoids," and properly classified as subholosteans and holosteans. They were of varied form and habits, and it is probable that they lived abundantly in many rivers and lakes. Certainly their remains are numerous in some Triassic horizons.

There were also large amphibians in the rivers and lakes, these being the stereospondyls, the last of the labyrinthodonts which had been so numerous and varied during the late Paleozoic time. The Triassic stereospondyls are represented in North America by *Eupelor,* a very

Fig. 16–28. Two fishes from the upper Triassic Newark beds of eastern North America (from Schaeffer, 1952). Above is *Diplurus*, a small fresh-water coelacanth, related to the living coelacanth, *Latimeria*, scientifically recognized in 1938 off the eastern coast of Africa. A shale layer containing thousands of specimens of *Diplurus* was encountered when excavations were being made some years ago for the Firestone Library at Princeton University. Below is *Turseodus*, a characteristic "ganoid." Fishes such as this inhabited Triassic fresh waters in great numbers.

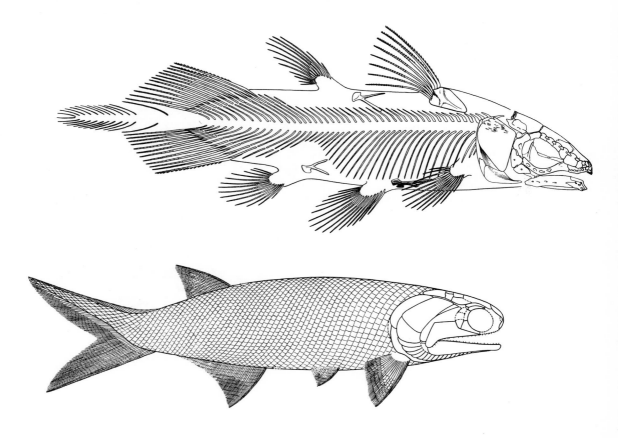

large, flat-headed amphibian, five or six feet or more in length, with extraordinarily weak limbs. Evidently this big amphibian spent its life in the water, where it fed upon fishes and was thus perhaps a competitor with the phytosaurs for this particular supply of food.

Although the phytosaurs were the largest and most aggressive of land-living predators, they were not as varied as the dinosaurs with which they were contemporaneous. These were the first of the dinosaurs, the ancestral theropods, from which many of the great meat-eating dinosaurs and the gigantic swamp-dwelling sauropods of Jurassic time were descended. Many of the Triassic theropod dinosaurs were small, lightly built reptiles. None was of more than medium size. They were all bipedal, walking on strong, bird-like hind limbs, using the fore limbs and hands for grasping and for aids in feeding. They were generally highly carnivorous—ate other animals. Perhaps the smaller theropods lived on insects and small reptiles, the larger ones very likely pursued game more in keeping with their size, possibly feeding to a considerable extent upon the lizard-like procolophonids and protorosaurs.

These dinosaurs are known in part from their bones, in part from their abundant footprints that are found in the Connecticut Valley. The bones show that there were various kinds of these dinosaurs (as do the tracks), such as *Coelophysis,* a light and very agile predator, and *Anchisaurus,* a relatively small but heavier animal, of the type that subsequently gave rise to the gigantic dinosaurs.

This recital of life on the land during late Triassic time gives us an impression of animal and plant associations rather different from those seen on the Permian delta of Texas. On the Permian delta the pelycosaurs were the dominant reptiles; in the Triassic forest the archosaurs—

Fig. 16–29. One of the last of the labyrinthodont amphibians, so characteristic of late Paleozoic and Triassic times (from H. J. Sawin, 1945). *Eupelor,* from the upper Triassic of North America, was an animal that lived in streams and ponds, and probably rarely came out on land. Although the skull and shoulder girdle are very large and heavy, the postcranial skeleton is weak—much of it evidently cartilaginous—and the limbs are small. Some of these amphibians are rather large, six or eight feet in length. The presence of very closely related genera in Eurasia reinforces the evidence for close continental connections across the northern hemisphere in Triassic times.

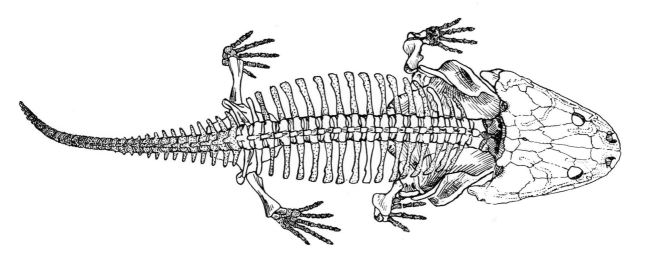

the pseudosuchians, phytosaurs and dinosaurs—are the dominant reptiles. On the Permian delta there were numerous varied labyrinthodonts and other amphibians; in the Triassic forest the labyrinthodonts are restricted to the large flat-headed stereospondyls, and as for other amphibians there is no evidence. Perhaps there were frogs.

As for the record, it is admittedly incomplete. The fauna of the North American upper Triassic gives us a partial picture of what life on the land was like two hundred million years ago, and this picture is repeated in northern Europe and in parts of Asia. In the eastern hemisphere are pseudosuchians and phytosaurs, dinosaurs and stereospondyls, and other backboned animals very closely related to those of the North American continent. So it is evident that there were paths of intercontinental migration from east to west and back. But in Europe there are some other upper Triassic tetrapods not found in North America, probably the results to a large degree in this latter region of the accidents of preservation and collecting. Thus there may be added to the list of animals described, primitive turtles, ancestral true lizards, and beaked rhynchosaurs, of which definite traces have recently been found in the Newark sediments. There are still fossils to be found. And when they are found we will gain additional insight into the new and vigorous animals that inhabited the Triassic world.

Fig. 16–30. The skeleton of the small, coelurosaurian dinosaur, *Coelophysis*, from the upper Triassic Chinle beds of New Mexico. This lightly built bipedal dinosaur, eight or ten feet in length, characterizes some of the first predatory dinosaurs that lived during late Triassic time. *Coelophysis* probably preyed upon other small reptiles, and such animals as it could catch.

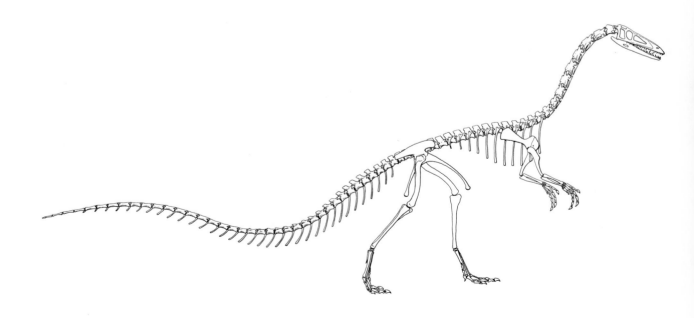

17
The Jurassic—
Advent of the Mexican Basin
and the Gulf Coastal Plain

The Great White Throne, Zion National Park, Utah, composed of lower Jurassic sandstone (J. Muench).

17 The Jurassic—Advent of the Mexican Basin and the Gulf Coastal Plain

The Rock of Gibraltar (Fig. 17–1) at the western entrance to the Mediterranean Sea is a great block of limestone, an erosion remnant of beds deposited in the seas of the Jurassic Period. Half Dome in Yosemite National Park (Fig. 17–2) is the face of a mass of granitic rock, granodiorite that has generally been considered to have been intruded in the Nevadan Orogeny, one of the principal mountain building times in North American history, late in the period; however, though Jurassic rocks are indeed intruded, recent age determinations by geochemists suggest that the granodiorite is younger than that period. The name Jurassic is from the Jura Mountains in the western Alps of France (Fig. 17–3), given to the limestones in the range by the noted explorer, Alexander von Humboldt (1769–1859) in 1799, a few years after he completed his studies under Werner at Frieberg. The stratigraphic position of the rocks was determined later and in 1829, the Jurassic System was so named by Ami Boué (1794–1881), the first system having a geographic name rather than the descriptive ones like Carboniferous and Cretaceous.

Fig. 17–1. Massive ledges of Jurassic limestone forming the Rock of Gibralter, rising 1400 feet above the Mediterranean Sea along its northern shore at the western end; the view is from the south, toward distant Spain (British Information Service).

Fig. 17–2.   *A*. Yosemite Falls on the north side of Yosemite Valley, on the western slope of the Sierra Nevada, California, and *B*, Half Dome, on the south (A. Greene and Associates). Yosemite Upper Falls is more than 1400 feet high, the stream dropping a similar distance over the lower cascades and Lower Falls. Half Dome has a nearly sheer face rising more than 4000 feet above the valley floor. The rock in the valley walls is granitic (granodiorite), intrusive into metamorphosed sedimentary rocks wherein Jurassic fossils were discovered nearly a century ago; prior to that, the metamorphic rocks were thought to be Precambrian. So the bathylith of the Sierra Nevada is post-Jurassic, and the Nevadan Orogeny, attributed to the post-Jurassic or latest Jurassic, gains its name from the range. However, age determinations on the ratio of potassium-argon isotopes show that some of the intrusives in the region are younger than some or all of the Cretaceous.

A

B

Though the name of the system came from southeastern France, it has the distinction of containing the rocks in southern England in which William Smith (1769–1839) recognized that fossils are arranged in a consistent order of succession. He made these observations near the close of the eighteenth century, and has been called the "Father of Stratigraphy." Smith was an engineer engaged in making surveys for canals and roads in the south of England. His profession led to his traveling extensively in the country, and he became interested in the distribution of the rocks and of fossils that he collected from them. About 1795, it came to his attention that the fossils were not indiscriminately distributed, but that each was in consistent association with others and these in systematic succession with different fossils above and below. It seems amazing that this simple observation had not been made so clearly before, for historical geology has been built on this foundation. Smith applied local names to the rocks that he mapped as outcropping in bands across the country, names such as Portland Rock, Purbeck Stone, and Coral Rag. In the course of time, these have become the standard of reference for some of the stages of the Jurassic System with alteration to forms such as Portlandian, Purbeckian, and Corallian.

AMMONITES AND CLASSIFICATION

The Jurassic rocks of western Europe have frequent specimens of ammonite cephalopods, forms with very contorted lines of juncture of the septa separating the chambers and the outer shell wall. Moreover, the forms change through the succession, hence give an exceptional basis for distinguishing the fossils in one faunal zone from those of another (Fig. 17–4). By the middle of the nineteenth century, the Jurassic rocks of northwestern Europe had been divided into some thirty or forty distinctive ammonite zones that were placed in about a dozen stages bearing place names that have become the Hettangian, Sinemurian, Pleinsbachian, Toarcian, Bajocian, Bathonian, Callovian, Oxfordian, Kimmeridgian, Portlandian, and Purbeckian stages; the first four are classed as lower Jurassic, and the Callovian is generally placed as the base of the upper Jurassic in North America, but in the top of the middle Jurassic by some European stratigraphers.

Paleontologists have recognized the stages and many of the zones throughout the world. As the period is thought to be thirty or forty million years long, the stages form recognizable spans of two or three million years. The subdivisions of the stages can be carried from place to place with varying degrees of confidence. The Jurassic Period is one

Fig. 17–3. *Folded sedimentary rocks in the Jura Mountains of southeastern France, the type region of the Jurassic System.* A. Anticline of Jurassic limestone near St. Wolfgang, from the southwest (Swissair).

B. Air view of the folded rocks in an area of about a square mile (Institut Geographique National, Paris).

A

B

in which events in distant places can be correlated as closely in time as within any period.

The reasons why such close correlation is possible between even distant areas are several. The ammonites are peculiarly suitable and useful. Precision in classification of organisms is enhanced when they have complex characters that are distinctive, constant in a single population, but changing recognizably with time. Correlation over great distances is enhanced when the animals were marine and free swimming. Ammonite cephalopods must also have been quite tolerant of varying conditions, or else the Jurassic conditions must have been rather uniform, or at least there must have been a great span of favorable conditions. After death some of their empty shells must have floated into depositional sites that the living forms could not inhabit. So in many ways, conditions are advantageous for the refined correlation of Jurassic rocks.

DETERMINATIONS OF TEMPERATURES

Jurassic fossils have seemed useful also in comparing the climates with those of the present, to conditions on the lands as well as in the seas. Temperatures and wind directions can be determined by direct observations today, and also in the Jurassic. A map of the average temperature of marine waters has lines of equal temperature, isotherms, that have only general similarity to those of latitude. Thus such well-established marine currents as the Gulf Stream carry warm water to much higher latitudes off northwestern Europe than along the east coast of North America. It may become possible to prepare maps of water temperature throughout the earth for past times. Calcium carbonate ($CaCO_3$) in marine shells has two oxygen isotopes, O_{16} and O_{18}. Shells that have grown in higher temperatures have a larger proportion of the lighter isotope, O_{16}, than those that have grown in lower ones. When the ratios of the two isotopes in marine shells have been calibrated against temperatures of the waters where they grew, it becomes

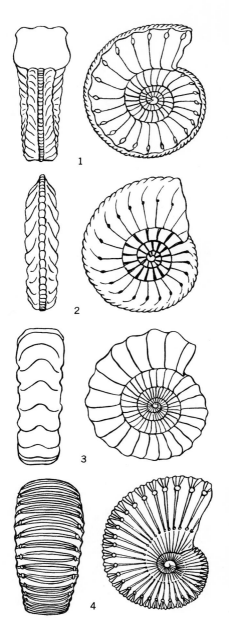

Fig. 17–4. *Jurassic ammonite cephalopods.* (1) *Liparoceras;* (2) *Oistoceras;* (3) *Amaltheus;* and (4) *Pleuroceras.* These forms are found in that stratigraphic order, and are related genera, though not in direct line of descent. They illustrate the variable ornamentation of the shells; complexities would be even more evident in their suture patterns.

The Jurassic was the first system in which the stratigraphic succession was divided into formal faunal zones, originally by Albert Oppel (1831–1865) in 1856–1858. He separated the Jurassic rocks in western Europe into some thirty zones that subdivided the larger time-stratigraphic units, stages, that had been named by Alcide d'Orbigny (1802–1857) in the preceding decade or so. The zone is a time-stratigraphic unit, one characterized by an assemblage of fossils, bearing the name of one of them—such as the Zone of *Ammonites oxynotus* of Oppel, which with revision in the nomenclature of the ammonite became *Oxynoticeras.* The fossil for which the zone is named

need not be restricted to the zone; the concept of the zone is abstract, for rocks elsewhere that are considered to fall within the span of time represented in the typical sequence are of that zone, whatever the basis for the interpretation. Zones, having fossil names, are of one or more stages, having names of places. We have referred to stages and zones in other systems, as in the Cambrian and Ordovician, but the terminology developed first in the studies in the Jurassic and Cretaceous of western Europe.

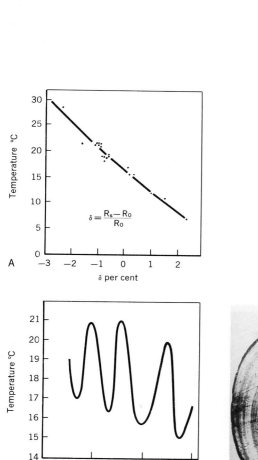

C

Fig. 17–5. *Oxygen isotopes determining the temperature of past seas.* The proportions of the isotopes of oxygen having atomic weights of 16 and 18 change with the temperature of the waters from which they precipitated. *A.* Isotopic temperature scale, showing the increasing proportion of O^{18} at increasing temperatures.

B. Graph showing the changing proportions of isotopes converted into temperatures for the successive summer and winter layers in a section of a belemnite cephalopod.

C. The belemnite cephalopod *Pachyteuthis* from the upper Jurassic of the Black Hills of South Dakota (see Fig. 17–14B): measurements were made on similar fossils from the Cretaceous of England (section from H. C. Urey and others: photo from the American Museum of Natural History). The winter (W) and summer (S) growth layers are darker and lighter, respectively. The numbers are on annual layers.

393

possible to learn the temperatures of seas of the past from the isotope analysis of ancient shells. Such a study made on Jurassic belemnites (Fig. 17–5), marine squid-like animals, showed reversing ratios from layer to layer representing seasonal variations, from which the average annual temperature can be computed. This is a physical chemical approach that has to be more fully tested and extensively applied, but there is promise of useful results.

Fossils give other means of judging temperatures, by analogy with present habitats of similar forms. For example, the Jurassic coral reefs in England have types of corals that live today only in latitudes twenty degrees south of England. Though judgments differ among paleontologists as to the significance of specific organisms, it seems well established that organisms that lived in warm waters extended farther north during the Jurassic than today. Moreover, assemblages of plants of the Jurassic that seem to have lived in temperate habitats are known in such presently frigid climates as those of East Greenland, and of Grahamland in Antarctica.

DINOSAURS AND CLIMATES

Of particular importance in this respect is the distribution of fossil reptiles in beds of Jurassic age. For example large dinosaurs of late Jurassic age, the dinosaurs so characteristic of the Morrison beds, together with turtles, crocodiles and other reptiles, are found as far north in North America as about the forty-fifth parallel. The closely related Kimmeridge dinosaurs occur in England above the fiftieth parallel of latitude, while in Central Asia dinosaurs and other reptiles of this age are found at localities equally far to the north. Modern reptiles have very definite temperature tolerances, and there is every reason to think that the same was true of extinct forms. Thus we can safely assume that large dinosaurs would have been limited to tropical and subtropical climates, as are large crocodilians in the modern world, and would not have been able to withstand even moderately severe winters. Consequently, the occurrences of these fossil reptiles indicate that climates similar to that of southern Florida of the present day must have been typical of, or extended to, the northern limits of the United States, northern Europe and northern Central Asia. Fragmentary remains of dinosaurs of this age in the southern part of South America indicate that these reptiles, and correlatively the mild climates in which they lived, seemingly extended as far to the south in the southern hemispheres as they did to the north in the northern land masses.

Fig. 17–6. Geomagnetic poles of the Jurassic. The symbols represent the recorded positions of the magnetic poles gained from the study of the remanent magnetism in rocks of Jurassic age on each of four continents. It will be seen that the determinations deviate far from the position of the present magnetic pole in northern Canada or of the rotational North Pole; the significance of these deviations will be discussed in Chapter 19.

Fig. 17–7. Granodiorite of the Sierra Nevada, California in Tuolumne Meadows, Yosemite National Park. The coarse-crystalline rock, principally of feldspar with some quartz and other minerals, has large crystals of potash feldspar that tend to project on weathered surfaces.

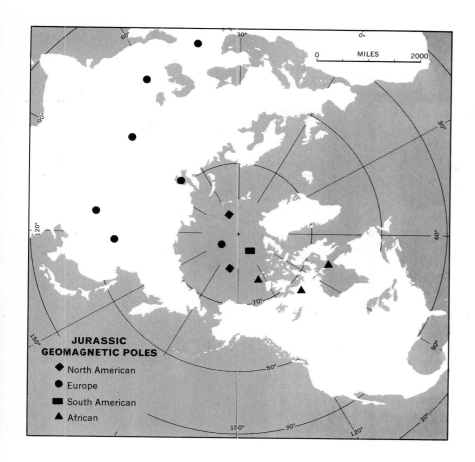

Fig. 17–8. Serpentine, "verde antique marble," in the base of the Alma Mater statue, Columbia University, New York City. The rock may have come from Paleozoic intrusions in Vermont, but is similar to that in serpentine belts in other parts of the world. The white bands are veins of calcite in fractures in the green serpentine, which is hydrous magnesian silicate.

395

If the Jurassic climates had been as of the present, we should expect to find evidences of glaciation in the higher latitudes, but none are known. The presence of warm water organisms and floras of temperate habitats in high-latitudes in Jurassic rocks may imply that earth temperatures were higher than today, so that temperature belts were relatively more poleward. An alternative is that the belts of Jurassic climate were at an angle to those of today, so that the equatorial belt was relatively northward in some longitudes but not in others. Thus it has been postulated that the poles were in the North Pacific and South Atlantic, so that there would be no continental ice caps (Fig. 17–6). Any placement of the pole on lands bordering the Arctic Ocean would seem improper because of the temperate climate faunas and floras around that ocean and the absence of evidence of glaciation—unless the average earth temperatures were higher as postulated above. But as will be mentioned in a later chapter, the problem is still more complex, for possibly the continents have not retained the same relative positions as of today. The positions of poles can be determined from the study of remanent magnetism in the rocks. The middle Jurassic pole relative to Europe seems to lie northeast of Lake Baikal, Siberia, but relative to North America, it lies to the southwest several hundred miles in China. The difference should represent the drift that

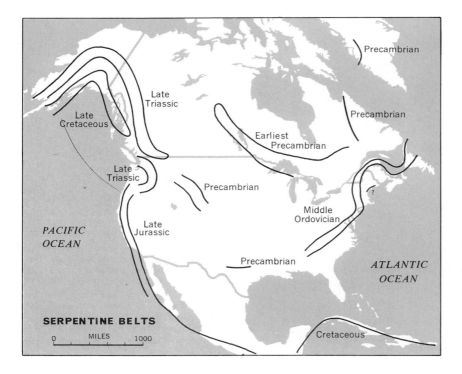

Fig. 17–9. Serpentine belts of North America. The map shows the distribution of ultrabasic rocks in North America, and the ages of the intrusives in the several belts (after H. H. Hess). Serpentine intrusions seem to have invaded geosynclinal belts at an early stage in their orogenic deformation. Their composition resembles that attributed to the subcrust below the Mohorovičić Discontinuity, which is shallow beneath ocean basins and trenches. Their presence in modern island arcs suggests the analogy to ancient tectonic belts.

B

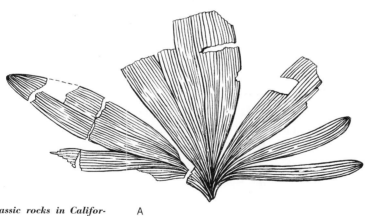

A

Fig. 17-10. *Jurassic rocks in California and Oregon. A.* Jurassic Ginkgophyta. The ginkgos or Maidenhair Tree is one of the most distinctive of present trees, having a tall straight trunk and fan-shaped leaves. Commonly planted in cities as a shade tree, it is a native of eastern Asia. The distinctive leaves are found in rocks as old as the Triassic, and are common in the Jurassic floras of central Oregon, from which the deeply dissected leaves of the genus *Ginkgoites* come.

B. Jurassic and lower Cretaceous rocks in east-dipping beds along the west side of the Sacramento Valley, central California; view to the northeastward (W. P. Irwin, United States Geological Survey).

has changed the relations between the continents since the Jurassic. On present lands, wind directions have definite relations to the climatic zones and latitudes. Information on the cross-lamination of wind-laid sands forming the Jurassic aeolian sandstones yields evidence corroborating that of other sources; studies are being made toward this end, as was discussed in Chapter 14.

CHANGES IN NORTH AMERICA

North America has a wealth of interesting and significant relations portrayed by its Jurassic rocks. From the standpoint of continental development a few seem most instructive. The present coasts on the Gulf of Mexico and the Atlantic Ocean have plains of Cretaceous and Cenozoic rocks that conform in main structure to those of the present geography. This tectonic form came into being by Jurassic and possibly in late Triassic along the Gulf. It did not prevail until the Jurassic in the East. The Nevadan Orogeny that affected the Pacific coastal belt in late Jurassic was one of the principal times of mountain forming and magmatic intrusion. Farther east, the Jurassic history of the Rocky Mountain region is marked by a return to tectonic relationships resembling those of the early Paleozoic. And the Jurassic is the first period giving an appreciable record in northern Mexico and in the states bordering the Gulf of Mexico.

INTRUSIONS IN THE SIERRA NEVADA

Returning to the granodiorite of Half Dome in Yosemite National Park (Fig. 17–2), on the west slope of the Sierra Nevada in California, the massive rock walls form the sides of the flat-floored glaciated valley. The age of the intrusion of the granitic rock is determined stratigraphically, because phyllites and schists that have yielded late Jurassic (Kimmeridgian) cephalopods to the east in the high Sierras are cut and altered by similar intrusive rock (Fig. 17–7). At other localities in northern California, similar intrusives into fossiliferous Jurassic rocks are overlain unconformably by lower Cretaceous conglomerates containing pebbles such as could have come from the erosion of the intrusives. It should be noted that there are assumptions that the rock in Yosemite, which is similar to intrusive rocks elsewhere, is of the same age. Thus great bathyliths, masses of granitic igneous rock, were emplaced within the latest Jurassic or earliest Cretaceous. Potassium-argon age determinations show that although some of the intrusions are late Jurassic, others are younger, so unfortunately it is not a warranted assumption that all the similar granodiorites are of the same age.

Fig. 17–11. Jurassic sedimentary rocks at Wide Bay, at the eastern base of the Alaska Peninsula, southwestern Alaska. *A.* Upper Jurassic (Shelikof) shale and sandstone west of Short Creek. Lower Jurassic (Kialagvik) shale is poorly exposed in the low foreground (M. A. Norton, Richfield Oil Corporation).

B. Conglomerates in the base of the upper Jurassic, lying with unconformity on middle Jurassic (L. B. Kellum).

Fig. 17–12. Conglomerate in the Jurassic southwest of Whitehorse, Yukon (J. Wheeler, Geological Survey of Canada). Triassic and Jurassic sediments in southwestern Yukon, thousands of feet thick, with some interbedded lavas, have conglomerates of increasing thickness and particle size southwestward. The conglomerate in the figure has well-rounded boulders of several rock types, is thus polymictic (in contrast to those having one type, called oligomictic). The dark volcanic rocks from lava flows are most common, the light, granitic boulders conspicuous; there are some fragments of sedimentary rocks. The size of boulders and their roundness, as well as the nature of the fragments, suggests that they have come from a land of older sediments and lavas, intruded, raised, and eroded, a tectonic land; they were laid on the margin of a deeply subsiding trough or geosyncline.

A

B

ULTRABASIC INTRUSIVE BELTS

The term Nevadan Orogeny is derived from the Sierra Nevada, the mountains extending northward through eastern California; Yosemite Park is on the west slope. Structures formed in the Nevadan Orogeny are Nevadides. In addition to the intrusions of granitic rocks in the bathyliths, the late Jurassic was the time of intrusion of masses of a very silica-poor rock, peridotite, and closely related serpentinite of similar composition, along the Pacific border in California. Serpentinite is a streaked green rock, such as verde antique (Fig. 17–8), frequently used as a decorating marble. Similar serpentinite is intrusive into rocks unconformably below the Jurassic in central Oregon. Serpentinized peridotites, ultrabasic rocks, have been found in narrow belts in mountains of many ages on all continents (Fig. 17–9). They seem to have been intruded into the central part of a major volcanic geosynclinal or eugeosynclinal belt at an early stage in its deformation, preceding other orogenies that led to its destruction. The ultrabasic intrusions along a belt are not all contemporaneous. It is suggested that the silica-poor ultrabasic rocks have penetrated into a very thin crust in a region that is marginal continental, not truly a part of the thick-crusted

A B

Fig. 17–13. *Navajo Sandstone, of early Jurassic age, in Zion National Park, southern Utah.* **A.** The Great White Throne, in which the lower sandstones are rather plane-bedded, the higher ones strongly cross-bedded (J. Muench).

B. Cross-bedded sandstone along the highway. The broad, sweeping cross-stratification is attributed to an eolian origin, the dipping surfaces being foreset beds on the lee slopes of a succession of dunes crossing a region that was accumulating sands. Although there is variation in the directions of dips, the dominant stratification in the exposure dips toward the right, eastward. Measurements of many such exposures determine the directions of winds in early Jurassic.

continent. The belts along the Atlantic Coast of North America are of the late Ordovician Taconian Orogeny. The verde antique "marble" from Vermont is a widely used serpentinite of that belt. The Pacific belt that had been mobile with deeply subsiding troughs and rising welts and volcanoes since early Paleozoic, was partially consolidated in the middle Mesozoic. But of course not all of it was—for the northwestern part, the Alaskan peninsula and Aleutian Islands, are still in an island arc stage.

Farther east, in Nevada, thrust faults of displacement of many miles involve sedimentary rocks as young as latest Triassic and early Jurassic (Toarcian). The faults and the folded thrust sheets are penetrated by granodiorites like those of the Sierra Nevada. Hence the thrusting has been attributed to the Nevadan Orogeny, though there are other thrusts that are known to be older, as discussed in the Carboniferous chapter.

PACIFIC COAST JURASSIC

The sequences of Jurassic rocks are widely preserved along the Pacific Coast and are quite varied. Some are difficult to determine, for they were deformed and metamorphosed in the orogenies that are called

Fig. 17–14. *Cephalopods in upper Jurassic limestones from the Black Hills, South Dakota (American Museum of Natural History).* *A.* Ammonite cephalopods, *Cardioceras.*

B. An accumulation of the shells of the belemnite *Pachyteuthis*, a dibranch cephalopod; see Fig. 17–5. The shells are internal supporting skeletons of the mollusc related to the modern squid and cuttlefish; there are a few associated pelecypods and small fragments of these and other organisms. The strong parallelism in orientation of the belemnites probably resulted from their being rolled by currents crossing from left or right.

A B

Nevadan within and at the close of the Jurassic and in later orogenies, and they have been deeply buried or eroded away over most of the area. Conditions varied greatly within rather short distances. Generally, there are miles of argillites with thousands of feet of lavas, agglomerates and tuffs, frequently with interbedded granite-pebble bearing conglomerates. For example, along the west of the Sierra Nevada in California near Yosemite, several miles of fossil-bearing medial and late Jurassic sediments, lavas, and fragmental volcanic rocks lie unconformably on Carboniferous slates; the volcanic centers must have been nearby, for the lavas change in thickness and character along the outcrop belt. In the northern Sierras, at a peak named Mount Jura by geologists nearly a century ago from their finding Jurassic ammonites, there are thousands of feet of marine and non-marine sediments, lavas, and agglomerates with fossils representative of each series and many of the stages of the system. But not far to the east in the higher Sierras, a land persisted

Fig. 17–15. *Restored sections of Jurassic rocks along the Western margin of the craton. A.* Section from southeastern Idaho across Wyoming to western South Dakota, showing the rapid decrease in thickness from the miogeosynclinal belt to the craton; detrital sediments from the rising highlands of the Nevadan Orogeny entered from the west in late Jurassic.

B. Similar section from southeastern Nevada across Arizona and New Mexico, in which there is more gradual convergence and overlap eastward.

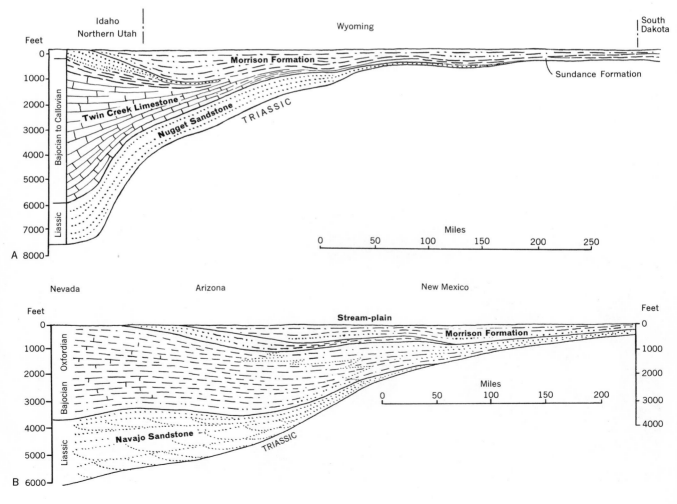

through most of the period, for very late Jurassic plant-bearing sediments lie unconformably on Paleozoic rocks (Fig. 17–10). Along the California Oregon border in the northern Klamath Mountains, very late Jurassic sediments having plants like those in the higher Sierras unconformably overlie sediments and volcanic rocks of late Jurassic age like those near Yosemite and Mount Jura. This unconformity within the upper Jurassic has been thought to date the Nevadan Orogeny; but it may not be of the same age as some of the folds, thrusts, and intrusions that are commonly attributed to the orogeny.

A few hundred miles northward in central Oregon, four or five miles of the section is of marine sediments without appreciable volcanics—one of the most representative records among marine faunas. Sequences of similar variety are known along the coast in British Columbia, Yukon, and Alaska. The Queen Charlotte Islands of western British Columbia have been long known for their well-preserved ammonites,

Fig. 17–16. Morrison shales and sandstones, outcropping in the slope to the left of the hog-back of east-dipping lower Cretaceous Cloverly or "Dakota" Sandstone, south of Morrison, Colorado (W. M. Cady). Mount Morrison on the left skyline is of Precambrian gneiss; it is flanked on the east by Pennsylvanian and Permian red sandstones and shales, including the Permian Lyons Sandstone forming the light-colored hog-back on the left. The sedimentary beds were deformed in the Laramian Orogeny late in the Mesozoic.

and a section of about four miles of fossiliferous sediments (Fig. 17–11) on the southern Alaskan coast has representatives of many stages of the system. The Jurassic in Yukon gives further evidence in its conglomerates (Fig. 17–12) of the presence of tectonic lands separating the geosynclines. The mobility of the Pacific margin produced subsiding troughs that permitted accommodation of thousands of feet of rocks while nearby rising welts yielded the terrigenous sediments; and lava and fragmental volcanic rocks issued from widely distributed fissures. Some of these troughs may have been folded and intruded by bathyliths in a Nevadan Orogeny late within or at the close of the Jurassic, but there probably were several phases in different areas in some regions, and consolidation may have been delayed into the later Mesozoic.

THE NAVAJO COUNTRY

Jurassic rocks are extensively exposed along the Rocky Mountains and in the plateaus to the southwest into Arizona. The Great White Throne in Zion National Park (Fig. 17–13), southern Utah is a spectacular bluff of sandstone having great, sweeping cross laminations (Fig. 17–13B) that have been attributed to preservation of the lower foreset beds of

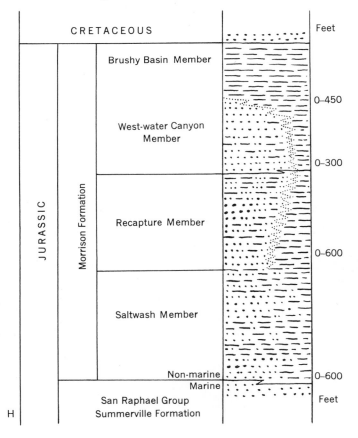

Fig. 17–17. *Morrison Formation in the states of Utah, Colorado, Arizona, and New Mexico, around the Four Corners where the states meet (L. C. Craig and others).* A. Isopach map of the Morrison Formation, showing the thicknesses of the whole.

B., C., D., and E., Isopach maps of the Salt Wash Recapture, Westwater Canyon and Brushy Basin members of the Morrison Formation; note that they are not time-stratigraphic units—have boundaries that are not synchronous planes.

F. Lithofacies map, showing the principal lithologies represented in the Salt Wash Member; similar maps could be presented of the other members.

G. Average direction of cross-bedding laminae in the Salt Wash Member; the directions are variable in the stream-laid deposits; their relative consistency is greatest in localities having the longest arrows.

H. Columnar section of the stratigraphic succession in the Morrison Formation, showing the names of the members and their intertonguing gradational relations.

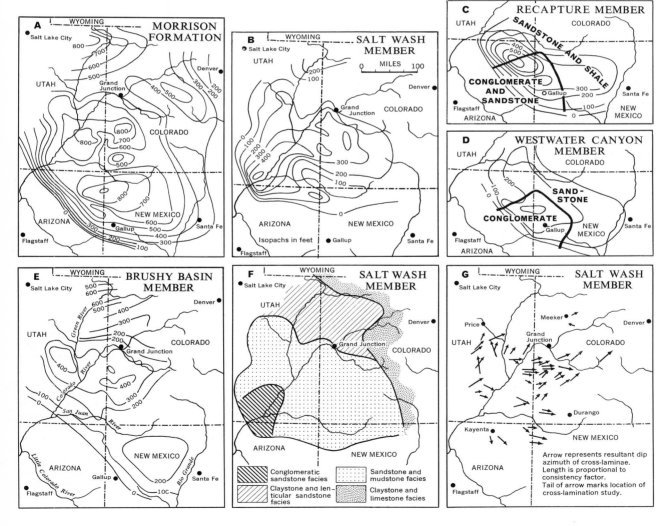

A WYOMING — MORRISON FORMATION
Salt Lake City
800 700 600 500
UTAH
Grand Junction
400 500 300 200
COLORADO
800 700 600 500
800 700 600
500
0
ARIZONA
300 Gallup
600
500 400 300
Santa Fe
Flagstaff
200 100

B WYOMING — SALT WASH MEMBER
Salt Lake City
0 MILES 100
UTAH
200 100
Denver
Grand Junction
COLORADO
0 100 200 300 400
300
200
100
0
ARIZONA NEW MEXICO
Isopachs in feet · Gallup Santa Fe
Flagstaff

C RECAPTURE MEMBER
UTAH COLORADO
SANDSTONE AND SHALE
400 500
CONGLOMERATE AND SANDSTONE
300 200 100
0 Gallup Santa Fe
Flagstaff NEW MEXICO
ARIZONA

D WESTWATER CANYON MEMBER
UTAH COLORADO
200 100 0
SAND-STONE
CONGLOMERATE
Gallup
Flagstaff NEW MEXICO
ARIZONA Santa Fe

E WYOMING — BRUSHY BASIN MEMBER
Salt Lake City
500 600
Green River
600 500 400 300
Denver
200 200
UTAH
COLORADO
Grand Junction
400
400
100
Colorado River
0
300 200
San Juan River
ARIZONA NEW MEXICO
Little Colorado River
Gallup 200
Rio Grande
100 Santa Fe
Flagstaff 0

F WYOMING — SALT WASH MEMBER
Salt Lake City
UTAH
Denver
Grand Junction
COLORADO
ARIZONA NEW MEXICO

Conglomeratic sandstone facies
Sandstone and mudstone facies
Claystone and lenticular sandstone facies
Claystone and limestone facies

G WYOMING — SALT WASH MEMBER
Salt Lake City
Price Meeker Denver
UTAH
Grand Junction
COLORADO
Durango
Kayenta
NEW MEXICO
ARIZONA
Flagstaff

Arrow represents resultant dip azimuth of cross-laminae. Length is proportional to consistency factor. Tail of arrow marks location of cross-lamination study.

405

dunes, built of wind-carried quartz sands. This spectacular sandstone, the Navajo, is known over a broad area extending northward to southeastern Idaho. From thicknesses of a thousand feet or more in the western exposures, it diminishes and disappears in New Mexico, central Colorado, and Wyoming. Evidence on the age of the Navajo Sandstone is sparse, but as it overlies rocks with Triassic reptile fossils and underlies rocks with marine middle Jurassic (Bajocian) ammonites, it is classified as lower Jurassic. This pattern of greater thickness in the west diminishing to the east is continued through the higher Jurassic beds, a change from the miogeosyncline in Utah and Idaho to the craton of Colorado and Wyoming (Fig. 17–15)—quite like that of the Paleozoic from Cambrian through Devonian times, but in marked contrast to the great differential movements that produced the troughs that received miles of sediments in Utah, Colorado, and New Mexico in the Carboniferous.

LATEST JURASSIC AND DINOSAURS

The latest Jurassic rocks in the region, the Morrison beds (Fig. 17–16) and their variants have conglomerates and sandstone from which uranium ores are mined in Colorado and southwestward. As the ores are related to organic material in fluviatile channels, the sedimentation has been studied in greatest detail. The sediments came from several sources and were carried in directions indicated by the inclination of the cross-laminae in the channel sands and gravels (Fig. 17–17). The youngest Jurassic

Fig. 17–18. Como Bluff, Wyoming, a ridge exposing shales and sands of late Jurassic, the Morrison Formation, from which many fine dinosaur skeletons were recovered (E. H. Colbert). When the Union Pacific Railroad was built across central Wyoming about a century ago, it passed along the north base of the hill, on the right. Dinosaur bones were discovered, and the skeletons were quarried and readily shipped away; moreover, the telegraph line gave a ready means of promptly giving the news of discoveries to the daily papers, for public interest was great.

Fig. 17–19. *Jurassic System in northern Mexico. A.* Outline map of the states of northern Mexico in which the Jurassic System is exposed.

B. Paleogeographic map with restored sections showing the relations between the Mexican Geosyncline, the Coahuila Peninsula, and the troughs and intervening lands to the east in Jurassic time (after Z. de Cserna). The geosyncline developed in early Mesozoic time over a basement of metamorphosed and intruded Paleozoic sedimentary and volcanic rocks that are now exposed only in a few small areas scattered widely through the northern states.

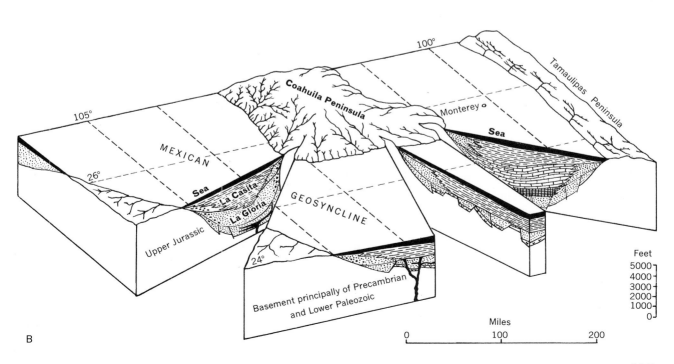

B

rocks in the western exposures in Idaho and Utah contain conglomerates formed from gravels laid in streams flowing eastward from the highlands raised to the west with the advent of the Nevadan Orogeny.

The widely distributed stream and lake-laid sediments of the Morrison formation have been a principal source of skeletons of the dinosaurs of the Jurassic that are so widely known and exhibited in museums. Como Bluff (Fig. 17–18), a ridge that lay along the route of the original transcontinental Central Pacific Railroad, was the site of the discovery of dinosaurs nearly a century ago. The regional constancy of the lithic and faunal units in the Rocky Mountain region demonstrates the relative stability of the area in comparison with the extremes in thickness and lithologies in the Pacific coastal region.

The easternmost exposures of Jurassic rocks in the United States are those in the east slope of the Black Hills, South Dakota, and in mountains along the Rio Grande River below El Paso, Texas, each near the longitude of 103°. Wells reveal the presence of Jurassic rocks below the surface in central South Dakota, western Nebraska, and northwestern Kansas. Wells have also found a thick wedge of Jurassic rocks that lies beneath the coastal plain of Louisiana, Arkansas, and Texas. And there are fine exposures of Jurassic in the mountain ranges of northern Mexico, the Texas localities being nearby.

MEXICAN GEOSYNCLINE

The Jurassic System is the oldest that gives substantial knowledge of the geologic history of northern Mexico. The state of Coahuila lies south of the Rio Grande River from the middle third of its course from El Paso, Texas to the Gulf of Mexico; Coahuila has an area of about 60,000 square miles, and is thus about the size of the state of Illinois (Fig. 17–19). Toward the Gulf lie the narrow states of Neuvo Leon and Tamaulipas, and to the west Chihuahua extends to the western boundary of New Mexico, and Sonora to the Gulf of California; these and other states are shown on the outline map of northern Mexico (Fig. 2–1). The base of the Jurassic is exposed in a number of small areas and is unconformable on rocks of great variety. Late Carboniferous and Permian sedimentary rocks, volcanic rocks, and subsequent intrusions underlie the Jurassic in southern Coahuila and adjoining Durango. In other areas eastward to Tamaulipas, the Jurassic lies on sediments of varied grades of metamorphism lacking distinctive fossils, and on intrusives into this basement. The lowest Jurassic beds commonly contain ammonites characteristic of the low upper Jurassic (Oxfordian), though some underlying non-marine beds intervene in Tamaulipas. The Jurassic sediments (Fig. 17–20), principally limestones, are

Fig. 17–20. Jurassic and Cretaceous in the Sierra de Parras, southern Coahuila, Mexico (Z. de Cserna). View to the southwest of an overturned succession, having Jurassic limestone on the crest of the range, succeeded stratigraphically by lower and middle Cretaceous limestones on the near face of the mountain, the youngest beds of late Cretaceous age being low on the right. The overturned Jurassic-Cretaceous contact is structurally below and stratigraphically above the limestone forming the left-descending light band left of the center.

Fig. 17–21. Map of the Gulf Coast of the United States, showing the distribution and thickness of Jurassic rocks that have been penetrated in the subsurface; they are not otherwise exposed, for the overlying Cretaceous rocks extend beyond them to the inner margin of the Coastal Plain; the isopach lines are at 1000-foot intervals (G. E. Murray).

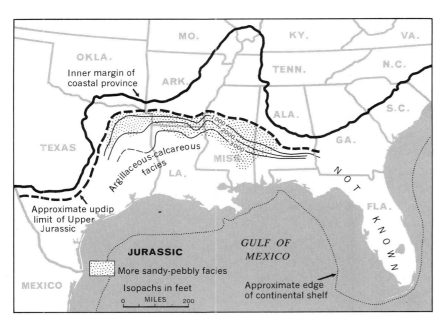

Map labels:

MO. KY. VA.
OKLA. TENN. N.C.
ARK.
Inner margin of coastal province
TEXAS ALA. S.C.
MISS. GA.
Argillaceous-calcareous facies
LA.
NOT KNOWN
Approximate updip limit of Upper Jurassic
FLA.
MEXICO
JURASSIC
GULF OF MEXICO
More sandy-pebbly facies
Isopachs in feet
0 MILES 200
Approximate edge of continental shelf

409

distributed through the ranges of eastern Chihuahua, barely extending eastward into Coahuila, southwestern Texas, southwestern New Mexico, and southeastern Arizona, with a maximum thickness of a few thousand feet. They become sandy as they approach a western shore toward Sonora and disappear eastward against a long peninsula that extended south through much of the area of Coahuila; hence it has been called the Coahuila peninsula or platform. This southeast trending trough in northern Mexico has been called the Mexican Geosyncline, continuing to be significant in the later Mesozoic. Lower and middle Jurassic rocks are widely known from more southern states of Mexico, so that the marine waters seem to have entered the Mexican Geosyncline from the south. Jurassic is also known in wells in the Tampico region, in coastal Tamaulipas, representing the sea on the eastern side of the Coahuila Peninsula and associated peninsular lands. The Tampico wells penetrate rocks that are probably continuous northward into those well known in the deep subsurface of the coastal states of the United States.

The paleogeography of northern Mexico suggests that the tectonic pattern of the late Paleozoic along the Gulf of Mexico, with a volcanic geosyncline curving westward to Chihuahua from coastal Texas, had been altered by folding and igneous intrusion in latest Paleozoic or early Triassic times. The Mexican Geosyncline and Coahuila Peninsula seem to have trends quite independent of the earlier tectonic pattern.

BEGINNINGS OF COASTAL PLAIN WEDGES

The geologic map of the Gulf Coast states from Texas to Florida does not show any exposures of Jurassic rocks, only Cretaceous and Tertiary formations in belts becoming progressively younger toward the Gulf. As the surface is gently rising inland, the exposed rocks must dip seaward. As we shall see subsequently, the sediments thicken toward the Gulf, so that the base of the Cretaceous must descend at an increasing rate, coming to lie tens of thousands of feet below the surface, perhaps several miles at the present shore line (Fig. 17–21). But the deepest wells that have penetrated the earth are oil wells about five miles or 25,000 feet deep—so the rocks that underlie the Cretaceous are known only toward the landward margin of Cretaceous outcrop. Some fifty years ago, important oil production came from some wells in southern Arkansas. As the techniques of petroleum engineering improved, and deeper wells became possible, rocks like those that had yielded oil in southern Arkansas were reached in extreme northern Louisana. Fossils that came from cores included ammonites of late Jurassic types, Oxfordian to Portlandian. The beds thicken from absence

Fig. 17–22. Restored section of the subsurface Jurassic sediments beneath the Gulf Coastal Plain of norther Louisiana and east Texas (after F. W. Swain). The Eagle Mills Salt, generally classed as Jurassic but possibly older, is the sediment that yields the salt domes that rose progressively throughout later geologic time, the salt being of lower density than the surrounding poorly consolidated rocks. The Jurassic is the oldest system laid in a coastal plain similar to that of the present.

Fig. 17–23. Reconstruction of a late Jurassic scene in North America, showing some of the reptiles that lived during Morrison time (American Museum of Natural History). On the left in the background are the herbivorous ornithischian dinosaurs, *Camptosaurus*, and in front of them two thirty-foot-long saurischian predators, *Antrodemus* (often known as *Allosaurus*). On the right are three individuals of the ornithischian, *Stegosaurus*. In the background, in the water, are several gigantic saurischians, the sauropod *Apatosaurus*, or *Brontosaurus*. These dinosaurs are shown inhabiting a tropical landscape in which marshes, ponds and rivers were abundant. Such Jurassic environments were widely distributed throughout the world.

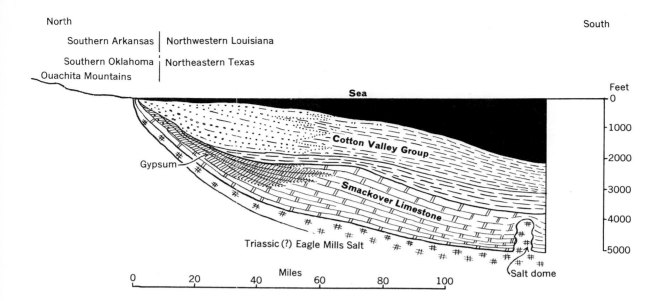

North

Southern Arkansas | Northwestern Louisiana

Southern Oklahoma | Northeastern Texas

Ouachita Mountains

South

Sea

Feet
0

Cotton Valley Group

Gypsum

Smackover Limestone

Triassic (?) Eagle Mills Salt

Salt dome

-1000

-2000

-3000

-4000

-5000

Miles
0   20   40   60   80   100

in southern Arkansas to a mile or more in northern Louisiana before going down to such depths, about fifty miles from the northern edge, that they are not penetrated by present wells (Fig. 17–22).

Coarse terrigenous sediments, conglomerates and sands, some of them red, pass southward into argillitic clays with limy interbeds that contain the marine fossils; stream-laid sediments on the north pass into the marine offshore deposits on the south. The oldest rocks in the sequence are salts that lie on argillites and siltstones thought to be Paleozoic, separated by unconformity. This Eagle Mills salt, though known in place only in wells at the northern margin of the Jurassic wedge, is thought to extend far to the south, perhaps to beneath the present shore, and to be the source of salt that rises into younger beds to form the salt domes that we will consider in the Tertiary; it may be older than Jurassic.

EVAPORITES UNDER THE GULF

Geophysical profiles made in the deep Carribean Sea suggest that salt domes, and thus source salt, extend clear across to Yucatan. As the seismic studies show shallow depths to the Moho Discontinuity, the Carribean seems oceanic. Thus there is the most surprising suggestion that an oceanic area may have been an evaporite basin in early Jurassic. Such an hypothesis is certain to be thoroughly tested in the coming years.

The Jurassic wedge has been drilled in wells to the west and southwest in Texas, and eastward into Mississippi and southwesternmost Alabama. These are the first deposits that are known to have been laid along a coast generally similar to that of the present Gulf of Mexico,

Fig. 17–24. Upper Jurassic dinosaurian faunas, closely related to each other, occur in the Morrison formation, which is widely exposed throughout the Rocky Mountain states of western North America, in the Kimmeridge and related beds of southern England, and in the Tendaguru beds of Tanganyika. There were obviously open migration routes between these areas during the late Jurassic that allowed a wide interchange of faunal elements between distant parts of the earth.

though in seas that extended farther inland than today's. The Jurassic shows marked changes from the Paleozoic, which in this belt was laid in rapidly subsiding troughs associated with rising structural welts that supplied the terrigenous sediments. Whether these rocks were folded and intruded toward the coast is not known, but the Carboniferous rocks of the Ouachita Mountains in Arkansas had been folded and thrust faulted before the Jurassic sediments were laid on them. The Jurassic seems to have introduced relations along the north of the Gulf of Mexico generally like those of today. But the relations to the west, within the Mexican Geosyncline, were to remain in contrast to those of the Paleozoic and of today through the rest of the Mesozoic Era.

ATLANTIC AND ARCTIC COASTS

We have summarized the Jurassic sedimentary and volcanic rocks on the Pacific Coast, and those deeply buried on the Gulf Coast. On the Atlantic Coast, too, Jurassic rocks do not seem to be exposed at the surface. Nor have they been recognized as present in wells. But quite possibly they lie at depths below the continental shelf, to be tested by wells that will surely be drilled there. On the Arctic Coast, the Jurassic forms part of the coastal plain of the Canadian Islands, much as does the Triassic that has been described.

JURASSIC CONTINENTS AND REPTILES

Reference has been made to the exposures of Morrison sediments at Como Bluff, Wyoming and the discovery of upper Jurassic dinosaurs there. The quarries at Como Bluff are perhaps the most famous of Morrison collecting localities, but they are in fact only one group of

Fig. 17–25.   Restoration of the sea bottom as in late Jurassic time in South Dakota, with swimming squid-like belemnite cephalopods *Pachyteuthis* (see Fig. 17–5) above a bank of oyster-like pelecypods (Chicago Natural History Museum).

numerous sites throughout the Rocky Mountain region, where Morrison reptiles have been collected in this widely exposed sequence of continental sediments. The type locality of the Morrison is near the little town of Morrison, Colorado, some fifteen miles west of Denver. Today, audiences in the beautiful Red Rocks outdoor theater north of Morrison may look beyond the stage where symphony concerts, operas, and ballets are performed by world-renowned artists, and see the long hogback, on the side of which Marsh's collectors excavated Morrison dinosaurs for Yale University in the decade after the Civil War. Morrison reptiles are likewise found in other areas as well, especially in New Mexico and Utah.

The Morrison fauna (Fig. 17–23) is composed of various dinosaurs, among which the gigantic sauropods, the great predatory theropods and the plated stegosaurians are particularly noteworthy. It contains also many turtles and crocodilians. And at one quarry along the front of Como Bluff, Quarry Nine, there were found many tiny Mesozoic mammals, which at one time lived in burrows, and among the vegetation through which the great dinosaurs wandered.

Various dinosaurs, or closely related ones, and other reptiles and mammals that characterize the Morrison sediments, are found in the Kimmeridgian beds of Europe—in England and in Portugal. Furthermore, some of these same animals have been excavated from the Tendaguru beds, near Lake Tanganyika, in Africa.

REPTILE MIGRATIONS

Such far-flung distributions of land-living reptiles must mean that there were broad avenues of intercontinental migration between the several land blocks during late Jurassic time (Fig. 17–24). The occurrence of *Brachiosaurus,* the greatest of the gigantic sauropods and the largest animal ever to walk the earth, in East Africa and in North America, must mean that this huge dinosaur, as well as many others, wandered widely during the final stages of Jurassic history. It must mean that much of the globe was a tropical and subtropical world at that time, in which lowlands, clothed in tropical forests, were widely connected. It was a time when climates and floras and faunas on the land were remarkably uniform. It was a time when environments were favorable to the evolution of reptilian giants, so that these giants flourished in abundance and dominated the continents.

18
The Cretaceous Record
in North America

Lower Cretaceous (Aptian) limestone standing vertically along the canyon in the Sierra
Madre Oriental west of Linares, Neuvo Leon, Mexico (Z. de Cserna).

18 *The Cretaceous Record in North America*

The name Cretaceous, from the Latin word creta, chalk, was given in 1822 by the Belgian geologist Omalius d'Halloy (1783–1875) to the system of rocks that is well exposed in the Paris Basin (Fig. 18–1), and which in western Europe has an abundance of chalk, particularly well shown along the English Channel (Fig. 18–2). The chalk is a rock consisting of fragments of marine organisms that had been detritus on the sea floor and that is now a porous, poorly cemented siltstone of calcite particles. Thus Cretaceous is a systemic name, like Carboniferous, that was gained from a rock type rather than from a place. A French geologist and paleontologist, Alcide d'Orbigny (1802–1857) arranged the Cretaceous rocks into seven stages (Fig. 18–3) named from localities with Roman names in France, Denmark (Danian), and Switzerland (Neuchâtel, Roman Neocomum), terms that have continued in use to this day. The Maastrichtian Stage was added below the Danian, and other names have been given for subdivisions, particularly of the Neocomian and Senonian. They are recognized by distinctive fossil assemblages, in which ammonite cephalopods (Fig. 18–4) are important constituents. Another kind of animal that is used in zonation is the echinoid (Fig. 18–5). It is conventional to call Cretaceous rocks lower and upper, and in North America to assign provincial series names for the rocks in several parts of the continent. But the names of the stages in northwestern Europe are universally applied.

We have considered the principles on which the interpretation of historical geology is based in discussing the preceding systems. The Cretaceous System in North America will be presented with respect to its present distribution, thicknesses, lithologies, and structures. The history of the period will be interpreted from these records, and those of the fossils that the rocks contain.

GENERAL DISTRIBUTION

The Cretaceous System in North America can be described in terms of several principal areas of its exposure and subsurface distribution (Fig. 2–6). Very gently dipping Cretaceous sediments underlie the Gulf and Atlantic coasts and extend seaward to undetermined distances. A very large area beneath the Great Plains from Kansas and New Mexico to northeastern British Columbia and the Yukon has gently dipping beds that become folded and faulted westward as an effect of the Laramian Orogeny that took place within the later Cretaceous and thereafter. A third principal area of Cretaceous rocks is in northern

Fig. 18–1 (right). Geologic map of western Europe, showing the Tertiary basin in central northern France where the stratigraphic sequences in the Cretaceous System and the Tertiary System were determined early in the nineteenth century.

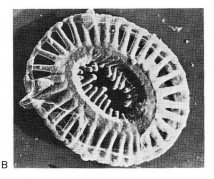

B

Fig. 18–2. *The typical chalk of the Cretaceous System. A.* Cliffs of Cretaceous limestone along the south coast of England on the English Channel—the Chalk Cliffs of Dover (Her Majesty's Geological Survey). The name Cretaceous comes from the association of the system with the chalks in France, across the channel. The rock is composed of minute tests of organisms, or dismembered parts.

B. An electron photograph of a coccolith, magnified nearly a thousand times (from Maurice Black). The structure formed in an alga that accumulated to build dominant portions of some chalks; other chalks are made of foraminifera and other organic detritus.

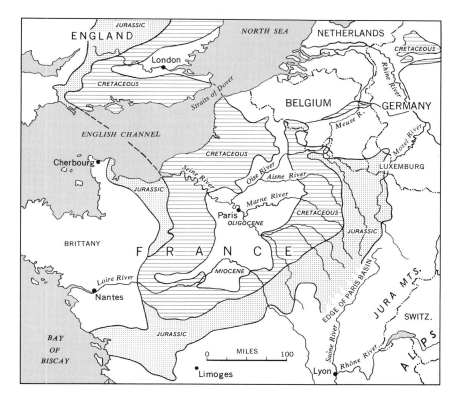

A

Mexico, extending into southeastern Arizona, in which the Cretaceous rocks are considerably folded and faulted. The Arctic Coast from Alaska to Ellesmereland has extensive Cretaceous rocks that are gently dipping, but locally the rocks form folded mountains. And another belt of Cretaceous of quite complex structure and distribution follows along the Pacific Coast from Baja California to Alaska.

GULF AND ATLANTIC COASTS

As can be seen on the geologic map, Cretaceous rocks are exposed almost continually from northern Tamaulipas and southern Texas to northern New Jersey; there are limited exposures on Long Island, New York, and rocks have been dredged northeastward to the Grand Banks of Newfoundland. The stratigraphy of the surface exposures and their correlations have been worked out in great detail along most of this belt of exposure. The interpretation of the history requires knowledge in three dimensions. As the Cretaceous is dipping seaward along this entire belt of outcrops the full thickness is still preserved in the down-dip. Yet much of it is beyond present drilling depths.

On the west Gulf Coast of Texas, Cretaceous rocks are exposed (Fig. 18–6) in a belt of a few tens of miles breadth extending northward barely into Oklahoma (Fig. 18–7). The top of the system underlies the Tertiary toward the Gulf, so the full thickness is preserved from erosion at that line of contact and southward. The complete section of the system is penetrated by wells southward to where the base of the system is deeper than the wells that have been drilled, generally about fifteen thousand feet. As the base of the Cretaceous descends at an increasing rate southward, it soon goes below the deepest present wells (Fig. 18–8); full knowledge of the system is limited to a narrow

Fig. 18–3. Generalized structure section across the Cretaceous of the eastern margin of the Paris basin, France, published by Alcide d'Orbigny in 1852. Seven stages were recognized at the time, all of common use to the present; the Maastrichtian Stage is generally now inserted below the Danian, and the Neocomian and Senonian divided into several stages or substages. The Cretaceous section is a mile or so thick and exposed in a breadth of some seventy miles.

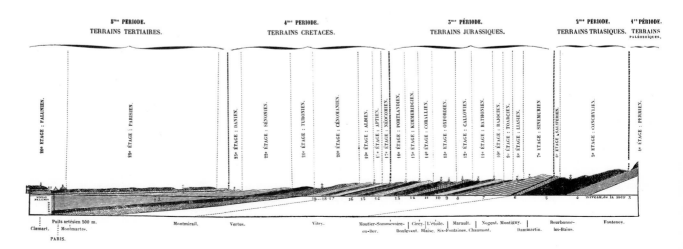

band between the contact with the Tertiary and the trend of the deepest wells. The upper part of the system is, of course, penetrated somewhat farther toward the Gulf. The surface and subsurface information shows the Cretaceous to be thickening in depth, the increase being partly by thickening of individual stratigraphic units, and partly by addition of beds, particularly to the base. As an affect of this basal overlap of the Cretaceous, the oldest Cretaceous Neocomian Stage is not found at the surface, where the lower part of the exposed section forms the Comanchean Series, the upper part, the Gulfian Series. The Cretaceous rocks older than the Comanchean have been called the Coahuilan Series, of Neolomian age. The problem of the downdip continuation below the deepest well will be considered in the discussion of the Tertiary of the region.

The Cretaceous section along the outcrop belt is separable into many formations (Fig. 18–9) and members that have disconformities and changes of facies evidencing the marine advances and regressions and the introduction of terrigenous sediments from stream systems such as that entering the Mississippi Basin (Fig. 18–10). Though there was general increase in subsidence toward the Gulf, with associated thickening of the sediments that are largely marly and calcareous, the deformation was not wholly uniform. There were areas of greater and lesser subsidence, forming troughs and domes in the southern interior of the continent. Thus eastern Texas had a south plunging subsiding

Fig. 18–4. *Cretaceous ammonite cephalopods.* (1) Smooth-shelled *Hantkeniceras;* (2) ribbed *Ptycholytoceras;* and (3) *Baculites,* a genus in which the first chambers are tightly coiled, but the shell becomes quite straight.

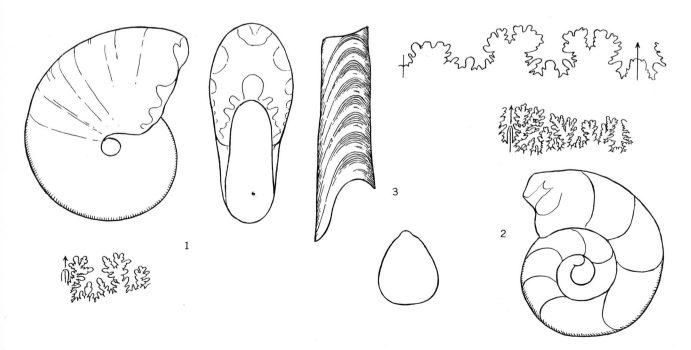

basin between a less subsiding area in northwestern Louisiana and that of central Texas (Fig. 18–11).

The Cretaceous System east of the Mississippi River outcrops in a belt that curves in an arc from the southern tip of Illinois through northeastern Mississippi into central Georgia. The rocks that outcrop are all Gulfian, upper Cretaceous. The thickest surface section, of fifteen hundred feet or so in central Alabama, has non-marine sandstones and clays overlain by marine sandstone and siltstones in the lower half, and a dominance of chalk in the upper. Toward the north and the east, along the outcrop, the chalks grade into sediments like those beneath them, near shore and stream-laid sediments (Fig. 18–12). Offhand, we could interpret the data from the outcrop as indicating that offshore chalks occupied the center of the north-south trending trough from which the sea advanced through later Cretaceous time. But because the data are from a line, even though a curved one, they do not adequately permit interpretation of the three-dimensional relations. Study of the sediments in wells shows that the section is first gradually and then rapidly increasing southward and southeastward. The shore in later Gulfian had a belt of shore sands passing through Alabama north of the present outcrop, joining the Georgia and Tennessee rocks of the same facies. The silts and muds drifted offshore southeastward, extending into northern Florida in the earlier Gulfian, but toward the close of the period, the area of the peninsula had a broad, gradually subsiding platform that accumulated hundreds of feet of calcareous shell detritus. The subsidence through Cretaceous time was greatest along the southwest coast, being restricted to that region in early Cretaceous; the shoal waters probably extended eastward beyond the Florida Strait into the Bahama Banks region. Thus the early Cretaceous geosyncline seems to have continued around the northeastern margin of the Gulf of Mexico (Fig. 18–13). When one

Fig. 18–5. *Echinoids.* A Cretaceous echinoid *Epiaster*, from Texas, (*A*) top and (*B*) bottom views. an—anal opening; m—mouth; the radiating rays of plates on the top side are the ambulacra(am). The living animal had many movable spines, like the sea urchin of the present.

Class Echinoidea. The echinoids are free-moving echinoderms, not attached by stems as are the crinoids and blastoids. The soft parts are within a somewhat compressed spheroidal shell of calcareous plates. Five radiating ambulacra or food-grooves have plates contrasting with those that intervene. The grooves in living echinoids have peculiar tube-feet that penetrate through pores; they facilitate the movement of the echinoid. In simplest forms, the test is round, with the mouth below and anal opening above. The surface of living forms has long projecting and moving spines; detached plates and spines found frequently as fossils, but rarely is the calyx found with the spines attached.

Echinoids are known from the Ordovician and later systems, are frequent in Devonian and younger rocks, but particularly useful in classifying Cretaceous and Tertiary. Many of the later forms are bilaterally symmetrical—oval, heart shaped, or even perforated, with the plates, mouth, and anal openings of the various genera distinctly and consistently arranged.

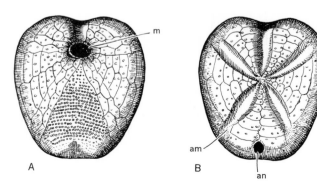

A   B

Fig. 18–6. *Lower Cretaceous (Coman-
chean) rocks exposed in Texas (Texas
Bureau of Economic Geology).* **A.** View
of the Trinity, Walnut, Goodland, and
Kiamichi formations in succession up-
wards along the Red River northwest of
Swells Bend, Cooke County, northern
Texas;

B. Glen Rose Limestone along the
Paluxy River near Glen Rose, Somervell
County, central Texas.

421

recalls that this general belt was one of active orogeny in the late Paleozoic, we can appreciate how great was the change toward the present continental outline.

The relatively simple seaward dip of sediments along the Gulf Coast is interrupted by one peculiar feature in western Mississippi. Reef limestones formed on several platforms of limited subsidence during the latest Cretaceous; the limestone near Jackson, Mississippi, is as much as fifteen hundred feet thick and has volcanic rocks beneath it. The Jackson Dome is evidenced by structure having relief in the lower Tertiary of about six hundred feet (Fig. 18–14) and on the top of the Cretaceous of about two thousand feet. The upper Cretaceous reef limestone is very thin on the crest of the structure, thickens on the flanks, and passes into thickening bedded limestone down the slopes. The lower part of the Gulfian is missing, represented in an unconformity that has the Comanchean sediments intruded by volcanic rocks below it. It seems that the Coahuilan and Comanchean series were invaded by lava which raised them into a dome that extended above the sea. The shoal was eroded in later Comanchean and earlier Gulfian time, then subsided sufficiently to be covered by later Cretaceous marine water, reefs forming on the crest of the platform. The structure on the younger, Tertiary rocks that cover the reef rock may be due in part to the compaction of the thick sections laid in the deepening water of the flanks of the platform as compared with the thinner cover on the crest of the dome.

Along the Atlantic Coast, Cretaceous rocks have quite continuous outcrops through the coastal states northward from Georgia to New

Fig. 18–7. Geologic map of the coastal plain of the Gulf of Mexico, from Texas to Florida. The map shows the gently southwardly dipping sediments of the Cretaceous and Tertiary systems. Although Oligocene rocks are not shown on the west Gulf Coast, they may be present within the rocks mapped as Miocene.

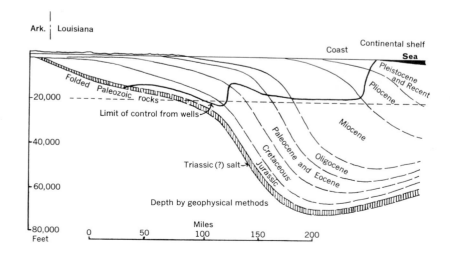

Fig. 18–8. Generalized structure section through the Mesozoic and Cenozoic rocks underlying the coastal plain of the Gulf of Mexico and the submerged shelf from eastern Texas southward (after G. C. Hardin, Jr.). The limit of penetration of wells restricts direct knowledge of each stratigraphic unit to a rather narrow band. Interpretation of the rest of the section has been largely inferred from geophysical, principally seismic, evidence. Techniques have been improved so as to give a fairly continuous profile on some horizons of discontinuity.

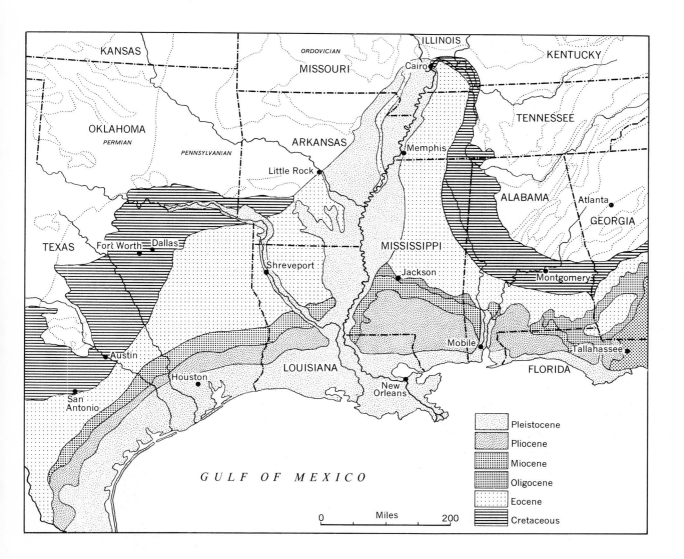

GULF OF MEXICO

0   Miles   200

Pleistocene
Pliocene
Miocene
Oligocene
Eocene
Cretaceous

Jersey. They barely appear on the surface of Long Island, New York, but are penetrated in wells; and they crop out as far to the northeast as in the canyons off the Grand Banks, south of Newfoundland. The exposed and drilled sediments in New Jersey are quite like those exposed along the east Gulf Coast, except that there are few chalks. The deepest wells along the Atlantic penetrate less than two thousand feet of upper Cretaceous fluviatile sands and clays overlain by marine marly and sandy strata, many with the potassium iron aluminum silicate glauconite which seems to form by alteration of other minerals beneath the sea and which seems to be distinctive of very slow deposition in marine waters. In Alabama, further information is gained by drilling wells in the hundred miles and more of the area south of the outcrop belt, but the New Jersey section does not have such a broad belt for exploration, for it is close to the Atlantic. Though methods have been devised that will permit drilling offshore, this has not been done along the Atlantic. However geophysical methods have obtained data that permit interpretation of the nature of the rocks beneath the sea floor (Fig. 18–15).

Fig. 18–9. Upper Cretaceous Austin Limestone along Little Walnut Creek; Travis County, central Texas, with the Burditt Marl at the top of the bluff (Texas Bureau of Economic Geology).

An explosion placed on the sea floor, recorded by instruments set at a distance, can determine the character of the rocks beneath the sea in terms of the speed with which they carry the explosion waves. Some waves will travel in the muds directly beneath the floor; others will go down, and travel more rapidly in more consolidated sediments. Rocks such as granites and strongly cemented or compacted sediments transmit the waves far faster than poorly consolidated muds or sands. It is possible to determine the thickness of layers beneath the sea that have properties of low, intermediate and higher wave-travel times, interpreting them as unconsolidated, somewhat consolidated and consolidated sediments. The same studies can then be made on the coastal plain of New Jersey or Virginia where drilled wells have shown what lies beneath the surface. In the former case, it has been found that the Tertiary and uppermost Cretaceous rocks behave as relatively unconsolidated sediments, most of the Cretaceous as semi-consolidated and the Triassic rocks of the terrane unconformably below the Cretaceous as consolidated.

The unconsolidated sediments, which are about five hundred feet thick on the shore, run out under the continental shelf with gradual increase to about a thousand feet. The semiconsolidated rocks also

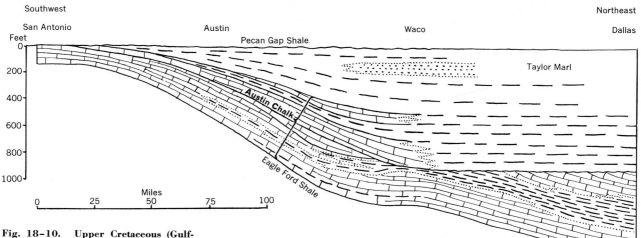

Fig. 18–10. Upper Cretaceous (Gulfian) Austin Group in Texas. Restored section of part of the Gulfian Series as it outcrops from San Antonio, Texas northeastward to Dallas (after C. W. Durham). The upper part of the chalk at Austin grades into argillaceous shales. The unconformity north of Austin is where the axis of the gulfward-plunging San Marcos Arch crosses the outcrop; when the sea retreated, sediments were eroded from a peninsula that projected along the sinuous shore while deposition continued longer in the embayments.

gradually increase in the shoreward part of the subsurface, but then they rapidly increase until their thickness is several miles. In form and proportion, the geosyncline containing these partially consolidated sediments is very like that of the Cretaceous and Jurassic along the Gulf Coast. It has been traced off the coast from the Carolinas northward to New Jersey. It seems not to continue northeastward, however, but to rise and disappear as it approaches the coast of New England. Thus the studies of the ocean bottom off the Atlantic Coast suggest that the Cretaceous history was similar to that along the coast of the Gulf of Mexico, where the flexure on the geosynclinal margin lies well within the present shore. In each belt of outcrop, the record is not nearly as full or as thick as that determined down the dip toward the sea by drilling and by geophysical methods. None of the rocks have been folded, though they have been tilted seaward. Cretaceous fossiliferous sedimentary rocks have been scraped from the walls of submarine canyons and abyssal scarps on the Atlantic continental shelf, suggesting that their continuation oceanward is broken by faults dropping them down toward the ocean basin.

ST. LAWRENCE REGION

Mount Royal (Fig. 18–16), from which the city of Montreal gains its name, rises from the St. Lawrence River for seven hundred feet within

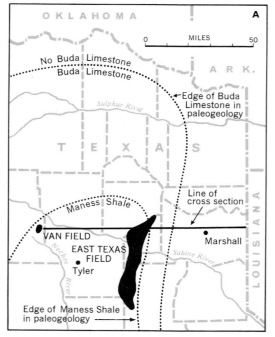

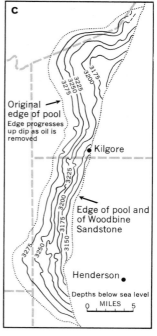

Fig. 18–11. *The East Texas Oil Field.*
A. Map of northeastern Texas showing the location of the field and the paleogeology of the top of the Comanchean Series that is unconformably overlain by the basal Gulfian Woodbine Sandstone that yields the oil (after T. L. Bailey, F. G. Evans, and W. S. Adkins).

B. Structure section through the field, showing also the position of the oil in the anticlinal Van Field to the west (after H. E. Minor and M. A. Hanna).

C. Structure contours on the top of the Woodbine Sandstone in the East Texas Field; and

D. structure section showing the relationship between the stratigraphy and the petroleum in the field (after J. S. Hudnall).

A porous quartz sandstone, the Woodbine at the base of the Gulfian Series,

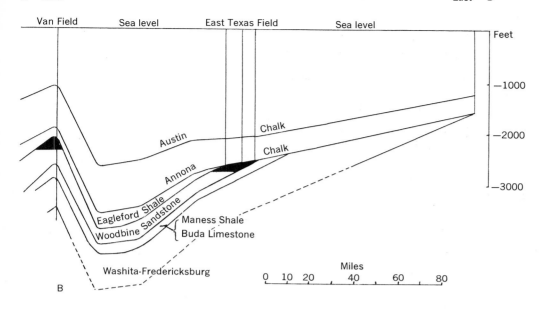

A West / East B

Van Field — Sea level — East Texas Field — Sea level

Feet
—1000
—2000
—3000

Austin Chalk

Annona Chalk

Eagleford Shale
Woodbine Sandstone
Maness Shale
Buda Limestone

Washita-Fredericksburg

Miles
0 10 20   40   60   80

B

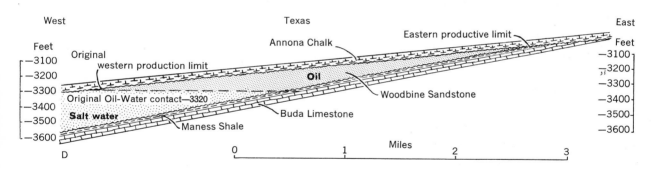

West — Texas — East

Annona Chalk — Eastern productive limit

Feet
—3100
—3200
—3300
—3400
—3500
—3600

Original western production limit

Oil

Woodbine Sandstone

Original Oil-Water contact—3320

Salt water

Buda Limestone

Maness Shale

Feet
—3100
—3200
—3300
—3400
—3500
—3600

D

Miles
0   1   2   3

traps the petroleum in the field because of a combination of original stratigraphic relations and later imposed structure. The oil was derived by the alteration of organic matter, probably the soft parts of marine animals, in associated sediments. The Woodbine Sandstone lies on a regional unconformity that bevels the upper formations of the Comanchean Series, and the sandstone is overlapped eastward by the relatively impervious Eagle Ford Shale and Annona (Austin) Chalk; the chalk extends far beyond to the east into Louisiana. Thus the Woodbine is a westwardly dipping thickening wedge of sandstone.

When oil was first discovered in 1930, it filled the sandstone from the updip line of extinction or edge down the dip to a level surface having oil above salt water, the latter driving the lighter oil up under pressure in such a way as to cause the oil to flow from the wells. The western edge of the field was where the elevation of the oil over water surface passed above that of the top of the Woodbine. Because the edge of the sandstone wedge on the east has been deformed into an arch, the southern limit of the field was where the edge passed below the oil-water surface, and the northern end of the field was similarly limited. As oil was removed, the oil-water surface rose, and thus the limit of production gradually moved opposite to the dip direction, thereby reducing the area of the productive field. The westernmost, northernmost, and southernmost wells came to yield salt water instead of oil, but the eastern wells should continue to produce quite to the end of the life of the field.

The Woodbine Sandstone thins to disappear eastward, because while east Texas was gradually subsiding the area of the Sabine Arch to the east in Louisiana rose somewhat and was eroded during latest Comanchean, and it subsided very little during early Gulfian. As can be seen in the structure contour map, the eastern edge was further warped or arched with the further rise of the arch.

The East Texas Field has produced and will produce several billion barrels of oil containing about 80 United States gallons or the equivalent 65 imperial gallons such as are used in Canada. It has been the largest single oil field in North America in terms of its production and reserves.

the city. The mountain is composed of basic igneous rocks intruded into Ordovician sediments and overlain by the glacial drift of the quite recent Pleistocene Epoch. On this basis, then, the age of the intrusion is not well defined. Beneath Jacques Cartier Bridge across the harbor is St. Helen's Island, which has a peculiar type of rock—a dark rock long taken to be an intrusive, like a basalt, containing peculiar limestone boulders with fossils of early Devonian age; thus the Devonian once extended over Montreal, and blocks fell into a volcanic channel, now thought to have been a vent having fused rock debris, formed sometime long after the Devonian. Mount Royal is but one of a number of volcanic necks, the Monteregian Hills (Fig. 18–17) scattered across southeastern Quebec and adjacent Vermont (Fig. 18–18). A second conspicuous feature of this region in Quebec and eastern Ontario is the high angle faults (Fig. 18–19) that define blocks that have been raised or have subsided—horsts and grabens. The Ottawa-Bonnechere Graben seems to have the Monteregian Hills within its southeastern extension—at least the hills lie in the trend of the graben, as though related in some way. To the northeast several hundred miles is the Saguenay Graben, which contains Lake St. John. Far to the north on the south shore of James Bay, stream and lake laid sediments occupy a small area in northern Ontario that is also bounded by a high angle fault. This latter sediment has yielded plants that are considered to be diagnostic of the early Cretaceous. The presence of pleochroic haloes in biotite, and of glass dikes in some of the rocks had previously suggested an early Tertiary age, as the haloes would normally become diffuse with the continued radioactive bombardment of the mineral biotite through very great time spans (Fig. 5–8), and glass is a rather unstable substance that will become crystalline in time. In recent years, with the improvement of isotopic methods of measuring geologic time, age determination on the biotites in the Monteregian Hills by the potassium argon method give ages of one hundred million years or more, which should evidence an early Cretaceous age.

The faults in the region have displacements of as much as two thousand feet, small as compared to those in the Triassic rift-blocks along the Atlantic Coast to the southeast. But within a region of very low relief such as the southern Canadian Shield, they produce some of the most impressive mountains, such as those in the Ottawa Valley and the front of the Laurentides (Fig. 2–13). Thus it had become evident that the Cretaceous Period was one in which taphrogeny and volcanism both produced substantial effects in eastern North America. The factors that produced the rifts and accompanying volcanism are, of course, not clear, but they have been attributed to tension related

Fig. 18–12. Diagrammatic restored section of the Cretaceous System along the belt of outcrop from western Tennessee through Mississippi and Alabama to Georgia (after L. W. Stephenson). As can be seen on the map (Fig. 18–7), the outcrop forms a northeasterly concave crescent. The coarser, terrigenous facies on the ends of the outcrop were stream-laid and shore-laid sediments, the chalks and limestones formed offshore. From the section alone it is not possible to reconstruct the three-dimensional form of the basin of deposition.

Fig. 18–13. Isopach map showing the gulfward thickening of the lower Cretaceous rocks along the northern margin of the Gulf Coast (after G. Murray and P. L. Applin).

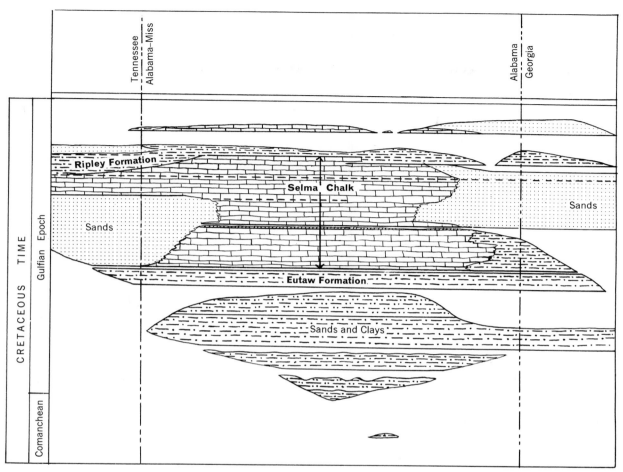

Ripley Formation

Selma Chalk

Sands

Sands

Eutaw Formation

Sands and Clays

CRETACEOUS TIME

Gulfian Epoch

Comanchean

Tennessee
Alabama-Miss

Alabama
Georgia

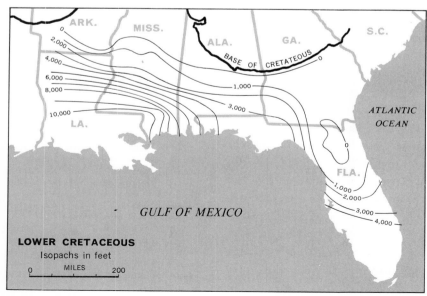

ARK.

MISS.

ALA.

GA.

S.C.

0
2,000
4,000
6,000
8,000
10,000

LA.

BASE OF CRETATEOUS

0

1,000

3,000

ATLANTIC
OCEAN

FLA.

0

1,000
2,000
3,000
4,000

GULF OF MEXICO

LOWER CRETACEOUS

Isopachs in feet

0   MILES   200

429

to the stress field that developed the Appalachian Mountains to the southeast in the late Paleozoic—a time so long prior to early Cretaceous that the relationship is hardly plausible.

ARCTIC COAST

Cretaceous rocks are known also in a coastal plain along the south of the Arctic Ocean, and their relations, from rather limited information, seem similar to those along the Gulf and Atlantic coasts (Fig. 16–22). In the extreme north, some of the rocks have been folded and faulted (Fig. 18–20). The stratigraphy in the northern coast of Alaska shows a northward thickening mass of sediments having tongues of coarse terrigenous rocks that were derived from rising lands in northern interior Alaska. These relationships (Fig. 18–21) are like those in the belt of the Rocky Mountains and Great Plains southward to Mexico.

ROCKY MOUNTAINS AND GREAT PLAINS

The stratigraphy of the Cretaceous rocks in the quite continuous broad belt of outcrops from southeastern Yukon and southwestern Mackenzie to Arizona and New Mexico evidences their position between interruptedly rising lands to the west and quite stable, gently subsiding crust to the east. The rocks reach a maximum thickness of some three miles (Fig. 18–22), and are generally of coarse, non-marine terrigenous rocks on the west, passing into argillites, or even non-terrigenous chalks and marls on the east; but at the extreme east is a narrow belt of shore sandstones and conglomerates. In the western Great Plains of Alberta and northeastern British Columbia, coarse-textured early Cretaceous (Neocomian) detritus is restricted to a belt within and directly in front of the present Rocky Mountains. They came from lands that rose to the west during the Nevadan Orogeny, from the Nevadides. Subsequent Cretaceous marine rocks spread southeastward to the American border (Fig. 18–23). By latest lower Cretaceous (Comanchean), marine sediments spread far to the east; basal shore sands and gravels on the east side of the seaway have consolidated to form the Dakota sandstones and conglomerates that make prominent narrow ridges or hogbacks where the tilted strata are exposed to erosion (Figs. 18–24, 17–16). The lower Cretaceous rocks attain a thickness of many thousands of feet from southeastern Idaho to northeastern British Columbia, then diminish rapidly across an original flexure at the margin of the craton, just as in preceding systems (Fig. 18–25); this Cretaceous is measured in hundreds rather than thousands of feet in Wyoming, South Dakota, and states and provinces to the north and south. Thus the lower Cretaceous shows both the effect of the Nevadan Orogeny in the in-

Fig. 18–14. *Map and section of the Jackson Dome, a structure lying directly below the city of Jackson, Mississippi (after W. Monroe). Porosity in the carbonate rocks on the crest and flanks of the dome have been the source of natural gas. A. Structure contour map, showing the depth to the surface of the Paleocene limestone that was laid on a submarine platform and flanking slopes.*

B. Structure section across the dome, showing the core of lower Cretaceous rocks that were intruded and eroded to form a platform that gained little sediment while thicker sections accumulated on its flanks.

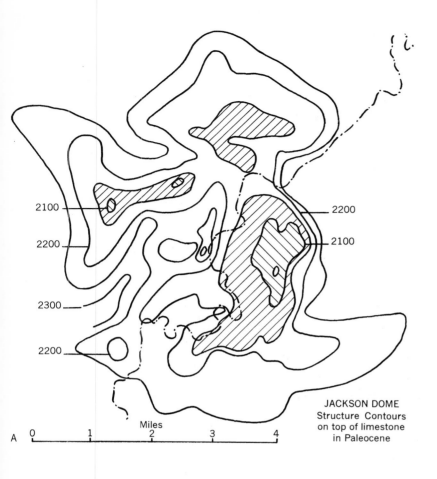

2100

2200

2100

2200

2200

2300

2200

Miles

0   1   2   3   4

A

JACKSON DOME
Structure Contours
on top of limestone
in Paleocene

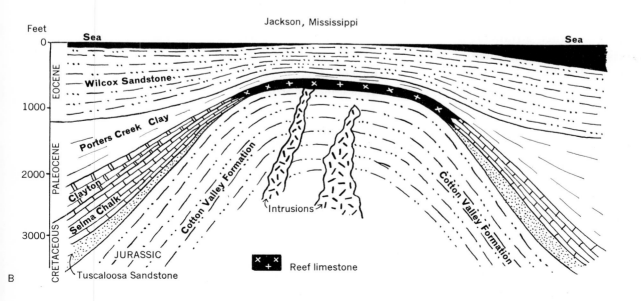

Jackson, Mississippi

Feet
0

Sea   Sea

EOCENE

Wilcox Sandstone

1000

PALEOCENE

Porters Creek   Clay

Clayton

2000

Cotton Valley Formation

Selma Chalk

Intrusions

Cotton Valley Formation

3000

CRETACEOUS

JURASSIC

Tuscaloosa Sandstone

B

× ×
+   Reef limestone

troduction of coarse detritus into a geosyncline west of the craton, and continued greater subsidence than in the interior region that has been relatively more stable from the beginning of the Paleozoic.

In the early Gulfian, rapid subsidence no longer was limited to the west of the continental interior; the earlier craton became the site of a geosynclinal belt that by the close of the period was buried by coarse terrigenous sediments derived from newly risen highlands to the west. Thick sections of lower Gulfian extend into the western part of the earlier craton, and tongues of coarser and non-marine sediment spread far eastward. Never before had such coarse sediments come into this belt from the west; now for the first time, streams with sediments from the orogenic belts raised in the earlier geosynclines in the west traversed the geosynclinal belt and distributed their loads within the earlier craton. A new geosyncline thus developed along the cratonal border.

Relations are analogous to those in the late Ordovician and the late Devonian in the East, where similar geosynclines, exogeosynclines, sank as complements to the lands rising outside ("exo") the craton, to the east, in the Taconian and Acadian (Shickshockian) orogenies. But the eastern Paleozoic geosynclines were semi-lenses that covered a very small area and contained a small volume of sediments as compared to those along the western side of the continental interior in the Gulfian. Early Gulfian coarse terrigenous deposits are in an east-thinning wedge (Fig. 18–26A) entering gradationally into progressively finer marine sediments to the east; they seem to represent a rejuvenation of uplands to the west that yielded a greater volume of sediments than could be accommodated below sea level in the adjoining subsiding geosyncline, so the detritus spread eastward in the stream-laid plains into the retreating and then advancing sea. Far to the east, in Nebraska, the Dakotas, and Manitoba much of the younger sediment of the early Gulfian is fossiliferous chalk, calcareous sediment (Fig. 18–26B). A second, even more extensive wedge of coarse, non-marine sediment extends into the later Gulfian; and the greatest extension of all caused seas to withdraw to the area of the eastern prairie states by the close of the period. It is in these stream-laid sediments that the great dinosaurs were buried and preserved. And the coastal plains on the west as well as the east supported forests that are the source of coal, as well as distinctive plants (Fig. 18–27). Thus a section through the system in northern United States or southern Canada has the early Cretaceous much thicker within the belt of the non-volcanic geosynclines that is to the east on the craton. Later Cretaceous rocks formed in a geosyncline subsiding east of this earlier one; they contain tongues of

Fig. 18–15. Structure sections of the coastal plain and the continental shelf along the Atlantic Coast. The section has been drilled on the coast but is interpreted from seismic geophysical investigations under the sea (after W. M. Ewing and others).

Submarine sections such as this have been compiled from data gained by setting off explosions on the sea bottom, receiving the time of travel of propogated waves in hydrophones suspended from one or more ships at distances of a few miles. Though some waves have a path directly to the hydrophones, others enter the sediments below the sea floor; their times of arrival at the receiving stations depend on the density of the rocks through which they pass in paths at differing depths, reflected from interfaces within the sedimentary layers. The interpretation of such records, as of those on the land, demands skill and experience of the geophysicist. In the past few years other methods of continuous acoustic reflection have been developed that give a more direct record of the subsurface interfaces (see Fig. 22–5).

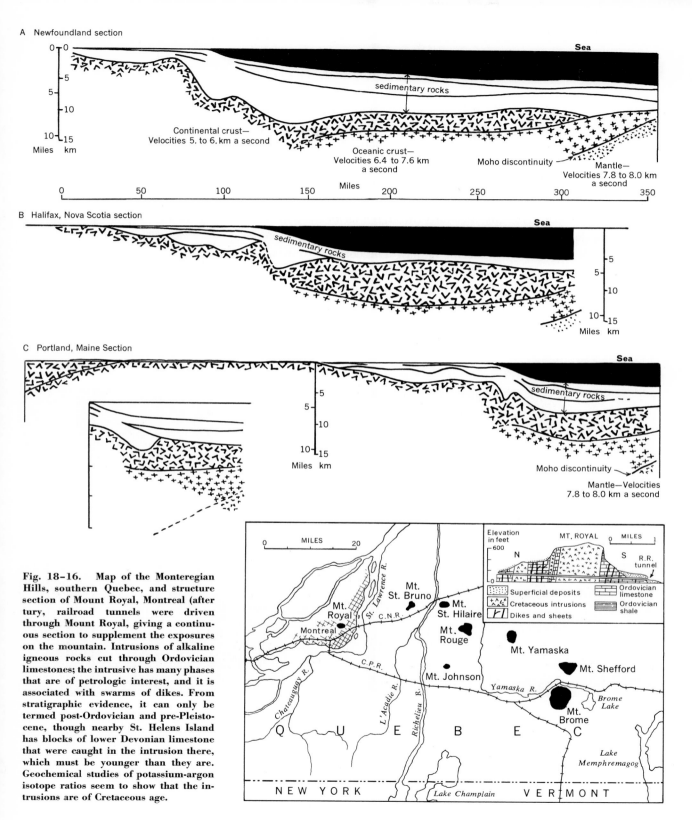

Fig. 18-16. Map of the Monteregian Hills, southern Quebec, and structure section of Mount Royal, Montreal (after tury, railroad tunnels were driven through Mount Royal, giving a continuous section to supplement the exposures on the mountain. Intrusions of alkaline igneous rocks cut through Ordovician limestones; the intrusive has many phases that are of petrologic interest, and it is associated with swarms of dikes. From stratigraphic evidence, it can only be termed post-Ordovician and pre-Pleistocene, though nearby St. Helens Island has blocks of lower Devonian limestone that were caught in the intrusion there, which must be younger than they are. Geochemical studies of potassium-argon isotope ratios seem to show that the intrusions are of Cretaceous age.

non-marine sediments derived from uplifts that rose to the west twice within the epoch, and from a culminating uplift near the close of the period.

The progress of the Laramian Orogeny, the term applied to movement in late Cretaceous and early Tertiary, is shown by evidence of erosion of beds deformed during Cretaceous time to the south of Utah, Wyoming, and Colorado. In east-central Utah, and northeastward into southwestern Wyoming, early upper Cretaceous (Gulfian) sediments along the cratonal margin are thousands of feet of coarse conglomerates and sands that pass laterally eastward through coal-bearing fluviatile and swamp-laid deposits into sands laid on the margins of the interior sea and shaly argillites with marine faunas. The coarsest westernmost sediments have angular unconformities within them; they were folded and eroded, then buried beneath later Cretaceous stream gravels. There are many phases of deformation, erosion, and deposition, and folding proceeding interruptedly along the cratonal margin and in the adjacent non-volcanic geosynclinal belt through Gulfian time.

Fig. 18–17. Map of the St. Lawrence Lowlands and adjoining mountains, showing the distribution of the Monteregian Hills, erosion remnants of alkaline intrusive rocks that stand higher because of their greater resistance to erosion. In addition to the seven principal intrusives in Quebec, there are other similar stocks in Vermont and dikes along Lake Champlain that may be part of the same petrographic province.

Fig. 18–18. Monteregian Hills, Quebec. Rougemont and Mount St. Hilaire are two of the stock-like masses of igneous rocks that intrude the lower Paleozoic rocks of the St. Lawrence Lowlands and the adjoining mountains of southeastern Quebec. The view is westward from the front range of the folded Ordovician rocks east of St. Hyacinthe; the mountains are eight and twelve miles away. The Lowland is blanketed by marine clays and sands laid there when marine waters drowned the region in quite recent time, at the close of the glaciation of the Pleistocene Period. The intrusions are known from stratigraphic relations to be younger than lower Devonian, for blocks of limestone of that age are contained in the volcanic rocks at Montreal. Analysis of isotopes of disintegrating elements suggest that they were intruded in about the early Cretaceous.

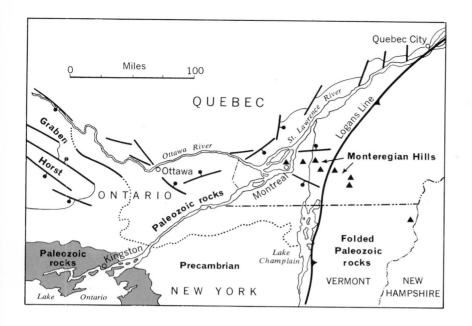

Fig. 18–19.   Fault-line scarp trending northwestward in the Ottawa-Bonnechère Graben west of the Ottawa River near Eganville, Ontario; (see Fig. 9–9*B*) (Royal Canadian Air Force).

The Precambrian gneiss with foliation dipping toward the lower right is cut off by a fault at the scarp; Ordovician sedimentary rocks lie in the downfaulted lowland to the left, the fault having a displacement of some 500 feet. The linear stream courses, on *lineaments*, in the middle distance are probably along other fault lines; the rather regular plains are essentially on the erosion surface over which the Paleozoic rocks were laid and subsequently stripped away. Other faults in this graben have displacements of more than 2000 feet. The Saguenay River and Lake St. John, north of the St. Lawrence River below Quebec City, lie in another graben in the Precambrian rocks of the Laurentide Mountains, with even greater fault displacements.

435

Structures rose as far east as central Colorado in late Cretaceous, becoming additional source lands for complementing more deeply subsiding parts of the generally subsiding region of southern Wyoming and eastern Colorado. From discussion of the Cretaceous sedimentation in the north, it would seem that any rock fragments, pebbles, or sand-grains came from rocks exposed by orogeny in British Columbia, western Montana, or Idaho. Late Gulfian conglomerates in southern Wyoming, however, have fragments of rocks of the facies found in the early Gulfian of the same region, as well as fine-textured detritus that may have come from the source lands of Idaho and western Utah. By the close of the period, Precambrian basement rocks of Colorado yielded pebbles to Cretaceous sediments. The folds of the present Rocky Mountains, the Laramides, began long before the end of the Cretaceous time. In southern Nevada, thrust faults cutting late Cretaceous rocks are buried in still younger Cretaceous sediments; the thrusts formed, ceased movement, and were buried in the later part of the period. Although much of the Laramian structure had developed by the close of the Cretaceous Period, some deformation continued in the succeeding Paleocene.

The rocks in South Dakota give evidence on the rate of advance of seas overlapping the continental interior and on the time significance of lithologic changes. And there are rocks that contain detritus eroded from inside the craton to the east. In the Black Hills in western South Dakota, the section from the top of the late Comanchean Dakota Sandstone to the base of the early Gulfian Greenhorn Limestone is about 1500 feet of shaly dark argillites (Fig. 18–28) thought to have been deposited quite continuously in water of such depth that waves and currents could not disrupt the delicate bottom layers. The argillites have paper-thin laminae, each foot averaging about 1500 alternations of units of alternate textures. The laminae seem to represent some seasonal influence, perhaps the response of organic life to seasonal water temperatures, so each foot seems to represent more than 1000 years of deposition, and the whole deposit, more than 2,000,000 years. Four hundred miles to the east in southeastern Dakota, the interval from top of the Dakota to the Greenhorn shrinks to 200 feet or so of quite similar shales; the Dakota seems to be the shore sand of the advancing sea. The material in the Dakota was carried by streams flowing from the interior to the eastern shore of the Cretaceous sea. In Minnesota, the Dakota is succeeded quite directly by Greenhorn-like sediments. So the advance of the sea across the state of South Dakota, some 400 miles, took place in about 2,000,000 years, or roughly at the rate of a mile in 5000 years, or a foot in a year. And if depth of water and sea

Fig. 18–20. Air photograph of a part of the west coast of Axel Heiberg Land on the eastern margin of the Sverdrup Basin, Queen Charlotte Islands, Arctic Canada (Geological Survey of Canada). The folded sediments shown in much of the area are Cretaceous and early Tertiary sandstones and shales that were folded within Tertiary time. The light colored mountain on the right is a mass of Carboniferous or Permian gypsum that has been intruded into younger rocks during folding, a piercement dome of diapiric structure. The white areas are glaciers or snow fields.

Fig. 18–21. Restored section of the Cretaceous rocks in northern Alaska (after G. Gryc). The coarser terrigenous sediments derived from lands to the south intertongue with marine shales and sandstones extending into the coastal plain and continental shelf of the Arctic Ocean.

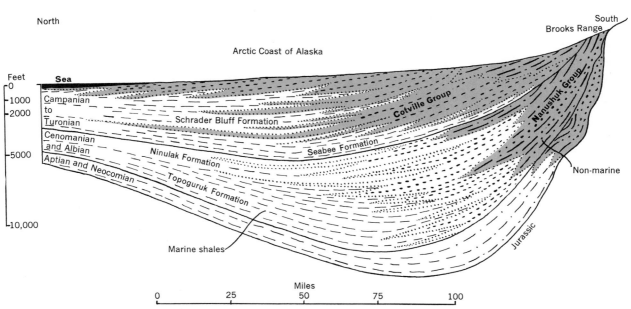

North

Arctic Coast of Alaska

South
Brooks Range

Feet
- 0
- 1000
- 2000

- 5000

- 10,000

Sea

Campanian
to
Turonian

Cenomanian
and Albian

Aptian and Neocomian

Schrader Bluff Formation

Seabee Formation

Colville Group

Nanushuk Group

Ninulak Formation

Topoguruk Formation

Non-marine

Marine shales

Jurassic

Miles

0   25   50   75   100

level remained constant in the west, the subsidence would have been at a rate of a foot in 1500 years. These figures give a measure of the probable rates of change in this relatively quiet part of the continent during the Cretaceous time.

In the western part of the same region of eastern Montana and Wyoming and the western Dakotas, the Greenhorn Limestone, or calcareous marl, is associated not only with black argillites but also has a few beds of altered volcanic ash, bentonite, spread from several outbursts of volcanism, perhaps in mountains far to the west, and settled in the marine waters of the Dakotas (Fig. 18–28B). The ash has been altered into a sticky white to buff clay that is used in clarifying oils, in forming a mud in drilling wells, and to some degree for cosmetic clays. From the standpoint of stratigraphy, the beds permit precise correlation of layers in time, but the changes in lithic character do not precisely parallel them. The sea bottom that at one place had calcareous muds at another not distant had argillaceous muds, though some of the sedimentary types spread rather widely at one time. Were it not for the ash beds, one might assume that the lithic contacts are synchronous through a greater area.

The late early Cretaceous (Comanchean) and late Cretaceous (Gulfian) seas spread continuously from the interior to the Gulf.

NORTHERN MEXICO

The Cretaceous stratigraphy of northern Mexico shows the influence of both the basin-forming subsidence that began in later Jurassic, and the general westward thickening that is an attribute of the late Cretaceous stratigraphy in the plains states to the north. The Early Cretaceous (Neocomian) limy sediments (Fig. 18–29) were laid in a subsiding basin extending southeastward from southeastern Arizona through Chihuahua toward southern Tamaulipas (Fig. 18–30). Because these sections have one of the best records of rocks of the Neocomian Stage, and are older than the Comanchean of Texas, they have been called the Coahuilan Series from the state of Coahuila. This trough and seaway was separated by a long peninsula in Coahuila from the downwarp along the northern and western margins of the Gulf that had originated in the preceding period. In medial Cretaceous Comanchean and early Gulfian, sediments in the Mexican Geosyncline continued to be dominantly calcareous, non-terrigenous, and of maximum thickness of a few thousand feet, but the seas spread over the peninsula in Coahuila and joined those along the Gulf. The platform between geosynclines was subsiding slowly so that deposition of carbonate sediments, perhaps partly of reefs, formed shoals that deterred

Fig. 18–22. *Lower Cretaceous of the Rocky Mountain region of western Canada and the United States. A.* Isopach map showing the thickness of lower Cretaceous sedimentary rocks in the region (Canada after R. P. Glaister; United States after R. J. Weimer and J. D. Haun).

B. Paleogeographic map showing the spread of the seas southeastward during Albian time (after C. R. Stelck).

free circulation and encouraged precipitation of salts in the shallow waters. Normal marine waters in the more rapidly subsiding areas on the flanks of the platform permitted deposition of fossiliferous limestones. By early Gulfian time, waters so deepened over the peninsula that the Mexican Geosyncline and Gulf Coast Geosyncline had direct marine connections. Fractured limestones along the Gulf Coast in Tamaulipas are the source of oil in fields that in the period after World War I were among the principal world sources.

A marked change affected northern Mexico in the later Gulfian; coarse terrigenous sediments began to enter from the west, and as the region subsided more than two miles, rising highlands to the west supplied coarsening silts and sands that graded into a few thousand feet of argillaceous shales near the Gulf (Fig. 18–31). These sediments are the continuants of those already described in the western interior states and prairie provinces of Canada, fluviatile plains spreading east-

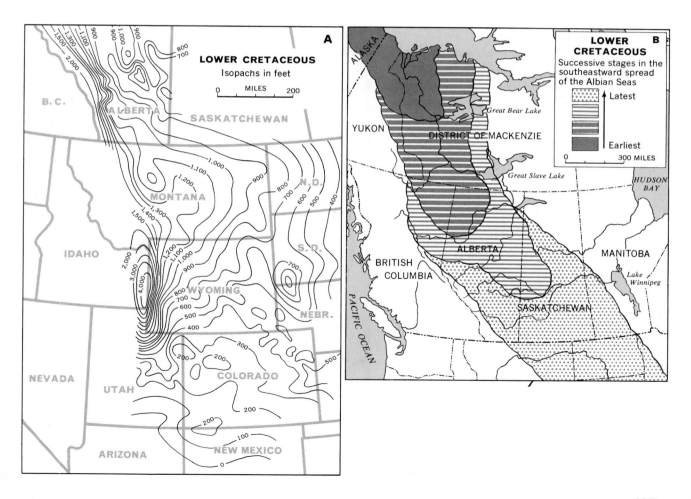

ward from rising folds in the belt to the west of the craton, passing into deepening marine waters eastward. They are the product of the earlier phases of the Laramian Orogeny. By latest Cretaceous, stream-laid sediments, derived perhaps also from anticlines rising in the region of the present Front Range of the Rocky Mountains in Colorado and New Mexico, spread westward into the Rio Grande Valley in West Texas.

PACIFIC COAST

The Nevadan Orogeny in the late Jurassic had greatly altered the distribution of highlands and seaways in western North America. The Cretaceous record is preserved both in a sedimentary volcanic sequence and as plutonic intrusions in large areas in the western interior states and interior British Columbia that had become land by the close of the period. The region bordering the Pacific Coast in California was the site of deeply subsiding geosynclines in which some four or five miles of Cretaceous accumulated (Fig. 18–32). It was not a single subsiding trough, nor one that retained the same form through the period. The early Cretaceous trough was partially folded into ranges that were eroded to yield part of the sediment in later Cretaceous (Fig. 18–33); and that was in turn further folded, raised and eroded before the latest Cretaceous was laid. Some sediments came from lands lying off the present Pacific Coast, and others must have been derived from the erosion of the Nevadides, the mountains raised in the orogenies to the east. In contrast to the Jurassic record, that of the Cretaceous is only sparingly volcanic; the broad belt of the Paleozoic and Mesozoic

Fig. 18–23. *Upper Cretaceous of the Rocky Mountain region of the United States.* *A.* Isopach map of the upper Cretaceous of the region (after W. C. Krumbein and F. G. Nagel).

B. Lithofacies map, the lines representing the proportions of coarser terrigenous sediments to the finer-textured sediments as ratios; thus a higher number shows coarser facies and a lower number or smaller fraction, a finer facies (after W. C. Krumbein and F. G. Nagel). Such lines represent the average of a stratigraphic thickness. The source is evidenced in the coarsening of the rocks toward the west. But the span of time is so great that a map at a specific time might differ considerably from that of the whole upper Cretaceous. Judgments of source are best made on paleolithologic maps for short time intervals or on paleogeographic maps.

C. Paleolithologic maps comparing the distribution of sediments in the lower Cretaceous Albian Stage with those in the middle of the late Cretaceous during deposition of the Campanian Stage (after R. J. Weimer and J. D. Haun). These maps representing short time intervals are essentially paleogeographic maps, maps of a time rather than stratigraphic maps averaging a span of time.

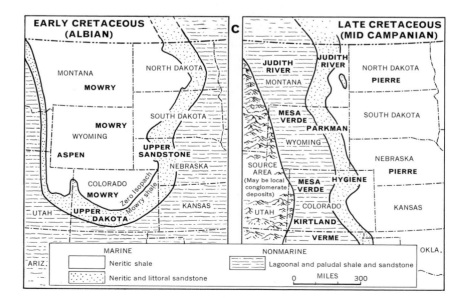

Fig. 18–24. Exposure of east-dipping lower Cretaceous Cloverly or "Dakota" Sandstone on the flank of the Medicine Bow Range, west of Laramie, southeastern Wyoming, with the characteristic hog-back trending southward. The Jurassic sandstones and shales are on the right, separated by the more resistant Dakota from the younger Cretaceous sediments on the left. The name Dakota is applied in the region to sandstones of somewhat variable age; the name is derived from Dakota City, Nebraska, far to the east. The rocks were folded in the Laramian Orogeny near the close of the Cretaceous period, though further deformed and uplifted in the Cenozoic Era.

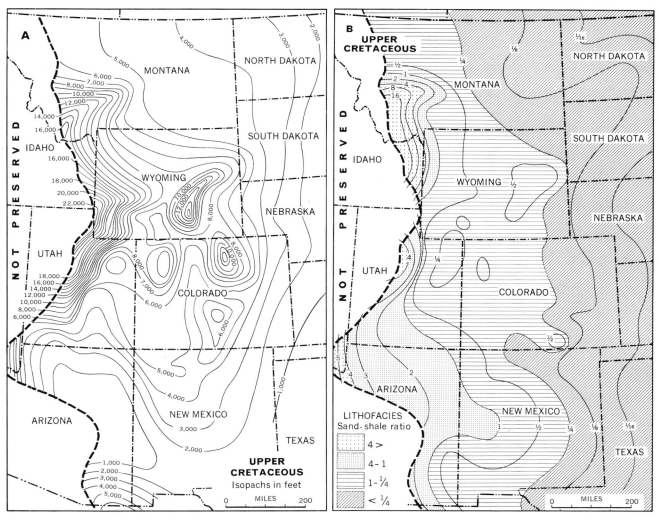

volcanic geosynclines and islands that spread eastward to central Nevada and Idaho has been reduced to a narrower belt along the ocean. A few areas are known to the east in which non-marine sediments, to thousands of feet, are remnants of deposits laid in deeply subsiding troughs within the Nevadides.

Similar rather narrow geosynclines subsided several miles on the western flank of the intruded earlier geosynclinal belt in Washington and British Columbia. And the whole of southern Alaska had a number of parallel arcuate belts of subsidence with rising, sediment-producing lands interrupting them. In some of the Pacific areas, excellent successions of Cretaceous fossiliferous sediments include representation of the Neocomian, so little represented in the Gulf-Atlantic and interior provinces, particularly the west side of the Sacramento Valley in California and the Queen Charlotte Islands in the Pacific off the mainland of British Columbia.

Fig. 18–25. *Cretaceous sections in the Rocky Mountains and Great Plains. A.* Restored section of the Cretaceous System from Idaho to South Dakota through southern Wyoming at the close of the period. The prevalence of coarse terrigenous sediments on the west is indicative of interrupted uplift of the Nevadides and the recurring folding in that region. Some of the detritus in southeastern Wyoming in later Cretaceous time came from rising welts in the emergent Rocky Mountains in central Colorado. The spread of the marine sediments in South Dakota over the basal unconformity is shown better in Fig. 18–28.

B. Diagram showing tongues of coarser terrigenous sediments on the West penetrating finer rocks on the east (after R. J. Weimer).

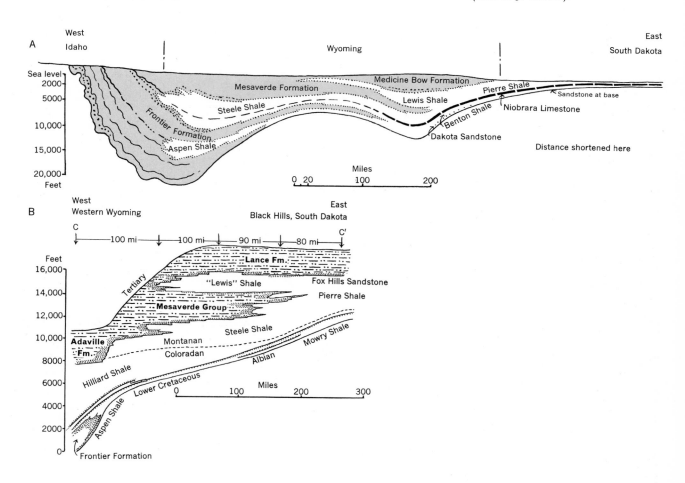

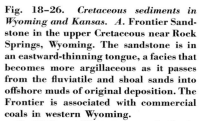

Fig. 18–26. *Cretaceous sediments in Wyoming and Kansas. A.* Frontier Sandstone in the upper Cretaceous near Rock Springs, Wyoming. The sandstone is in an eastward-thinning tongue, a facies that becomes more argillaceous as it passes from the fluviatile and shoal sands into offshore muds of original deposition. The Frontier is associated with commercial coals in western Wyoming.

B. Niobrara Limestone, a chalk, in Logan County, western Kansas; sediment was deposited in the relatively clear waters of the great inland sea, far from the lands that spread sands and clays into the western margin of the sea in the Rocky Mountain states (Kansas Geological Survey).

A

B

443

SUMMARY OF THE CRETACEOUS HISTORY

The Cretaceous was a period of about sixty or seventy million years, a span long enough to show appreciable changes in the character of the North American continent and in the life that inhabited it. At the beginning of the period, the lands were quite as extensive as today on the continent, if not more so. If any seas remained on the present continent, they were in the limited areas that retain records of the latest Jurassic and earliest Cretaceous, the Neocomian Stage, and they were not extensive. Much of the continent was a land of low relief. Nearly level plains remained in areas from which Jurassic seas had withdrawn—in the central and northern Rocky Mountain region, in Louisiana and coastal Texas and Mississippi, in Arctic Canada, and northern Mexico.

Fig. 18–27. Fossil *Cycadophyta* from the Cycad National Monument of South Dakota (G. R. Wieland, after H. N. Andrews, Jr.). The cycadophytes are represented in modern floras by nine genera of tropical plants, such as the sago palm of Australia and the Orient. They have a columnar trunk that may reach tens of feet high, crowned by whorls of long leaves having a central rib and many margining pinnules. One class of cycadophytes is dioecious, having separate trees with two types of cones, whereas the other has flowers with male and female organs borne on the same axis. The cycads were widely distributed in the Mesozoic, being dominant in some areas. Though cycadophytes are widely distributed, they are found with the conifers of the Petrified Forest and in many Jurassic rocks, but they are best known in American in the late early Cretaceous of the Cycad National Monument near the Black Hills, South Dakota. The silicified trunks are of the monoecious group, with male and female organs on the same axis. Though this locality is the best known, the cycadophytes are perhaps more representative of the Triassic and Jurassic periods.

Cycadeoidea, a silicified, many branched trunk of a low-growing cycad only a yard or two high; the pits are leaf scars.

Though older rocks were exposed throughout the most of Canada and of central and eastern United States, they were of low relief. Mountains such as the Appalachians, as well as the block mountains that formed to the east in Triassic time, had been reduced to a peneplane by early Cretaceous, for the streams and advancing seas of the period laid thin, rather uniform deposits, derived from the extensive low coasts.

Only in the west, in the region from central Nevada to the Pacific, and the coastal part of British Columbia and Alaska, were highlands associated with subsiding troughs in the same coastal belts, the effects of the Nevadan Orogeny which was to continue interruptedly in that region until it rather merged with the Laramian movements of the later Cretaceous. The widespread volcanism that had prevailed in the sub-

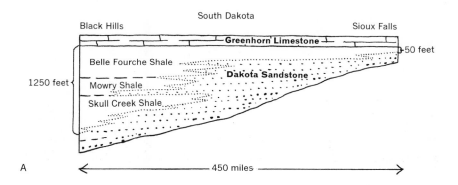

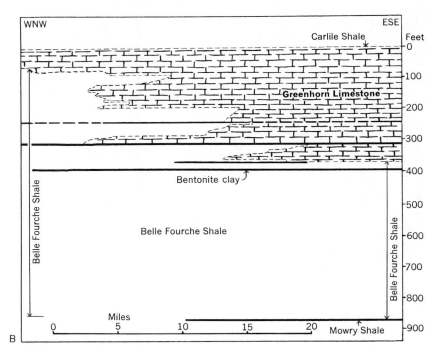

Fig. 18–28. *Sections of Cretaceous rocks in South Dakota and Montana (after W. W. Rubey).* **A.** Restored section of the lower part of the Cretaceous System in South Dakota; the Mowry Shale is thought to be at the top of the Albian Stage, the Belle Fourche to be Campanian. The Dakota Sandstone is younger in its type section at Dakota City, Nebraska, opposite Sioux City, Iowa at the east end of the section, than similar rocks in the Black Hills, western South Dakota. The sequence has been considered to represent the changing distribution of facies with the advance of the Cretaceous Sea. The 1250 feet of shale in the west has about 1500 laminae in each foot, which if annually laid would represent some 2,000,000 million years of deposition. The section in the east is reduced to 50 feet below the Greenhorn Limestone that is thought to be essentially time-stratigraphic, synchronous. If the thin section represents deposition at comparable rate, it was laid in less than 100,000 years. Thus the advance of 450 miles across South Dakota from the Black Hills would be in less than 2,000,000 years, or at a rate of one mile in 4000 years in round numbers, less than a foot a year. This gives us some insight into the enormous span of geologic time.

B. Section showing the relations in upper Cretaceous sections in southeastern Montana. The bentonite beds, altered volcanic ash falls, are horizons of constant age, isochronous surfaces; the associated lithic units, the formations, have boundaries that shift in age from place to place.

Fig. 18–29. *Cretaceous sedimentary rocks in Mexico (Z. de Cserna).* *A.* Lower Cretaceous limestones in the frontal folds of the Sierra Madre Oriental west of Ciudad Victoria, Tamaulipas, northeastern Mexico, the view is northward. The massive cliff-forming Albian Aurora Limestone contains reefs of rudistids, peculiar pelecypods having dissimilar shells, one of them extremely thick, large, and conical.

B. Upper Cretaceous (Coniacian) deformed sandstone and shale on the Mexico City–Acapulco Highway near Iguala, Guerrero, southern Mexico. Terrigenous sedimentary rocks are widespread in the younger Cretaceous northward into the Rocky Mountain region of the United States.

Fig. 18–31 (right). *Restored section of the Cretaceous System from western Durango to the Gulf of Mexico in latest Cretaceous time (after R. Imlay).* *A.* The upper Cretaceous, in which the younger beds are terrigenous sediments coarsening toward the rising highlands to the west.

B. Lower Cretaceous carbonate rocks deposited in the Mexican Geosyncline a subsiding trough extending southeastward from southeastern Arizona through Chihuahua, with the Coahuila Platform of gypsiferous carbonate rocks on the east.

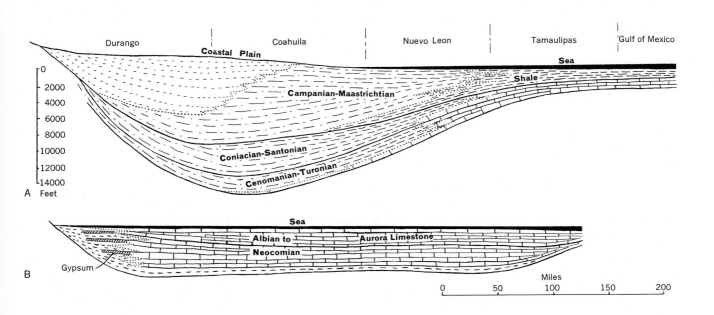

Fig. 18–30. Paleogeographic map showing the interpreted limit of spread of the earliest Cretaceous Coahuilan Series, the Neocomian Stage, in northern Mexico (after L. B. Kellum). The Coahuila Peninsula separated the Mexican Geosyncline, trending southeastward from Chihuahua, from the Gulf Coast Geosyncline of Tamaulipas and Texas; the peninsula was inundated in later Cretaceous and became the Coahuila Platform. The geography to the west of the Mexican Geosyncline is more conjectural, for there are few areas exposing Mesozoic rocks because of the thick cover of Cenozoic volcanic rocks. (*See* Fig. 17–19).

EARLY CRETACEOUS

Neocomian seas

MILES
0   300

siding troughs of the Pacific belt since early Paleozoic virtually ceased in much of the belt but continued in more limited areas into the Cenozoic. Thus there are thick submarine lavas in the early Tertiary of Vancouver Island and adjacent Washington.

We could construct many maps of successive stages in the Cretaceous, if the records were assimilated. The seas generally spread from the minimum at the beginning of the period, in successive and varied incursions and retreats such as would be suggested on a correlation chart, until they covered very much of the continent. Perhaps more of the continent was beneath the sea in the Ordovician Period of the early Paleozoic; there was little land at either time.

The lithologies of the later Cretaceous rocks in the Rocky Mountain region show that the Nevadides rose interruptedly during the period, as is confirmed in the dated Cretaceous intrusions in the Pacific coastal region. There were also intermontane valleys, as in Nevada, that filled with thousands of feet of stream and lake-laid sediments as they subsided, later to be folded. The principal structure of the present Rocky Mountains came into being in the Laramian Orogeny, which in the first phase had rising anticlinal belts in Colorado that were stripped of earlier Cretaceous rocks even within later Cretaceous time; the last phase of the Laramian Orogeny was in the Tertiary. Along the west of the orogenic belt in Utah, Cretaceous beds were folded and eroded, only to be succeeded by later Cretaceous sediments, those again to be folded and eroded. So orogenic movements took place interruptedly.

Along the eastern margin of the orogenic belt in western Wyoming, Idaho, Montana, and the Canadian Rockies, latest Cretaceous and the underlying succession to the Precambrian were folded and thrust eastward over the Cretaceous foreland (Fig. 18–34). Yet on the plains to the east, beyond the direct effect of the movements, deposition continued with little interruption from the dinosaur-bearing latest Cretaceous into the mammal-bearing earliest Paleocene. The causes and manner of movement of the thrust sheets, such as in southeastern Idaho and western Wyoming, have had varying explanations. The most apparent source of energy for the long-distance travel of such thrust sheets would seem to be gravity, but to produce a gently east sloping surface on which the thrust could glide, there should have been highlands to the west from which the Paleozoic and later sediments were to slide away (Fig. 18–35). There could have been an area devoid of the sediments, floored by Precambrian or lower Cambrian rocks that formed the plane of movement, or possibly the younger rocks could stretch by flow or by minor faults that would extend them sufficiently to permit their eastward spread. This is a subject of study and speculation, in which the care-

Fig. 18–32. Isopach map of the Cretaceous System in California (after O. P. Jenkins).

Fig. 18–33. Restored section of the Cretaceous System from the Pacific Ocean west of San Francisco Bay to the Sierra Nevada (after N. L. Taliaferro and A. Safanov). Sediments came principally from lands on the west that were rising during the period and from the highlands formed on the east in the Nevadan Orogeny in late Jurassic. Although lower Cretaceous is well developed in western sections, the seas did not spread over the margin of the Sierras until late Cretaceous, when tectonic lands rose in the region of earlier Cretaceous subsidence.

Fig. 18–34. Chief Mountain, Montana, east of Glacier National Park (R. Rezak, United States Geological Survey). The upper part of the mountain is of Precambrian sedimentary rocks of the Belt Group. These lie structurally on Cretaceous shales, the lighter colored, more rounded slopes in the lower part of the mountain face, the Chief Mountain thrust fault separating the two. The deformation took place in the Laramian Orogeny in late Cretaceous and early Tertiary times.

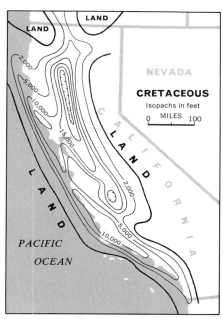

NEVADA

CRETACEOUS
Isopachs in feet
0 — MILES — 100

2,000
5,000
10,000
15,000

CALIFORNIA

2,000

5,000
10,000

LAND

LAND

LAND

PACIFIC
OCEAN

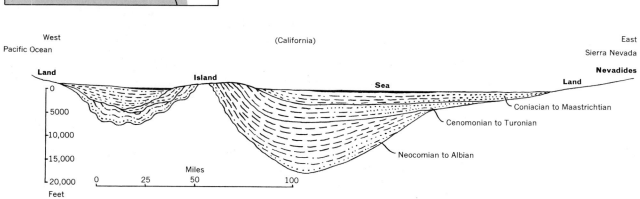

West
Pacific Ocean

(California)

East
Sierra Nevada

Land

Island

Sea

Nevadides

Land

0
-5000
-10,000
-15,000
-20,000
Feet

Coniacian to Maastrichtian

Cenomonian to Turonian

Neocomian to Albian

Miles
0   25   50   100

ful observations of field relationships must precede their understanding. In any case, the Cretaceous Period closed with the land of the North American continent again exposed or with marine tongues of limited extent and doubtful presence. The differences from the late Jurassic was principally in the more easterly extent of the western mountains.

THE CLIMAX OF DINOSAURIAN EVOLUTION

One of the striking features of Cretaceous stratigraphy in North America is the extensive development of broad floodplain deposits in the Rocky Mountain states, the detritus from ancient mountains that were rising in this part of the continent during the early phases of the Laramian Revolution. As has been noted on a preceding page of this chapter, the skeletons of many dinosaurs were buried and preserved in these stream-laid deposits. And these skeletons in the Rocky Mountain upper Cretaceous sediments are numerous and varied beyond anything seen in previous Mesozoic deposits, even in the productive Morrison beds of late Jurassic age. Here is a magnificent record of the dinosaurs and of the animals that lived with them, a record that is paralleled only by the recently explored upper Cretaceous continental beds of Central Asia.

The fossil evidence of land-living, backboned animals that lived during the early part of Cretaceous history in North America is incomplete, to say the least. A fragmentary fauna from the Arundel Formation in the lower part of the deposits of the eastern coastal plain, a much better fauna, as yet undescribed, from the "Dakota" Cloverly Formation of Montana, and some enormous footprints in the basal marls of the Trinity beds of Texas (Fig. 18–36), constitute the bulk of our information about the dinosaurs that followed in time the widely distributed late Jurassic faunas. During the years when Cretaceous seas had spread extensively across the continental platform, the habitats for land-living animals were restricted.

But with the succession of events which made up the history of late Cretaceous time, with the rise of high lands, and the erosion of vast quantities of materials from these lands to form the great flood plains that bordered them on the east, the habitats for the latest of the dinosaurs were greatly expanded. Dinosaurian populations expanded to fill the land.

The record is beautifully preserved in many areas, but particularly in Alberta and the states to the south (Fig. 18–37), where a sequence of upper Cretaceous formations contains quantities of bones which show, frequently with dramatic emphasis, the evolution of many dinosaurian lines through time. In these beds, of which the Belly River, Edmonton, and Hell Creek succession of Alberta and Montana makes

Fig. 18–35 (right). Interpretation of the development of thrust faults in southeastern Idaho and northwestern Wyoming in the Laramian Orogeny (after W. W. Rubey and M. K. Hubbert). Such thrust sheets move on fault planes that pass along relatively less competent shales and rise through more competent beds on the faces of crustal folds. Fluids under great pressures in the rocks have been thought to reduce friction to such low degree that rocks may travel on extremely low slopes through the gravitational force.

Fig. 18–36. Trackways made by giant dinosaurs in the Lower Cretaceous Trinity marls, along the Paluxy River near Glen Rose, Texas (American Museum of Natural History). The overlying shales have been removed to expose a large area containing the trackways; on the right side a trench has been cut through the limestone, preparatory to removing a section of it for museum display. The large, round depressions, each some three feet long, are the footprints of a giant brontosaur that was obviously wading through shallow water. To the left are the three-toed tracks of a carnivorous dinosaur that followed the brontosaur.

West

Idaho

East

Wyoming

A

Middle Cretaceous

Triassic–lower Cretaceous

Precambrian

Carboniferous and Permian

Cambrian to Devonian

B

MK

Tr–LK

C–P

€–D

C

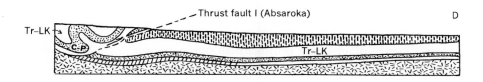

MK

Tr–LK

D

Thrust fault I (Absaroka)

Tr–LK

C–P

Tr–LK

E

Thrust fault II (Darby)

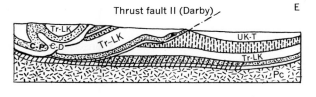

Tr–LK

UK–T

C–P   €–D   Tr–LK

Tr–LK

Pc

UK–T   Upper Cretaceous–lowest Tertiary

MK   Middle Cretaceous

Tr–LK Triassic–lower Cretaceous

C–P   Carboniferous and Permian

€–D   Cambrian to Devonian

Pc   Precambrian

F

Idaho–Wyoming Thrust fault III (Prospect)

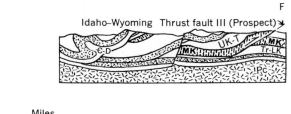

€–D   MK   UK–T   MK

Tr–LK

Pc

Miles

50   40   30   20   10   0

451

a first-class example, are the remains of dinosaurs in profusion and in great variety. This last point should be stressed.

Although the dinosaurs of the late Jurassic Morrison beds are numerous, they show nothing like the variety to be seen among the dinosaurs of the upper Cretaceous of North America. Suborders, families, genera, and species of dinosaurs appear as newcomers to the late Cretaceous scene, representing lines of evolution that developed after the close of Jurassic time. Of particular significance is the fact that these newcomers are all plant-eating dinosaurs. And here we see very plainly the chain of cause and effect in environments and among animals.

During early Cretaceous times there was a virtual explosion in the increase in numbers and kinds of the angiosperms, the flowering plants and the broad-leafed trees (Fig. 18–38). The flora of the earth became variegated far beyond its Jurassic expression, and landscapes which in late Jurassic time had a primitive aspect became by late Cretaceous time truly modern in appearance. Vistas became colorful with the bright hues of flowers, and somber forests of conifers and other primitive trees (Fig. 18–28) were replaced by deciduous woods in which oaks, willows, and sassafras trees flourished. In short, there was a new food supply available for animals that could utilize it.

So there arose new lines of plant-eating ornithischian dinosaurs, the duck-billed hadrosaurs, the dome-headed pachycephalosaurs, the armored ankylosaurs and the horned ceratopsians, all of Cretaceous origin and development. They evolved quickly to fill the ecological niches open to them, and to populate the late Cretaceous plains and forests in great numbers. And of course where there are herbivores, there are carnivores to prey upon them. Consequently, the late Cretaceous was a time of gigantic meat-eating dinosaurs, too, the predators that lived upon the numerous vegetarians. This was indeed the climax of the dinosaurs. There were other reptiles, also, turtles and crocodilians, lizards and snakes. There were toothed birds, and primitive mammals, many of them closely related to modern opossums and shrews. There was, in short, a great expansion in the variety of backboned animals on the land of the late Cretaceous North American continent.

The details of dinosaurian evolution during their late Cretaceous climax can be nicely followed in the sequence of western formations that have been named, and in other formations as well. Indeed, the particular development of dinosaurs that is here being discussed would seem to have taken place in an Eastern Asiatic–Western North American theater. In Central Asia, especially in Outer Mongolia, there are upper

Fig. 18–37. A comparison of the number of genera of Cretaceous dinosaurs in North America and eastern Asia. The silhouettes indicate the suborders represented by each bar, which are; the ceratopsians or horned dinosaurs, the ankylosaurs or armored dinosaurs, the hadrosaurs or duck-billed dinosaurs, the sauropods or giant brontosaurs, and the theropods or carnivorous dinosaurs.

Fig. 18–38. Cretaceous flowering plants. Late Cretaceous (Cenomanian) forest in Bohemia, after Saporta (1879) in Darrah, *Principles of Paleobotany.*

The *Anthophyta* or *Angiosperms*, the flowering plants, consist of the *monocotyledons*, of which the grasses and palms are examples, and the *dicotyledons*, containing the principal trees and shrubs of the present. The origin of the phylum is obscure, but flowering plants are known from seeds found in Jurassic rocks, and an assemblage of leaf impressions in the upper Triassic of southwestern Colorado would surely be called those of a palm—were there not reluctance because of its antiquity.

A marked development of life is that of the angiosperms in rather late early Cretaceous, late Aptian and Albian. Floras have been found in many localities that were much like those of the present in having leaves of dicotyledons like modern willows, sycamores, oaks, laurels, magnolias, poplars, and figs. The associated wood is similar to that of the present plants. The cause of such rapid evolution is only conjectured. Perhaps their susceptibility to insect pollination may have been a favorable factor for flowering plants. If we could revisit the lands of the Cretaceous, plants would seem quite familiar, except perhaps for the sparse representation of monocotyledons, particularly the grasses. These were to thrive in the Tertiary.

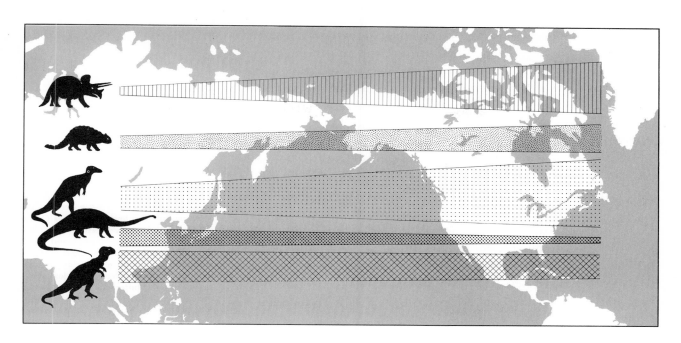

Cretaceous dinosaurs showing very plainly that there was essentially a succession of widely distributed faunas through time, each of which stretched across eastern Asia and western America. And these successive faunas contain the evidence of evolution in detail. Within them we can see the increase in size and predatory adaptations of the gigantic carnivorous dinosaurs, of the adaptive radiation and the various specializations in the armored dinosaurs, the duck-billed dinosaurs, and the horned dinosaurs. It is the record of animals becoming variously adapted in different directions, sometimes in parallel directions, to a varied environment. It is the record of expansion through time.

And this record of expansion continued to the very end of the Cretaceous. The dinosaurs reached their climax; they enjoyed this state of varied dominance for some millions of years, and they then became extinct. They would seem to have been supremely successful to the end of their reign. Why they should have become extinct is a large question, one that will be explored in a little more detail in Chapter 20.

Fig. 18–39. Reconstruction of Cretaceous life as it might have been in the later part of the period in Texas (Texas Technological College). The large coiled shell is an ammonite cephalopod which from the outside gives no suggestion of the intricate pattern of the internal sutures; the long, slender conical shells are other ammonites, the straight shelled *Baculites*. Coiled gastropods or snails, and pelecypods or clams, lie on the sea bottom.

19

The Permanence or Drift
of Continents

A Lower Permian reptile, *Mesosaurus* from the state of São Paulo, eastern Brazil (American Museum of Natural History). The presence of such mesosaurs on the two sides of the South Atlantic has suggested that the continents were connected.

19 *The Permanence or Drift of Continents*

"There are many bonds of union which show that Africa and South America were formerly united," wrote F. B. Taylor (1860–1938), an American geologist, in 1910. Two years later, Alfred Wegener (1878–1928), a German, published a provocative book, stating that

> . . . he who examines the opposite coasts of the South Atlantic Ocean must be somewhat struck by the similarity of the shapes of the coast lines of Brazil and Africa. Not only does the great right-angled bend formed by the Brazilian coast at Cape San Roque find its exact counterpart in the reentrant angle of the African coast-line near the Cameroons, but also, south of these two corresponding points, every projection on the Brazilian side corresponds to a similar shaped bay in the African. (Fig. 19–1).

This similarity or coincidence had attracted attention long before the beginning of the twentieth century, but it remained for Wegener to develop the hypothesis most fully. He believed the present continents to be but fragments of a once greater continent that broke apart in the late Paleozoic, and the continents to be relatively light, of sial, floating on and drifting through the heavier sima that lies below them and below the floors of the ocean basins. Several types of observations led Wegener to believe in drift.

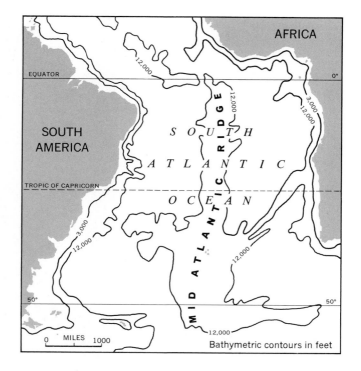

Fig. 19–1. Map of the South Atlantic Ocean showing the marked similarity in the shapes of the shores and continental margins of South America and Africa that suggested that the continents have drifted apart. The Mid-Atlantic Ridge that lies in the middle of the ocean basin has a narrow graben or rift zone within its crest.

456

Fig. 19–2. Reconstruction of the map of the world for the late Carboniferous by Alfred Wegener, showing the original single continental mass that became separated in his Displacement Theory. The heavy lines roughly outline the present continents and larger islands. Wegener conceived of the displacement having continued through the Mesozoic and Cenozoic eras.

Fig. 19–3. *Permian Dwyka Tillite* (G. W. Bain). *A.* Tillite lying on a grooved glaciated surface near Kimberley, South Africa. The glacial scratches show that the ice moved from right to left, northeast being on the right, a direction that seems to make it impossible to derive the ice from a center that also produced glaciation in Australia.

B. Surface of glaciated rock with overlying boulder-bearing tillite, which has in turn been grooved by a later glacial movement.

457

Wegener thought that the Americas have been close to Europe and Africa, and that India and Madagascar were along the east coast of Africa, Antarctica being to the south of them, and Australia adjoining Antarctica and Southern India. The remarkable similarities of the geology on the two sides of the South Atlantic in late Paleozoic and early Mesozoic seemed to require their junction during those periods (Fig. 19–2); so the South Atlantic would have originated in middle Mesozoic time, Newfoundland, Greenland, and Ireland forming a continuous block that split asunder at the end of the Tertiary in his hypothesis.

Several decisions are possible with respect to continental drift. The evidence might be so overwhelming as to convince of the validity of drift as theorized by Wegener. Or the evidence for some parts of the theory might be strongly suggestive, though with details that seemed in error. The theory might be applicable but for times different than Wegener postulated. His evidence might not be significant, but drift in some other form might have been a reality. Continents would drift apart if they retained their size while the volume of the earth expanded. And the whole history of the oceans and continents might be essentially one of permanence in their present positions.

Late Paleozoic consolidated glacial deposits, tillites, are widely present in South America—Argentina, Brazil, and the Falkland Islands —and in South Africa (Fig. 19–3). Similar tillites are present also in the peninsula of India and in western Australia (Fig. 19–4). Marine beds associated with the glacial rocks show them to range from lower Carboniferous into the Permian, though there are some differences in age assignments in the several regions. Some of the glaciers that formed the tills seem to have flowed from beyond the borders of the present lands as shown by the orientations of grooves and striations. The glaciers are generally thought to have been broad ice sheets rather than restricted valley glaciers. The presence of the glacial deposits in the scattered regions, of course, shows that the climates were quite alike in each.

A distinctive kind of doubtfully seed-bearing fern-like plant, *Glossopteris* (Fig. 19–5), is associated with the sequences having tillites in each of the regions, in Permian and earliest Triassic rocks. The sediments seem to have been laid in one or more great subsiding basins within the continents and to have reached thicknesses of three or four miles, though of only a few thousand feet in some places. The fact that the plants are remarkably similar in the latest Paleozoic and Mesozoic in these southern areas—Australia, southern India, south Africa, and southeastern South America—has led to the belief that they are relics of forests on parts of the great continent of Gondwana (Fig. 19–6). There must have been environmental similarities, but plants on separate

Fig. 19–4. Map of the late Carboniferous glacial deposits of the theoretical Gondwanaland, with the area of the ice cap on the interpreted continent. Arrows mark deduced ice flow (after A. L. DuToit).

Fig. 19–5. Leaves of the seed-bearing plant *Glossopteris*, whose distribution in the southern continents has suggested that these continents were once one land, Gondwana. *A*. The specimen is from the Permian rocks near Newcastle, New South Wales, Australia (Princeton University collections, from E. Dorf).

B. Detail of frond from South Africa (after H. Thomas).

C. Seed-like reproductive organ (after E. P. Plumstead in H. N. Andrews, Jr.).

The term *Glossopteris* flora is applied to an assemblage of plants that was widely distributed in the southern hemisphere in the late Paleozoic and Triassic. The plants of the genus *Glossopteris* have long, slender leaves, each with a prominent midrib and delicate veins spreading to the leaf-margins; in some forms, the leaves are arranged in whorls. The plants and similar forms have what seem to be seed-bearing structures, which would place the glossopterids in closest association with the Anthophyta, or angiosperms. But some have interpreted the parts as comparable to those in the pteridosperms. In either interpretation, they are quite distinctive plants that are believed to have grown in a cool temperate climate during times of extensive glaciation.

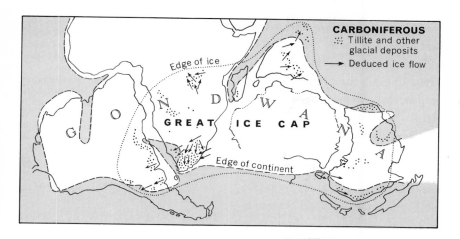

Edge of ice

GONDWANA

GREAT ICE CAP

Edge of continent

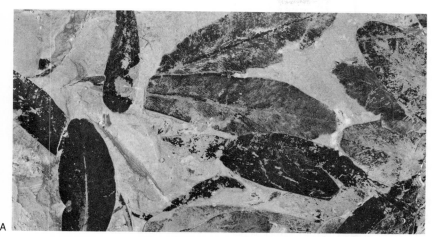

A

B

C

continents would be similar if climates were comparable and if they had means of traveling across intervening water barriers. Seeds can be drifted by sea and spores carried by the winds. There were no birds in the late Paleozoic to transport seeds or plants. Though the chance of a seed crossing the South Atlantic seems remote, the dominant winds, tremendous number of individuals available, and enormous duration of geologic time make such a crossing conceivable if not quite possible.

The non-marine rocks not only have this flora, but also a variety of animals, among which vertebrates are of particular interest. The small, fresh-water, fish-eating reptile, *Mesosaurus*, (Fig. 19–7), is found in South Africa in white sandstones directly above the late Carboniferous tillites and an overlying tongue of marine shales; it is in sediments in a comparable stratigraphic position in Brazil. The fact that the plants are similar may not seem compelling evidence for drift. And marine animals such as those in associated strata could have drifted across a sea or followed its circuitous shores. But it does seem unlikely that small reptiles could swim or be carried across such a great expanse of ocean as the South Atlantic. They might have traveled by roundabout land routes across thousands of miles, or they might have crossed the ocean on an island belt. But forms like *Mesosaurus* have not been found in intervening regions that would be possible land routes; hence it has been thought that they did not migrate along the course of present lands. The factors of the accidents of preservation and of discovery cannot be completely discounted in explaining their absence from regions other than South Africa and Brazil. And it is quite impossible that the forms on the two sides of the Atlantic that are so alike could have developed independently.

Fig. 19–6 (right). *Several interpretations of the relative positions of the continents under theories of continental drift. A. A polar centered Gondwanaland in the late Carboniferous (after R. Maack).*

B. and C. Interpretations of the distributions of the southern continents in Permian and late Cretaceous times (after F. Ahmad).

D. Distribution of continents in early Carboniferous on the basis of study of fossil corals; those living in temperate climates were thought to show seasonal growth records, whereas those living in the tropics did not (after T. Y. H. Ma).

Fig. 19–7 (below). A restoration of a reptile of the genus *Mesosaurus* from the Triassic rocks of eastern Brazil. The presence of such reptiles, land living, both in Africa and South America presents a problem of migration from a common ancestor, whether by some unascertained route along which similar forms have not been discovered, or by direct travel on a single land that has been severed by formation of the South Atlantic Ocean (American Museum of Natural History); also see the chapter opening.

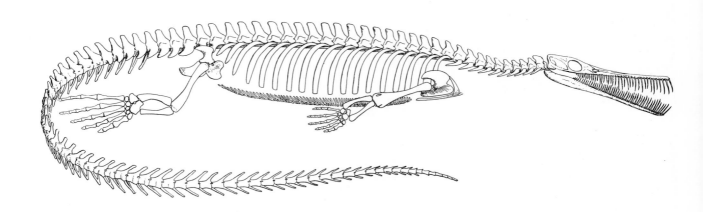

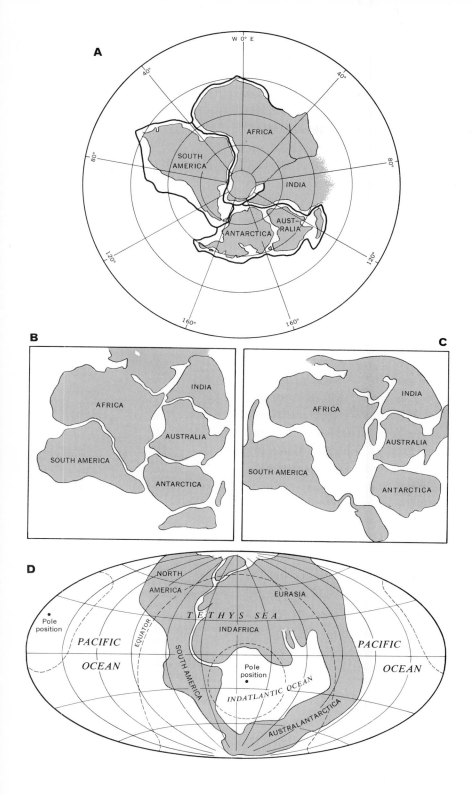

461

The similarities between South America and Africa are not only those of coast line, stratigraphic sequence, and life (Fig. 19–8). Some of the structures in the one continent seem to be cut off abruptly at the Atlantic Coast and to be like those in the corresponding position on the opposite coast, as though both were continuous before a splitting apart. The folded mountains of the Cape Province of South Africa seem to find their continuents in the ranges in the region of Buenos Aires in Argentina. The similarities between northern Africa and the Carribean region of America are not as clear. The great mountain belt that extends from the Pyrenees and Alps into the Himalayas of Asia has been attributed to the pressing of the Gondwana continent on a broad geosynclinal belt, the Tethys, that separated it from the Eurasian continent. The western extension of this belt does not find a counterpart in the Antillean region of the Americas, for that island system forms a loop emerging from North America in Yucatan and re-entering South America in northwestern Venezuala; there is no loose end to join the Tethys Belt. Wegener conceived of these island festoons as marginal ranges that lagged behind the drifting continents. They were compared to the island festoons of the East Indies, which he thought to have been drawn out of southeastern Asia by the drift of Australia.

There are not strong similarities in coastal form between North America and Europe, as there are on the two sides of the South Atlantic. Yet there are many comparable geologic features. Thus the Cambrian System has strong resemblances in the trilobite-bearing sequences in Newfoundland and the British Isles, discussed in earlier chapters. Trilobites, such as *Paradoxides* (Fig. 6–12), are very like those found in south Wales (Fig. 19–9), associated in the so-called Atlantic Province. Cambrian faunas in western Newfoundland resemble those seen in northwestern Scotland, both being in the Pacific Province. It would not be surprising if there were appreciable differences, perhaps, for the present lands in Newfoundland and Ireland would have been separated by four or five hundred miles of present continental shelves even if an original single mass has separated along a break at the present continental slopes (Fig. 19–10). And as Cambrian faunas have not been found in Ireland, South Wales would be another two hundred miles away from eastern Newfoundland.

The Ordovician in western Newfoundland and northwestern Scotland, as well as in islands in the northernmost Atlantic, Spitzbergen and Bear islands, is mostly of limestone and quartzite, whereas that of central Newfoundland and of Ireland and the rest of Great Britain has quite variable successions of terrigenous and volcanic rocks, graywackes and argillites, some with graptolite faunas. The Silurian in these

Fig. 19–8. Correlation of late Paleozoic and early Mesozoic sequences in South Africa and Brazil on the opposite coasts of the South Atlantic Ocean through times when the two continents have been interpreted as being contiguous parts of Gondwanaland (after K. E. Caster). The numbers inserted in the charts record the positions of significant fossils that are reported from the two countries; for example, 4 refers to the plant *Gangamopteris*, of the *Glossopteris* flora, and 10 to the *Mesosaurus* fauna, which, though not recorded in Brazil, is known in middle Permian in Argentina. Most of the records are of plants or floras, some, such as 16, 17, and 18, of reptiles, and a few, such as 8, refer to a marine invertebrate fauna.

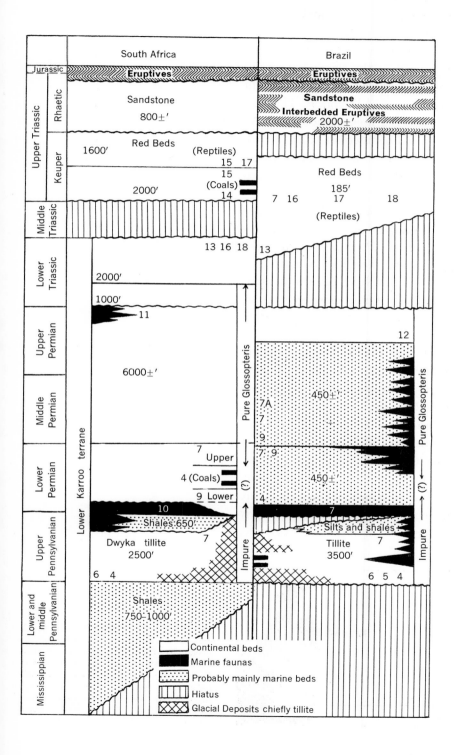

South Africa
Brazil
Jurassic
Upper Triassic
Rhaetic
Keuper
Middle Triassic
Lower Triassic
Upper Permian
Middle Permian
Lower Permian
Upper Pennsylvanian
Lower and middle Pennsylvanian
Mississippian
Eruptives
Eruptives
Sandstone
800±′
Sandstone
Interbedded Eruptives
2000±′
Red Beds
1600′
(Reptiles)
15 17
15
(Coals)
14
2000′
Red Beds
185′
7 16 17 18
(Reptiles)
13 16 18
13
2000′
1000′
11
12
6000±′
Pure Glossopteris
Pure Glossopteris
450±′
7A
7
9
7 9
7
Upper
4 (Coals)
9 Lower
(?)
450±′
4
10
7
Shales 650′
Silts and shales
Dwyka tillite
2500′
7
Tillite
3500′
7
Impure
Impure
(?)
Lower Karroo terrane
6 4
6 5 4
Shales
750–1000′
Continental beds
Marine faunas
Probably mainly marine beds
Hiatus
Glacial Deposits chiefly tillite

areas in Newfoundland has many coarse conglomerates suggesting a late Ordovician orogenic event. Silurian rocks in western Ireland lie on regional unconformities, but seem not to be as coarse; but coarse sediments change into fine ones along trends in each region. A strong orogenic time affected the British Isles in the late Silurian and early Devonian—the Caledonian Orogeny, whereas the somewhat similar history of the American Acadian or Shickshockian Orogeny is thought to have been at a somewhat later time, in medial Devonian. Thus the early Devonian Old Red Sandstone, non-marine sedimentary rocks and lavas in the British Isles, finds a partial counterpart in the medial and late Devonian beds and occasional volcanic rocks of the maritime provinces and eastern New England, but they cannot be considered identical.

The coal-bearing sediments of the Carboniferous of Britain resemble the thick, economic deposits in Nova Scotia and New Brunswick, and those of many other regions. Red beds in the Permian are like those in the late Paleozoic and succeeding Triassic New Red Sandstone of the British Isles. Few of the comparisons that have been mentioned are particularly diagnostic of the need of original close geographic position, and there is the broad intervening water-concealed continental shelf in any case.

The evidence may be compatible with drift from proximal positions. But are there alternate hypotheses that will explain the facts? Though there is similarity between the stratigraphic sequences and structural events on the two sides of the North Atlantic, there is not identity. For instance, the Cambrian and Ordovician of northeastern Newfoundland is as similar to that along the coast of Norway as to that in Scotland, and also has close similarity to successions in northwestern Argentina. The Cambrian section in eastern Newfoundland is not very like that of Wales, resembling more that of Sweden. Possibly all are along a belt of similar character, an eugeosynclinal belt of the early Paleozoic that was deformed by orogenies in the later Paleozoic;

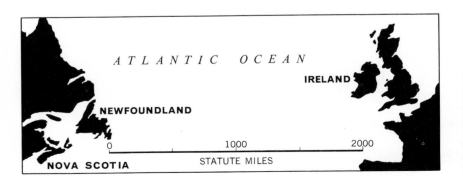

Fig. 19–9. Map of the present North Atlantic Ocean showing the relations between Newfoundland and the British Isles. The islands off Newfoundland and Ireland are now somewhat less than 2000 miles apart, but about 200 miles of sea on each side lies above continental shelf having depth of less than two hundred fathoms.

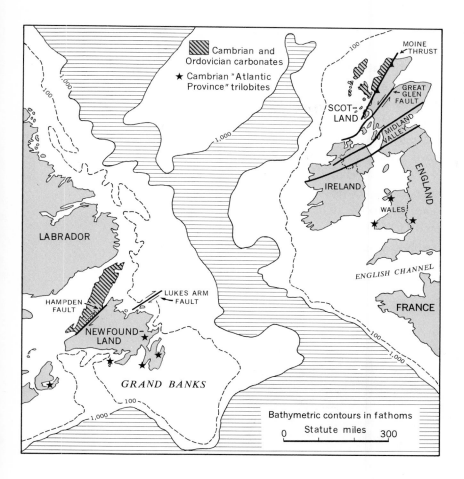

Fig. 19–10. A palinspastic map of the North Atlantic Ocean and of Newfoundland and Ireland made on the hypothesis that the continents were once together (M. Kay). The map shows them as separated a little at the margins of the present continental shelves. The shelves, with water depths of less than 200 fathoms, extend for about 200 miles from the islands on each side of the Atlantic. The shaded area of western Newfoundland has Precambrian gneiss overlain by two thousand feet or so of fairly constant Cambrian and Ordovician limestones (see Fig. 9–19); similar rocks outcrop in northwest Scotland. Central Newfoundland has thick and varied sequences of Ordovician and Silurian argillites, graywackes and volcanic rocks, conglomeratic in the upper part and intruded by Paleozoic intrusive rocks; this deeply subsiding eugeosynclinal belt does not expose Cambrian and older basement. Similar lower Paleozoic rocks are present in many areas in the British Isles. Eastern Newfoundland has one or two thousand feet of rather constant Cambrian argillites laid on an eastern relatively more stable platform of metamorphosed sedimentary and volcanic Precambrian rocks, folded and faulted later in the Paleozoic. The stars are in localities having Cambrian trilobite successions like that in Sweden of the so-called Atlantic province, contrasting with other forms found in the shaded areas. Some transcurrent faults, such as the Hampden Fault, displace Carboniferous rocks; Carboniferous rocks continue across other high-angle faults, such as those south of the Midland Valley in Ireland. Although there are suggestive similarities in the geology of Newfoundland and Ireland, there are differences as well.

465

perhaps this belt continues beneath the present North Atlantic, in which case the ocean north of the belt might differ from that to the south, being essentially a depressed part of a continental block rather than truly a simatic ocean basin. Islands in the northern-most Atlantic, Spitzbergen and Bear Islands, have Ordovician sections like those found on the continents. There are other alternatives.

There are suggestive structural features. The Great Glen Fault crosses central Scotland (Fig. 19–10); it is a transcurrent fault, one in which a more northerly block moved easterly relative to a more southerly one, called right lateral because as one stands on one block, the other block that he faces moved relatively to the right. In eastern Newfoundland, an important right-lateral fault, the Lukes Arm Fault, enters from the sea in central Newfoundland (Fig. 19–11); and there is suggestion that another major fault may run along White Bay on the east of the main northwestern peninsula of Newfoundland. Some

A

B

Fig. 19–11. *Lukes Arm Fault, New World Island, northeastern Newfoundland.* This transcurrent or strike-slip fault has relative movement to the eastward on the northwest side and is thus right lateral, as determined by deformational effects along the fault. *A.* Northeastward from the hills south of Indian Arm; the fault follows the depression and the left side of the arm of the sea. Probable Ordovician sedimentary and volcanic rocks are on the left, Silurian conglomerates in the foreground and in the distant right, south of Pikes Arm.

B. Locality near Herring Neck where the fault passes into the sea; Silurian conglomerates in the foreground and on Herring Head on the right; older lavas and sediments on the left and in islands in the distance.

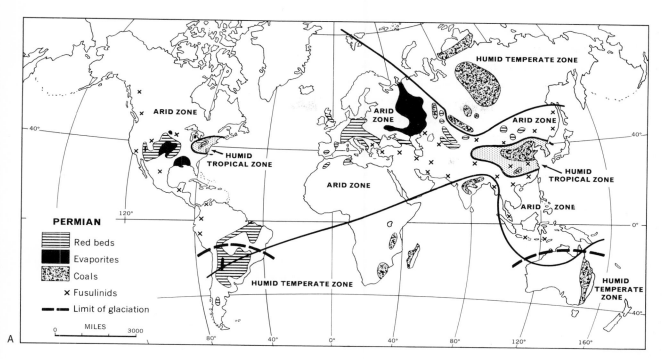

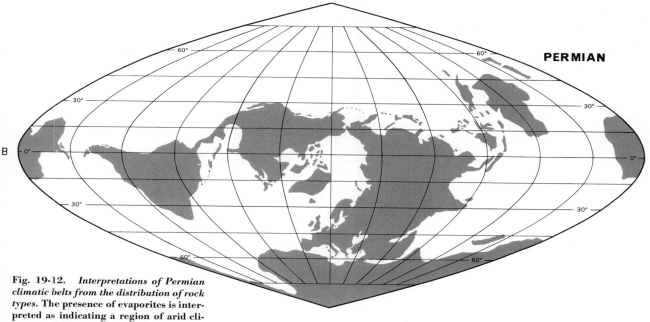

Fig. 19-12. *Interpretations of Permian climatic belts from the distribution of rock types.* The presence of evaporites is interpreted as indicating a region of arid climate. *A.* Interpretation of the climatic zones of the Permian as relating to the present arrangement of latitudes and poles (after A. A. Ronov).

B. Interpretation of the zones in the belief that the continents have retained relative positions but have shifted relative to the poles (after G. W. Bain).

467

comparable faults in Britain such as the Highland Border Fault along the Midland Valley (Fig. 19–8), moved during the Paleozoic Era, for they are succeeded unconformably by Carboniferous rocks, and some may have been scarps during deposition of Ordovician and Silurian rocks. Some transcurrent faults in the maritime provinces may be of late Paleozoic age, and others seem to cut through Triassic. Just as in the case of the stratigraphic evidence, there are many interesting points of similarity, but few that might not have developed in distant lands having similar histories. Any attempt to match these records on the two sides of the Atlantic Ocean as evidence of a once continuous continent is quite speculative, as our knowledge is limited to lands that are widely separated not only by sea, but by broad continental shelves.

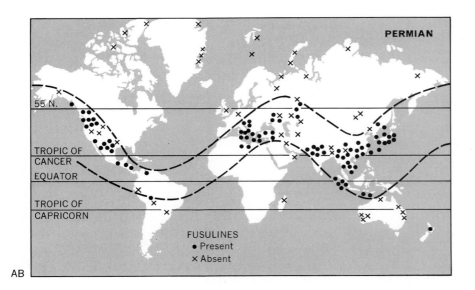

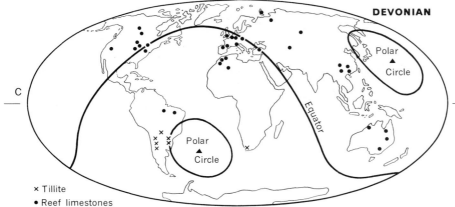

Fig. 19–13. *Interpretations of climatic belts from the distribution of marine fossils.* **A.** Localities in which Permian rocks contain or lack fusulinid foraminifera, which are considered to have lived in warm marine waters; the map interprets the records as evidencing stability of poles and latitudes (after F. G. Stehli).

B. Superimposed on the same map are the equatorial and tropical belts, bounded by broken lines, under another interpretation of the same data, that the poles had a different position than the present ones (after G. W. Bain).

C. Devonian geography based on the distribution of reef corals, taken as indicating the changed positions of poles and latitudes (after M. Schwartzbach).

We need evidence that is more definitive. If we know that the rocks on the margin of the continental shelves on the opposite sides of the Atlantic had similar, matching structures and that these were interrupted by the Atlantic Ocean, the evidence would be much more than suggestive.

In time, the nature of rocks beneath the continental shelves of Newfoundland and Ireland will be known. For instance, the rocks in a small and perhaps representative area in northeastern central Newfoundland are in fault-separated belts having Ordovician with frequent lava flows contrasting with Silurian conglomerates and argillites that are volcanic-poor. The different rocks affect the earth's magnetic field so that their distribution sometimes can be recognized from records of flights with airborne magnetometers, just as has been discussed for the iron ranges of northern Minnesota (Fig. 3–1). In the future, similar data will be gained from flights over the shelf, permitting better interpretation of the structural nature of the submarine rocks; perhaps some of this information ultimately will be gained by direct observation. Critical seismic surveys may determine the trends of the structures. And having established something of the character of the shelves of Newfoundland and Ireland, there should be abrupt truncation of the structures at the edge of the continent if there has been drift, for these two areas are exceptional in that the oceanic margins cut across the continental structures rather than parallel them. Thus the present is one of promise and opportunity rather than of conviction from the

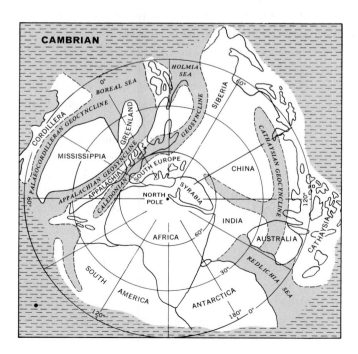

Fig. 19–14. An interpretation of a single continent, Pangaea, centered at the north pole in Cambrian time; as in many of the maps, the boundaries of the continents are only crudely sketched. The map has an interpretation of the extent of Cambrian seas (after A. W. Grabau).

stratigraphic and structural knowledge along the North Atlantic.

A number of geologic criteria may be useful in establishing the original latitude of rocks from place to place. Salt deposits that are the result of the evaporation of marine embayments having restricted access should be limited to regions of limited precipitation and high temperatures, generally within 30 degrees or so of the equator. Deserts are similarly present only in the more equatorial parts of the earth, extending to about 45 degrees in the present climatic conditions. Desert deposited sands have cross-stratification reflecting dominant wind directions, which are of predictable character in the successive climatic belts of the earth (Fig. 19–12). Marine organisms are affected not only by water temperatures, which are generally higher in lower latitudes, but also by depth of light penetration. Thus the distribution of coral reefs in the Paleozoic and Mesozoic, differing from that of the present, suggests changes resulting from continental drift, polar shift, or general earth temperatures. Unfortunately, rocks of similar age are rarely of such wide distribution, and preserve enough records of these climatic indicators to permit unequivocal conclusions as to the relations among continents. Distributions that seem to some to demonstrate that continents and poles in the late Paleozoic were as today, seem to others to show constant relations among the continents with a shift of poles (Fig. 19–13), and to still others to represent changing relative position of continents (Figs. 19–2, 19–6, 19–10, and 19–14). As we shall see, the importance of transcurrent faults, on which blocks move laterally, has become recognized; possibly such movements and the spreading aside of the blocks have been significant (Fig. 19–15).

The nature of the earth's crust beneath the oceans should have a bearing on theories of drift. Drift is conceived of as rather resistant sial masses floating through a weak sima layer that underlies the continental sial plates as well as the floors of the ocean basins. Only in the past decade or two have we come to know the nature of oceanic crust, which is quite thin compared to that of the continents. Such a weak suboceanic crust must have means of gaining excessive depth, for the deepest troughs of the oceans, like the Puerto Rico Deep north of that island, Tuscarora Deep off Japan, and Mindinao Deep off the Phillipines, are along the borders of ocean basins. If the oceanic crust is so weak that the continents can pass through it, it must nevertheless have such strength that it can retain these great crustal downfolds. Perhaps there are subcrustal forces that are drawing the dense earth substance down along these belts; the deeps do not seem to have constant positions relative to the front or the lea of drifting continents.

The mountain ranges were conceived by Wegener as developing in

Fig. 19–15. Map interpreting the displacement of parts of the northern polar regions as by drift and displacement on transcurrent faults (from S. W. Carey).

Fig. 19–16. Observations to test the time when the remanent magnetism in rocks was established (after J. W. Graham). As the pebbles in a conglomerate bed are derived from older rocks, and have been rearranged during their travel, they should have rather random orientation of poles in the deposit unless magnetism has been imposed on the rock after its consolidation. If a rock such as a lava flow gained uniformly oriented poles on solidification, these poles should be systematically rearranged with later folding of the lava.

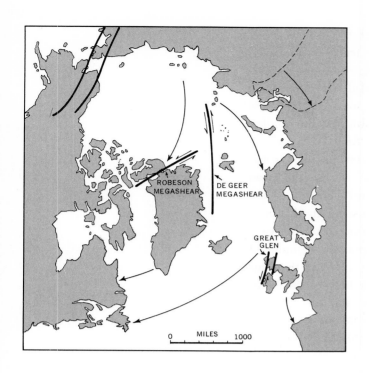

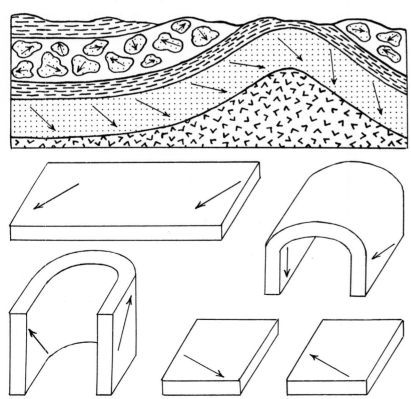

the fronts of the continents as they drifted through the oceanic crust. Though too weak to retard drift, the oceanic crust has been attributed the strength to deform the margins of the advancing continental blocks! Such folded, eugeosynclinal volcanic belts are found all along the Pacific shores of both Asia and the Americas, as well as southerly into New Zealand, presumably in the front of the drifting blocks. But similar, older belts also lie on the opposite shore of North America, where the Atlantic Coast northeastward to Newfoundland is on the "lee" of the continent; but the similar belt is in the "front" of the European continent from Scotland to northern Norway and Lapland. If the structures of the West Coast of the Americas must be attributed to drift westward, those of the East Coast must be attributed to drift eastward at an earlier time—in fact long before the supposed beginning of drift of Wegener; but westward drift would be required at the same earlier time in northwestern Europe. The orogenic belts of the coasts of the continents are apparently of quite the same character and presumably, therefore, of similar origin as those found within the continents, not only in the Paleozoic of such belts as the Urals but also in the Precambrian going far back toward the beginning of earth history. If we attributed the drift to time earlier than Wegener's dating as of medial Mesozoic, the very evidence that was thought most significant in the comparison of the southern continents becomes anachronous. However, this simply shows that the mountains were not formed on the fronts of drifting continents; it does not disprove drift.

Fig. 19–17. *Postulated paleomagnetic polar-wandering curves.* ***A.*** **Successive positions for Europe and North America; map centered in the south Pacific Ocean. *B.* Successive poles for Europe, North America, Australia, India, and Japan; map centered at North Pole (after R. R. Doell and Alan Cox).**

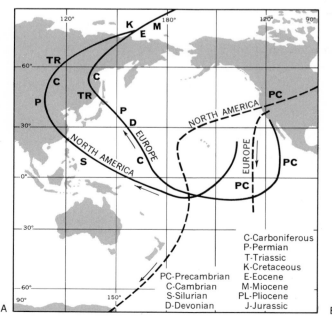

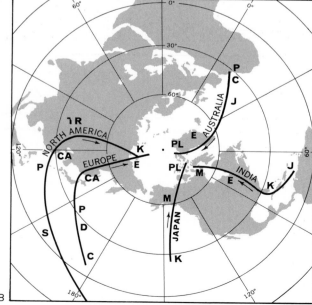

PC-Precambrian
C-Cambrian
S-Silurian
D-Devonian

C-Carboniferous
P-Permian
T-Triassic
K-Cretaceous
E-Eocene
M-Miocene
PL-Pliocene
J-Jurassic

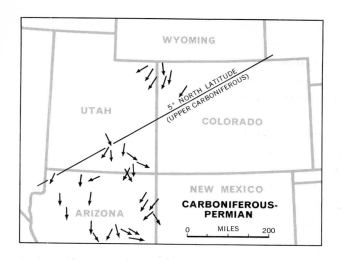

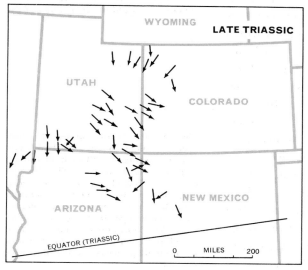

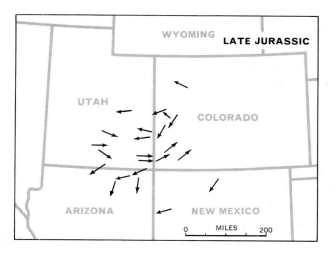

Fig. 19–18. Maps showing wind directions during the late Paleozoic and earlier Mesozoic eras for the Colorado Plateau, Utah, Colorado, Arizona, and New Mexico based on the prevailing direction of cross-lamination in wind-blown sands (after F. G. Poole).

473

Though it has been argued that the strength of earth materials is too great to permit the slight forces to produce drift, the validity of such objections is not acceptable to those who are convinced of drift—they can but say that the forces are of some additional cause or have been misjudged. If the continents did drift, there were sufficient forces to move them!

The theory of continental drift can be tested quite independently by the study of paleomagnetism. The earth rotates on an axis emerging at the North and South poles. The earth has a magnetic field, but the magnetic poles are not identical with the rotational poles; the magnetic pole in the Northern Hemisphere is now in northern Canada about Somerset Island, 20 degrees from the rotational pole. The magnetic pole moves rapidly through time and is now several degrees nearer the North Pole and east of its position of a century ago. The reasons for these progressions are not clear. Moreover, a compass needle points toward the magnetic pole, but the field also causes a compass with a horizontal axis to have a dip which is vertical at the magnetic pole. When a large number of readings is taken of the declination or direction of the compass and of its dip, in rocks of about the same age and area, it is found that although they scatter over a considerable range of values, they average a direction toward the rotational pole. Though the magnetic pole rarely coincides in position with the rotational pole, the average of its positions through time does seem to coincide.

Remanent magnetism is a permanent or irreversable magnetism produced in a rock. Certain minerals become magnetized in the direction of the earth's field at the time of their formation. Thus a cooling lava has minerals that fix the magnetization as they cool below the Curie point, a temperature of some 600 or 700 degrees Celsius or centigrade. Some minerals that crystallize, as cement within a rock for example, or that form through recrystallization in metamorphism, gain a magnetism paralleling or approaching that of the earth's field at the time. Fine magnetized mineral particles are deposited with a preferred orientation determined by the earth's magnetic field. The changing magnetic field of the earth through time may impress itself if the rocks are placed under certain stresses and temperatures, but there are means of determining whether the original magnetization has been destroyed. For instance, if in a conglomerate bed each pebble has a distinct magnetic field orientation, or if the magnetism in rocks on the two limbs of a fold differ by the amount of their inclinations, the remanent magnetism is that gained before deposition or folding (Fig. 19–16). The techniques are somewhat sophisticated, but the results are so consistent in rocks of the same age within a region that they permit

Fig. 19–19. The prevailing surface winds of the world for the month of July.

Fig. 19–20. Section across the Mid-Atlantic Ridge in the North Atlantic Ocean, showing the rift or trench in the ridge (after B. C. Heezen). The rift, bordered by tilted blocks of oceanic crust, suggests that the Atlantic is separating along the ridge. Lavas of basaltic nature have been dredged from the slopes of the ridge. The ocean-floor profile was produced and recorded continuously by an echo-sounding device as the ship progressed.

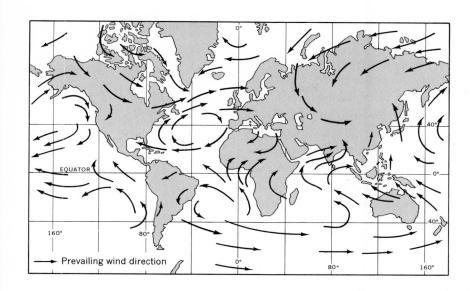

→ Prevailing wind direction

Northwest   Atlantic Ocean   Southeast

Fathoms 2700 | (16,200 ft)   **Water**   **Water**

                                        Oceanic crust and mantle
2800 | (16,800 ft)

                                              Atlantic Ocean

2900 | (17,400 ft)   **Water**   **Water**

3000 | (18,000 ft)   Oceanic crust and mantle

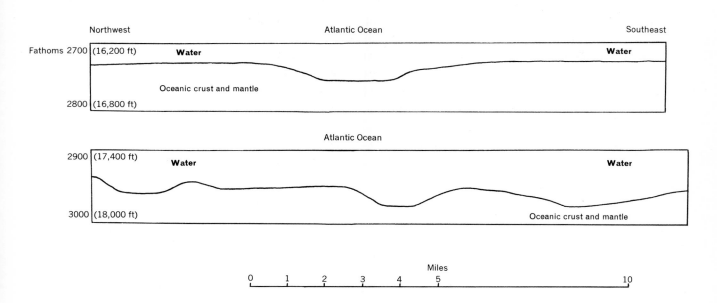

Miles
0   1   2   3   4   5   10

the determination of the positions of the magnetic and approximate rotational poles when there are many readings.

Present readings taken from rocks of relatively recent age would average directly north. But when observations are made of rocks in the distant past, it develops that the position of the magnetic pole has changed through time as measured on a single continent, relative to the present rotational pole (Fig. 17–6). Thus for North America the relative position of the pole in past times on the present globe is progressively southward in Asia through the Mesozoic and then eastward into the Pacific from southern China in late Paleozoic time and going toward the center of the Pacific at the equator in early Paleozoic (Fig. 19–17). This does not mean that the pole of the earth's rotation has changed, but it may mean that the crust of the earth has moved over the mantle and core through time.

Another anomaly appears when observations are extended to Europe. From observations in North America, the pole for the late Paleozoic should have been in southeast China; but we find on making a similar study in Europe (and presumably the same would be true of studies made in Asia) that during the same time the pole was somewhere in eastern Siberia (Fig. 19–17)! So, when the earth's pole relative to North America should have been in southeastern China, that area was not itself in that position! This can be explained on the assumption that the poles for the time were the same, but the relative positions to each other of the rocks of the crust of the two continents have changed.

Studies are now progressing in widely separated parts of the earth on rocks of many ages. When the records have been made and analyzed, the present opinion, by no means unanimous, is that they will show that there has been separation of continents in the Western Hemisphere from those in the Eastern and some separation among Australia, Africa, and southern Asia. It remains to be seen whether the pieces will be found to form a single block as envisaged by Wegener; the present impression is that they will not. Moreover, the present judgment is that some of the continents have rotated relative to others. And there should be some changes in relative position as an effect of orogenic movements if we interpret them correctly. The objections that were raised against the drift hypothesis seem to have been valid, in that drift is not responsible for some of the earth features that were attributed to it. Arguments that seemed to oppose drift were really criticisms of the effects that were attributed to drift, such as having it form folded mountains and other earth features. Some of the observations that relate to climate seem to be consistent with paleomagnetic results (Fig. 19–17).

One of the recent postulates is that the Mid-Atlantic Ridge has a central rift or graben (Fig. 19–20) that represents the widening gulf between the continents. Earthquakes are frequent along the rift, and volcanic rocks have been recovered from bottom dredging there. Such rifts could develop not only if the continents were drifting from an original single mass, extending and separating the ocean area, but also if they retained their size and the whole earth were to expand. The later would not produce the relative change in positions shown in the paleomagnetic data. The present is a time of many new discoveries that are giving better understanding of the relations of oceans to continents.

20
Mesozoic Life

Skeleton of the carnivorous dinosaur *Tyrannosaurus* from the Cretaceous Hell Creek beds of Montana. The animal has a height of about eighteen feet (American Museum of Natural History).

The Mesozoic Era was a time of new life on the earth, a time when plants and animals evolved in new directions and attained in many instances heights of development far beyond any levels reached by Paleozoic organisms. It was a time of wide adaptive radiation and the abundant occupation of ecological niches in the sea and on the land. Indeed, it was a time when various organisms invaded environments which hitherto had been closed to them, when some air-breathing vertebrates went back into the sea and others conquered the air. It was a time of much change, when many plants and animals spread throughout broad expanses of oceans and continents.

At the opening of the Mesozoic, at the beginning of Triassic history, it would appear that the animals of the oceans were just beginning to recover from the extensive depauperization that had taken place as a result of the wide-spread extinctions marking the close of Permian history. Of course, many evolutionary lines had continued in the ocean from the Permian into the Triassic Period, but to a considerable degree this was a time when new adaptations were established which were to determine in many respects the nature of Mesozoic marine life. And although it was a time of new beginnings, these beginnings did not become formulated all at once. There was a certain amount of evolutionary lag in the development of marine life after the advent of Triassic time.

Perhaps this was owing in some measure to the fact that seas would seem to have been cold then. The close of the Permian was evidently a time of comparatively harsh climates in many parts of the world, and such conditions would appear to have continued into the beginning of Triassic history. Thus there is a notable lack of corals in lower Triassic marine deposits, a very probable index as to the lack of tropical waters in which these organisms might have lived. But as time continued, this hiatus in marine faunas disappeared as a result of the evolution of new types of corals, the scleractinians, appearing during middle Triassic time, these being the first of the reef-building corals which in the course of time increased to populate Mesozoic seas.

Brachiopods were rare in Triassic seas. So were crinoids, but various other echinoderms proliferated, with the result that during Mesozoic times the mobile forms, the starfishes and sea urchins and their relatives, became very numerous indeed. These were successful lines of development, and today such echinoderms populate the oceans throughout the world.

Fig. 20–1.   The marine reptiles of the Mesozoic era (E. H. Colbert, 1955).

Fig. 20–2.   Restoration of an ichthyosaur, after a restoration by Lois Darling, a reptile similar to a modern porpoise in size and habits (E. H. Colbert, 1955). Here is a prime example of convergence in evolutionary development: the ichthyosaurs of Jurassic and Cretaceous times quite evidently played the same ecologic role in the oceans of those days as do the small cetaceans in today's oceans.

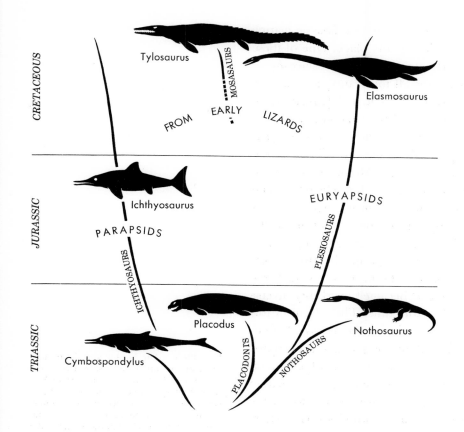

Two additional developments in the burgeoning of Triassic marine invertebrate life deserve particular notice at this place. One of these was the increasing development and importance of crustaceans of modern type (the fossil record of which is all too scanty) to take over the niches occupied by trilobites and other ancient Paleozoic types, and the other was the remarkable radiation of the ammonites.

Ammonites have been the subject of much study because of their successive and wide radiations through Mesozoic time. Their evolutionary history was not, however, limited to the Mesozoic. The comprehensive group of ammonoids appeared in the Devonian and developed rather abundantly through later Paleozoic time; but this group of invertebrates suffered heavily during the extinctions at the close of Permian history, so that of the various ammonoid groups which occupied late Paleozoic seas only one small group of rather simple morphology continued into the Triassic. From this remnant the ammonoids experienced a renaissance and became remarkably diverse and complex during the course of Mesozoic time. The advent of this wide range of adaptive radiation took place during the Triassic Period, with the appearance of some four hundred diverse genera that occupied numerous ecological niches. Here is a striking case of evolution progressing at a rapid rate and resulting in wide adaptive radiation among a certain group of animals, enabling them to occupy many habitats—to take advantage of the ecological opportunities with which they were confronted.

The several evolutionary events that have been mentioned, and others as well, filled the seas of the Triassic with new invertebrates,

Fig. 20–3. The hunter and the hunted in Morrison times, as restored by Alexander Seidel (American Museum of Natural History). This restoration shows some giant, swamp-dwelling sauropods, *Apatosaurus*, escaping from the predators, *Antrodemus*, by taking refuge in the deep water of a swampy lake. Compare this picture with the trackways from the Cretaceous of Texas, shown in Fig. 18–36.

giving to oceanic faunas a rather modern look. Competition was ever on a higher plane, with correlative increases in the organization and complexity of animal life. This fact is particularly reflected in the development of the marine vertebrates during the early years of Mesozoic history.

The bony fishes which had evolved primarily in fresh waters and lakes during the initial stages of their history, became increasingly adapted for marine life with the advent of the Mesozoic Era. Triassic seas were invaded by subholostean and holostean fishes which, even though primitive as compared with the fishes of later geologic ages, were none the less widely adapted for life in the oceans. Some of these Triassic fishes were of generalized appearance, some were elongated, some were deep-bodied like reef fishes of today. Their lines of adaptive radiation suggest specializations in many cases to rather narrow habitats—this is an index to the severity of competition for life in the seas.

Of particular importance in the history of Mesozoic life is the invasion of the oceans by air-breathing reptiles. All through Paleozoic time the trend of reptilian evolution had been predominantly away from the water in the direction of ever-increasing independence on the land. In the large sense this trend continued through the Mesozoic, too, with land-living reptiles becoming increasingly diverse as Mesozoic history progressed from beginning to end (Fig. 20–1). But at the same time there was a countertrend, and tetrapods for the first time became adapted for marine life. Here again is an index to the severity of evolutionary competition. The reptiles of the Triassic Period were more diverse

Fig. 20–4. Vertebrate flight in Jurassic times; after restorations by Lois Darling (E. H. Colbert, 1955). The flying reptile or pterosaur, *Rhamphorhynchus*, and the ancestral bird, *Archaeopteryx*, both from the Upper Jurassic Solnhofen limestone of Germany, represent two independent adaptations for flight, one of which was destined to succeed, the other eventually to fail.

Ramphorhynchus

Archaeopteryx

than ever before; they were adapted to a large array of habitats on the land. In the sea there was a new food supply—a supply consisting of the new invertebrates which were increasing at such rapid rates during the middle and later years of Triassic history and of the new types of marine bony fishes. Consequently, certain groups of reptiles became specialized as marine animals, and thus were able to feed upon animals unavailable to their reptilian relatives. They became adept swimmers, and this was brought about by changes in form and function. The legs, which for so many millions of years had been propelling organs on the land, became transformed into paddles, in some types for locomotion, in others for controlling balance. The tail, which had for so many million years been a balancing organ (among other things) became transformed in some types into a powerful scull, for propelling the animal through the water. The nostrils commonly retreated to the top of the skull for more efficient breathing. Because evolution is irreversible, the gills of the fish ancestors, once having been lost, could never be regained, so breathing was by means of lungs, and these marine reptiles must perforce come to the surface for their oxygen, as do modern whales and crocodiles.

These Triassic marine reptiles were the nothosaurs and plesiosaurs, the placodonts and the ichthyosaurs. The nothosaurs are small reptiles that in life pursued fishes. They were in a sense "reptilian seals" that could come out on rocks or sandy shores at the edge of the ocean. They were the ancestors of the plesiosaurs, these latter first appearing in middle Triassic time as completely marine reptiles, and they also were predators upon fishes, rowing through the water with great oar-like paddles. The placodonts were obviously shallow-water paddlers with huge crushing teeth in the jaws with which they broke open mollusc shells. The ichthyosaurs were completely fish-like in form and, like modern porpoises, were evidently fast swimmers in the open ocean.

Such was life in the sea during the Triassic Period, life that showed new activity and new directions. What was life like on the land? As was remarked at the end of Chapter 15, the record of Permo-Triassic land animals is an unbroken one in two areas, South Africa and Russia. In South Africa, for example, we can follow the fortunes of the many mammal-like reptiles, as well as of certain other groups, across the boundary between Lower and Upper Beaufort beds, which is coincident with the boundary between the Permian and the Triassic. Indeed, this continuity between Permian and Triassic sediments and fossils is so notable that the rocks and fossils of South Africa are placed by many geologists in a single Permo-Triassic entity, the Karroo System. Similarly, the Permo-Triassic rocks of India are grouped into a Gondwana System, and those of South America into a Santa Catharina System. The fact that the Permo-Triassic rock sequence of the Southern Hemisphere and of India crosses a major division between two eras of the standard geologic column is an illustration of the complexity of geological evidence. What is missing in one place may be present at some other locality. It was an inevitable historical accident that the founders of geology should have worked in northern Europe, and thus developed a scale of succession of rocks and fossils and the breaks between them upon the basis of the evidence with which they were familiar. If the first geologists had worked in the southern hemisphere, it is most probable that the arrangement of geologic eras, periods and systems would have been somewhat different from what it is.

With this digression in mind, we need only to cite very briefly some of the evidence. In South Africa the therapsids, the mammal-like reptiles which so dominate the Permo-Triassic faunas of that region, continued from the one period into the next. The same was true for some other reptiles, notably the cotylosaurs which are primitive types, and the eosuchians, which are essentially ancestors of the dominant Mesozoic reptiles. Thus reptilian life was unbroken in its large aspects. But when we look at the record in detail we see that the succession is not as simple as would seem evident from the foregoing remarks. Thus there were various changes in faunas between Permian and Triassic times. The gorgonopsians, a large group of mammal-like reptiles with saber-like canine teeth, the ecological prototypes of the saber-tooth cats of the Cenozoic, became extinct with the close of the Permian in South Africa. And other groups of mammal-like reptiles showed sharp reductions in their numbers; at this time these were the predatory therocephalians and the bizarre, tusked dicynodonts. But the most mammal-like among the therapsids, the cynodonts, did continue strongly from the Permian into the Triassic.

Added to these continuing lines of development there were new lines of reptilian evolution, appearing in the Triassic, those lines which taken together founded the Mesozoic dynasties of ruling reptiles. Thus as the Triassic Period progressed in Africa the faunas became ever more weighted with the elements which were characteristically Triassic and Mesozoic; reptilian assemblages became increasingly dominated by the archosaurians, the thecodonts and their lineal descendants, the croco-dilians and the dinosaurs. So it was that by late Triassic time the land-living animals of South Africa showed many resemblances to those in other parts of the world. These animals, exemplified by the vertebrates of the Chinle beds of the southwest, have been described in Chapter 16.

As is evident from the Chinle fauna, and from contemporaneous faunas in other parts of the continent and of the world, late Triassic time was a phase in earth history when the lands were inhabited by large, almost gigantic flat-headed, stereospondyl amphibians, by primitive turtles, by a few tusked dicynodonts (the last of a long line of strange plant-eaters), by very advanced mammal-like reptiles—animals on the very threshold of mammalian organization, by strange, beaked rhynchocephalians, the rhynchosaurs, and finally and above all by a host of archosaurians—phytosaurs in northern lands and armored pseudosuchians, ancestral crocodilians, and early dinosaurs. In the streams and ponds were heavy-scaled fishes.

These animals lived far and wide in forests of conifers and cycads, among ferns and other primitive plants. The presence of such associations of land-living plants and animals, in North America and northern Eurasia, in the southern reaches of South America and South Africa, gives some indication as to the wide extent of tropical climates during the final days of Triassic history. The world was to a large extent a tropical world, as was so often the case during past geologic history.

The end of the Triassic Period was marked by some extensive and rather significant extinctions, a fact that has not been sufficiently appreciated by many students of earth history. These extinctions took place in the seas and on the land, and brought about changes in faunas that were to make the animal associations of the Jurassic very recognizably different from those of the Triassic. These extinctions were particularly marked among the backboned animals.

But they were not confined to the vertebrates. In the oceans the ammonites, which had multiplied in such spectacular fashion during Triassic history, were almost wiped out. Only one group of these animals survived into the Jurassic Period, a group that was to give rise to the diverse and abundant ammonites of later Mesozoic time. In the oceans, too, some of the marine reptiles that had been so characteristic of Triassic history reached the end of their existence. These were the small, fish-eating nothosaurs and the rather massive, molluscivorous placodonts. On land various tetrapod lines came to their separate ends. Among these were the large stereospondyl amphibians, the passing of which marked the end of labyrinthodont evolution. Likewise the little, lizard-like procolophonid reptiles disappeared with the close of Triassic history, and their exit concluded the long history of the primitive cotylosaurian reptiles. The dicynodonts similarly became extinct at this time. All of these groups were "holdovers" from the Paleozoic, animals more typical of that earlier geologic era than of the Mesozoic. That they were able to continue through Triassic time in competition with the more "progressive" reptiles that had arisen during this period is some indication of their evolutionary vitality.

But it must not be thought that all of the extinctions which took place at the end of the Triassic Period affected only the old remnant lines, the persisting animals from a more ancient age. Some of the newcomers also suffered extinction, the most notable among these being the thecodonts. All of the thecodonts, which had arisen with the advent of the Triassic, disappeared with the end of the period. These were the pseudosuchians of various types and the phytosaurs. They vanished from the scene to be replaced by their more modern relatives; the phytosaurs by the crocodilians and the pseudosuchians by the various dinosaurs. There were still some other tetrapod extinctions at the close

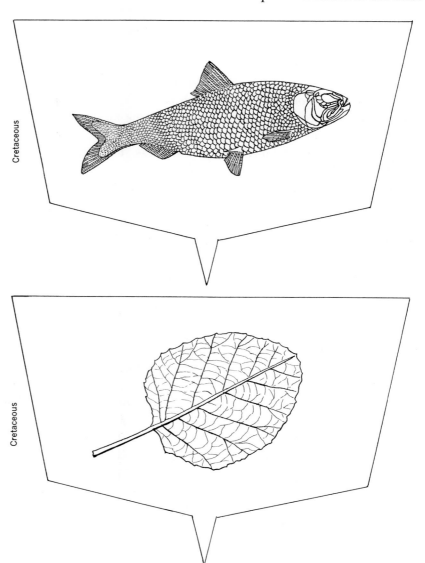

Fig. 20–5. The sudden evolutionary expansion of the angiosperms or flowering plants, and the teleosts or bony fishes, in early Cretaceous times. The flowering plants which dominate modern lands, and the teleosts which populate modern oceans, became firmly established long before the close of Mesozoic history. They lend a modern aspect to Cretaceous lands and seas, which otherwise were the habitats for many forms of life long established and soon to become extinct.

487

of Triassic time. Taken together, four tetrapod orders with ten families out of a total of eleven orders with thirty families living in the Triassic, became extinct at the end of the period. This compares to a rather surprising degree with the tetrapod extinctions characterizing the end of the Cretaceous Period, so famous in the literature of geology. These later extinctions were marked by the disappearance of eight orders containing twenty-four families, out of a total of fourteen orders and fifty-seven families of Cretaceous tetrapods.

As the Jurassic Period developed, the oceans spread widely throughout the world, so that extensive habitats were available for the evolution of marine life. This trend in the physical evolution of the earth, with the correlative radiation of marine animals, was well under way by the beginning of the Jurassic, as is shown by Liassic rocks, with an abundance of marine life, in Europe and in other regions. As time progressed the world became increasingly a tropical world—much more so than had been the case during the Triassic Period—and benign climates and environments reached far and wide from north to south. Tropical seas washed the shores of lands in high latitudes, seas in which coral reefs were extensive and coral lagoons were the locales where many animals lived and died.

Among the animals inhabiting Jurassic seas, the ammonites were particularly abundant and varied. The presence of these interesting cephalopods in such variety and numbers in Jurassic marine deposits indicates an evolutionary rebirth among these molluscs, for it will be remembered that with the close of the Triassic the ammonites suffered extinctions of such vast extent that only one phylogenetic line remained to perpetuate the group into later geologic history. But this one line was enough. From it the Jurassic ammonites evolved into two basic stocks, one of these being the lytoceratids, with rather loosely coiled shells, weakly ornamented or lacking ornamentation, and with an acute pattern characterizing the lobes and saddles of the sutures that joined the walls or septae between the chambers and the outer shell, the other being the phylloceratids, in many respects similar to the lytoceratids, but with a leaf-like pattern of sutures. It would seem probable that the conservative representatives of these two basic stocks inhabited deeper waters of the great geosynclines, and from them there evolved the remarkably diverse, highly specialized, and frequently highly ornamented ammonites that flourished in the shallower seas beyond the geosynclinal deeps. This pattern of evolution was supremely successful during the middle and late years of the Mesozoic, as is obvious from the numbers and the broad phylogenetic spectrum of these cephalopods to be found in Jurassic strata. Mention has already been made in Chapter 17 to

Fig. 20-6. The elasmosaurs of late Cretaceous times were long-necked plesiosaurs that flourished in shallow seas (Skeleton after S. P. Welles; restoration after Lois Darling, from E. H. Colbert, 1955). They evidently swam near the surface, rowing themselves with large paddles, and caught fishes by complex movements of the long, supple neck, which gave extraordinary mobility to the head.

the fact that it is possible to divide the Jurassic of Europe into some forty successive zones upon the basis of evolutionary details in ammonite phylogeny.

One more point may be made in connection with the Jurassic renaissance of the ammonites, namely that this affords a first-class example of parallelism in evolutionary history. From a single ancestral stock the middle and late Mesozoic ammonites developed along numerous parallel lines, lines in which the constituent members reveal their common ancestry by their basic resemblances and their separate adaptations to a variety of habitats and modes of life by their detailed differences. Here is evolutionary parallelism on a grand scale.

The presence of abundant ammonites, great coral reefs and the host of invertebrate life which inevitably lives in tropical seas was matched by the richness of marine back-boned animals of this age. Oceanic fishes were abundant as never before, and there were diverse types adapted for many kinds of habitats and many kinds of diets. Of particular interest is the marked deployment of the sharks and their relatives in Jurassic seas. Not only were the waters inhabited by aggressive long-bodied, fast-swimming predatory sharks, many of them not greatly different from the sharks of modern seas, but also there were highly

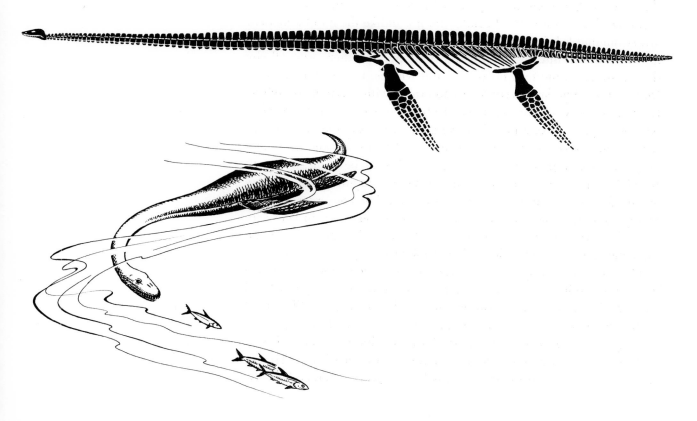

489

specialized bottom-living sharks, similar to the modern chimaeras and to some extent to the modern skates and rays, fishes that lived upon diets of molluscs, which they gathered from the floors of shallow seas. The oceans were filled also with many bony fishes, almost all of them holosteans. These fishes, like the highly evolved teleosts of the present day, were adapted for many oceanic habitats, for life in the open water and for life among the crannies of coral reefs. The first teleost fishes, the ancestors of the overwhelming majority of modern fishes, appeared near the close of Jurassic history, these being descended from Jurassic holostean ancestors.

The reptiles which had invaded the oceans in the Triassic Period, to live off an abundant supply of invertebrates and fishes, became very highly developed and numerous in Jurassic oceans. The small, seal-like nothosaurs and the mollusc-eating placodonts of the Triassic had become extinct with the close of that period, but the plesiosaurs and the icthyosaurs, which were of comparatively primitive form in the Triassic seas, reached high levels of adaptation in Jurassic oceans (Fig. 20–2). Jurassic ichthyosaurs were as completely specialized for life in the open ocean as are modern porpoises. In these reptiles the body was of fusiform shape, perfectly streamlined for slipping through the water with a minimum degree of friction and turbulence. The tail was specialized as a vertical caudal fin, not unlike the fin of a modern fast-swimming fish in shape. The legs were modified into paddles, not for locomotion, but for steering and braking, and there was a fleshy dorsal fin, as in modern porpoises and whales, to prevent the body from rolling. The jaws were long and armed with sharp, fish-catching teeth, and the eyes were enormous, the better to see in dim waters. Here we see a prime example of convergence in evolution, the similar development of unrelated forms (such as ichthyosaurs and fishes and porpoises) for a particular mode of life, this to be contrasted with parallelism, (as in the ammonites) in which there are strong resemblances among descendants from a common ancestry. The convergence of the ichthyosaurs to fishes and, to be precise, of the porpoises in later ages to the ichthyosaurs, is striking proof of the strict limitations placed on form and function among fast-swimming animals.

The plesiosaurs of the Jurassic had become of very large size. These were large-bodied, slow-swimming animals that evidently inhabited shallow waters, where they paddled or rowed with long, oar-like paired limbs among schools of fishes. They were able to catch their prey by darting the small head, at the end of a long, sinuous neck, this way and that. The ichthyosaurs and plesiosaurs, descended from Triassic ancestors, were joined in Jurassic seas by one group of crocodiles, the geosaurs, that also became highly adapted for marine life. These reptiles, like the ichthyosaurs, had a fish-like tail fin and paddle-like limbs. They were the only crocodilians to become strongly specialized for marine life.

Most of the crocodilians of the Jurassic Period were animals of fresh-water lakes and rivers, as they are today. At this stage of earth history the crocodilians showed a wide range of adaptive radiation— here again, as is so often the case in evolutionary history—similar to the crocodiles of later ages in general aspect but different in basic details. For the Jurassic crocodiles were mesosuchians, representative of a suborder that preceded the eusuchian crocodiles with which we are now familiar, just as the holostean fishes of the Jurassic preceded the teleosts of later geologic ages. Nevertheless, the Jurassic crocodiles, so numerous and widely adapted for life, give a modern look to Jurassic faunas on the land.

So do various other land-living animals of Jurassic age, notably the frogs, the turtles, and the lizards.

With the opening of Jurassic history lands were restricted, as has been mentioned, and life on the land obviously was also restricted. Consequently our record of Liassic (early Jurassic), land-living tetrapods is a thin record indeed and is based to a large degree upon the fossils found in the lower Jurassic deposits of northern Europe. Even though this fossil record is woefully incomplete, it does give some inkling as to the beginning of the dinosaurian dominance so characteristic of Jurassic and Cretaceous history. Dinosaurs were common enough in the late Triassic, as we have seen, but they certainly did not rule the lands which they shared with various thecodonts and other reptiles. Nor were they particularly diverse. In lower Jurassic deposits, however, we see not only the carnivorous theropods, which were the dinosaurs of the Triassic, but also the first of the great giants, the sauropods, and some armored dinosaurs or stegosaurians, as well. Undoubtedly, there were others.

That this must have been so is evident from the array of upper Jurassic dinosaurs entombed in the sediments of the Morrison Formation of North America, the Kimmeridgean and related beds of Europe, and the Tendaguru Formation of Africa. The implications of the distribution of dinosaurs of this stage of Jurassic history has already been briefly noted in Chapter 17. As for the dinosaurs themselves, they show a considerable range of adaptive radiation. There are small coelurosaurs such as *Ornitholestes* of the Morrison, little dinosaurs that preserve without a great deal of change the adaptations of their Triassic forebears. These were the hunters of small game. Then there are the giant carnivores, such as *Allosaurus* (more properly *Antrodemus*) of the

Morrison, and *Megalosaurus* of Europe, which preyed upon other giants, probably upon the gigantic, swamp-dwelling sauropods, *Brontosaurus, Diplodocus, Brachiosaurus,* and many others (Fig. 20–3), and upon the plated stegosaurs. The pressure of predation upon the sauropod giants obviously was not of such proportions as to inhibit the evolution of these great dinosaurs. It was a good time for giants on the land.

The wide distribution of the upper Jurassic sauropods is paralleled by that of the stegosaurians, represented by *Stegosaurus* in the Morrison, *Dacentrurus* in the Kimmeridgean, and *Kentrosaurus* in the Tendaguru. These are ornithischians, belonging to the more specialized of the two dinosaurian orders, and with them in the Jurassic habitats were other ornithischians, too. These were more primitive types, hypsilophodonts and camptosaurs, preserving the bipedalism and in general the small size of their ancestors. The presence of these variously adapted dinosaurs in the upper Jurassic is some indication as to the amount of dinosaurian evolution, as yet imperfectly known, that must have taken place in early and middle Jurassic time.

The predominance of sauropods and other giant dinosaurs in upper Jurassic continental deposits is an indication too of far-flung tropical climates and a very abundant supply of plant food. Evidently the low continents of late Jurassic time were clothed with a green mantle of tropical plants, conifers and cycads, ferns, horsetails, and other plants that gave to the forests an archaic look. As has been pointed out in Chapter 17, it was a time of climatic and environmental uniformity. Is was a time of abundant life on the land and in the seas.

It was a time, also, when backboned animals conquered the air. The vertebrates had made the first essays at aerial locomotion during late Triassic times, when certain small reptiles possessed elongated ribs supporting a membrane that would permit these animals to glide from one tree to another, as do the "flying lizards" of modern oriental jungles. But true flight, by means of flapping wings, was a development among the vertebrates that took place in the Jurassic. The pterosaurs, or flying reptiles appeared in the Liassic and established a line of evolution that continued through the remainder of Mesozoic time (Fig. 20–4). In the upper Jurassic Solenhofen limestones of Germany are found the famous skeletons of *Archaeopteryx,* the first bird. This is a truly annectent type, best described as a feathered reptile, for if the imprints of the feathers of *Archaeopteryx* had not been discovered in the fine-grained limestone along with the skeleton, the bones would surely have been classified as of reptilian affinities. This is an unusual example in the fossil record of evolution from one large taxonomic category into another.

Fig. 20–7. The late Cretaceous carnivorous dinosaur, *Tyrannosaurus,* attacking the horned dinosaur, *Triceratops.* Restored by Charles Knight (American Museum of Natural History).

We can be sure that with the appearance of the birds during late Jurassic times, the "warm-blooded" or endothermic animals had become well established on the earth. Actually they had appeared antecedent to this stage of earth history, because the first mammals in the fossil record are of middle Jurassic age, the descendants of Triassic mammal-like reptiles. And in late Jurassic time the mammals undoubtedly inhabited the lands in great numbers, even though the fossil record is sparse. The warm-blooded vertebrates eventually were to dominate the earth; their first appearances were evolutionary events of the utmost significance.

The final stages of Jurassic history were marked by strong tectonic movements in western North America, the Nevadan Orogeny that has already been described. Lands were uplifted along the western border of the continent, one result of this uplift being the deposition of sediments to the east in which the Morrison dinosaurs were entombed. This mountain-making revolution was an incident in Mesozoic history, but seemingly not an incident that greatly affected life. The Morrison dinosaurs thrived throughout this phase of continental disturbance, and from what little evidence we have in lower Cretaceous sediments on this continent, they continued with no appreciable changes into the opening years of Cretaceous history. The same was probably true of life in other parts of the world, on the land and in the seas. There were not the wide-scale extinctions to separate life of the Cretaceous from

that of the Jurassic that so definitely separate Jurassic and Triassic life, or Triassic and Permian life. There was a continuity of evolution that carried animals and plants across the boundary which has been set up between the middle and the last of the Mesozoic periods.

Although the procession of life continued from the final years of Jurassic history into the opening ones of Cretaceous time without any extensive breaks and extinctions such as appear in the transition from the Permian into the Triassic period, or the Triassic into the Jurassic, there were nevertheless some changes that marked the opening of the last Mesozoic period. Certain groups of organisms would seem to have experienced vicissitudes that caused them to decline at the close of the Jurassic Period, and among these the ammonite cephalopods are of particular interest. These molluscs, which had proliferated so remarkably during Jurassic history from a very limited group of Triassic survivors, were again drastically cut—this time to about one-third of their former taxonomic abundance. But again, as in Jurassic time, the ammonites quickly recovered, once more to populate the seas in remarkable diversity and profusion.

But the changes of greatest significance that took place with the advent and the development of Cretaceous history were physical ones, especially the extensive invasion of continental areas by shallow seas. These events repeated in many aspects the happenings that took place during Jurassic history, so that once again as in the earlier period marine habitats were enlarged and terrestrial habitats reduced. Of course such happenings had correlative effects on the life of the time, and during the early Cretaceous there was an enrichment of marine life and certain limitations of life on the land. Tropical seas, many of them shallow, epicontinental seas, extended far and wide, and these warm oceans were the homes of abundant invertebrate faunas, in which the ammonites, once they had recovered from the drastic limitations that had cut them down at the end of the Jurassic, were exceedingly numerous and diverse. The belemnoids, cephalopods closely related to the modern squids, each with an internal skeletal rod, were also numerous in the seas, as they had been in Jurassic time. Pelecypods were very widely distributed and varied during this stage of earth history. And as the tropical seas of the Cretaceous spread, there was, after some regression at the beginning of the period, a spread of coral reefs. The seas were good environments for fishes and for the marine tetrapods that fed upon fishes, and so these animals also developed abundantly during the extent of Cretaceous history.

Since the early years of the Cretaceous, like those of the Jurassic, were marked by the restriction of lands, our records of lower Cretaceous terrestrial faunas and floras are indeed rather sparse—perhaps not as sparse as those of the lower Jurassic, but none the less very incomplete in many respects.

To turn from the imperfection of the lower Cretaceous fossil record to its more positive aspects, two very significant evolutionary events are depicted in rocks of this age. One of these was the establishment of the teleosts as the dominant fishes of the seas. It will be recalled that the first teleosts had appeared late in Jurassic history, but during the early years of the Cretaceous there was what may aptly be termed an "evolutionary explosion" of these fishes, whereby they displaced their more primitive holostean ancestors, and filled the oceans (Fig. 20–5). From that time until today the teleosts have been the overwhelmingly dominant fishes of the world, the most numerous of the vertebrates, not only in species but also in individuals. The teleosts owed their sudden success in early Cretaceous waters to the fact that they were far better adapted in structure and physiology for an aquatic life than were the holosteans. In the teleost fishes the skeleton is highly ossified, thus affording a strong framework for the attachment of powerful muscles, and the scales though thin and pliant are none the less strong enough to give ample protection. This is a great advance over the heavy, bony scales of the holostean fishes. There are many other characters in addition to these by which the teleosts are more efficient as fishes than the Jurassic holosteans, but perhaps enough has been said to give some impression as to the reasons for the early Cretaceous explosion of the highest of bony fishes.

The other evolutionary event of great significance that took place in early Cretaceous time was again an explosion—this time of the angiosperms, or flowering plants, on land (Fig. 20–5). This development, briefly discussed in Chapter 18, established broad-leaved trees and flowers that gave variety to the landscapes, so that forest and field took on a decidedly modern appearance. It can be assumed that with this explosion of the angiosperms there was also an evolutionary proliferation of the insects, the agents of pollenization. We may say that the Age of Angiosperms and Insects had begun, an age in which we are still living. Here we see the interdependence of life. Organisms are adjusted to each other, wherever they live, and frequently the existence of one organism depends upon the simultaneous existence of another—for instance of flowers upon insects and insects upon flowers.

The lower Cretaceous continental deposits of the world which have yielded the record of angiosperm beginnings, have also given us a glimpse of the animals that were living among the first of the modern forests. As in late Jurassic time the dinosaurs were dominant, most of them the lineal descendants of Jurassic forebears. This was the time when the stegosaurians or plated dinosaurs were nearing the end of their existence; these dinosaurs, alone among the dinosaurian suborders, became extinct prior to the close of Cretaceous history. They died out at the end of early Cretaceous time.

Perhaps the most important and famous of lower Cretaceous horizons and localities is the Wealden, a succession of sands and clays on each side of the English Channel. Here are found the dinosaurs and the associated reptiles and other animals that give us a graphic idea of what life was like on the land throughout large parts of the world.

In the sea the marine reptiles, the ichthyosaurs and plesiosaurs continued their evolutionary progress. With a new food supply of teleost fishes available they prospered through the remainder of Cretaceous time, especially the plesiosaurs, which grew into giants of diverse forms. Some of these plesiosaurs were long-necked giants, such as *Elasmosaurus,* a reptile forty feet in length, of which length more than fifty per cent was taken up by the sinuous neck (Fig. 20–6). But some of the plesiosaurs had, instead of long necks and small skulls, short necks and large skulls. The extreme in this evolutionary trend was reached by *Kronosaurus* of the lower Cretaceous of Australia, a reptile fifty feet or more in length, with a skull more than twelve feet long. Here was a reptile that performed the role in Cretaceous oceans fulfilled by the great sperm whales in the oceans of the present day.

One interesting development taking place in late Cretaceous time was the invasion of the oceans by still another group of reptiles, the mosasaurs. These are nothing more nor less than gigantic marine lizards, closely related to the Old World monitor lizards of the present day. The mosasaurs evolved rapidly from their terrestrial ancestors and populated the oceans far and wide, where they shared the abundant food supply of teleost fishes with the other marine reptiles of that age.

The late Cretaceous is notable in geologic history as the time when the dinosaurs reached the climax of their evolutionary development (Fig. 20–7). This climax of dinosaurian evolution, dependent in turn upon the angiosperm explosion of early Cretaceous time, has already been discussed in Chapter 18 and need be only mentioned at this place. As has already been shown, the dinosaurs lived in far greater variety than ever before and this variety was owing to the wide adaptive radiation of herbivorous dinosaurs, dinosaurs that wandered among forests of modern aspect, where they fed upon the abundant available supply of angiosperms. And with the dinosaurs on the land were turtles and lizards and snakes, frogs, birds, and great flying reptiles in the air, and everywhere a host of small mammals underfoot. This was the time

Fig. 20–8. Above: the continuation and extinction of reptiles during the transition from Cretaceous to Tertiary times. Below: the reduction of reptilian families during the transition from Cretaceous to Tertiary times, compared with reductions during the Triassic to Jurassic transition and the Permian to Triassic transition. As will be seen, the extinction that took place at the end of the Cretaceous was proportionately somewhat less than those which occurred at the end of the Triassic and the end of the Permian.

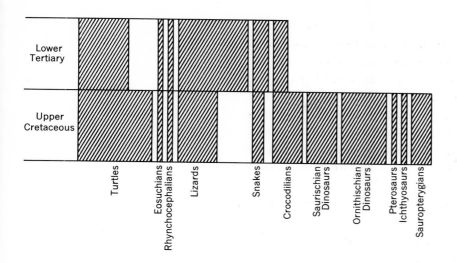

Turtles
Eosuchians
Rhynchocephalians
Lizards
Snakes
Crocodilians
Saurischian Dinosaurs
Ornithischian Dinosaurs
Pterosaurs
Ichthyosaurs
Sauropterygians

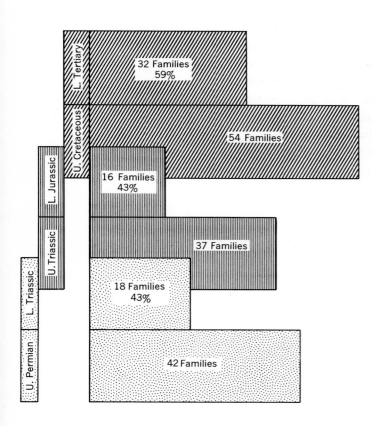

L. Tertiary
32 Families 59%

U. Cretaceous
54 Families

L. Jurassic
16 Families 43%

U. Triassic
37 Families

L. Triassic
18 Families 43%

U. Permian
42 Families

when the ancestors of the modern mammals had made their entrance into the long drama of life on the earth, an event that is recorded in upper Cretaceous deposits by the bones and teeth of opposum-like marsupials and of insectivores related to modern shrews and hedgehogs.

There was a fullness of life on the land and of life in the sea. Then there came the great extinction, which marked the close of Cretaceous history (Fig. 20–8). With the end of Cretaceous time all of the dinosaurs disappeared, as did the flying reptiles, the great marine reptiles, and the ammonites, the belemnites, and numerous large, thick-shelled, Cretaceous pelecypods known as rudistids. The reasons for these wide exterminations which brought to an end many of the dominant animals of the Mesozoic have long been debated, without any satisfactory answers having been reached. It has been thought that the rising of lands that took place as a result of the Laramian Revolution may have been a factor in bringing about the disappearance of the dinosaurs and other land-living animals. But the emergence of modern mountains was a slow process in terms of years, and it seems hard to think that the great reptiles of the Mesozoic might not have become adapted to new conditions brought about by this tectonic event. It has been suggested that the Laramian Revolution instituted colder climates, which in turn affected the plant life of the world. But the evidence of the fossils would seem to show that climates in early Cenozoic times were not greatly different from those in late Mesozoic times. Conversely, it has been suggested that climates became warmer and this caused the extinction of the reptiles of the Mesozoic. But there is no good geologic evidence for this. It has been suggested that the dinosaurs were killed off by world-wide epidemics. But epidemics are usually limited in a taxonomic sense—they do not wipe out great numbers of species but are generally restricted to a single species or a group of closely related species. The dinosaurs died out at the end of the Mesozoic, but their close archosaurian cousins, the crocodilians, did not. If some crocodilians could live through the transition from Mesozoic to Cenozoic times, why could not some of the dinosaurs? Moreover, the great extinction affected life in the sea as well as on the land.

Extinctions are difficult to understand and probably very complex. All we can say is that the world changed, and some of the life of late Cretaceous time failed to adapt itself to a changing world. So the extinctions took place. And it is these extinctions which, among other things, mark the Mesozoic as distinct from the succeeding Cenozoic Era. The Mesozoic was the Age of Dinosaurs; the Cenozoic was not.

That this is so, is amply proven by the fossil record. Despite a century and more of geologic exploration there have been no discoveries of dinosaur bones and teeth in their place of death in Tertiary sediments. An interesting illustration of the termination of the dinosaurian story is afforded by the North Horn Formation of Utah and its contained fossils. This is a sequence of continental sands and siltstones which extends in an unbroken series through a considerable thickness. In the lower part of the formation are the bones of dinosaurs and other vertebrates, including mammals. Then there is a relatively thin interval of barren sediments, above which there are mammal bones but absolutely no dinosaur bones. There are no physical criteria in the sediments to denote the passage from Cretaceous into Paleocene time. But the fossils give evidence of this passage. The rocks show the continuity of geologic time, the fossils record the disappearance of the dinosaurs and the end of Cretaceous history, the succeeding mammals are such as are invariably found above dinosaur-bearing beds.

And so ends a chapter in the history of the earth, a chapter of particular interest because of the spectacular evolution of life throughout its extent. At the close of the chapter great segments of life that had been so very characteristic of the Mesozoic disappeared from the face of the earth. But life continued in many new aspects. The era of recent life on the earth, the Cenozoic, was at hand.

21
The Tertiary History of Western North America

Bryce Canyon National Park, central Utah, southeast of Salt Lake City (A. Duvaney). The early Tertiary, Eocene sandstones and siltstones are beautifully tinted with reds and buff shades, and erosion along fractures and of less resistant beds produces the intricately sculptured and beautiful erosional columns.

21 *The Tertiary History of Western North America*

The Cenozoic (originally Kainozoic and still so spelled in Europe) was named from the Greek *kainos*, recent, because the fossil shells were seen to be much like those of the present sea. Classification and subdivision of the era dates back to times when it was not the practice to give geographic names to rock and time divisions. The Cenozoic Era is divided into the Tertiary and Quaternary Periods; the first of these periods is a name that persists from the eighteenth century, when Primary, Secondary, and Tertiary were terms for the classification of rocks based on their induration or resistance; thus a granite or a gneiss was Primary and a sand or a gravel, Tertiary. As most youngest rocks are little consolidated, most of the Tertiary rocks in this sense were also very young; thus the name came to have time connotation. Quaternary was applied in 1822 to such substances in the Tertiary of that time as the alluvium along streams, and the diluvium, conceived as laid by the biblical flood or "deluge," but now interpreted as of glacial origin.

CLASSIFICATION OF THE TERTIARY

The subdivisions of eras are periods. The Cenozoic embraces only about seventy million years, as measured from the disintegration of radioactive minerals in igneous rocks of the close of the Mesozoic. The span of time is no longer than that represented in the earliest Paleozoic systems, or no longer than two average periods. So it seems reasonable that the Tertiary and Quaternary are classed as periods of time, and their subdivisions as epochs, with the corresponding rocks forming series. Originally there were three epochs, Eocene, Miocene, and Pliocene, named by Sir Charles Lyell (1797–1875) in 1839 on the percentages of recent shells within their known faunas. Subsequently, the term Paleocene was introduced to replace the lowest part of the Eocene, the Oligocene similarly for the uppermost Eocene and lowest Miocene, and the Pleistocene to replace the upper Pliocene.

In Europe, the series have many stages named from geographic localities and having type sections. The succession of rocks and their representative fossils was established in the Tertiary of the Paris Basin (Fig. 18–1) by Georges Cuvier (1769–1832) and Alexandre Brongniart (1770–1847) at the beginning of the last century (Fig. 21–1), one of the first demonstrations of the use of fossils in determining stratigraphic successions. Many of the stages (Fig. 21–2) are named from localities

Fig. 21-1. Diagram of the section of Tertiary rocks in the Paris Basin as published by Georges Cuvier (1769–1832) and Alexandre Brongniart (1770–1847) in Paris in 1811. They described the sequence of rock units and recognized fossils that distinguished each.

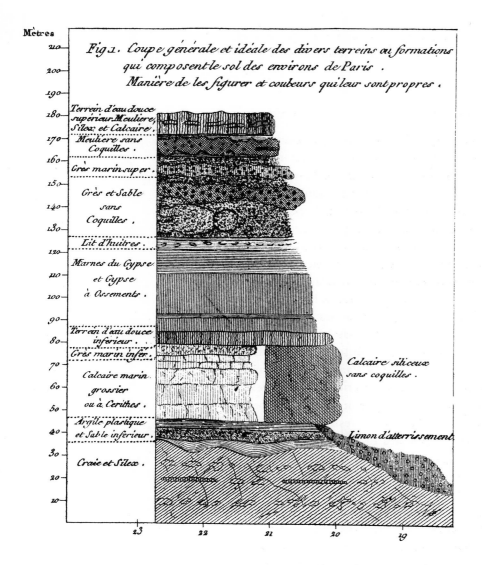

Fig 1. *Coupe générale et idéale des divers terreins ou formations qui composent le sol des environs de Paris.*

Manière de les figurer et couleurs qui leur sont propres.

Métres

Terrein d'eau douce supérieur, Meulière, Silex, et Calcaire.

Meulière sans Coquilles.

Grès marin super.

Grès et Sable sans Coquilles.

Lit d'huîtres.

Marnes du Gypse et Gypse à Ossements.

Terrein d'eau douce inférieur.

Grès marin infer.

Calcaire marin grossier ou à Cerithes.

Argille plastique et Sable inférieur.

Craie et Silex.

Calcaire siliceux sans coquilles.

Limon d'atterrissement.

503

in the Alps, a system of mountain ranges noted for its peaks and glaciers, and in geology for the deformation that developed in the early Tertiary Period (Fig. 21–3).

BASES OF CLASSIFICATION

It is difficult to reach agreement on the boundaries of deposits representing series and stages in terms of the European standard in places distant from Europe. The definition of boundaries based on the percentage of present species contained in an ancient fauna has shortcomings. It was originally applied to molluscs, particularly pelecypods. Even in the original region, the number of recent species counted in a fauna by one paleontologist differs from that made by another because species do not have sharply defined limits. Clearly, the method cannot apply to rocks having different sorts of fossil organisms or laid in contrasting environments. To judge the rocks in a distant place, such as the Pacific Coast, on the assumption that the percentages of species in all parts of the world at a particular time were identical leads to further error. There are very few species of molluscs common to collections from the Atlantic and Pacific sides of the Isthmus of Panama, for instance. The similarities between European forms and American will be greater among organisms that are free-floating and tolerant of variable conditions than among those that must limit their lives to restricted conditions. It has been more practical in North America to apply provincial, that is, local regional names to the stages and to recognize the uncertainties in the series-epoch classification of each (Fig. 21–2). Separate stage names have also been applied to the non-marine sediments, for they are largely isolated from the marine beds and are classified independently on vertebrate organisms rarely found in the marine strata.

The history of North America through the Tertiary Period differs considerably from that of the preceding eras, and is of peculiar intrinsic interest. The record of the Tertiary is distinctive, unique in that the continent had developed until marginal volcanic belts that had prevailed along the continental borders in the Paleozoic and along the Pacific in the Mesozoic were retained only in the Alaskan peninsula and islands. Nevertheless, the Pacific Coast was a belt of active deformation. The Gulf and Atlantic coasts had passed to a condition where gradual warping movements were predominant. The western interior, the region of the Laramian Orogeny of the late Cretaceous, records the later phases of that revolution; and it preserves a much fuller record of the events that succeed such an orogeny than could be shown for earlier orogenies. Tertiary history can explain the manner of the development of some of the most striking features of the present land-

Fig. 21–2. The stages of the Tertiary epochs as arranged in several regions.

Left panel:

Epoch	European	Atlantic and Gulf Coasts
PLEISTOCENE	Sicilian	
PLIOCENE	Calabrian	
PLIOCENE	Astian	
PLIOCENE	Plaisancian	
PLIOCENE	Pontian	
MIOCENE	Sarmatian	
MIOCENE	Tortonian	
MIOCENE	Helvetian	
MIOCENE	Burdigalian	
MIOCENE	Aquitanian	
OLIGOCENE	Chattian	
OLIGOCENE	Rupelian (Stampian)	
OLIGOCENE	Tongrian (Sannoisian)	
EOCENE	Ludian	Jacksonian
EOCENE	Bartonian	Jacksonian
EOCENE	Auversian	Claibornian
EOCENE	Lutetian	Claibornian
EOCENE	Cuisian	Sabinian
EOCENE	Ypresian (Sparnacian)	Sabinian
PALEOCENE	Thanetian	Midwayan
PALEOCENE	Montian	Midwayan

Right panel:

Epoch	Pacific Coast		North American (Vertebrates)
PLEISTOCENE	Hallian	Upper	Rancholabrean
PLEISTOCENE	Hallian	Lower	Irvingtonian
PLIOCENE		Wheelerian	Blancan
PLIOCENE		Venturian	Blancan
PLIOCENE		Repettian	Hemphillian
MIOCENE		Delmontian	Clarendonian
MIOCENE		Mohnian	Barstovian
MIOCENE		Luisian	Barstovian
MIOCENE		Relizian	Hemingfordian
MIOCENE		Saucesian	Arikareean
OLIGOCENE		Saucesian	Whitneyan
OLIGOCENE		Zemorrian	Orellan
OLIGOCENE		Refugian	Chadronian
EOCENE		Marizian	Duchesnean
EOCENE		Marizian	Uintan
EOCENE		Ulatisian	Bridgerian
EOCENE		Penutian	Wasatchian
PALEOCENE		Bulitian	Clarkforkian
PALEOCENE		Bulitian	Tiffanian
PALEOCENE		Ynezian	Torrejonian
PALEOCENE		Ynezian	Dragonian
PALEOCENE		Ynezian	Puercan

A

NNW

SSE

Oligocene-Miocene Molasse

Eocene Flysch

Cretaceous Flysch

Jurassic } Limestones
Triassic } and shales

Mount Pilatus
6475 feet

Stanserhorn
6070 feet

Helvetic Nappes

Molasse

Sea level

B

Length of section: 10 miles

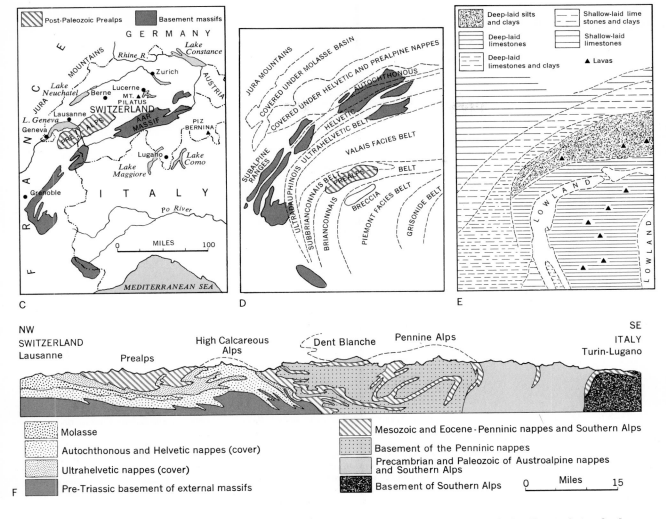

C (map):

Post-Paleozoic Prealps | Basement massifs

GERMANY
Rhine R.
Lake Constance
JURA MOUNTAINS
Lake Neuchatel
Lucerne
Zurich
AUSTRIA
Berne
MT. PILATUS
SWITZERLAND
Lausanne
AAR MASSIF
L. Geneva
PIZ BERNINA
Geneva
PREALPS
Lugano
Lake Como
Lake Maggiore
FRANCE
Grenoble
ITALY
Po River
MILES 0 100
MEDITERRANEAN SEA

D (map):

JURA MOUNTAINS
COVERED UNDER MOLASSE BASIN
COVERED UNDER HELVETIC AND PREALPINE NAPPES
AUTOCHTHONOUS
HELVETIC
ULTRAHELVETIC BELT
SUBALPINE RANGES
ULTRADAUPHINOIS
VALAIS FACIES BELT
SUBBRIANCONNAIS BELT
PREALPS
BRIANCONNAIS BELT
BELT
BRECCIA
PIEMONT FACIES BELT
GRISONIDE BELT

E (map):

Deep-laid silts and clays | Shallow-laid lime stones and clays
Deep-laid limestones | Shallow-laid limestones
Deep-laid limestones and clays | ▲ Lavas
LOWLAND
LOWLAND

F (structure section):

NW
SWITZERLAND
Lausanne
Prealps
High Calcareous Alps
Dent Blanche
Pennine Alps
SE
ITALY
Turin-Lugano

Molasse
Autochthonous and Helvetic nappes (cover)
Ultrahelvetic nappes (cover)
Pre-Triassic basement of external massifs

Mesozoic and Eocene - Penninic nappes and Southern Alps
Basement of the Penninic nappes
Precambrian and Paleozoic of Austroalpine nappes and Southern Alps
Basement of Southern Alps
0 Miles 15

Fig. 21–3. *The Alpine region.* A. Mount Pilatus (6475 feet) a few miles south of Lucerne, Switzerland, an eroded remnant of a mass of late Cretaceous and Tertiary sandy shale of the Flysch that lies discordantly on early Tertiary Flysch, far in front of the crustal folds that form the backbone of the Alps; the rising folds yielded mountainous masses of the covering sediments, which glided down northward into the subsiding troughs in which Tertiary sediments were accumulating, these "overthrust" masses, the Helvetic nappes, being intricately folded as they moved. The southwestward view shows the distant high peaks of the Bernese Alps. Eiger (13,040 feet) and more distant Jungfrau (13,668 feet) on the far left, are composed of pre-Mesozoic gneisses of the Aar Massif; in the middle distance are the high calcareous Alps with Helvetic Nappes (Swissair).

B. Structure section through Mount Pilatus and adjoining ranges, suggesting the intricate deformation of these frontal, floating masses, and their underlying surfaces of displacement or "thrust faults." The section is conventionally oriented with northwest at the left, so is reversed relative to the view in the preceding figure (after Albert Heim).

C. Map of the western Alps, showing the present distribution of the Prealps, and the basement massifs.

D. A palinspastic map showing the facies belts and the relative original relationship of the Prealps to the basement massifs.

E. A paleolithologic and paleobathymetric map of the Cretaceous of the western Alps, with the complex paleogeography (parts C, D, E, and F after R. Trümpy).

F. Generalized structure section through the Alps, in Switzerland.

The Alps are a system of mountain ranges extending for six hundred miles in an arc from the Mediterranean coast at the French-Italian border through Switzerland, northern Italy, and Austria to northern Yugoslavia; the highest peak, Mont Blanc (15,781 feet) is in the west in southeastern France.

The highest mountain ranges in the world, such as the Himalayas in Asia, the Caucasus in southeastern Russia, the Andes in South America, and the Alps, were raised during the Tertiary Period; because of their youth, they have not been reduced, so have remained high. All are called "Alpine Mountain Chains" from the Alps, which have been the object of intense study by several generations of scientists, and whose development is still a subject of speculation. A

(continued on next page)

507

scape. The end of the Tertiary, a million years or more ago, is so recent that events of the succeeding Quaternary or Pleistocene time did little more than sculpture and smooth, accentuate or moderate features gained in the preceding epochs.

WESTERN INTERIOR REGION

The Tertiary deposits of the western interior give us an exceptional opportunity to reconstruct the history of the later stages and aftermaths of an orogeny. In orogenies in the distant past, parts of the earth have been raised as well as deformed. With the rise, the surface has come to be far above sea level. Records of the higher parts of the mountains were to a considerable extent removed, as erosion reduced regions toward sea level. Thus although marine deposits frequently persist, the original surficial rocks laid on the lands often are eroded. If the present continent were similarly reduced to the level of the sea, much of the Tertiary non-marine sediment and Quaternary glacial and alluvium material would be destroyed; the areas where the present basement lies below sea level are shown in Fig. 2–8. Much surficial material was laid above sea level, and some has been raised subsequently; only that which had been depressed would be retained. With the removal of

Fig. 21–3 (continued)

brief discussion of the western Alps, the most thoroughly investigated young mountain system, can but suggest the complicated history that has produced such mountain ranges.

The transformation of the Alps from a region of gently subsiding troughs and intervening stable or gradually rising lands progressed through the Mesozoic and Cenozoic eras. The oldest rocks in the Alps, Precambrian and Paleozoic metasedimentary and igneous rocks that were in the great fold belt of central Europe, had been reduced to a virtual plain by the end of that era. The earliest Mesozoic Triassic System formed rather a constant plate of moderate thickness in a region that had but little crustal deformation before the end of the period. Warping caused differences in water depths and resultant thicknesses of deposits along certain belts, initiating much greater contrasts. In the Jurassic, deeply subsiding geosynclinal troughs developed, some of them with volcanic rocks, cherts, and little detritus, laid in increasing depths and separated by platforms or lowlands. The geosynclines were not purely downfolds, for their margins seem to have been partly fault controlled. Thus

the Alpine mid-Mesozoic geosynclines were in an eugeosynclinal belt, yet were formed on a base that had been deformed earlier and had become rather stable; they do not seem to represent conversion from a pre-Mesozoic oceanic crust; moreover, they seem to have been little influenced by associated tectonic welts.

In late Mesozoic, later Cretaceous, tectonic lands rose in the area, producing terrigenous sediments, argillaceous silts and sands with some gravels, that were swept into the complementing subsiding troughs; the resulting rocks, known as the Flysch, commonly have sole-markings and internal graded bedding that reflect transportation into fairly deep water. Welts continued to rise in the Paleocene and Eocene, and with compression, some of their rocks traveled as thrust sheets and nappes toward the outer margins of the higher ranges, much as in the fashion that has been discussed for some American ranges; large masses seem to have been detached from the rising folds, and to have slid or glided slowly into the deepening troughs on the flanks of the ranges, gaining internal folds and faults as they were transported. By Oligocene, the debris from the rising ranges spread into foredeep troughs sub-

siding on the north of the earlier geosynclines; these can be compared to the exogeosynclines that subsided in the margin of the American craton in orogenic times. The coarser materials, carried from the raised mountains, comprise the Molasse of the Oligocene and Miocene that extends beyond the folded Alpine ranges. By the Pliocene Epoch, the Alpine region had been eroded to rather a low terrain, which was uplifted and eroded prior to the coming of Pleistocene glaciation. Thus the present elevations of the Alps are a reflection of rather recent epeirogeny, as is so frequently the case with our most elevated ranges.

Study and interpretation of the Alps emphasize that certain generalizations can be made, but many of the stratigraphic and structural features of a single mountain system may be peculiar to itself. Such ranges have been affected by a succession of tectonic changes in many phases; the conception of a single great trough being deformed in an orogeny of great magnitude has to be replaced by many more complex sequences of events and of resulting paleogeographic features.

this veneer that covers the older folded and faulted rocks, the depositional history would have to be surmised.

END OF THE LARAMIAN OROGENY

The folds and faults of the Laramian Revolution developed about the close of the Cretaceous though some of the deformation continued well into the Tertiary. The details of the record and the interpreted history differ from area to area, so that only representative records can be used in illustration. The story is well portrayed in the Rocky Mountains of southern Wyoming and northern Colorado. The resistant conglomerates and sandstones of the Cretaceous form distinctive hogbacks along the mountain fronts tilted in the Laramian folding (Figs. 17–16, 18–24). As remarked in the discussion of the Cretaceous, some of the late Cretaceous conglomerates have fragments of the older rocks indigenous to the area, transported from folds that rose nearby late in the Cretaceous Period—hence the conclusion was reached that the ranges started to rise in Colorado before the close of the Cretaceous (Fig. 21–4A).

The steeply dipping Cretaceous beds near the mountain fronts are overlain with angular unconformity by Paleocene gravels and sands (Fig. 21–4B). Distant from the ranges, the succession is quite continuous and conformable (Fig. 21–5). The ranges were deformed and eroded while the adjoining subsiding basins continued to receive the products of erosion. In some instances, Eocene sediments include very coarse granite-boulder bearing conglomerates near low-angle faults that cut the older rocks (Fig. 21–4C); the fragments decrease in size and grade into sands and clays within a mile or so. Eocene rocks lie unconformably above the Paleocene near the ranges. Thus some folds developed later than Paleocene and prior to Eocene. But this is not the close of the tectonic history. After a time of erosion, Oligocene and Miocene gravels and sands spread across the basin and over peneplanes that bevel older rocks high into the mountains. The region was eroded to low relief within the earlier Tertiary perhaps after the Eocene, and then upwarped so that these mid-Tertiary gravels and sands have been dropped into the Precambrian and Paleozoic rocks in fault blocks having throws of hundreds and thousands of feet (Fig. 21–4F). The Laramian Orogeny seems to have terminated about the end of the Paleocene, but there were later epeirogenic (warping) and taphrogenic (block-faulting) movements, and there may have been folding in some regions. The structures of the orogeny, the Laramides, are late Cretaceous and early Tertiary. The further elevation of the rocks has taken place later, the present ranges stand high both for this reason and because of the presence of resistant rocks. As was discussed in Chapter 2, the ages of rocks and structures in a mountain system are several.

While the mountains were rising in the early Tertiary, some areas of the region below present plains were subsiding. Where subsidence was rapid, as in the Denver, the Laramie and Powder River Basins (Fig. 21–5), the stream-carried debris was laid near the source. The surface of the Precambrian rocks in southeastern Wyoming rose to more than three miles in the area of the present ranges, and sank below four miles in some parts of the basins (Fig. 21–6). The subsidence of the Powder River Basin in northeastern Wyoming preserved a record of rather continuous deposition of Cretaceous to Tertiary sediments. The Paleocene stream-laid rocks grade eastward into marine sediments of the last invasion of seas into the interior of the continent (Fig. 21–7).

DEPOSITS ON THE GREAT PLAINS

To the east of the Black Hills in the Great Plains of South Dakota, the surface of the Cretaceous sediments did not subside during early Tertiary. Sediment was being carried to the Gulf of Mexico by streams flowing southward in the Mississippi Valley region, the gradients were sufficiently high in the Eocene in the region southeast of the Black Hills that none of the sediment was retained. In the Oligocene Epoch, some combination of gradient decrease, increase in the quantity of ter-rigenous and volcanic detritus and decrease in water discharge led to the resumption of deposition in great alluvial plains in South Dakota and Nebraska; thus Oligocene lies on the eroded surface of the Creta-ceous. The little-consolidated Oligocene rocks, exposed by erosion in the Badlands (Fig. 21–8), have yielded some of the finest fossil verte-brates of Oligocene and Miocene times.

UPLAND SURFACES AND CANYONS

The surface of the higher mountains of Colorado and Wyoming is re-markably smooth. We can travel over the Laramie Range west of Cheyenne on the highest elevation reached by the Union Pacific trans-continental line without appreciating that we are on the top of a

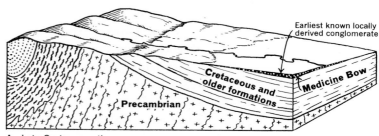

Fig. 21–4. Block diagrams illustrating the development of the Front Range of the Rocky Mountains and the adjoining Laramie Basin in southeastern Wyoming through latest Cretaceous and Tertiary time (after S. H. Knight). The Laramian Orogeny gains its name from this area.

A Late Cretaceous time

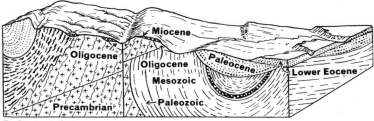

F Present time

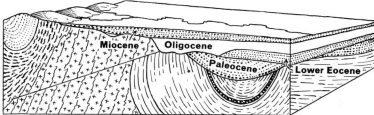

E Late Miocene time

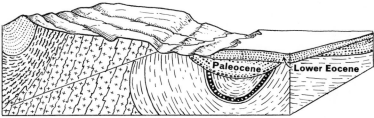

D Late Eocene time

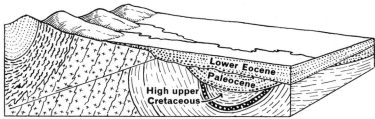

C Early Eocene time

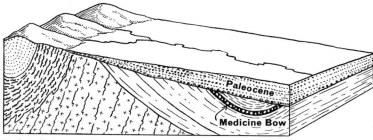

B Late Paleocene time

range, for this is a surface originally reduced by late Eocene and Oligocene streams to a gently tilted erosion surface. To the west, the Medicine Bow Range has a similar smooth upland, with the monadnocks of the Snowy Range rising abruptly above, just as Laramie Peak forms a distinctive erosion remnant in the northern Laramie Range (Fig. 21–9). Pikes Peak is one of the monadnocks on a similar high peneplane in Colorado, and there are many other such unreduced small remnants of the formerly lower surfaces. Sediments spread over these pediment slopes and over the intervening structurally lower areas (Fig. 21–4C). The surfaces were not peneplaned to sea level, for the region is far back from the sea, and the gradients of streams would not permit denudation to such low elevation.

Streams flowing over the alluvial surfaces in the late Tertiary were oblivious to the ranges that were concealed. The many gorges through the ranges (Fig. 21–10) are the effect of the rivers gaining courses on the higher levels of concealing sediments and retaining the courses as

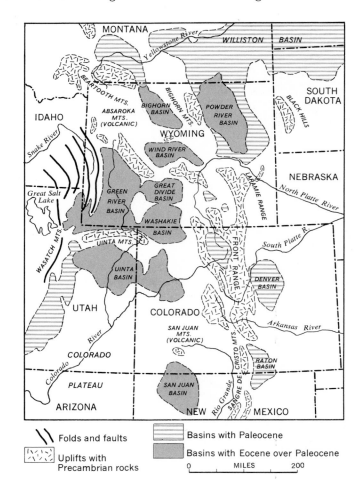

Fig. 21–5. Map of the central Rocky Mountains in Wyoming, Utah, and Colorado showing the principal mountain areas and basins in the early Tertiary Period.

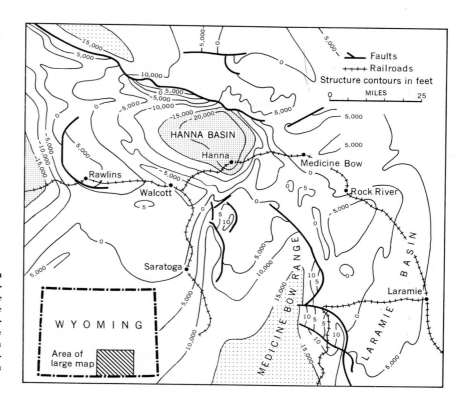

Fig. 21-6. Map of part of southeastern Wyoming to show the elevation of the surface of Precambrian rocks by structure contour lines (after S. H. Knight). The differential elevations between the mountains and the basins are principally the result of the deformation in the Laramian Orogeny in late Mesozoic and early Cenozoic. The actual elevations in the region were gained in later Cenozoic time.

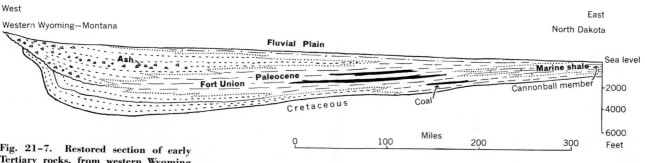

Fig. 21-7. Restored section of early Tertiary rocks, from western Wyoming to North Dakota, showing the Paleocene sediments and volcanic rocks grading eastward into coal-bearing (Fort Union) streamlaid sediments, and the marine (Cannonball) member, the last known marine invasion of the central part of the continent.

erosion reduced the surface and sculptured out the areas underlain by the less resistant Tertiary, Mesozoic, and (to some degree) Paleozoic sediments. The major streams were superimposed on the underlying rock pattern. Middle Tertiary sediments seem in some areas to have been normal faulted and tilted in the Wyoming ranges. Further rejuvenation of the Rockies took place toward the close of the Tertiary, perhaps in the early Pliocene. The movements must have raised the rocks to about their present elevations. Pleistocene glaciation shaped some of the higher ranges and their valleys, and streams and weathering still alter the details.

LAKES AND OIL SHALES

The sediments in the early Tertiary basins of southwestern Wyoming and eastern Utah have great thicknesses of oil-shale laid in fresh-water lakes (Fig. 21–5). Southward, the Colorado River and its tributaries drain the area of the Colorado Plateau, deeply dissected by their canyons. To the west lies the Basin and Range province of western Utah, Nevada, and eastern California, marked today by the interior drainage of the semi-arid basins. The history of this region was complex and varied through the Tertiary and Pleistocene periods.

Fig. 21–8. The Badlands, western South Dakota, in a national monument southeast of the Black Hills (J. Muench). The sedimentary rocks are relatively unconsolidated Oligocene silts and clays containing volcanic ash. The area has been the most prolific source of distinctive Tertiary mammals. The present deeply dissected character is the result of erosion in rather recent times.

The bluffs along the highway and transcontinental railroad that follow the course of the Green River in southwestern Wyoming are of very regularly bedded gray rocks that on close examination are seen to have very thin alternating laminae of dark and light clay and silt (Fig. 21–11). The darker bands have organic matter, plant spores, insects, and other organisms; occasionally there are well-preserved fossils of fishes such as live in fresh-water lakes (Fig. 21–11*B*). There are hundreds of laminations in each foot, and the sequence is of hundreds of feet. If each dark layer represents the abundant life of a summer season, the clays took several million years to accumulate. The coarse gravels (Fig. 21–12) that are interlayered can be followed along the miles of dissected cliffs; they thicken southward at the expense of the clays and silts, as though carried northward by streams from highlands to the south in the Uinta Mountain region into the margin of a deepening lake. Interior lakes covered large areas in the Eocene Epoch, margined by plains receiving the bedded silts and clays from the nearby mountains (Chapter 21 opening photo). Some of the sediments contain Eocene mammals.

Folds started to rise in the position of the present Rocky Mountain ranges in the late Cretaceous. Ranges in Colorado reached such eleva-

Fig. 21–9. Erosion surface on the Laramie Range in southeastern Wyoming, with erosion remnants, monadnocks, of more resistant rocks forming higher hills; the linear features in the latter are fractures in the granitic rocks (Washington National Guard).

Fig. 21–10. Platte River canyon through the Seminoe Mountains in eastern Wyoming, with a dam in the northern end (P. T. Jenkins and L. P. House). The region was once buried in Tertiary sediments, from which the stream has been superposed on the resistant rocks of the mountains. The two views show the canyon in Precambrian crystalline granite rocks (A) and in Paleozoic sedimentary rocks (B), respectively.

A

B

Fig. 21–11. *Eocene Green River Shale, deposited in a fresh-water lake in southwestern Wyoming. A.* Exposures along Highway 30 and Union Pacific Railroad west of Green River. The laminated, carbonaceous shales filled a basin surrounded by mountains raised in the Laramian Orogeny. The sediments are rich in organic matter, and form the greatest potential petroleum reserve in the United States, though the removal of the bituminous matter is not economically feasible at present. The top of the bluff has masses of sandstone, some of them contorted, that slumped into the deeper waters of the lake from shores to the south. The Green River beds pass laterally into tongues of the Wasatch formation. They contain a rich fauna of fossil fishes and other organisms.

B. A herring of the genus *Diplomystus* from the Green River beds. The illustration is somewhat reduced in scale (American Museum of Natural History).

C. Restoration of a lake bottom in western North America during Eocene, or Green River, time (Chicago Natural History Museum). Seen here is a school of *Diplomystus*, and on the bottom, are two fresh-water sting-rays.

C

tion that the accumulating snow formed ice fields that were the heads of glaciers. So long as the streams were able to pass around the rising folds or to cut through them, the sediment could be carried away toward the sea. Thus non-marine, stream-laid earliest Tertiary Paleocene deposits in Montana and Wyoming grade into the marine beds in the western Dakotas that represent the last time that the sea spread so far into the interior of North America (Fig. 21–7).

In other areas, the rising ranges prevented the flow to the sea. In southwestern Wyoming, a mountain-enclosed undrained topographic basin formed north of the Uinta Range that runs eastward through northeastern Utah (Fig. 21–5). Melting snows and summer rains fed mountain streams that flowed into the basin, carrying detritus from their courses. If the evaporation had been very high, the sediment might have been laid entirely in streams and saline lakes as in western Utah and Nevada today. But water came in more rapidly than it evaporated. A lake without an outlet rose within the depression. Streams built deltas out from the south, now represented by the coarse gravels that margin the southern exposures. Laminated clays accumulated in the deeper waters. And from a count of the layers, it is judged that the lake remained for thousands, even a few million years, for the

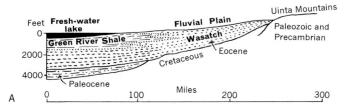

A

B

Fig. 21–12. *Early Tertiary of Utah and Wyoming. A.* Restored section of the early Tertiary from the Green River Basin of southeastern Wyoming southward to the highlands raised in the Laramian Orogeny in Utah. Fluviatile sediments were carried northward into the rising waters of the lake in which the Green River shales were accumulating. The lake-laid sediments have layers that may be each of a year's deposition—a type of varve.

B. Wasatch conglomerate along a highway and the Union Pacific Railroad east of Ogden, Utah.

sediments are a half-mile thick. The lake deposits that had accumulated within this interior basin form a great reserve of bituminous shale that will in the future be refined to yield tremendous quantities of petroleum. Perhaps the water continued to rise until it flowed over the lowest elevation on the brim, thus establishing the course of an outlet stream in the late Eocene. The basin is now drained by the Green River flowing through a deep gorge in the Uinta Mountains in northeastern Utah.

TERTIARY DEFORMATION AND THE COLORADO

By the close of the Eocene Epoch, the elevations of the mountains formed in the early Paleocene had been reduced, their rocks carried in fragments to fill the intermountain basins and spread in plains to the east, and similar basins were filled in other areas. Much of the region came to be deeply buried beneath sediments that were eroded from the adjacent mountains. The erosion surfaces that underlie Oligocene or

Fig. 21–13. Oliogcene sediments in the Wind River Mountains, Sublette County, western Wyoming (J. Muench).

Miocene stream-laid sediments extend far up on the slopes of the present mountains (Fig. 21–13). In central Colorado, hundreds of feet of lava flows (Fig. 21–14) issued and spread over some of these surfaces; they form terraced slopes in the mountains that lie above the narrow and steep-walled Black Canyon of the Gunnison River. The Colorado River now drains much of this region. The front of the highlands that produced Cretaceous sediments was in western Utah and southwestern Arizona. This region gained early Tertiary fluvatile sediments and the mountains were of low relief by Miocene, when the region was raised (Fig. 21–15). With the increased gradients, streams incised their courses into the underlying sediments. The vastness of geologic time can be appreciated with realization that the erosion of the Grand Canyon has been proceeding only in the later Tertiary and Pleistocene epochs (Fig. 21–16).

Fig. 21–14. San Juan Mountains near Ouray, southwestern Colorado, a plateau of rather horizontal Tertiary lava flows and associated sediments that rise to elevations of more than 14,000 feet (Monkmeyer photo).

The precise times of folding and thrust faulting in an orogeny such as the Laramian are not always readily determined, if determinable at all. Rarely are folds preserved in rocks of one stage eroded and bevelled by sediments of a stage but little younger, so as to define closely the time of folding. The rocks below and above such unconformities are not always readily dated; frequently they are rather barren stream-laid sediments. One can state that a fault is younger than the youngest stratum that it cuts but rarely does one find a deposit that lies directly on a thrust fault, so rather circumstantial evidence must be used to limit its minimum age. However, there are known to be folds produced within the Cretaceous and formed as late as the Eocene in many regions all the way from Alaska to southern Nevada in the trend of the Laramides. In addition to those already mentioned, sediments in southern Nevada in late Cretaceous seem to have been laid in front of an advancing thrust plate containing earlier late Cretaceous sediments. In southern Alberta, far to the north, early Paleocene conglomerates lie unconformably with angular discordance on late Cretaceous sediments (Fig. 21–17). In the Absaroka Mountains east of Yellowstone Park in Wyoming, the principal folding took place in two or three disturbances that started prior to the Paleocene and that deformed Paleocene beds prior to their erosion and burial by lower Eocene; many of the rocks are lavas and volcanic fragmental sediments (Fig. 21–18). Other spasms of folding and thrusting of local and relatively minor nature continued in the Absarokas in medial and later Eocene. Also several phases of folding and thrust faulting took place in mountains in southwestern Wyoming and northeastern Utah. So the Laramian Orogeny was one that progressed through millions of years. The Paleocene and Eocene epochs have been judged to exceed thirty million

Fig. 21–15. Representative paleogeographic maps of the Colorado Plateau and adjoining areas in Utah, Colorado, Arizona, and New Mexico (After C. B. Hunt). The Cretaceous seas had spread over the northeastern part of this region, gaining their sediments from highlands in the western and southwestern part; by early Paleocene time, the Laramian Orogeny had raised ranges throughout the area. In undrained depressions the waters rose to form intermontane lakes such as the great Green River lakes in which tremendous volumes of bituminous shale accumulated (Fig. 21–11). Elsewhere, volcanism prevailed, forming such great plateaus of lava as those in the San Juan Mountains (Fig. 21–14). Gradually the basins became filled with sediments, the lakes were drained by streams deepening their outlets, and perhaps their waters lowered through evaporation. By Miocene time, the Arizona area developed the beginning of the Colorado River drainage, while high angle faults produced Basin and Range block mountains in western Utah. Volcanism was particularly widespread in later Miocene and Pliocene times, producing some of the conspicuous present peaks in Arizona and New Mexico. Subsequently the Colorado River system developed in the lower plains, from the headwaters of the Green River in Wyoming and of the Colorado in that state, to form the present topography, shown in Fig. 23–17. The stages in the development of the Grand Canyon are shown in Figure 5–12.

(continued on next page)

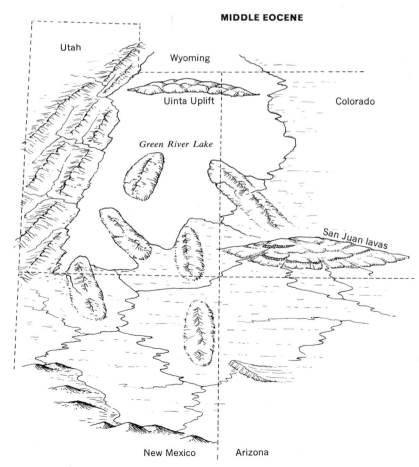

MIDDLE EOCENE

Utah

Wyoming

Uinta Uplift

Colorado

Green River Lake

San Juan lavas

New Mexico | Arizona

521

years duration, and Laramian structures had developed in the preceding later Cretaceous. The deformation in this long span of time progressed strongly in different areas at different times.

THRUST FAULTS AND THE PALEOZOIC GEOSYNCLINE

The principal thrust faults of the Laramian Orogeny are in a belt from the eastern side of the Canadian and Montana Rockies through the southwestern border of Wyoming and across central Utah to southern Nevada (Fig. 21–19). The folds extend to the east in southern Wyoming, Colorado, and New Mexico, and there is the one prominent fold of the Black Hills further east. The generalization has been made that mountains form in geosynclinal belts. Why was the Laramian folding in the Rocky Mountain region?

The geosynclines of the earlier Mesozoic and Paleozoic were limited

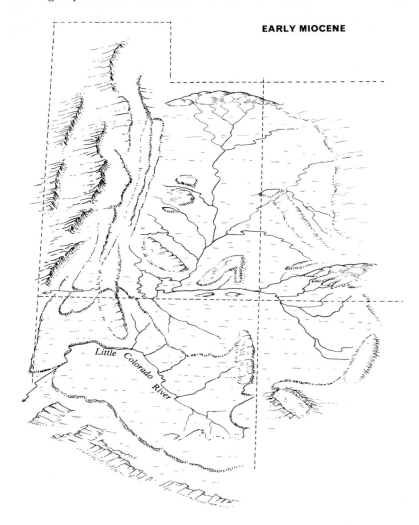

EARLY MIOCENE

Little Colorado River

Fig. 21–15 (continued)

on the east by a line that passed from central Alberta through south-western Wyoming and across Utah to southern Nevada. The thickness of sediments reaches many miles to the west decreasing to a few thousand feet to the east. It is at once apparent that the belt of great overthrusts in the Laramian Revolution is that of the thick sections in the mio-geosynclinal belt of Paleozoic and early Mesozoic. For whatever the reason, the thrusts developed at this zone of earlier flexure and west-ward down-bending. The Cretaceous exogeosyncline, on the other hand, lay largely to the east; it extended throughout the region that has Laramian folds, though Cretaceous is also thick in some regions that did not gain Laramian folds of any great consequence, such as the eastern part of the Alberta plains. Smaller belts subsided a mile and more in the late Paleozoic troughs in central Colorado and northern New Mexico, the zeugogeosynclines. This region started to rise again before the

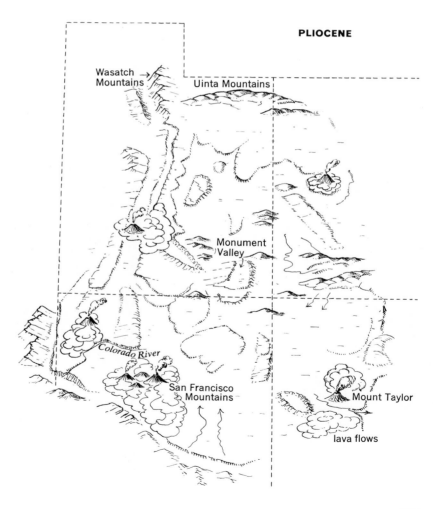

PLIOCENE

close of the Cretaceous, and is the highest and most severely deformed part of the earlier craton that was affected by the Laramian Orogeny. So the Laramian Orogeny involved great lateral displacement of crustal rocks through the production of great thrusts, in the belt of Paleozoic and earlier Mesozoic non-volcanic geosynclines, the miogeosynclines. Folds and smaller faults rose in the region that had thick Cretaceous, particularly upper Cretaceous, sediments. The deep subsidence of the belt to the west need not have caused the thrusts to develop there, but the striking coincidence suggests cause and effect. The Laramian Revolution involved crustal shortening within the belt of the western interior. It is possible that the whole western part of the continent moved relatively eastward against the stable interior, thus widening the Pacific; but the displacement of crustal sedimentary rocks in thrust sheets may be largely surficial sliding, as was discussed in an earlier chapter (Fig. 18–35).

BLOCKS OF THE BASIN AND RANGE REGION

The Basin and Range region to the west seems to have been principally land through the Cretaceous, the source of the sediments laid in the Cretaceous geosynclines to the east in eastern Utah, Colorado, and Wyoming and west in California. The few known Cretaceous deposits are stream- and lake-laid ones, formed in basins within the land. Moreover, there is a very limited record of the early Tertiary. It seems that the region was one of moderate relief by early Tertiary, and that

Fig. 21–16 (right). *Canyons in the Colorado Plateau, Utah, and Arizona.* **A.** Canyon of the San Juan River east of the Colorado River at the Goosenecks in southeastern Utah. The stream is entrenched in some 2000 feet of upper Carboniferous (Pennsylvanian) and Permian sedimentary rocks, the lower part of limestones and interbedded shales, the upper of red sandstones and shales. The dissection has taken place in the latter part of the Tertiary and the Pleistocene periods, concommitant with that of the Colorado River which it joins in the foreground (T. Nichols).
B. Grand Canyon of the Colorado River from the South Rim at Museum Point, showing the broad surface of the Kaibab Plateau to the north.

Fig. 21–17 (left). Paleocene Paskapoo Sandstone lying on shales and sandstones of the upper Cretaceous Edmonton Formation along Little Red Deer River west of Carstairs, north of Calgary, Alberta (J. Gleddie). The contact, disconformable in this section, becomes an angular unconformity to the westward in the front ranges of the Rocky Mountains.

A

B

the present arrangement of ranges and basins is the result of Tertiary and Pleistocene deformation.

The upper Humboldt River in northeastern Nevada preserves much of the record of events. West of Carlin, the Southern Pacific and Western Pacific railroads pass through the narrow Palisades Canyon of the Humboldt. To the south, west of Pine Valley, is the Cortez Range (Fig. 21–20), having Paleozoic sediments unconformably overlain by lavas that dip toward Pine Valley. Conglomerates and sandstones overlying the lava contain silicified or "petrified" wood of probable Oligocene age (though possibly older). These sediments and associated volcanic ash are separated by unconformity from gravels, sands, and volcanic rocks that dip at a lower angle and contain distinctive late Miocene vertebrate fossils such as three-toed horses. The sediments contain increasing quantities and sizes of fragments of Paleozoic rocks as they approach the Pinyon Range of Paleozoic rocks east of Pine Valley. As that range is bounded by a linear fault-line scarp, it seems that the faulting in the region began at least as long ago as Miocene. The lavas of Palisades Canyon, somewhat tilted and gently folded, lie unconformably on the Miocene sediments and so are latest Miocene or early Pliocene. Deposits that have Pliocene vertebrate fossils lie quite horizontally on the somewhat tilted Miocene; these younger sediments were laid in a basin having much the area of the present Pine Valley basin, for fluviatile fans along the margins grade into lake deposits within the central part.

Fig. 21–18. View of the east face of the Absaroka Mountains, on the eastern border of Yellowstone National Park in northwestern Wyoming (D. Muench). Thousands of feet of Eocene and Oligocene lavas accumulated in the Park during the early Tertiary.

THE SIERRA NEVADA

High-angle faults bound ranges westward to the Sierra Nevada. Some of the ranges are of Paleozoic and Mesozoic rocks, folded and faulted in late Paleozoic and Mesozoic orogenies. These are capped in other ranges by Tertiary volcanic rocks, many of them tuffs, some welded—like those that were spread as a blanket over underlying topography by the hot gasses emanating from volcanic explosions, *nuée ardente.* Lavas of Eocene age are known, and some are so young as to be nearly fresh. The ranges were reduced to low-relief within the Tertiary, and were re-elevated. The Sierras, now two miles or more high, were probably less than a mile high in eastern California in the Pliocene. The gentle relief extended southeastward into Arizona, broken not only by the eroded earlier ranges, but by volcanoes and mountains raised by the intrusion of laccoliths beneath them, as in southern Utah. The Miocene seems to have initiated the formation of many of the fault-bounded block mountain ranges, but tilting continued, and is still proceeding.

Fault scarps are still being produced and earthquakes related to faulting are recorded from time to time in the Great Basin of Nevada (Fig. 21–21), Oregon, and eastern California.

DEPOSITION ON THE PACIFIC COAST

The Tertiary sections along the Pacific Coast have an abundance of marine sediments. The volcanic islands of the Aleutians curve westward from the tip of the Alaska Peninsula. The mainland has scattered volcanoes that continue to Mount Edgecumbe in the southeastern islands (Fig. 21–22). The floor of the ocean has the Aleutian Trench (Fig. 21–23), like those margining the volcanic island arcs in the Indies, and there are other similarities, such as the distribution of deep-focus earthquakes. In earlier eras, the volcanic archipelagoes continued southward along the Pacific Coast to Mexico but most had been destroyed by orogenies that consolidated them and made them a part of the continent. Only the northwesternmost part remains as a volcanic island-submarine trench belt. The changes of the Cretaceous continued through the Tertiary. Though perhaps in a late stage of its mobile tectonic development, the belt still is the most active part of the continent, as it is in a stage of mountain building.

During the Cretaceous, the belt between the Sierra Nevada in California and land that lay to the west of the coast became deeply subsiding geosynclines that were somewhat deformed within the period and at the close. The Tertiary records progressive breaking of this belt into essentially rising and sinking, or relatively more and less rapidly sinking blocks and belts, bounded in some cases by normal faults which moved both vertically and laterally. The deformation still continues. Though there has been some orogeny in the sense of lateral shortening, much of the deformation has been of a warping and breaking nature, epeirogenic and taphrogenic. A geosyncline developed during Paleocene and Eocene between the Sierras and islands of somewhat folded Mesozoic rocks along the coast. Through the southern Great Valley, the sinking progressed more rapidly during Miocene and achieved a total of four miles or so by the end of the Tertiary Period (Fig. 21–24). Seas that had covered much of the region west of the Sierra Nevada in early Tertiary time became progressively reduced as sedimentation exceeded subsidence, and deformation raised parts of earlier basins.

FOLDS AND LATERAL DISPLACEMENTS

Several phases of folding were interspersed, affecting limited parts of the section, some producing anticlinal ridges that were beveled during succeeding stages (Fig. 21–25). The magnificent Kettleman Hills

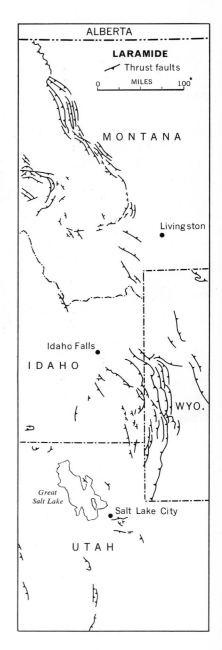

Fig. 21–19. Map of the Rocky Mountains and adjoining region, showing the distribution of thrust faults formed in the Laramian Orogeny near the close of the Mesozoic Cretaceous Period (from Geologic Map of the United States).

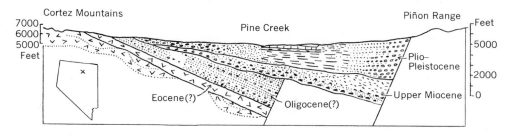

Fig. 21–20. Section of Tertiary rocks in Pine Valley, south of Carlin in northern Nevada, showing the progressive tilting of the sediments in the Basin and Range region (after J. Regnier).

Fig. 21–21. *Fault scarps in the Basin and Range region of Nevada and California.* A. View westward across the Owens Valley to the front of the Sierra Nevada and Mount Whitney (14,500 feet), highest point in conterminous United States; Death Valley, having elevations below sea level, lies to the left.

B. East front of the Sonoma Range in central Nevada, with the face of a fault scarp having formed in 1915 with a resultant earthquake displacement of tens of feet (from Eliot Blackwelder).

529

anticline (Fig. 21–26), which is an oil reservoir, exemplifies the folding. The lateral displacement along the San Andreas Fault (Fig. 21–27) has brought the west side northward scores of miles. This is shown in the offset of lithologies that were originally contiguous (Fig. 21–28) on the two sides of the fault. There was very little volcanism, evidenced by ash deposits in parts of the sequence and perhaps contributing silica in the extensive and abundant cherts in the Miocene. These cherts of California seem to have formed by the accumulation of siliceous minute plant skeletons (Fig. 21–29). The opaline silica is unstable under the load of a mile or so of burial, changing to porcelaneous chert.

The deformation in the Tertiary of California resembles that of the Carboniferous of the Maritime Provinces of Canada, deeply subsiding somewhat linear basins separating sediment-producing lands. In each region there has been dispute as to the importance of normal faulting and increasing demonstration of the presence of lateral slip on wrench faults; those on the Pacific are right lateral—that is as one looks across the San Andreas Fault the opposite block has moved relatively to the right. The geography resembled what would pertain if the Pacific Ocean were now to rise a thousand feet, permitting broad estuaries to invade the interior California lowlands. At times the waters were as much as a half-mile deep. In the Ventura Basin northwest of Los Angeles, sediment of the Pliocene contains foraminifera such as are now present at a depth of several thousand feet in the Pacific; intercalated gravels and muds have been carried in dense flows of mud that, dislodged from

Fig. 21–22. View westward along volcanoes in the Aleutian Mountain chain of southeastern Alaska, rising to Mount Shishaldin, nearly 10,000 feet high (Monkmeyer Photo).

Fig. 21–23. *Faults along the Pacific Coast. A.* The southern coast of Alaska and the Denali Fault, interpreted as a right-lateral fault having displacement of about 150 miles; that is, the southwestern, seaward side of the fault is thought to have moved westward relative to the North American continent on a transcurrent fault (after P. St. Amand). *See* Fig. 2–15.

B. Palinspastic map moving the geography back to its interpreted original position before the fault displacement. The earthquake of March 27, 1964 that caused such destruction in Alaska was centered on the northern margin of the Aleutian Trench near Anchorage.

C. Map showing the relationship of the Denali Fault to the San Andreas Fault of California and the submarine faults in the Pacific off the coast, which also have some transcurrent movements as judged from the displacements of submarine geophysical magnetic belts along them (after P. St. Amand).

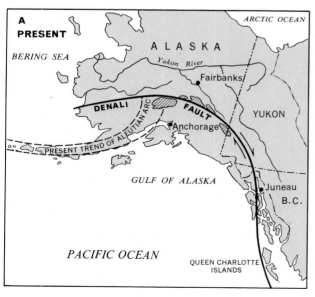

the shores, swept down the steep submarine slopes to accumulate amid the deep water-laid muds. Depths of water varied greatly as a resultant of varying rates of subsidence and sedimentation (Fig. 21–30).

NORTHWESTERN STATES AND BRITISH COLUMBIA

The volcanic geosynclinal belt persisted in the Tertiary farther north. Thousands of feet of Eocene submarine lava accumulated in a sinking trough extending from Vancouver Island, British Columbia, southward beneath coastal Washington and Oregon. As vulcanism ceased in late Eocene (Fig. 21–31), it was succeeded by continued great subsidence, but with deposition of several miles of sands and silts (Fig. 21–31*A*), eroded from lands in the belt of the Nevadan Orogeny in central and eastern Washington and Oregon. As in California, some of the sediments were carried in density currents to bottoms at a depth of hundreds of feet or more. By the Miocene, a new sort of volcanism flooded the interior, plateau basalts and other lavas spreading widely over the Columbia River Basin (Fig. 21–32 and Chapter 2 opening), not in deeply

A

Fig. 21–24. *Tertiary rocks in California* (Mary Hill, California Division of Mines). *A.* Quarry in the Miocene Monterey Formation in Monterey County, California. The siliceous rocks of the Monterey are consolidated diatomaceous earth that became porcelaneous chert on deep burial.

B. Badlands in Furnace Creek Formation (Miocene-Pliocene) along Twenty Mule Team Drive in Death Valley National Monument, eastern California, overlain by the Funeral Conglomerate in the middle distance; the rocks are nonmarine sediments. The footpath descending the slope in the foreground serves as a scale.

subsiding geosynclines, but flowing as plates over a rather stable cratonic area. Later Tertiary volcanism extended into Nevada, as previously noted, and eastward into Idaho and Wyoming. Some of the most striking lavas and tuffs in Yellowstone Park (Fig. 21–32), as in the Absarokas (Fig. 21–18) are Eocene, however, and others, Pliocene. The great volcanoes of the Cascades, mountains like Shasta (Fig. 21–33), Rainier, Hood, and the cauldron of Crater Lake (Fig. 21–34) formed at the close of the Tertiary and in the Quaternary. The lavas perhaps of the Miocene, are widely broken into great fault blocks of the Basin and Range Region of southern Oregon, continuing into the ranges of Nevada and western Utah that have been discussed.

Tertiary sedimentary and volcanic rocks are present northward along the Pacific Coast to the Alaska Peninsula. In the region about Anchorage Alaska, for example, Cretaceous rocks are succeeded with little, if any, structural discordance by Paleocene and Eocene marine and non-marine sedimentary rocks having a thickness of one or two miles, including some coals. The principal angular unconformity is between middle

B

and upper Eocene. Similar and younger sedimentary rocks are widespread along the south coast of Alaska, but as marine rocks are restricted to areas within fifty miles of the coast, it is clear that the Pacific shore did not extend far within its present limit. Volcanoes and volcanic rocks on the other hand, are present far within the state, though not beyond the Arctic Circle. Tertiary sediments of the Arctic Coast are north-thickening terrigenous rocks becoming finer textured and marine northward. Interest in these regions has increased in recent years through exploration for petroleum resources. Volcanic rocks are absent. So Alaska in the Tertiary retained a separation into a volcanic southern and non-volcanic northern belt, as it had since the early Paleozoic. But the proportion of land to sea had very greatly increased.

The examples that have been cited are but glimpses of the record of Tertiary history in western North America. In the prevalence of early Tertiary orogeny and subsequent normal faulting over broad areas, it contrasts with the history of the southern and eastern coastal regions.

MAMMALS IN NON-MARINE ROCKS

From what has been written in the foregoing pages, it is abundantly evident that North America during Tertiary times was a single continental region, within the western portion of which there occurred a long history of mountain uplifts with concomitant faulting and erosion, and

Fig. 21–25. *Sections of the Tertiary rocks in southern California. A.* Restored section of the Tertiary on the east slope of the Sierra Nevada westward across the southern San Joaquin Valley near Bakersfield, California (M. L. Hill and R. Eckis). Oil fields have been developed in sands in the Tertiary that form traps where they underlie unconformities and the crests of folds, as in the Kettleman Hills (Fig. 21–26).

B. Columnar sections in the Cuyama Valley and in Caliente Mountain, a few miles apart southwest of Bakersfield. The marked contrast in the sections within such short distance has been attributed to sharp downflexure into a more rapidly subsiding trough, or to contrast in rate of subsidence on two sides of a high-angle fault; in some instances, such contrasts are brought about along transcurrent faults moving together sections that were some distance apart at times of deposition.

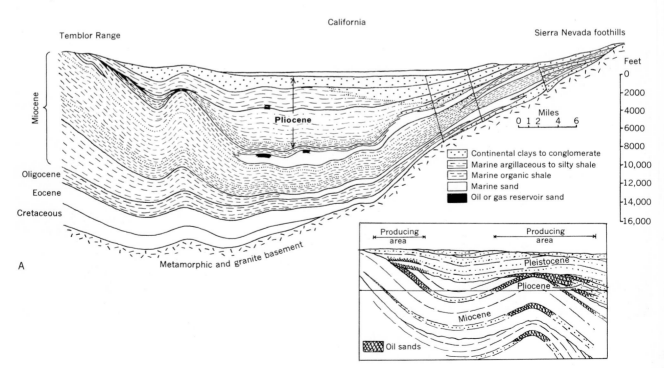

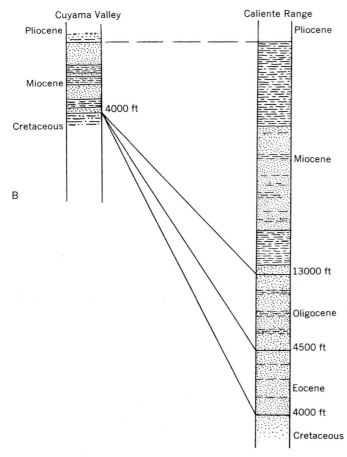

Cuyama Valley — Caliente Range

Pliocene

Miocene

Cretaceous

4000 ft

B

Pliocene

Miocene

13000 ft

Oligocene

4500 ft

Eocene

4000 ft

Cretaceous

Fig. 21–26 (above). Northward aerial view of the Kettleman Hills anticline, near the southern end of the Great Valley of California west of Bakersfield. The Tertiary sedimentary rocks have been folded to form the structure with an axis trending toward the upper left to the right of the center of the view. The anticline is an important petroleum reserve, the oil being trapped in reservoirs where sandstones wedge out on the flank of the anticline (Fairchild Aerial Surveys).

with correlative deposition of sediments in the basins and lowlands contiguous to the rising mountain masses. In these sediments is an unexcelled fossil record showing the sequence of mammalian faunas, from those of exceedingly primitive aspect to those with very modern features. It is a record of unusual completeness, because it contains some twenty or so successive faunas, these, one after another, documenting in a reasonably full fashion the evolution of mammalian life on the North American continent through a time span of approximately seventy million years. There is perhaps no Tertiary record elsewhere in the world to equal it, for which reason, among others, it is of particular significance not only to students of mammalian evolution and to students of past life in general, but also to those whose concern is with the broad principles that underlie the facts of organic development through time. For the Tertiary fossil record of western North America, accumulated to a large degree in sediments deposited by streams and lakes within upland regions far removed from the sea, is more than the documentation of many faunas through time, it is an array of materials showing the details of evolution among various developing lines of phylogenetic development.

Some mention was made (Fig. 21–2) of the various Tertiary stages that have been set up on the basis of the North American sequence of mammalian faunas. Although these stages are not to be correlated point by point with the European Tertiary stages, they may be compared in a general way with the well-known Tertiary divisions of the Old World (Fig. 21–35).

AMERICAN STAGES

The Paleocene was first named upon the basis of sediments in Europe, but as is evident from the foregoing list, the oldest Paleocene deposits are the Puerco beds of western North America. As was shown in the preceding chapter, the lowest Paleocene of this continent, sediments of Puercan age whether or not they belong to this particular formation, succeed without sedimentary interruption the latest known dinosaur-bearing beds. Here is the sedimentary record of the transition from Cretaceous to Paleocene time.

It should be explained that the North American Tertiary stages have been placed in order, in part upon the basis of their stratigraphic succession and to a large degree upon the basis of the evolutionary progress shown by the faunas typical of each. In no one locality are there more than two or perhaps three or four sedimentary sequences representative of these stages present, one on top of the other. But each of such limited sequences gives a portion of the Tertiary record, and by

Fig. 21–27. Trace of the San Andreas Fault zone through San Mateo County, California (Clyde Sunderland). The fault is a transcurrent right-lateral fault, one in which the block to the west, on the left, moved northward relative to that on the right. San Francisco Bay on the far right passes San Francisco in the right distance, where it joins the Pacific at the Golden Gate, to the right of where the fault zone meets the sea. In the far distance on the left of the fault is Point Reyes, separated by the fault zone from the mainland to the right; the fault lies offshore northward. Movements along the fault have been associated with earthquakes that have caused great damage, such as the San Francisco quake of 1906. The hills and mountains in the view are principally of later Mesozoic, Jurassic, and Cretaceous rocks. The movement along the fault has been interpreted as of several hundred miles since mid-Mesozoic.

carrying correlations from one locality to another, and especially by establishing the evolutionary progress of the several faunal assemblages, it has been possible to arrange the North American Tertiary stages so that they present a remarkably complete record of evolutionary development—stratigraphic and organic—on the continent.

The Pliocene, also named (as were all of the Tertiary epochs) upon the basis of European evidence, is represented in North America by the Blancan Stage, based upon sediments and fossils which in some places verge upon the very border of the Pleistocene record. And so the continental Tertiary record in North America runs from the very beginning to the very end of the system.

PHYLOGENY AND THE HORSES

Something might be said about the record of mammalian phylogenies preserved in the Tertiary of North America, to which reference already

has been made. There are many such phylogenies known, some of them in much detail, others only partially preserved. None is better known than the phylogeny of the horses, an example that is repeatedly set forth in books. This evolutionary story is frequently cited because it is so very complete that it gives us a vast amount of evidence as to the facts, and in many ways the workings, of evolution through time.

The facts, very briefly stated, are about as follows. The first horses were members of a family of perissodactyls or odd-toed hoofed mammals that appeared in North America with the advent of Eocene history, during the Wasatchian Stage. They made their entrance into the Tertiary history of Europe at the same time, but in the Old World they failed to continue, whereas in North America their evolutionary development carried on through the remainder of Cenozoic time.

These first horses were small animals, about the size of terrier dogs, with four toes on each front foot and three on each hind foot. The

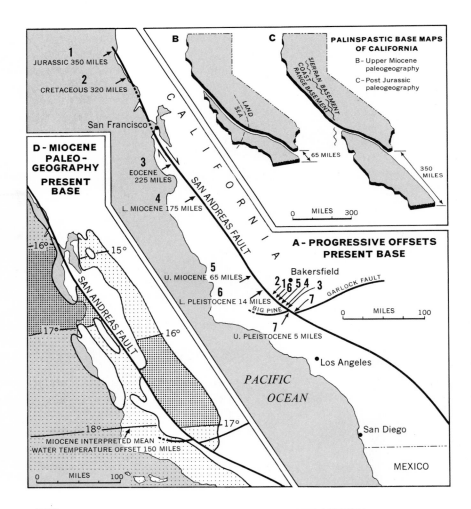

Fig. 21–28. *Palinspastic maps showing interpretations of movements along the San Andreas Fault, a right-lateral transcurrent fault in California. A.* Present map showing by numbers places that were interpreted as on opposite sides at several times in Mesozoic and Cenozoic time, with the approximate displacements in miles since each time.

B. and *C.* Palinspastic maps, showing the relationships across the fault in post-Jurassic and late Miocene (after M. L. Hill).

D. Miocene paleogeography, showing interpreted mean water temperatures based on studies of the organisms on the present base map; the western temperature bands have moved northwestward some 150 miles relative to the eastern (after C. A. Hall).

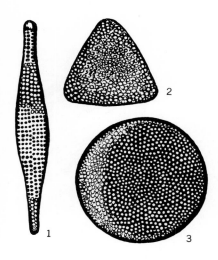

Fig. 21–29. *The Diatoms.* Diatoms from the Miocene of Maryland (after Maryland Geological Survey): (1) *Rhaphoneis*; (2) *Biddulphia*; and (3) *Coscinodiscus*.

Phylum Chrysophyta Diatoms are microscopic, one-celled or colonial, marine and fresh-water plants. They occur in such large numbers in the oceans that they form an important part of the plankton which is the source of food for marine life. Their cell walls consist of two siliceous, elaborately ornamented valves, one of which partly overlaps the other like the lid of a box. The cells exhibit either radial or bilateral symmetry. Because of their abundance, their valves accumulate on the ocean floor in the form of diatomaceous ooze. In the past they were dominant in rocks called diatomaceous earth—a material which is used for filtering and as an abrasive. The opal in diatoms is sufficiently unstable under pressure of depths of a mile or so that the rock alters to porcelaneous chert—a dense, glassy substance which hardly preserves the evidence of its origin. Such diatomaceous rocks and chert are prevalent in some of the Tertiary rocks of California. Diatomaceous earth also forms persistent beds in the Miocene in Maryland and adjacent states. Like radiolarian-bearing rocks, the abundance of diatoms requires siliceous waters such as are sometimes associated with volcanism. Probably the instability of the siliceous valves is responsible for the fact that diatoms are reported frequently only in Cretaceous and younger rocks. They are definitely known from Jurassic and quite doubtfully in beds as old as Silurian. In as much as they store food in the form of fatty oil, they probably were important contributors in the formation of petroleum deposits.

skull was rather low, and the cheek teeth had low crowns. From such a beginning the horses evolved through time, this evolutionary development being characterized by certain very definite trends. One trend was toward increase in size; another was in the direction of reduction of toes so that in the end the horses had a single functional toe, the third digit; still another trend was that of increase in the height and complexity of the cheek teeth, so that they became in the end very tall crowned, with very complicated patterns of loops and crenulations in the enamel band exposed on the worn grinding surface of each tooth. Such evolutionary trends were, of course, correlated with the evolution of the Tertiary environment, namely with the development of high plains and hard grasses. The development of tall-crowned, complex cheek teeth was an adaptation that enabled the horses to feed upon the grasses as they evolved. The reduction of the toes to a single hoof was an adaptation for running over hard ground. And the increase in size was among other things an adaptation for life in the open, away from the protection of low underbrush.

One can follow the development of these trends nicely through the successive stages of the North American Tertiary. In each stage the

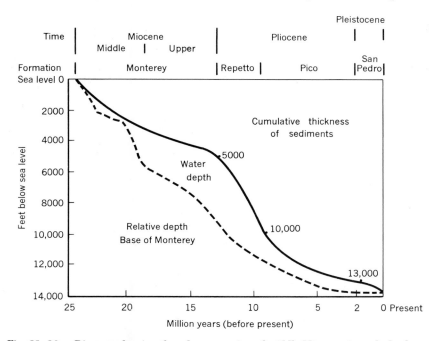

Fig. 21–30. Diagram showing the relative effects of deposition and subsidence on water depths in the Tertiary sediments in the Los Angeles Basin, southern California (after J. W. Durham). The solid line represents the cumulative thickness of sediments from the beginning of middle Miocene time; the broken line is below the solid line by an amount equal to the interpreted water depth, which reached a maximum of some 4000 feet in late Miocene and earliest Pliocene, as interpreted from the ecology of the foraminifera.

A

Fig. 21-31. *Tertiary rocks in Oregon and Washington. A.* Eocene conglomerate and sandstone, looking north along the Pacific shore near Coos Bay, Oregon (R. H. Dott, Jr.).

B. Early Eocene pillow lava in a sequence of an estimated 50,000 feet of Eocene in the eastern Olympic Mountains, northwestern Washington, near the Big Quilcene River; the lavas are associated with a variety of volcanic rocks, argillites and graywackes. In the northward view, the beds are nearly vertical with the tops of beds toward the right, eastward, as is suggested by the rather regularly convex surfaces normally upward in pillows; See Fig. 3-8 (R. W. Tabor, United States Geological Survey).

B

Fig. 21–32. *Tertiary volcanic rocks of the Columbia Plateau and Yellowstone Park. A.* Miocene or Pliocene lavas and associated fragmental volcanic rocks exposed along the canyon of the Snake River at Twin Falls, Idaho; westward view; see also the opening of Chapter 2.

B. Eocene sediments and volcanic fragmental rocks at Yellowstone Falls in Yellowstone National Park, Wyoming (A. Green); see also Fig. 21–18 of the Absaroka Mountains.

541

horses generally are larger, have higher crowned cheek teeth, and they tend to show more reduction of the lateral toes than in the preceding stages. Evolutionary development among the horses is strongly *oriented,* and by some students, especially those of an earlier scientific day, this story of evolution has been cited as an example of *orthogenesis,* that is to say, of evolution in a straight line in the direction of some definite goal. But the evolution of the horses, although oriented, is not orthogenetic, it is not straight-line evolution.

The usual presentation of the evolution of the horses (Fig. 21–36), as based upon modern evidence, shows the progress from the Eocene *Hyracotherium* (very commonly designated as *Eohippus*) through the Oligocene *Mesohippus,* the Miocene *Merychippus,* the Pliocene *Pliohippus* to the Pleistocene *Equus,* the horse of the ice age and of modern times. This is a perfectly valid line of evolutionary development and is an oriented line, with trends always in the direction of increased size, reduction of the lateral toes and increased complexity of the cheek teeth. But it is not an orthogenetic line. There was no "goal" toward which horses were evolving. Their development along the certain trends that have been outlined was the genetic response to the evolution of the environment. The evolution of the continent and consequently the vegetation of the continent determined that genetic mutations in certain directions would be favorable; thus the horses evolved along the lines that they did.

The progression that has been cited was only one of various lines of horse evolution, which were variously successful. It so happens that the line culminating in *Equus* is the one that has survived to the present day, but during much of the Cenozoic there were other lines of horse evolution parallel to the one with which we have been concerned. In some of these there was only a limited increase in size, in others only a limited increase in the heights and complexities of the cheek teeth, and in these same lines and others only limited reductions of the lateral toes. The evolution of the horses was indeed a rather complex and diversely branching example of development through time. It seems "straight line" to us merely because we commonly look only at one line of this extensive phylogenetic history.

CORRELATING WITH THE MARINE STAGES

Although the epochs of the Tertiary Period were originally defined upon percentages of molluscan species supposedly like those of today, this period of geologic history is more significantly subdivided and these subdivisions defined by its mammalian fossils. This introduces, however, a basic difficulty, namely the comparison of marine and continental

Fig. 21–33. Mount Shasta (14,162 feet) in the Cascade Mountains of northern California, a volcano built during eruptions in the Pliocene and Pleistocene periods (J. Muench). The lavas and fragmental rocks were dissected by radial streams. Glaciers, now somewhat reduced from their earlier limits, shaped and further deepened the valleys in the late Pleistocene.

Fig. 21–34. The west wall of Crater Lake, in the national park in southwestern Oregon, with Wizard Island in the middle of the lake. A high volcano, like mounts Shasta and Rainier, once rose over the locality; the top collapsed to form a caldera now occupied by the water of the lake. Crater Lake is 2000 feet deep, deepest of any lake in North America; it is about five miles across and the bluffs rise 2000 feet and more to elevations of nearly 9000 feet.

deposits and their contained fossil remains. By the nature of circumstances the marine sediments must be dated largely by fossil invertebrates, and fortunately the evidence of the foraminifera, which as fossils are the hard parts of protozoans, is admirably suited and much used for this purpose. Nevertheless, the fossil mammals remain for the evolutionist the most important of the Tertiary fossils.

THE AGE OF MAMMALS

The Tertiary was above all the "Age of the Mammals." Within the confines of this geologic period by far the greater part of mammalian evolution can be traced, from very primitive to almost modern types. Indeed, many lines of mammalian evolution developed within only a fraction of the extent of Tertiary time.

At the close of the Cretaceous period the mammals, though probably very numerous, were according to the fossil record all of small size and of limited taxonomic range. These were, so far as we know them, the rather rodent-like multituberculates, the marsupials related to the modern opossums, and the insectivores related to modern shrews and hedgehogs. With the opening of Paleocene time the mammals had expanded remarkably, although they were still for the most part of rather small size. But as the Tertiary Period progressed the mammals also progressed, and proliferated at ever accelerated rates to become exceedingly varied as time went on. Almost thirty orders of mammals arose and evolved during Tertiary history, to become the dominant animals of this age.

The wide adaptive radiation of the mammals on the land was equalled or even surpassed by the adaptive radiation of the teleost fishes in the waters of the earth, where certain groups of mammals also prospered, especially in the oceans. It will be remembered that these fishes went through an explosive phase of evolutionary development in early Cretaceous time, so that by the close of the Cretaceous Period fish faunas were essentially modern in their composition. This evolutionary progress continued through the Tertiary, to make the teleost fishes the most numerous of all vertebrates, both in species and in numbers of individuals. The Tertiary is in a sense an Age of Teleosts as well as being an Age of Mammals.

OTHER VERTEBRATE ANIMALS

And during this period the birds also evolved as the dominant animals of the air, while many of them became well adapted to land and water as well. Birds had reached essentially their modern stage of evolution by early Tertiary time; they did not enjoy the long continued progress

Epoch	European	North American (Vertebrates)
PLEISTOCENE		Rancholabrean
PLEISTOCENE	Sicilian	Irvingtonian
PLIOCENE	Calabrian	Blancan
PLIOCENE	Astian	Blancan
PLIOCENE	Plaisancian	Hemphillian
PLIOCENE	Pontian	Hemphillian
MIOCENE	Sarmatian	Clarendonian
MIOCENE	Tortonian	Barstovian
MIOCENE	Helvetian	Hemingfordian
MIOCENE	Burdigalian	Hemingfordian
MIOCENE	Aquitanian	Arikareean
OLIGOCENE	Chattian	Whitneyan
OLIGOCENE	Rupelian (Stampian)	Orellan
OLIGOCENE	Tongrian (Sannoisian)	Chadronian
EOCENE	Ludian	Duchesnean
EOCENE	Bartonian	Uintan
EOCENE	Auversian	Bridgerian
EOCENE	Lutetian	Bridgerian
EOCENE	Cuisian	Wasatchian
EOCENE	Ypresian (Sparnacian)	Wasatchian
PALEOCENE	Thanetian	Clarkforkian
PALEOCENE	Thanetian	Tiffanian
PALEOCENE	Montian	Torrejonian
PALEOCENE	Montian	Dragonian
PALEOCENE	Montian	Puercan

Fig. 21–35. Comparison of the stages of the standard section of the Tertiary in Europe with those applied to the nonmarine, vertebrate fossil-bearing succession in the interior of the United States.

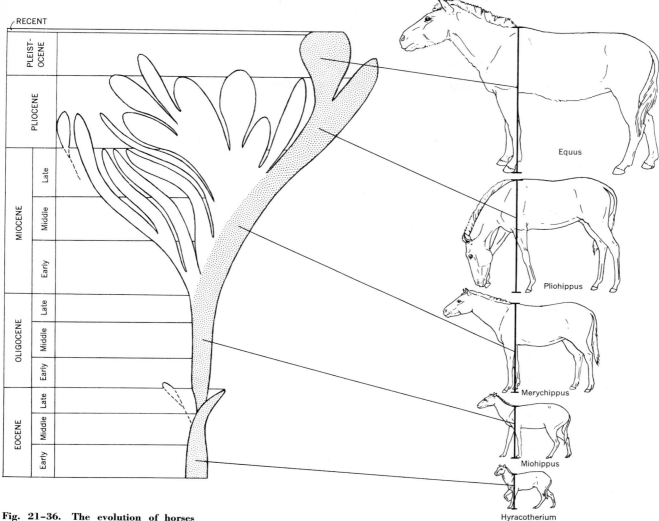

Fig. 21–36. The evolution of horses during Cenozoic history. In its totality, it was a many-branched phylogeny, showing wide adaptive radiation through some fifty million years or more of earth history. Modern horses, zebras, and asses are the end products of one particular trend, shown by the shaded part of the phylogeny. This trend was marked by many correlated adaptations, of which a progressive increase in size, here demonstrated by shoulder heights, is particularly apparent.

545

in form and function so characteristic of mammalian evolution. But these vertebrates, so highly evolved in structure by the opening of Cenozoic history, did become widely adapted in form and in behavior through the course of the Tertiary. Their evolutionary history during the Cenozoic was equally as successful as that of the mammals and of the teleost fishes.

As for the amphibians and reptiles, the Tertiary Period was the time when they became established in great specific variety even though limited in their larger taxonomic expressions. The Tertiary, like the present, was an age of frogs and toads, of lizards and snakes. The turtles and the crocodilians were also successful, as they have been up to the present time, but were never so numerous in taxonomic diversity as the lizards and snakes. The rhynchocephalians can be regarded as survivors from the Mesozoic, just managing to hold on through Tertiary time.

From this it can be perhaps seen that the Tertiary record of vertebrate life is a rich one, so rich indeed that any detailed consideration of it would necessarily run to much greater length than can be accommodated in a book such as this. More will be said, however, about Tertiary vertebrates in Chapter 24.

22
The Tertiary of the Gulf and Atlantic Coasts

Oil well in the Choctaw Field, Atchelefelaya Swamp, Plaquemine Parish, Louisiana (Standard Oil Company of New Jersey).

The Tertiary history of the western interior and Pacific Coast is one of mountain building and its aftermaths. In contrast, the record of the Gulf and Atlantic coasts is one of warping and subsidence within a region that had not been active orogenically since the late Paleozoic and early Mesozoic. Much of the record is buried, reached only by wells that penetrate apparently horizontal, but generally slightly dipping loosely consolidated sediments along the margin of the sea. The present is but a continuation of conditions that have prevailed throughout the Cenozoic, so that one can reconstruct past events in terms of present geography.

THE GULF COAST

The Gulf Coast is a region (Fig. 18–7) that has had great subsidence within the last one hundred and fifty million years, the later Mesozoic and Cenozoic eras. Wells have penetrated almost five miles, and geophysical evidence suggests that the Tertiary rocks are no more than seven or eight miles thick (Fig. 22–1). The sediments are principally poorly consolidated sands, silts, and clays. Dips are exceedingly low

Fig. 22–1. Structure section through the younger Tertiary rocks on the East Texas–Louisiana border, interpreted from the wells shown by vertical lines (after G. C. Hardin, Jr.). The successive solid lines connect points where well cuttings or samples contained similar assemblages of fossil foraminifera, or of a similar distinctive lithic marker bed. For example, the top of the Oligocene Series is interpreted as being above a zone with a genus *Heterostegina*. The section also shows the maximum depth of penetration by wells.

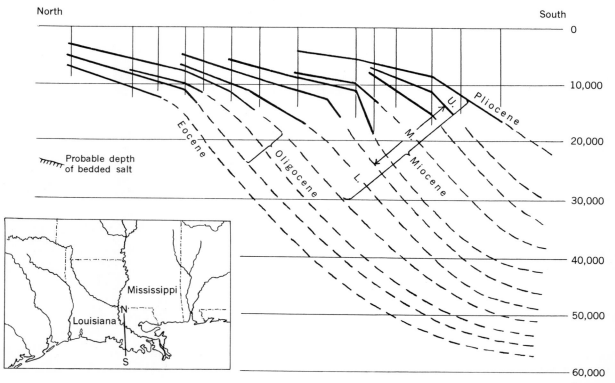

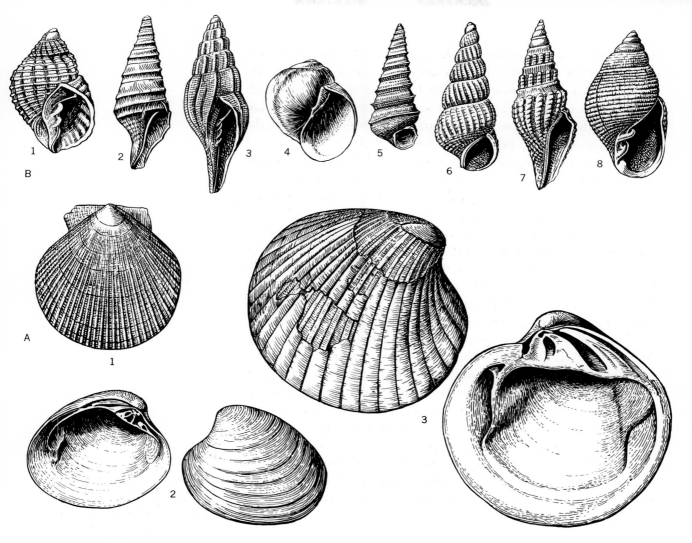

Fig. 22-2. *Invertebrate life of the Tertiary.* *A.* Pelecypoda. Eocene pelecypods from Maryland and Virginia. (1) *Pecten* of the subgenus *Chlamys*; (2) Exterior and interior of left valve of *Dosiniopsis*; and (3) exterior and interior of right valve of *Venericardia*.

B. Gastropoda. Tertiary gastropods from Maryland and Virginia. (1) *Pleurotema* and (2) *Cancellaria* from the Miocene. (3) *Mitra*; (4) *Lunatia*; (5) *Turritella*; (6) *Turbonilla*; (7) *Pleurotoma* and (8) *Tornatellaea* from the Eocene.

The series of the Tertiary were originally three, named in reference to the increasing similarity of the molluscs in the successively younger fossil faunas to those living today. Thus, the Eocene originally was considered to have 3½ per cent of the shells of species that are still living, the Miocene, 17 per cent, and the Pliocene, 36 to 95 per cent. Of course,

such a figure is affected by many variables: for example, individual scientists differ on whether a particular fossil shell is enough like the recent related one to be considered of the same species. The original studies were based on collections in Britain, France, and Italy so other regions may have differing factors controlling how many similar descendants were retained. The Tertiary shells are very much like those of the present beaches, whatever the measure of likeness.

The Tertiary Mollusca are dominantly of the classes Pelecypoda and Gastropoda, the clams and snails, though Cephalopoda and other molluscs are occasionally found. As has been stated, the Pelecypoda are bivalved molluscs having two similar valves, symmetrical with respect to the plane of junction. The importance of pelecypods in reaching

certain stratigraphic and biologic interpretations was discussed with respect to the Carboniferous rocks.

Gastropoda, the snails, are nearly all coiled molluscs, commonly in the form of a twisted spire but occasionally in a plane. The latter resemble some cephalopods but lack the internal partitions or septa forming the chambers of cephalopods. Gastropods live in marine and fresh water, and even on the land. They are among the oldest known fossils and are found with trilobites in early Cambrian rocks. Although they have been frequent and useful from time to time (Fig. 11–19), they became a most important means of classifying rocks in time in the Cenozoic.

The similarity of early Tertiary fossil clams and snails is illustrated by examples of the Eocene of Maryland (after Maryland Geological Survey).

and exposures are limited to such places as the banks of streams where erosion actively cuts into the deposits. When these exposures are plotted on maps, and classified in time by means of their fossils (Fig. 22–2), it is seen that the dip is gently seaward, for the older rocks are exposed farther inland than the younger. When wells are drilled, the elevations of correlated beds decline toward the Gulf. Correlation of Tertiary sediments is accomplished by comparing lithologies as sampled in cuttings and cores, by the fossils in such samples, and by several geophysical prospecting techniques. To a great degree, the comparison of the rocks is by electric logging—sliding electrodes down along the walls of each well, recording the electric properties such as those of resistivity to the passage of current—and "self-potential," the measure of the fluids in the rock as compared to those in the muds (Fig. 22–3). These physical characters can be interpreted in terms of lithologies on the basis of experience with similar measurements in known rocks. Other physical properties, some of the measurements of the radioactivity of the sediments and of the effect of directing radioactive particles into the rocks are also applied within wells. The logs are a continuous record, which, when plotted to uniform scale, can be compared directly with those of other wells. Similar properties on the log reflect similar properties of the sediment, correlation of likeness that may suggest time-equivalence (Fig. 22–4). In exploration prior to drilling and afterward, the seismograph is of great importance; we have learned of this sort of use in the study of the coastal Cretaceous sediments that are beneath the Atlantic. Some recent developments permit the production of a record that is essentially like a subsurface structure section (Fig. 22–5).

MICROFOSSILS AND ECOLOGY

The microfossils, particularly foraminifera, are recovered with the fine particles in the cuttings (Fig. 22–6). Experience has shown that some of the foraminifera are quite constant in their appearance within certain parts of the sequence, thus useful in recognizing such strata in new wells. Foraminifera of the same kinds as those found in the wells live and die today on the bottom of the Gulf. Collections from the sediments on the floor of the sea show the different kinds to appear as bands on the bottom, each form having upper and lower levels of tolerance of such physical and chemical conditions as temperature and salinity. Thus, a faunule from a shallow-water environment may have entirely different fossils than one from deep water (Fig. 22–7). Similar interpretations by foraminiferal analysis are applied in California (Fig. 21–30). The same genera and some of the same species in the Tertiary

Fig. 22–3. Electric log of a drilled oil well and explanations of the records (after H. Guyod). Depths are recorded on the left. The left irregular line represents the millivolts of current induced by the relative salinity of the water in the rock to that of circulating water in the well. The right-hand line shows the resistance to passage of current between two electrodes that can be placed at variable distances apart as they descend the wall of the well bore. The design of probing equipment and the interpretation of the results are professions that employ skilled scientists; the records are a principal means of study of petroleum prospects from subsurface records.

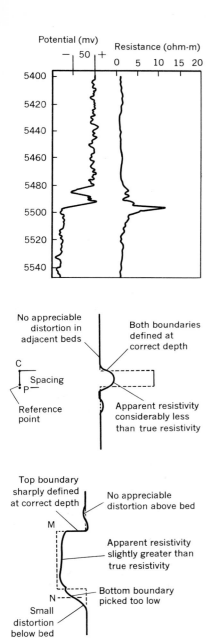

Potential (mv)

Resistance (ohm-m)

No appreciable distortion in adjacent beds

Both boundaries defined at correct depth

C

Spacing

P

Reference point

Apparent resistivity considerably less than true resistivity

Top boundary sharply defined at correct depth

M

No appreciable distortion above bed

Apparent resistivity slightly greater than true resistivity

N

Bottom boundary picked too low

Small distortion below bed

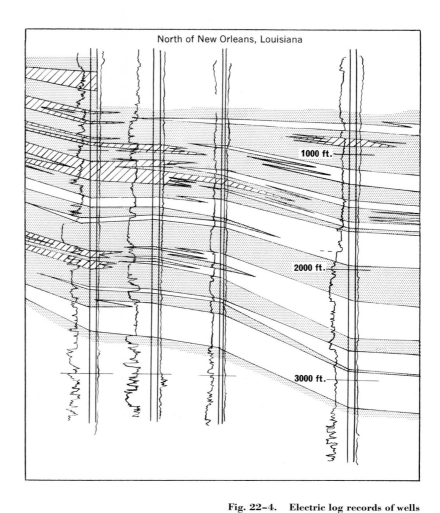

North of New Orleans, Louisiana

1000 ft.

2000 ft.

3000 ft.

Fig. 22–4.   Electric log records of wells penetrating the Pleistocene sediments in the Gulf Coast of Louisiana (after W. H. Akers and W. J. J. Helck). The self-potential and resistivity records in each log can be interpreted in terms of the sediment and its fluid contents. Thus the correlation of wells is essentially a method of lithic correlation that pertains to time insofar as the lithic units are time-stratigraphic.

sediment thus can be useful in judging whether correlation in time is warranted in two wells having the same fossils but also in giving guidance on whether a sediment is near shore in origin, or was laid far off shore. Many of the sediments are non-marine; they have, for the most part, very few recognizable organic remains within them, but within recent years they have yielded a moderately rich harvest of fossil mammals. Some use is made of the quite resistant plant fossils, such as spores and pollen (Fig. 22–8).

LIMITATIONS IN DIRECT KNOWLEDGE

Wells are numerous, for this is a very important region in the production of petroleum. The reserves of petroleum in the rocks of the Gulf Coast of Texas, Louisiana, and Mississippi have been estimated as about 25 billion barrels of 42 gallons each (36 imperial gallons each) and those of coastal Mexico as another 5 billion barrels. Of this total about half is Tertiary, the rest principally Cretaceous, including the five billion barrels in the East Texas field (Fig. 18–11). It might seem that the result of the great economic importance would be great knowledge of the sediments throughout a broad region. But the youngest rocks, the Pliocene, are exposed only in a belt near the Gulf, and hence they would be penetrated completely only between that belt and the coast. The Paleocene sediments which outcrop farthest inland should be preserved under the greatest area, fully in wells drilled from outcrop of Eocene and younger strata. Successively older beds would be missing from the top as one proceeded from the Eocene to the Cretaceous on the surface, until only the basal Paleocene bed would remain (Fig. 22–1). The full section of the Paleocene will be known then from its surface contact with the Eocene as far south as wells are drilled through it. This is not a great breadth. The deepest wells are about five miles, but few approach that depth, and those only in limited areas. The seaward dip of the Paleocene carries its base rapidly deeper and deeper. Within twenty or thirty miles, it goes so deep that wells do not pass through it. So our knowledge of each series is limited to the exposures in the outcrop belt and through wells in a belt of twenty or thirty miles toward the Gulf (Fig. 22–9).

CHANGES IN SECTION TOWARD THE GULF

Each stage and series increases in thickness toward the Gulf. Within the limited belt of observation, this tendency is nearly invariable. As a result, oldest Tertiary formations increase in dip as they become more deeply buried. They are at a lower elevation than would be the case if the surface dip continued unchanged to the Gulf. For the most part,

Fig. 22–5. Continuous reflecting profiling of structure beneath the Gulf of Mexico continental shelf off Louisiana (from P. Oxley, Signal Oil Company). The records are interpreted as showing a mass of salt that has invaded the subsurface Pleistocene sediments.

Continuous reflecting profiling is a method of exploring the subsurface by recording the intervals of time taken by acoustic waves to reach successive interfaces at the bottom of the sea and within the underlying rocks, and, reflecting, to return to hydrophones at the surface; the principle is like that of echo sounding, so widely used by ships at sea to determine water depths. There are many methods of producing sounds, such as explosion of charges, of gas mixtures, and discharge of electric arcs (sparkers); each produces waves of differing characteristics. Waves of higher frequency record finer details, but do not penetrate as deeply, so are not as useful to record deep features. Interpretations require skill and experience, for there are many reflections that can produce false bands suggesting surfaces or interfaces that do not exist.

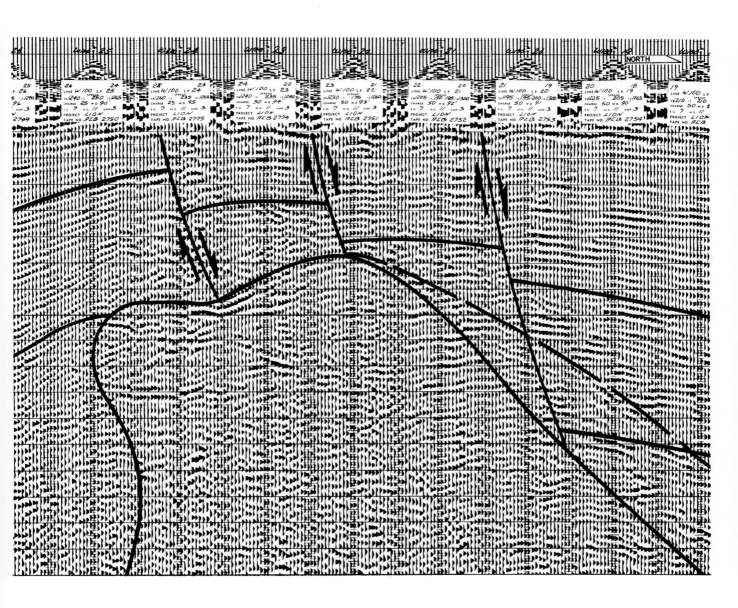

they thicken at an increasing rate of downdip as they cross a flexure. The increases also take place abruptly along fault lines and zones that are more or less parallel to the Gulf shore. In an extreme case, the thickness of a unit of a few hundred feet increases as much as seven times across a fault. This indicates that the down dip block was subsiding more rapidly than the updip block as the deposits were forming. The increasing rate cannot continue indefinitely; the rate must be reversed, for otherwise the volume would become limitless! The sediments become finer textured toward the south, a statement of an average condition rather than a description of a single unit. Formations tend to be non-marine at the outcrop, passing from stream and delta laid into offshore facies. To say that the stream-laid sediment becomes finer gulfward is a generalization that could be demonstrated only by means of large numbers of samples. As the sediments pass from stream environment into marine, the change of agents tends to result in sorting of coarse from the fine, the former remaining at the shore and the latter drifting seaward; thus the average texture may actually increase, for removing the finer particles from a mixture results in an average that is coarser than that of the mixture (Fig. 22–10).

Coarsening does not take place uniformly, but as extensive wedges of non-marine rocks penetrating between oppositely faced wedges of marine, with bands of shallow and brackish water sediments facing them (Fig. 22–11). Each penetrating non-marine tongue represents a retreat

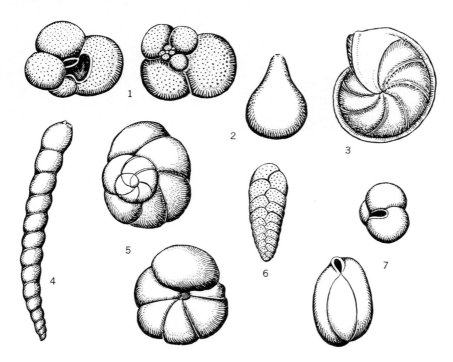

Fig. 22–6. Foraminifera from the Pleistocene of California. (1) *Globigerina*, (2) *Oolina*, (3) *Robulus*, (4) *Dentalina*, (5) *Globorotalia*, (6) *Bolivina*, and (7) *Globobulimina*; each is enlarged about ten times (after J. J. Galloway and M. Murrey).

Fig. 22–7. **Diagram to show the changing assemblages of foraminifera that live in the bottom of the Mississippi Delta and the Gulf of Mexico at the present time (after S. W. Lowman).** The names on the lower diagram represent different genera or families of foraminifera, the proportions of each sort being shown by the proportion of each vertical line drawn at successively deeper water points. Thus *Ammobaculites* is prevalent in waters of ten feet or so depth, but Buliminidae, in waters of 1000 feet. Knowledge of these ecologic ranges in present foraminifera permit interpretation of probable sites of life of ancient similar forms. Note how strongly the ecology controls the distribution of the organisms, all of which live at the same time, the present; this emphasizes the difficulties inherent in carrying correlations by study of fossil organisms.

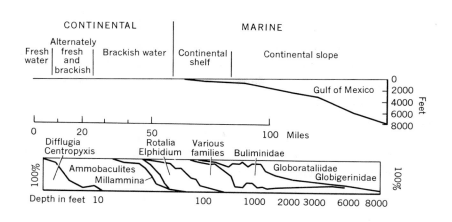

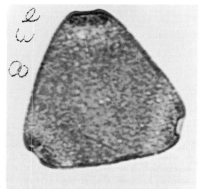

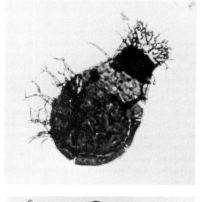

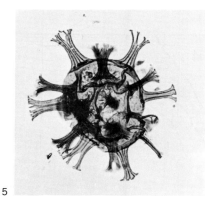

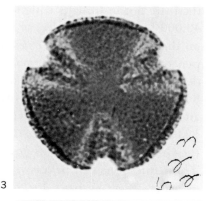

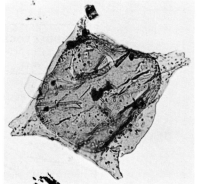

Fig. 22–8. **Several kinds of microfossils from Paleozoic and Cenozoic rocks. These forms have been found useful in correlating sedimentary rocks (S. A. Levinson, Humble Oil and Refining Company).**

Chitinozoans: (1) *Angochitina* and (2) *Sphaerochitina*, 350 times enlarged, from middle Deonian shale, Arkona, Ontario.

Pollen: (3) *Porocolpopollenites* and (4) *Triatriopollenites*, 1000 times enlarged, from the Paleocene Wilcox Group, Sabine Parish, Louisiana.

Hystrichosphaerid: (5) *Hystrichosphaeridium*, 500 times enlarged, from the Eocene Claiborne Group, Clarke County, Alabama.

Dinoflagellate: (6) *Wetzeliella*, 500 times enlarged, from the Eocene Claiborne Group, Clarke County, Alabama.

of the sea, because of eustatic change, fall in sea level, or the building forward of the non-marine sediments into a sea with stationary level. Each extension of marine beds shows rise in the sea level, or subsidence at a greater rate than the sediment-building. In each marine advance, the sediments gradationally overlap the stream laid deposits, and the latter become more extensive with each retreat. In Texas, only the Paleocene and Eocene are marine at their outcrops, but thick wedges of marine Oligocene and Miocene enter the downdip, partly as a result of non-marine sediments passing into marine, and partly as deposits in the sea at times of non-deposition in the place of the present outcrop.

THE GULF COAST GEOSYNCLINE

The Tertiary deposits have been considered as laid in a great trough or geosyncline lying beneath the coast and continental shelf, the Gulf Coast Geosyncline. Unlike some of the geosynclines we have considered, these are not in troughs complementing adjoining highlands. Perhaps 40,000 feet, nearly 8 miles, of Tertiary sediments are in the sequence on the shore in eastern Texas and Louisiana, a figure based on the penetration to the Eocene, and judgment of the thickness below the deepest well. Thus the base of the Paleocene may be as much as 8 miles below sea level at the Gulf margin (Fig. 22–12) but this is only an estimate. The Sigsbee Deep of the Gulf of Mexico is only about 12,000 feet deep 300 miles to the south (Fig. 22–13). How far out into the Gulf do thick sediments go? What is the shape of the geosyncline? These questions can be answered qualitatively from the geophysical evidence; there is basis for some conclusions. The thickness must

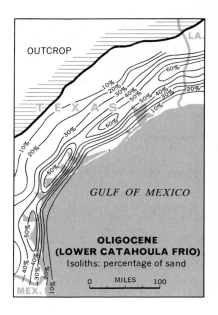

Fig. 22–10. Map showing the percentage of fragments of sand-size in the Oligocene of the Texas Gulf Coast (after J. A. Waters, P. W. McFarland and J. W. Lea). The sediments to the northward are poorly sorted sands and silts. Sorting along the shore zone caused the finer materials to be taken into suspension and transported gulfward, so that the remaining sediment is more sandy than that on either side; the average size particle in a mixture can be increased if the finer particles are removed and transported elsewhere.

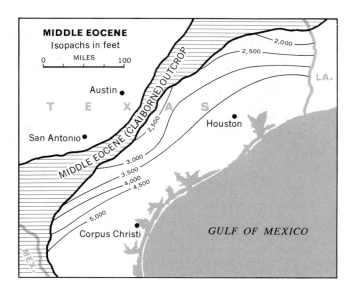

Fig. 22–9. Map to show the limited belt in which rocks of one stage are fully penetrated. The belt of outcrop of the middle Eocene Claibornian Stage in Texas, with the isopachs drawn within the belt in which wells passed through both the top and the base of the time stratigraphic unit. Seaward, the base of the Claibornian is so deep that its thickness is not determinable; landward, the upper part is reduced by erosion until it is no longer preserved (after D. Malkin and D. J. Echols).

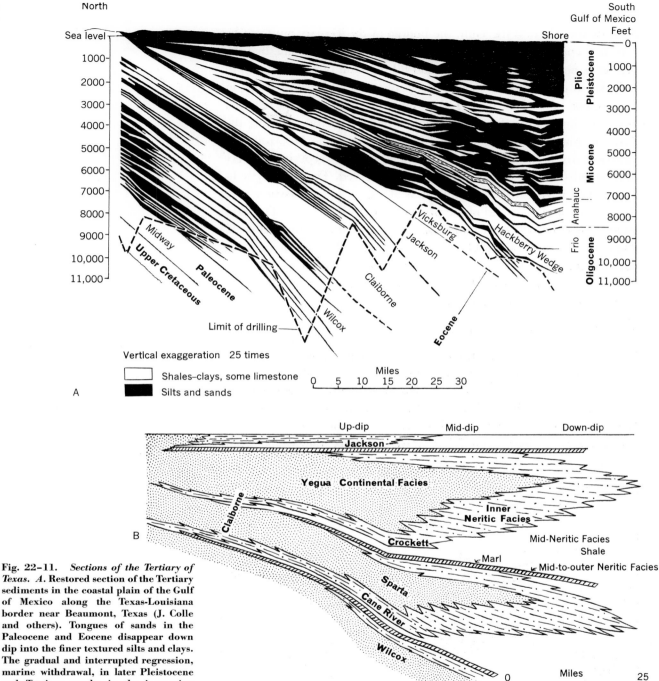

North   South
                                                                                                           Gulf of Mexico
                                                                                                                    Feet

Sea level   Shore

A

Vertical exaggeration 25 times

☐ Shales–clays, some limestone
■ Silts and sands

Miles
0 5 10 15 20 25 30

Labels on section A: Midway, Upper Cretaceous, Paleocene, Wilcox, Limit of drilling, Claiborne, Jackson, Vicksburg, Eocene, Hackberry Wedge, Frio, Anahauc, Plio Pleistocene, Miocene, Oligocene

B

Up-dip   Mid-dip   Down-dip

Jackson

Yegua Continental Facies

Claiborne

Inner Neritic Facies

Crockett

Mid-Neritic Facies
Shale

Marl

Mid-to-outer Neritic Facies

Sparta

Cane River

Wilcox

Miles
0   25

Fig. 22–11. *Sections of the Tertiary of Texas. A.* Restored section of the Tertiary sediments in the coastal plain of the Gulf of Mexico along the Texas-Louisiana border near Beaumont, Texas (J. Colle and others). Tongues of sands in the Paleocene and Eocene disappear down dip into the finer textured silts and clays. The gradual and interrupted regression, marine withdrawal, in later Pleistocene and Tertiary results in the increasing prevalence of sands in the higher parts of the section. Downwarping of the Gulf is shown in the divergence in the time-stratigraphic units and increasing dip with depth. Wells have penetrated nearly all of the sandy facies, so that their character and thickness are fully controlled.

B. Diagram of facies in the Tertiary of the Gulf Coast of Texas (after S. W. Lowman). The lines on the diagram are symbolic of the facies distributions; there is progressive gradation across the positions of the lines, and the gradation is complex, with far-penetrating intertonguing of the sedimentary types. Yet within the intricacies of local details, there are larger general patterns such as are represented in the diagram.

557

decrease quickly into the Gulf; the bottom of the Tertiary must not be very far beneath the floor of the Sigsbee Deep. If we assume that the base does not rise, the Tertiary would fill the Gulf of Mexico basin to a thickness of 6 miles or more, from a plane at 40,000 feet depth to the present bottom surface. The volume of sediment thus required would be impossible; for if that volume had been eroded during the Tertiary from the interior of the continent, all of the Mesozoic and Paleozoic rocks would have been eroded away except in the very deepest subsided areas; yet they are still preserved. The Tertiary sediments must have a maximum thickness along an axis somewhere near the position and the trend of the present shore (Fig. 22–14). The sediment thins gulfward, most rapidly at the edge of the continental shelf. Thus, the Tertiary of the Gulf Coast seems to fill a geosyncline that parallels the coast, perhaps going into the Gulf in Louisiana, for there is diminishing thickness of Tertiary south of an axis trending from northern Louisiana toward central Georgia.

THIN CRUST BENEATH THE GULF

Submarine geophysical studies by seismograph have shown that the base of the crust, the Mohorovičić Discontinuity, is only about 12 miles below sea level at the Sigsbee Deep and thus some 10 miles below the sea as compared to 20 miles or so at the shore. The crust in the Sigsbee Deep is of no more than half a mile of sediments some of them presumably Mesozoic or older. Even if they all were Tertiary, they would be far

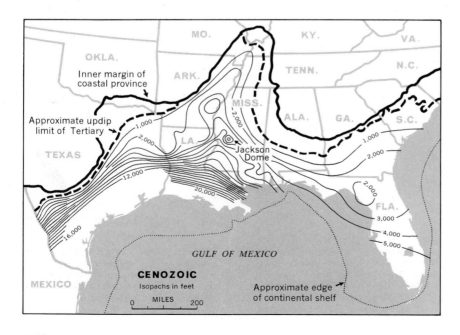

Fig. 22–12. Map giving the thickness of Cenozoic deposits along the northern margin of the Gulf of Mexico (after G. A. Murray and Lyman Toulmin, Jr.). The northern limit is that of preservation, the southern that of penetration. Note the Jackson Dome in western Mississippi.

thinner than those on the coast, and their base is at a higher elevation than the base of the Tertiary at the coast. They probably are partly Mesozoic, for there are salt dome-like structures such as developed from salt, perhaps of Jurassic age, on the coast, as will be discussed.

SALT DOMES

The Gulf Coast is the site of a great many salt domes (Fig. 22–15), features that are of limited area, but of great interest in the light they throw on fluid mechanics within the earth, and of great economic importance as the source of rock salt, sulphur, and petroleum that has

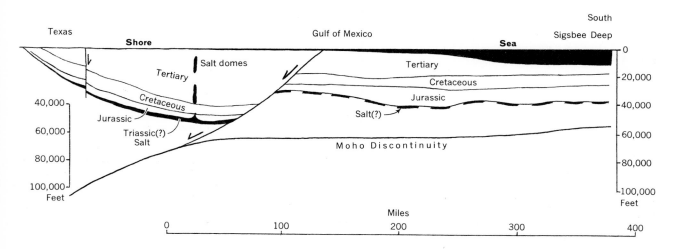

Fig. 22–13 (above). An interpretation of the present structure of the northern Gulf of Mexico and Texas coast (after J. L. Worzel, J. L. Shurbert, and P. Lyons). Salt domes rise from layers of salt that may be Jurassic, that possibly extend beneath the breadth of the Gulf. The Gulf Coast Geosyncline, a term applied to the belt along the coast, is interpreted as having subsided on a fault from the more stable floor of the Gulf, with the Mohorovicic Discontinuity or Moho becoming shallow beneath the Gulf, as shown in geophysical studies.

Fig. 22–14. Map of the coast of the Gulf of Mexico showing an interpretation of the thickness of Paleocene and Eocene sediments along the Gulf of Mexico from Texas to Alabama (after M. Kay and M. Bornhauser). Since much of the area lies beyond the limit of penetration in wells, the thicknesses are hypothetical in such areas.

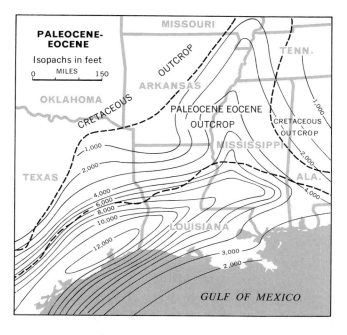

559

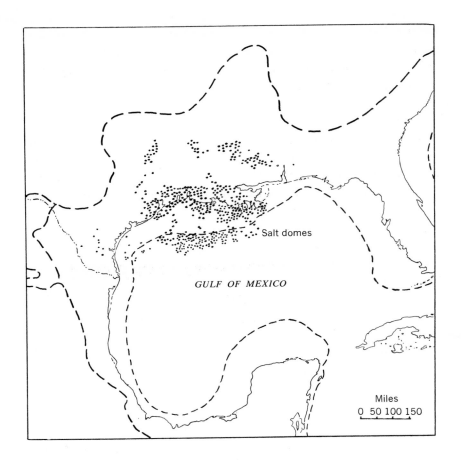

Salt domes

GULF OF MEXICO

Miles
0 50 100 150

Fig. 22–15. The distribution of salt domes along the Gulf Coast of the United States (after *Tectonic Map of the United States*). Most of the domes are in Tertiary sedimentary rocks, and contain rock salt or halite that has risen through the sediments from source beds that underlie the upper Jurassic. The salt, being of lower specific gravity than the overlying sediments, tends to flow through them toward the surface somewhat as a layer of grease may rise in a blob through water if the temperature rises so as to make it fluid.

Fig. 22–16. *Salt domes that yield oil from Tertiary rocks on the Gulf Coast of Texas and Louisiana* (from W. H. Bucher, Humble Oil and Refining Company). *A.* High Island Dome, Galveston County, Texas, along the shore of the Gulf of Mexico; the structure is evidenced by roads, and wells that have been drilled on the margins of the dome.

B. Avery Island Dome, Iberia Parish, Louisiana; although artificial drainage outlines the dome, there are contrasts in vegetation produced in the gently dipping surrounding bands of sediments. The pictures are made from an assembly of vertical air photographs into an aerial mosiac.

A

B

accumulated along their flanks. There is little evidence of these structures on the surface; occasionally, one flying over a coastal region in a plane might see a circular pattern of vegetation (Fig. 22–16) and exceptional drainage. In a few cases, quarries or mines opened in the center of such a structure revealed rock salt, quite pure and with intricate folds (Fig. 22–17). Drilling and geophysical methods show that these masses of salt are somewhat circular (Fig. 22–18). The salt being lighter than the adjoining sediment, its mass exerts less pull on a gravimeter than the heavier rocks. Waves reflected from its surface can be recorded on seismographs and interpreted. Where the salt is not exposed on the surface, the frequent situation, a cap is of carbonate rock and gypsum, and frequently sulphur is on the flanks. Hot water pumped into the latter melts the sulphur, which is pumped to the sur-

100 Feet

Fig. 22–17. Sketch of a room in the salt mine in Grand Saline Dome, east of Dallas in north central Texas (after R. Balk). Such salt, rising gradually through somewhat heavier sediments, shows structures that related to its flow, suggesting the way in which other metamorphic rocks developed their folds of similar character. The ceiling of the room is about one hundred feet above the floor.

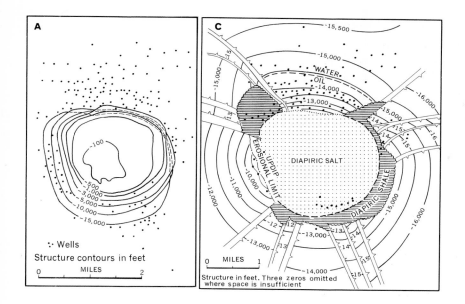

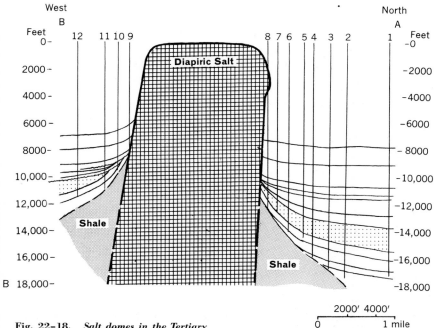

Fig. 22–18. *Salt domes in the Tertiary of the Gulf Coast of Texas. A. and B. Structure contour map on the top of the salt core and structure section along lines A and B of the Weeks Island salt dome.*

C. Structure contour map on a sand stratum that has been displaced by faults radiating from the salt dome; some of the blocks have been dropped or raised relative to others; oil has been trapped in the rocks on the north side of the dome (after G. I. Atwater and M. J. Forman).

face to crystallize on cooling. Geophysical exploration shows that the salt masses extend far into the subsurface, with rather vertical or overhanging walls (Fig. 22–18*B*). The Tertiary sediments approaching the domes are in some cases simply cut off by the salt. But in others, they thin as they approach, or even have a thickened circular ring with thinning within it; frequently broken by high-angle faults (Fig. 22–18*C*).

Salt is of lighter gravity than the surrounding sediments, which are generally fine clays and silts near the Gulf, rocks that are not hardened and are rather plastic when wet. A mass of salt buried in heavier muds tends to rise, just as a wooden ball placed under water comes to the surface. Far below the present ground at the base of the upper Jurassic sediments is a formation of bedded rock salt that had variations in original thickness; the age is uncertain and is referred to as Jurassic, Triassic, or even Permian. After it had been buried by younger deposits, elevations rose where the salt was thickest, gradually piercing the overlying muddy sediments in the late Mesozoic and Tertiary coastal geosyncline. Salt within the formation flowed toward the rising node. Some of the salt columns rose only part of the distance to the surface, but others have reached the present coastal plain. The salt, exposed in shallow mines, reveals the flow structures resembling those seen in deformed marbles and gneisses (Fig. 22–17). The salt and the muds were so viscous that the movement was very gradual.

As the salt invaded the sediments, they subsided into the surrounding source of flow, producing encircling troughs. Anticlines and synclines may also have formed as the rising column forced the sediments aside. The stresses also resulted in the formation of a network of high-angle faults, most of them of small throw (Fig. 22–18*C*). Oil and gas

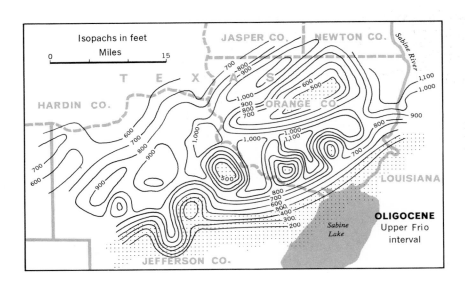

Fig. 22–19. Isopach map showing the thickness of a stratigraphic interval in the Oligocene near Beaumont, Texas (after F. Reedy). The fact that the sand is thinner around the domes than in the intervening areas indicates the presence of rising mounds at the time of deposition. Spindletop, the dome in the center of the map, was the site of a great gushing oil well in 1901 that initiated the industry in the region; the field has produced nearly 150 million barrels of oil since then.

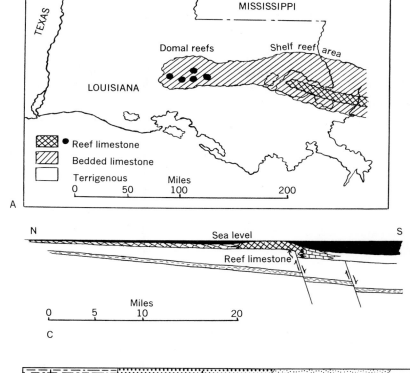

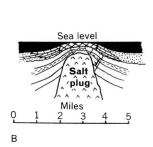

Fig. 22-20. Tertiary reef limestones in Louisiana (after G. I. Atwater and M. J. Forman). A map and sections of the distribution of reefs and associated limestones on salt domes and on flexures along fault scarps in the Oligocene rocks of southern Louisiana.

Fig. 22-21. *Eocene of Alabama and Florida.* *A.* Paleogeographic map showing the distribution of the bottom sediments of the late Eocene (Jacksonian) along the Gulf Coast of Alabama and northern Florida, which at the time was a broad shoal or bank on which carbonate sands and muds accumulated (after A. Cheatam).

B. Late Eocene (Jacksonian) Ocala Limestone from near Ocala, Florida, having specimens of the large orbitoid foraminifer *Lepidocyclina*; the specimen is somewhat reduced in size (K. Waage, Peabody Museum, Yale University).

C. Similar limestone from the core of an oil well in Libya, containing specimens of the foraminifer *Nummulites*. Similar Eocene rocks in Egypt were quarried in ancient times to construct the great Pyramids along the Nile (American Museum of Natural History and American Overseas Petroleum Company).

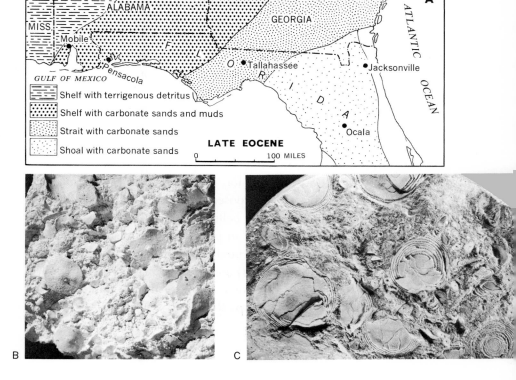

565

have been trapped in reservoirs in surrounding anticlines, against the faults, and in dome structures above the top of the rising columns, as well as in the reservoirs against the salt columns, perhaps coming from original organisms such as were the source of oil, or possibly from the alteration of calcium sulphate, gypsum, such as is present in some of the sediments. The thickening and thinning of sediments in concentric bands around the margins of the columns (Fig. 22–19) show that rise of salt has progressed since early Tertiary and presumably longer; Paleocene and older beds are too deep to be penetrated in areas of abundant domes. The presence of dome-like structures in the bottom of the Gulf of Mexico, suggesting salt domes, implies that the area has salt widely spread beneath the Gulf; salt precipitated from marine water in basins of restricted ingress in several prior geologic periods, and it may be that the Gulf area was such a basin in the early Mesozoic. Salt domes are also prevalent in the Tertiary in an area in Vera Cruz, Mexico, and a few are in Mississippi.

REEFS AND SHOALS

The gentle seaward dip of the Tertiary is interrupted by structures additional to the salt domes. In discussing the Cretaceous, the Jackson Dome in Mississippi was shown to be on an intruded structure of early Cretaceous age that was eroded and covered by a reef limestone platform in latest Cretaceous. The Paleocene and Eocene rocks are also thin on the structure as compared to its flanks, about twelve hundred feet on the dome corresponding to several thousand feet away from the dome. Reefs also mantled the platform on the Monroe Arch in northeastern Louisiana. The Sabine Arch (or "Uplift," though it is a subsiding structure in that its basement sank progressively lower than the level of the sea) in northwestern Louisiana is a much broader structure that is on the flank of the East Texas or Tyler Basin; the structure was instrumental in producing the Cretaceous stratigraphic relations forming the reservoir for the great East Texas oil field (Fig. 18–11). The oldest rocks exposed on the platform are Paleocene, margined on the geologic map by belts of younger Tertiary rocks; thus differential movements continued in post-Paleocene Tertiary. In addition to the reef limestones that formed platforms in the late Cretaceous on the Jackson Dome (Fig. 18–14) and Monroe Platform, reefs formed in the Tertiary on the crests of some of the rising salt domes, and on the margin of the submarine scarps that faced the subsiding downdip slope into the Gulf (Fig. 22–20).

Fig. 22–22. Andros Island, the Bahamas (N. D. Newell). The Bahama Islands are low land areas on a bank of calcium carbonate mud and sand extending some seven hundred miles from fifty miles east of Florida to north of Haiti. Andros, the largest island, lies near Florida, and is surrounded by shallow seas that serve as an excellent laboratory for the study of the deposition of carbonate rocks, such as have consolidated to form limestones.

Fig. 22–23. The Bahama Banks, to the east of Florida Straits, gradually subsiding shoals on which thousands of feet of carbonate sediments accumulated in Cretaceous and Tertiary times. A. Map showing the islands and surrounding shoals, Andros Island being in the center of the map.

B. Lithologic map showing the distribution of calcareous sands of several types reflecting such factors as depth, intensity of wave and current action and organic growth, these have been further subdivided in surveys (after J. Imbrie and E. G. Purdy).

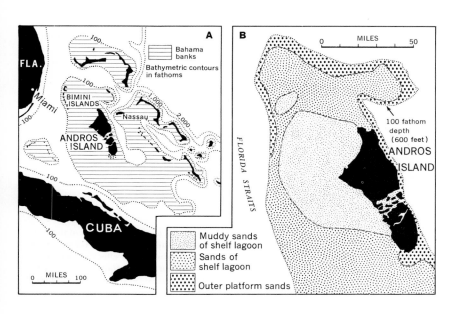

What caused the great thickness of rocks along the Gulf Coast? Possibly the sediment that was laid on the sea floor caused the crust to sink. This is the principle of isostasy—that parts of the crust tend to attain equilibrium relative to their masses. A cubic foot of sediment after compaction would not be as heavy as an equal volume of the rock that would have to be displaced deep within the earth by the load; it is thought that it would have only about five-sixth the mass: 6000 feet of sediment would cause about 5000 feet of subsidence. If water were 1000 feet deep to begin with, if the sea retained its level, and if the subsidence balanced the load, 6000 feet of sedimentary fill would cause 5000 feet of subsidence, and the surface would rise to sea level. The

Fig. 22–25. Correlation table to show the representation of the several series of the Tertiary Period in the surface exposures along the Gulf and Atlantic coasts. Horizontal lines represent equal time, and vertical ruling represents absence of deposits of the corresponding time. Most correlation tables have separated columns to show sections at a succession of places.

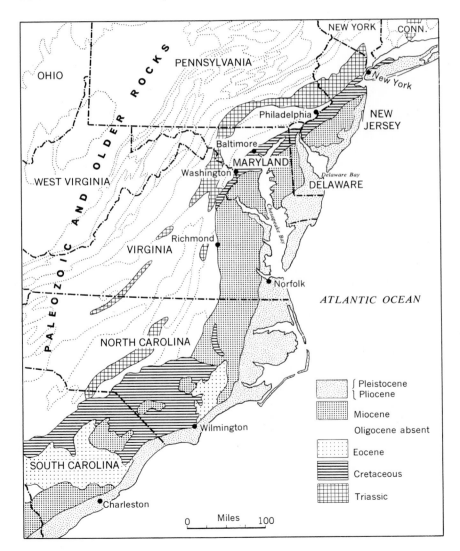

Fig. 22–24. Geologic map of the coastal plain of the Atlantic from Georgia to New York, showing the effect of unconformities within the Tertiary on the distribution of the several series. Although the most extended shore is at Cape Hatteras, the principal structure brings the Cretaceous nearly to the sea at Wilmington, North Carolina.

	Texas	La.	Miss.	Ala.	Fla.	Ga.	S.C.	N.C.	Va.	Md.	Del.	N.J.
PLEISTOCENE												
PLIOCENE			Terrace gravels and sands									
MIOCENE		Oakville	Catahoula			Tampa		Yorktown	St. Mary's			
OLIGOCENE		Vicksburg										
EOCENE		Jackson			Ocala							
		Claiborne										
		Wilcox										
PALEOCENE			Midway									

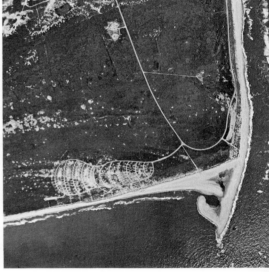

Fig. 22–26. Views of Cape Hatteras, North Carolina, taken from the air in 1955 (*A*) and 1962 (*B*), to show how rapidly the minor features along the Atlantic Coast are changing (W. D. Athearn, from the United States Coast Guard). The earlier picture shows banding to the left, west of the cape that represents older beach ridges that have been eroded away. In the millions of years represented in the Tertiary sedimentation beneath the coastal plain, the preserved sediments represent but a small fraction of the beds that were continually being laid and lost in the shifting shores of the advancing and retreating seas.

sediments would build forward and displace the deeper water; they would show in their lithologies and faunas that the water shoaled with time. But the sediment could not have been laid in the beginning if there had not been initial subsidence. A thickness of perhaps eight or ten miles would require an initial depth of a mile or two. The early Tertiary sediments do not seem initially to have been laid in such depths, nor to have formed in progressively shallower water; some of the lower Eocene sediments are of shallow origin or even non-marine, whereas some of the later Eocene in the same localities seem to have been deposited in depths of hundreds of feet, judging from their organic content. The same general Gulfward tilting proceeded, as in the Jurassic and Cretaceous.

Thus there has been a cause of subsidence independent of loading; subsidence preceded the first deposition, for it caused the placement of the initial sediments. It has been stated that the sediments are thicker off the mouth of the Mississippi than where a major stream did not enter. It is possible that the greater thickness was an effect of added load, but subsidence had been going on, making it difficult to confirm that the cause of initial subsidence was not also the cause of all continued subsidence. The geologist is not able to give a satisfactory answer as to what causes the initial subsidence. Some have attributed it to convection flow in the earth's interior; such hypotheses are difficult to evaluate from the evidence of the geology. Something more fundamental than loading accounts for the presence of the continent and the Gulf, and the great thickness of sediment near their junction.

MISSISSIPPI DELTA AND FLORIDA

The most conspicuous coastal projections along the Gulf are the Mississippi delta and the peninsula of Florida. The Mississippi has built its distributaries into the Gulf in quite recent time, subsequent to the Tertiary. An embayment entered the site of the lower valley during most of that period. Florida, rising about 400 feet above the waters of the Gulf and Atlantic, is the land along a platform that subsided less than its surroundings even in late Mesozoic. Carbonate rocks accumulated in the shallow, clear, and agitated waters. Only in the north was there terrigenous detritus drifted from the lands of the Appalachian region, so there are shifting zones of intertonguing in the Tertiary strata (Fig. 22–21). Florida had peninsular lands and small islands from time to time during the Tertiary. Warping raised the early Tertiary strata that traversed the area, so that the oldest exposed rock is the late Eocene Ocala limestone in the core of the domed anticline of northeastern Florida. Some beds are replete with large calcareous foraminif-

Fig. 22–27. The Bermuda—New England seamounts (after J. Northrop, B. C. Heezen, and others). This chain of submarine volcanic peaks extends southeastward from Georges Bank on the continental shelf south of New England for nearly a thousand miles, with scattered mounts in the direction of Bermuda. Only Bermuda is an island, carbonate rocks crowning a subsurface rise having volcanic rocks. The other seamounts rise to within about five hundred fathoms of the surface, more than a half-mile in depth, though most are much deeper. Although the volcanic rocks have not been dated, fossils of Tertiary age are reported to have been dredged from the crests of the mounts. The lines on the map are in fathoms of depth, with seamounts shaded; a fathom is equal to six feet, so the west Atlantic abyssal plain has a depth of some three miles, the extremely flat surface being formed of deposits carried by turbid flow from the continental margins.

Fig. 22–28. Susquehanna River viewed northwestward from Harrisburg, Pennsylvania, showing water gaps through Silurian and Devonian sandstones below the even-crested mountains preserving remnants of a Tertiary peneplane (G. Heilman). Paleozoic sedimentary rocks, folded in the Appalachian Revolution, had been reduced to a peneplane late in the Tertiary Epoch; as the peneplane was raised, the Susquehanna continued in its course, cutting the water gaps through the more resistant ridges, while tributary streams sculptured out the less resistant intervening rocks to form the lowlands. The nearest ridge is of north-dipping Silurian quartzite (Fig. 14–18B).

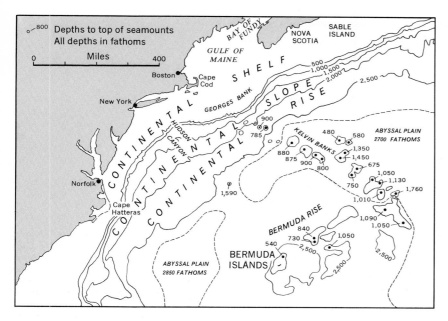

Depths to top of seamounts
All depths in fathoms

Miles

BAY OF FUNDY
NOVA SCOTIA
SABLE ISLAND
GULF OF MAINE
Boston
Cape Cod
New York
GEORGES BANK
CONTINENTAL SHELF
CONTINENTAL SLOPE
CONTINENTAL RISE
HUDSON CANYON
Norfolk
Cape Hatteras
CONTINENTAL

500
1,000
1,500
2,000
2,500

900
785
KELVIN BANKS
480 580
1,350
1,450
880
875 900 800
675
1,050
1,130
1,760
750
1,010
1,090
1,050
ABYSSAL PLAIN 2700 FATHOMS

1,590

BERMUDA RISE
840
730 1,050
540
2,500
BERMUDA ISLANDS
2,500
2,500

ABYSSAL PLAIN 2850 FATHOMS

era, nummulites; similar nummulites are important constituents of the rock quarried in Egypt long ago to build the Pyramids (Fig. 22–21C).

Florida in its present form developed during the subsequent epochs of the Tertiary (Fig. 22–12). In the Miocene, for instance, there was an area of land in northeastern Florida that was inhabited by mammals that are preserved in the non-marine deposits. Eastward for seven hundred miles from Florida are the Bahamas (Fig. 22–22), low islands and shallow submarine flats having a few deep channels that pass northeastward into the deep ocean basin (Fig. 22–23). The Cenozoic rocks are almost completely of carbonate rocks, having a thickness of more than 8000 feet in the single well that has penetrated them on Andros Island some 150 miles southeast of Miami, Florida—a thickness greater than the maximum of some 6000 feet known in Florida. The Bahamas basement must exceed 15,000 feet, the well approaching that depth having Cretaceous limestones such as must have been laid in rather shallow water. Hence, it seems that the Bahama Platform is an area in which the organic growth through later Mesozoic and Cenozoic was at as great a rate as the subsidence of the crust. The quite precipitous descent into oceanic depths marks the limit wherein organisms were able to keep pace with the subsidence.

ATLANTIC COASTAL WARPING

The seaward extension of Tertiary and Cretaceous rocks has been discussed. Along the Coastal Plain, marine sediments of some stages outcrop fairly continuously from the south to the north, but thicken and thin along the belt of outcrop. Thus Miocene lies on Cretaceous along a tongue of outcrop of the latter extending toward Cape Fear in North Carolina (Fig. 22–24). To appreciate the significance of these distributions, one must think of each series as in the form of a wedge with a sinuous irregularly curving landward termination. Where this termination lies inland from the present outcrop, the sediments are exposed at the surface, but where the edge does not approach so far landward, the sediments do not reach the surface. Thus, the absence of Eocene on the northern side of the Cape Fear Cretaceous outcrop reveals that the edge of the wedge swings sharply seaward, passing southward beneath the surface seaward from the present shore, coming back beneath the land in Georgia. If this represents the limit of original marine deposition, the sea swung out around a promontory; and if it represents the bevelling of post-Eocene, pre-Miocene time, there must have been a seaward plunging upwarp in that area. Thus the distribution along the outcrop (Fig. 22–25) reflects the trends of the edges of the sedimentary tongues. The coast was warping irregularly

Fig. 22–29. *Block diagrams showing the development of the Atlantic Coast through Pennsylvania and New Jersey (after D. W. Johnson).* **(1) In the Jurassic Period, the fault block of westerward dipping Triassic sediments and the folded Paleozoic rocks of the Appalachian region, with the intervening belt of Precambrian metamorphic and intrusive rocks, had been raised and eroded to a land of ridges and intervening valleys.**

(2) The erosional reduction of the area in later Jurassic and earlier Cretaceous had reduced the elevations and produced a peneplane of very low relief by the time that Cretaceous seas advanced along the Atlantic coast. The principal streams were flowing gently toward the sea.

(3) As the land was raised, the streams reduced their valleys rapidly enough to keep their courses, now entrenched within the raised peneplane.

(4) In time, the tributaries removed the belts of less-resistant rocks to form linear ranges such as are seen in the present folded Appalachian Mountains.

(5) Further relatively smaller uplifts and continuing erosion in the later Tertiary produced the present topography (6), with the Appalachian Plateau and Mountains bordered by the crystalline rocks, such as those in the Highlands of the Hudson River and the Reading Prong extending into northeastern Pennsylvania; to the east is the Triassic Lowland, overlapped by the sedimentary rocks of the Cretaceous and Tertiary of the Atlantic Coastal Plain.

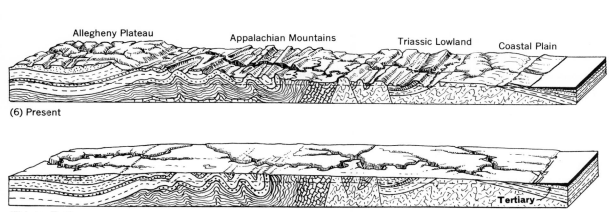

Allegheny Plateau Appalachian Mountains Triassic Lowland Coastal Plain

(6) Present

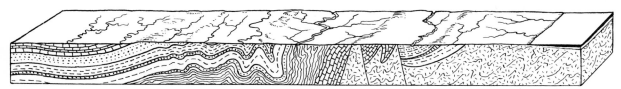

Tertiary

(5) Later Tertiary

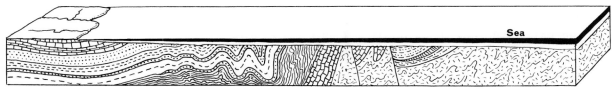

Cretaceous

(4) Early Tertiary

Sea

(3) Late Cretaceous

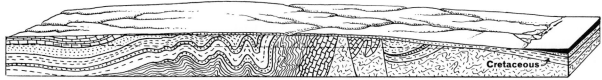

(2) Early Cretaceous

Paleozoic

Triassic

Precambrian and Paleozoic

(1) Jurassic

during the Tertiary. At times, the sea retreated far off the present shore; there must have been relative uplift of the Appalachian belt. At other times, the sea advanced inland beyond the present outcrop, and the fluvial plains covered much of the Piedmont, if not the more inland parts of the mountain system.

The relations along the Carolina Coast (Fig. 22–26) exemplify the rather ephemeral significance of the distribution on the geologic map. The general relations that pertain to the Tertiary all along the Gulf and Atlantic coasts have been described. Generally it seems that rocks of any stage should be found in offshore marine lithologies if one continues far enough down dip. The differential movements, epeirogeny, along the continental margin and the eustatic levels from time to time are principally the determinants of the distribution of rock types. The elevation of regions as far away as the Rocky Mountains will have affected the quantities and textures of terrigenous sediments along the western Gulf Coast, as will the deformation of the Appalachian System along the eastern Gulf and Atlantic coasts. The former has been discussed and the Appalachian record will be described briefly. The distribution of rocks of the many stages of the Tertiary along the outcrop from New Jersey to southern Mexico can be shown on a correlation chart (Fig. 22–25). Briefly this shows the West Gulf Coast to have marine Paleocene and Eocene, the east Gulf Coast to have additionally marine Oligocene and Miocene, and the Atlantic Coast to have principally Paleocene and Miocene.

SUBMARINE EXTENSION OF INFORMATION

The submarine extent of these rocks is determinable by means of geophysical methods, such as seismic reflection and refraction, just as it was in the Cretaceous. Similar means establish relations seaward into the deep Atlantic. The edge of the Continental Platform, the oceanward face of the several banks (Fig. 18–15), has yielded Cretaceous and Tertiary fossils in several localities from Florida to Newfoundland, suggesting that strata terminate in the slope, possibly through slumping or faulting away of the face. The deep floor of the ocean is remarkably flat (Fig. 22–27), and seems to have sediments that have flowed in turbid currents from the banks. In recent times, many seamounts, submarine peaks of volcanic rocks, have been found. The belt of seamounts that extends southeasterly from New England may be related to a zone of faulting of that trend; the relatively enormous and much more extended volcanic chain of Hawaiian Islands is perhaps analogous. Fossils in sediments gained from the summits of the mounts seem of early Tertiary age, so the volcanic history should be early

Fig. 22–30. A diorama, a reconstructed view showing a group of mammals such as might have been seen in western South Dakota in the Pliocene Epoch (the Smithsonian Institution).

Among the forms gathered about a swamp or water hole are a variety of bizarre antelope-like animals, some with forked horns on the snout and others with a horn protruding from the back of the head.

Standing in the water are *Amebelodon*, a mastodon with shovel-like tusks in its lower jaw, and the short-legged rhinoceros *Teloceras*. Peccaries appear in the foreground. On the hillside in the background are ancestral horses and camels.

Tertiary or earlier; the suggestive similarity to the volcanic Monteregian Hills that seem Cretaceous in southeastern Quebec may not be significant. Bermuda (Fig. 22–26) is a carbonate-rock crowned seamount, more isolated to the south. The Bahamas east of Florida, are banks of thousands of feet of Tertiary and Cretaceous sediments that have been drilled more than three miles.

APPALACHIAN SUMMITS

The summits of Appalachian ranges tend to have rather uniform levels —to be peneplanes (Fig. 22–28). When projected seaward, these pass below Cretaceous sediments; they are surfaces formed prior to the cutting of earlier Cretaceous streams and the overlap of Cretaceous seas. Other surfaces seem related to periods of stability and erosion during the Tertiary period (Fig. 22–29). The interpretation has been made that the courses of the major streams began on the surfaces of the sediments that covered the peneplane of the pre-Cretaceous, or over the peneplaned rocks. As the Appalachian region rose quietly during the Tertiary, the streams cut through the veneer of sediment, notching the buried ridges that obstructed their channels. Ultimately, with continued denudation, the covering sediments were wholly removed, and the streams now course through ridges that once lay at lower elevations than their channels. This superposition of streams from the consequent course on an old sea floor or sedimentary cover is analogous to that of the western streams that we have described as cutting their gorges through the Rocky Mountain ranges in later Tertiary times.

23
The Quaternary
or Pleistocene Period:
The Great Ice Age

Slope of Mount Shasta, California, a glaciated volcano.

23 *The Quaternary or Pleistocene Period: The Great Ice Age*

The term Quaternary was applied about 1820 to the loose rocks that were at the surface of the earth in northern Europe, the alluvium along the banks of streams, and the "diluvium," tumbled mixtures of boulders and clay that formed the hillocks rising above the small lakes in widely scattered areas. Pleistocene was a name assigned by Charles Lyell a little later, in 1839, and with greater precision to a sequence in northern Italy which had been part of the Pliocene, but with fossil shells so recent in aspect that they seemed to warrant separate designation. Within the past few years some students have endeavored to separate the Recent from the Pleistocene, including both in the Quaternary—and have even attached an era name (Holocene) to time that seemed to mark a new milestone in earth history because of the dominance of man. The more conservative have not followed this practice. The Pleistocene has come to be associated with glaciation, for glacial ice was the most distinctive feature of the epoch in the northern part of northern continents. The name Pleistocene was applied, however, before it was recognized that the Quaternary deposits were partially glacial. The beginning of the glaciation is not a distinct time, so the definition of the system is by comparison with a standard marine section of rocks in Italy, where the Pleistocene has fossils that lived in colder water than those in the underlying Pliocene, as judged by the life environment of the most similar present forms, and in part by the appearance in this same region of modern horses (*Equus*), cattle (*Bos*), and elephants (*Archidiskodon*). The end of the Pleistocene has gradually seemed closer and closer to the present as new discoveries and techniques have developed. On the contrary, the record of the beginning of man has extended back into the past. We seem to be in the waning stages of the ice age, for the glacial ice of Greenland is gradually retreating. So the Pleistocene has come to include the Recent as a formal time designation.

DISCOVERY OF CONTINENTAL GLACIATION

Great glaciers occupy some of the higher valleys in the Alps (Fig. 23–1). Swiss travelers long ago recognized that the debris at the lower end of the ice, constituting the terminal moraine, was deposited as a direct result of the transportation of rock fragments on and in the ice, and their release with the subsequent melting. When they observed ridges of similar material even at a distance in front of the glaciers, they

Fig. 23–1. Valley glacier in the Alps. Morteratsch Glacier south of St. Moritz in eastern Switzerland; from the northwest, with Bernina Mountain (13,300 feet) on the right, in ranges of intruded metamorphic rocks of early Mesozoic age. The glacier has retreated from the terminal and lateral moraines as melting is in excess of the accumulation of ice. The presence of such moraines along the margins of present glaciers suggested to geologists in the nineteenth century that similar tills found farther to the north were of glacial origin (Swissair Photo).

Fig. 23–2. Maximum spread of Pleistocene ice sheets in North America and Europe (after R. F. Flint). The Laurentide ice sheet was quite continuous, a continental glaciation; in western mountains, glaciers such as those that occupied individual valleys in the more southerly ranges tended to coalesce into continous ice fields in the more northern mountains. Similarly, in Europe valley glaciers occupied the higher ranges in the Alps and Pyrenees.

surmised that the ice had been more extensive than it is now. The striation and grooving of the rock surfaces, and of the stones in the moraine, were readily associated with glaciation. In the third decade of the nineteenth century, the Swiss (subsequently American) naturalist, Louis Agassiz (1807–1873), traced the extent of these features that seemed indicative of glaciation, confirming the impressions of a colleague that ice that once covered the plains to the north of the Alps had melted back to the present restricted valley glaciers in the mountains. When he found similar smoothed and grooved surfaces beneath clays with striated pebbles widely distributed over the British Isles, the conclusion was inevitable that the ice had extended far over the lands. Subsequently, he realized that glaciers that spread over much of Northern Europe were separate from those in the Alps. By a century ago, it was recognized that glaciers in the rather recent past had occupied a great part of the lands of northern Europe and northeastern North America (Fig. 23–2).

RECOGNITION OF MULTIPLE GLACIATION

In the eighteen seventies, in northeastern Iowa, beds of peat and clay were found beneath thick boulder-bearing clay and overlying similar glacial till. The peat was interpreted as requiring that the glacial ice, after depositing the lower moraine, melted back for a long enough time so that a forest could grow, and then advanced again, burying the forest beneath the till it deposited. In southwestern Iowa, a gravel pit yielded remains of vertebrate fossils; the gravels lay above glacial till, and similar till was above them. In many localities, a peculiar sticky clay or "gumbo" seemed to separate boulder clays. Rolling, hummocky hills

Fig. 23–3 (right). *Pleistocene glacial deposits in central Iowa.* **A.** The section of about 70 feet is along the Milwaukee Railroad near Marshalltown, Iowa. The two lower layers are of Kansan till: the lower, 10-feet thick, being oxidized and leached of calcium carbonate; the upper, darker layer, 8-feet thick, is gumbotil—the product of weathering of till during the succeeding Yarmouth interglacial and Illinoian glacial times. The overlying 25 feet is of early Wisconsinian (Iowan) loess, wind-transported silt. The upper 25 feet of the slope is of middle or late Wisconsinian till, oxidized and shallowly leached in the uppermost part (W. C. Alden, United States Geological Survey).

B. Diagram to show the ideal sequence of zones of weathering on a drift sheet.

C. Idealized section of the succession of deposits as from west to east in Iowa (after G. F. Kay).

Fig. 23–4. *Glaciation in the middle states.* **A.** Maximum extent of each of the principal ice sheets that invaded the central part of the United States.

B. Map of Iowa, showing the distribution of successive Pleistocene tills in greater detail. Much of the moraine left by each retreating glaciation was covered by a veneer of loess, silt blown from the glacial till and associated fluvial deposits. The Driftless Area in northeastern Iowa is a small part of a much larger area, principally in Wisconsin; it may have been covered by earliest glaciers, but erosion removed virtually all trace of any deposited till (after G. F. Kay and J. B. Graham).

The Nebraskan till is exposed in the southwest in areas where streams have eroded through the succeeding Kansan glacial till.

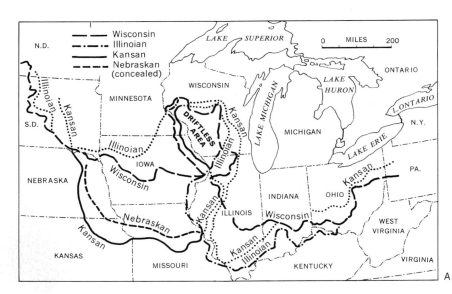

A

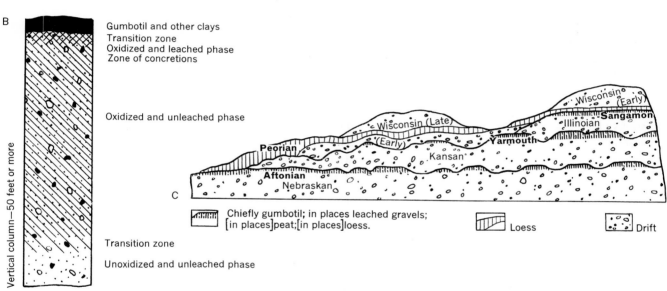

B

Gumbotil and other clays
Transition zone
Oxidized and leached phase
Zone of concretions

Oxidized and unleached phase

Vertical column—50 feet or more

Transition zone

Unoxidized and unleached phase

C

Peorian

Aftonian
Nebraskan

Wisconsin (Late)

(Early)

Kansan

Yarmouth

Wisconsin (Early)

Illinoian

Sangamon

Chiefly gumbotil; in places leached gravels;
[in places]peat;[in places]loess.

Loess

Drift

Fig. 23–5 (right). Loess along the east bluff of the Missouri River in western Iowa across from Omaha, Nebraska. Pleistocene winds blew silty dust from the floodplain to the west and deposited them with diminishing thickness on the east side of the Missouri. The dark spots are the burrows of swallows that build nests in the easily excavated loess.

of till and gravel with scattered lakes could be seen to end along a rather definite line, having beyond it southward other boulder-clay with a well-developed drainage system on its surface; evidently an older moraine had been eroded before deposition of a moraine so young that it is not dissected.

Peat and gravels indicate retreat and readvance of the ice, but do not give very substantial indication of how long a time elapsed between the advances. Plants and animals in the interglacial sediments seemed to show that the climate ameliorated considerably. The dark sticky clay or gumbo that separated the tills passed downward into yellow or buff silts and clays having some fresh boulders, such as quartzite, and others badly disintegrated, such as granite; there were no limestone boulders. Occasionally, similar very resistant boulders remained in the upper clay; and at times we might recognize the "ghost" of a granite boulder, a regular outline within which the resistant quartz remained but the other minerals had been altered (Fig. 23–3). So it was judged that the clay layer was actually a residual sediment, the product of the decomposition of material like the till below it; the deeply weathered sticky clay became known as gumbotil. The buff lower clay that lacked carbonate boulders was in turn judged to be the leached residue of the calcareous, carbonate-boulder bearing, but otherwise similar, buff till below it. And at the base, the till that was gray or blue had not undergone the oxidation that altered iron minerals to hydrous oxide, giving the rusty color to the succeeding zone. Thus a sequence was recognized from dark gumbotil through leached and unleached oxidized till zones into unaltered glacial till. Some of the dark clay has been found to be the product of sedimentation, so not true gumbotil. The profile is like that of soils in regions of moderate rainfall where water removes the surface lime carbonate.

As mapping continued, several tills in Iowa and adjacent states were found. Above the oldest till with the weathering profile or gravels (Nebraskan) came another till that had a similar weathering profile (Kansan). It was succeeded in the region of the Mississippi River by a third till (Illinoian) having a profile of deep weathering, and this was succeeded by the fourth till (Wisconsin). The Mississippi Valley was judged to have had multiple glaciation. When all the deposits were mapped, the limit of the ice at one time or another seemed to come toward or beyond the Missouri and Ohio rivers (Fig. 23–4). Winds blew silt from the surfaces of recently laid tills and outwash that was laid as loess, separating some of the tills and overlying the margins of the last deposits (Fig. 23–5).

Fig. 23–6. The predecessors of the Great Lakes (after J. L. Hough). A succession of lakes formed south of the ice front of the late Wisconsinian glacier in the last 20,000 years or so. Though the ice margin generally retreated, there were times of readvance that restricted the expanding basins. An important factor shaping the lakes and the courses of their outlets was the gradual rise of the crust as the load of ice was released.

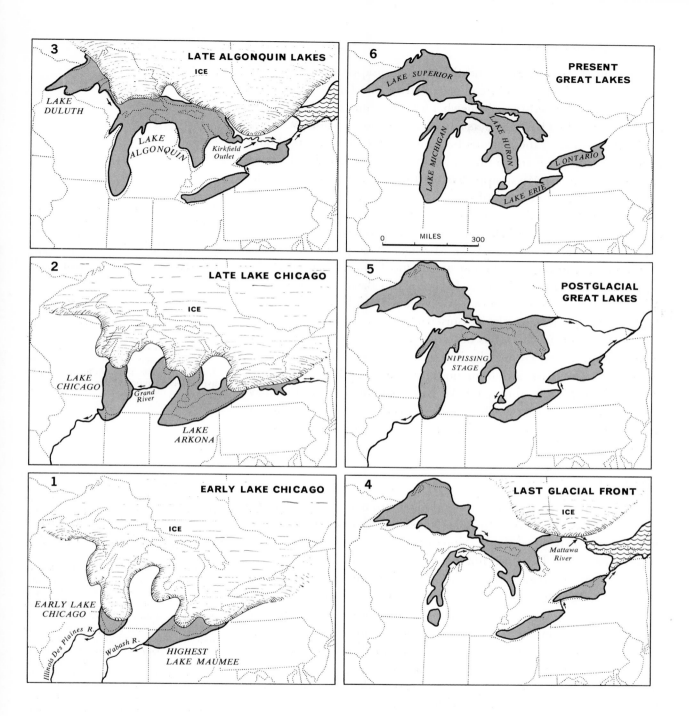

3 LATE ALGONQUIN LAKES

ICE

LAKE DULUTH

LAKE ALGONQUIN

Kirkfield Outlet

6 PRESENT GREAT LAKES

LAKE SUPERIOR

LAKE MICHIGAN

LAKE HURON

L. ONTARIO

LAKE ERIE

0 MILES 300

2 LATE LAKE CHICAGO

ICE

LAKE CHICAGO

Grand River

LAKE ARKONA

5 POSTGLACIAL GREAT LAKES

NIPISSING STAGE

1 EARLY LAKE CHICAGO

ICE

EARLY LAKE CHICAGO

Illinois Des Plaines R.

Wabash R.

HIGHEST LAKE MAUMEE

4 LAST GLACIAL FRONT

ICE

Mattawa River

583

POSTGLACIAL GREAT LAKES

The ice finally melted away and the glaciers disappeared. There had been broad valleys in the areas occupied by the Great Lakes. The glacial ice had shaped these somewhat by smoothing irregularities. As the ice front melted back from its advanced position south of the lakes, the southern tips of their basins became exposed (Fig. 23–6). Water from the melting ice accumulated in these smaller lakes and flowed southward into the Mississippi System through such streams as the Illinois from the lower end of Lake Michigan and the Wabash from the western tip of Lake Erie. As ice retreated farther, other outlets opened. The water from Lake Erie for a time flowed westward through the Grand River Valley into Lake Michigan. Georgian Bay of Lake Huron drained through a winding stream, the Trent River (Fig. 23–7), across southern Ontario into the Lake Ontario Basin and then southward through the Susquehanna. A readvance can be dated at about 11,000 years ago, because the ice buried a forest in southern Wisconsin, and the wood in the forest has been dated by the radiocarbon analysis method (Fig. 23–8). Ultimately, in a very late stage, the waters of the lakes all flowed eastward from northern Georgian Bay, the eastern part of Lake Huron, through Lake Nipissing into a marine embayment of the Ottawa valley. The sea also extended through the Hudson and Champlain lowlands, for there are marine shells in some of the terraces (Fig. 23–9).

TILTED TERRACES

But why should these waters have drained from Lake Erie through Lake Huron to leave through Lake Nipissing? Elevations now would not permit such flow there. Moreover, the beaches of the levels of the last Great Lake are not now level; they are far above Lake Huron on the northeast and pass below southern Lake Huron and southern Lake Michigan, lakes that are now of the same level (Fig. 23–10). The water in the lake must have been level in the time following glaciation, so it is clear that the continent has been tilted southward since the retreat of the ice. Study of shores of successively younger lakes show that the crust rose progressively northward as the ice retreated. And studies of present precisely determined elevations indicate that the southern end of Hudson Bay is still rising relative to the Great Lakes. The inevitable conclusion is that either the land happened to subside in the region that was being glaciated, and to rise thereafter, or that the glacial ice placed a load on the crust that produced isostatic adjustment, subsidence during glaciation followed by elevation after glaciation (Fig. 23–11). This is substantiated by similar

Fig. 23–7. Channel of the Trent River outlet of the upper Great Lakes, at Healey Falls, northeast of Peterborough, Ontario. This is one of two channels at the locality that discharged the waters of the upper lakes when the ice had melted away between lakes Huron and Ontario, prior to the rise of the land that followed the disappearance of the ice. The flow of water, probably greater than that of the present Niagara River, formed a falls over the rather resistant ledges of Ordovician lower Trentonian limestone. The old channel of the small modern Trent River has been dammed to gain hydroelectric power from the flow.

Fig. 23–8. Determination of age of carbon-bearing matter. Carbon of atomic number 14 comprises an infinitesimal part of the carbon in living organisms— only about one millionth millionth as much as other isotopes of carbon, but the isotope carbon 14 is radioactive, half of it breaking down to nitrogen 14 in somewhat less than 6000 years with loss of a beta ray (electron). As the number of radioactive atoms in a given mass of carbon of unit weight will decrease through time, therefore, and as it is possible to record electronically the atomic disintegration by means of sensitive counters, the isotope is useful in determining the age of carbon-bearing substances, such as carbonate in shells or carbon-bearing organic matter. We know there is a rather constant quantity of carbon 14 in living organisms, because over a long time there seem to be rather constant numbers of cosmic rays reaching the upper atmosphere—where they meet nitrogen 14 atoms and change them to carbon 14 by entering the nuclei of the nitrogen atoms.

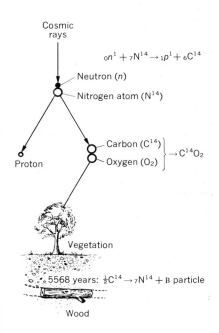

Cosmic
rays

$$_0n^1 + _7N^{14} \rightarrow _1p^1 + _6C^{14}$$

Neutron (n)

Nitrogen atom (N^{14})

Proton

Carbon (C^{14})
Oxygen (O_2) $\Big\} \rightarrow C^{14}O_2$

Vegetation

5568 years: $\frac{1}{2}C^{14} \rightarrow _7N^{14} + \text{B particle}$

Wood

Fig. 23-9. Raised marine terraces of the late Pleistocene in Sudbury, west of Brandon in western Vermont. As the continental glacier retreated from the New England–New York region in the latest Pleistocene, the sea came in over the depressed crust; the level of the water in the oceans was somewhat lower than at present, since much water was still contained in the waning glacial ice. The terraces, now more than 400 feet above sea level, have marine shells of cold-water habitat. The sea rose as water from the melting ice was added to its volume, but the marine terraces rose more rapidly in response to the unloading of the crust, so that they now are far above sea level. The late Pleistocene Champlain Sea extended into the Lake Ontario basin and far up the Ottawa River, the expanded Great Lakes flowing into the Ottawa from northeastern present Lake Huron, as shown on the maps of the Great Lakes (Fig. 23-6).

studies in the northern part of Europe (Fig. 23–12). Sea beaches far above the present sea level are conspicuous features in the Arctic regions (Fig. 23–13); yet melting of the Pleistocene glaciers should have produced a general rise in sea level of tens of feet and should have drowned any late Pleistocene terraces rather than leaving them far above the sea.

WIDESPREAD ICE—MARGINAL LAKES

The Red River Valley of North Dakota, Minnesota, and Manitoba drains north into Lake Winnipeg and ultimately into Hudson Bay. The divide separating it from the Mississippi drainage is through south-eastern North Dakota. When the ice retreated from this region, a lake formed between the ice and the height of land, Lake Agassiz. Similarly, a great lake formed north of the divide between the Great Lakes and Hudson Bay as the ice retreated northward in that region. In New York, lakes rose in the Mohawk Valley (Fig. 23–14) between ice in the Lake Ontario basin and that in the Hudson Valley; these waters flowed south over divides into the Susquehanna drainage. And in the valleys of southern New England, irregular melting back of the ice from Long Island Sound produced small lakes ponded between standing masses of ice in the valleys, or moraine filling drainage channels. Post-glacial lakes were scattered widely over eastern North America through the late Pleistocene.

Fig. 23–10. The rate of crustal uplift since the retreat of Pleistocene glaciers in centimeters in a century: *A*, for northeastern North America (after B. Gutenberg), and *B*, for northwestern Europe (after R. A. Daly and S. M. Sauramo).

Fig. 23–11. *Postglacial uplift in northeastern North America, showing the rise in the elevation of the land surface since the retreat of the ice. A.* Map showing the present elevation of marine beaches formed as the land was freed of ice (after W. R. Farrand and R. T. Gajda). The lines are *isobases,* contour lines drawn through the present elevations of a succession of points that were originally at sea level; those in the Great Lakes are of a level of a late glacial lake that was above the contemporary sea level. The map of isobases is further complicated in that the marine levels are not of exactly the same time, for the beaches are successively younger northward where the ice remained latest. The level of the sea rose as the glaciers melted, and the waters flowed to the ocean—sea level was changing through time.

B. Graph showing the progressive uplift of places near the Great Lakes; Cape Rich is near the southern end of Georgian Bay, Lake Huron. The ages were determined by radiocarbon dating (after W. R. Farrand).

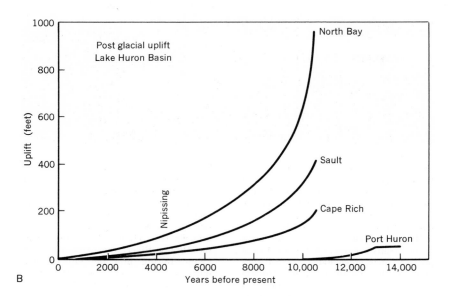

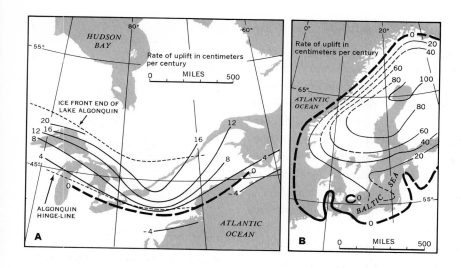

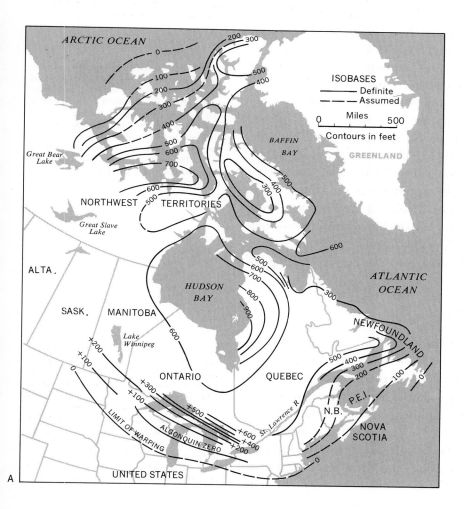

The Yosemite Valley is distinguished by its precipitous rock walls rising from the flat valley bottom (Fig. 23–15). The walls are of grano-diorite; though generally thought to be intruded during the late Jurassic Nevadan Orogeny, it seems from studies of the potassium disintegration that they are younger. The valley form is a result of glaciation. Near the mouth of the valley, a ridge of moraine represents deposition at the front of the last glacier. High up on the north wall on the road to Tuolumne Meadows is much older glacial till, deeply weathered in the long period of exposure. The forested slopes of the western Sierra do not present as favorable sites for the observation of the phenomena of glacial history as the relatively treeless semi-arid eastern slopes (Fig. 23–16). There the oldest and most deeply weathered till is preserved on the crests of ridge spurs separating the deep east-flowing stream canyons. The ice spread over a plain in front of early Pleistocene mountains; the mountain face has been raised by Pleistocene faulting, and deep valleys have been cut in the increased relief. The last glaciers are represented by the lateral and terminal moraines along the valleys, the uppermost and youngest being little eroded. The older phases have somewhat weathered and there are stream-sculptured moraines. Glaciers of this sort sculptured the valleys in all of the higher ranges of the west, the Sierras, the Ruby, Independence, and Snake ranges of northern and eastern Nevada, the Uinta and Wasatch mountains of Utah, the Rocky Mountains of Colorado, and ranges northward into Canada. A few of the mountains in the United States have glaciers today, such as Absaroka Glacier in the Colorado Front Range, Dinwoody Glacier in the Wind River Mountains of Wyoming, and those of Mount Ranier in Washington. Glaciers were much more extensive in the Pleistocene, as they are today in the Coast Range of northern British Columbia.

LAKES BONNEVILLE AND LAHONTAN

The region between the Sierra Nevada and the Wasatch Range of Utah, the Basin and Range region (Fig. 23–17), now receives very little precipitation and is semi-arid. In the Pleistocene, there must not only have been colder climate, but higher precipitation and smaller evaporation. The face of the Wasatch Range has conspicuous wave-cut terraces and gravel-capped benches, shore features of a lake that withdrew so recently that these features have been little dissected (Fig. 23–18). Similar terraces appear on the west side of the Great Salt Lake Desert in the ranges by the highway at the Nevada state line at Wendover. The great lake that filled this entire region drained southward into the

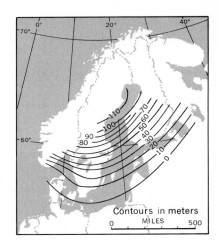

Fig. 23–12. Postglacial uplift in north-western Europe since the retreat of glaciers (after E. Fromm).

Fig. 23–13. Air photograph of the raised beaches along the west coast of Boothia Peninsula, on the mainland south of Queen Elizabeth islands, northern Canada (Geological Survey of Canada). The beaches are associated with cliffs cut in Paleozoic sedimentary rocks, successive sea levels in the late Pleistocene. The terraces have risen in more recent times as the glacial ice has melted from the region.

Fig. 23–14. Postglacial lake terraces along West Canada Creek, a tributary of the Mohawk River (the valley in the background) northeast of Utica, New York. The terraces were formed during two stands of water in lakes that formed between lobes of ice in the Mohawk Valley to the east and to the west; the waters flowed over the lowest divide in the plateau to the south into the drainage of the Susquehanna River. As the ice melted, the lake level dropped because the waters could drain eastward into the Hudson Valley.

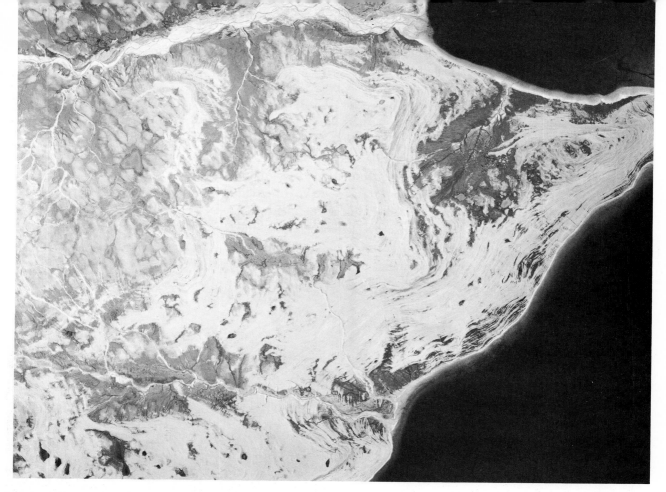

Colorado from an outlet far to the southwest beyond Sevier Lake in southwestern Utah. The waters have been evaporated, and the largest part of the floor of the Pleistocene lake is under Great Salt Lake and the surrounding desert. This lake has been called Lake Bonneville (Fig. 23–19) after the French explorer who visited the region early in the nineteenth century. The Humboldt River flows southwestward across northern Nevada to a desert evaporation basin or salt lake, and the rivers from the Sierras, such as the Carson, flow eastward into another interior basin that contained an extensive Lake Lahontan. Other basins, too, had large fresh-water lakes during the Pleistocene, for there are extensive beaches and lacustrine sediments. The tilted surfaces of some of the beaches are means of determining the warping that has gone on since the waters evaporated to their present shallow remnants.

Fig. 23–15 (left). The Sierra Nevada upland surface eastward from the lower Merced River Valley in Yosemite National Park (Sarah Ann Davis, California Division of Mines and Geology). The crystalline rocks of the mountains were denuded to a rolling plain in about Miocene, with elevation of three thousand feet or so on the divides. The region became a great, westward-tilted fault block with development of the rifts that persist along the eastern face (Fig. 21–21). The rise of the range to the distant elevations of 13,000 feet and more formed a barrier that affected the climate eastward into the Great Basin. The valley of the Yosemite in the foreground shows the effects of the glaciers that descended from higher snowfields during the Pleistocene Period.

When the glaciers melted, they returned to the oceans water that had been removed and retained by glacial ice. The water level in the sea subsided, an eustatic movement, as each glacier advanced. The Mississippi River and its ancestors have been draining the interior since early in the Tertiary. It has been possible to trace the succession of courses in the lower Mississippi in the past two thousand years by picking out the progression of meander patterns shown on air photos, with knowledge of changes within historic time. The river has had very nearly the present course for more than five hundred years (Fig. 23–20). For a similar time prior to that, it flowed somewhat west of the present course that is now occupied in considerable part by the smaller, levee-bordered Lafourche River. From about A.D. 100, it flowed for eight hundred years or so in a course farther to the west, the Teche channel.

In the past half century, the water of the Red River has become diverted to the Atchafalaya Basin, between the Teche and Lafourche channels; as much as 40 per cent of the Mississippi flow is in that channel during highest stages of the river. Far below the present surface channels are the sands and gravels of other channel fillings that have subsided. Several hundred feet below sea level is an irregular surface with decreasing elevation seaward, one of erosion on a deeply

Fig. 23–16. Moraines formed by Pleistocene mountain valley glaciation at the mouth of Bloody Canyon on the east side of the Sierra Nevada, in eastern California south of Lake Tahoe. The linear ridges are lateral moraines (E. Blackwelder).

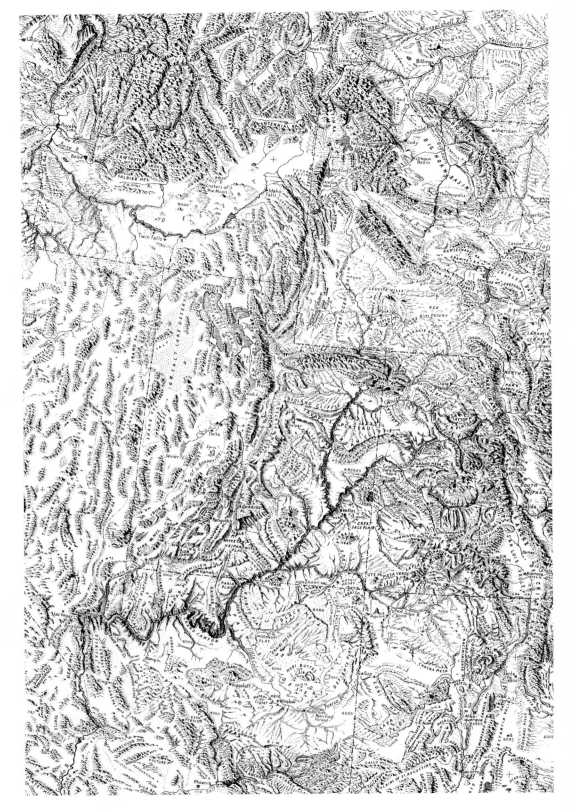

Fig. 23–17 (left). Physiographic diagram of Utah and surrounding states (through the courtesy of Erwin Raisz). The Basin Ranges in western Utah and eastern Nevada border the Colorado Plateau on the southeast, with the entrenched Colorado River system. Block-faulting in western Utah and eastern Nevada produced a succession of northerly elongate ranges having intervening semi-arid lowlands. They are generally of Tertiary volcanic and sedimentary rocks lying uncomformably on deformed Paleozoic sedimentary rocks. The Sierra Nevada, rising above fourteen thousand feet in eastern California lies on the western margin of the province beyond the map area.

Interior drainage enters intermittent lakes of high salinity, playa lakes; with greater precipitation and lower evaporation in the Pleistocene, the depressions filled to form large lakes, the largest, Lake Bonneville, of which Great Salt Lake is a relic, extended over a large part of western Utah, rising until its waters could flow southward into the Colorado River drainage.

The Rocky Mountains are to the east, ranges of folded sedimentary rocks having Precambrian crystalline rocks in the cores of structures. The Snake River Plain through southeastern Idaho has Tertiary lava flows, such as those in Craters of the Moon National Monument.

Fig. 23–18. Wave-cut and current-built terraces of the Pleistocene Lake Bonneville near Tooele, Utah. In relatively recent time, the volume of water that reached the basin in which Great Salt Lake lies was so great that water rose hundreds of feet, lapping against the ranges. Successive stages in the reduction of the lake produced several prominent terraces. Not only are these of interest in recording the extent of the lake stages, but surveys of their elevations show that there has been tilting of the region in rather recent time. Carboniferous limestones are visible in the range.

Fig. 23–19. *Late Pleistocene lakes in the Basin and Range region of Nevada and Utah. A. Boundaries of the Great Basin, with areas occupied by lakes in post-glacial time. Lake Lahontan and Lake Bonneville were two large lakes that have the present Pyramid Lake, Nevada and Great Salt Lake, Utah, as their remnants (after W. S. Broecker and P. C. Orr).*

B. The extent of Lake Bonneville, with the upper terrace some three hundred feet higher than the elevation of the remaining Great Salt Lake (after A. J. Eardley). The lake at its highest level may have drained southward into the Colorado River drainage from its southwestern bays.

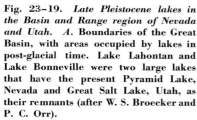

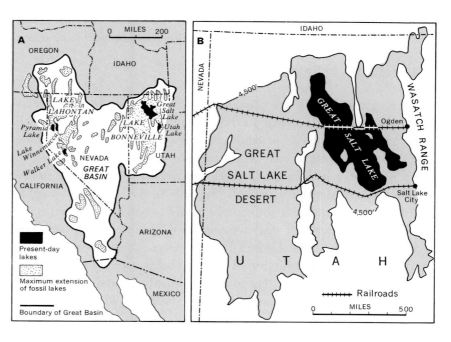

weathered soil surface (Fig. 23–21). The bottoms of the valleys in this buried surface are as much as two hundred feet lower than the intervening undissected weathered upland. Sands and silts are prevalent in the lower deposits on the surface; within them, plant material has been determined as about nine thousand years old by means of radioactive carbon studies, to be discussed shortly. On the other hand, wood from below the surface is too old to be measurable by the method, limited in its application to carbon not as old as thirty thousand years; improved techniques are extending the limiting age.

Beneath this eroded weathered surface, others have been traced at greater depths, five in all. The upper two are attributed to marine withdrawal corresponding to the two main phases of advance of the fourth glacial age, the Wisconsin. The lower three are correlated with the removal of oceanic water to form the ice of each of the older glacial ages, the successively older Illinoian, Kansan, and Nebraskan.

The deposits between the erosion surfaces, laid as the seas rose in the interglacial and postglacial times, are not of uniform thickness, nor do they form regularly seaward-thickening wedges. They thicken into elliptical, lenticular bodies of sediment, which are interpreted as corresponding to the positions of the Mississippi deltas in each of the glacial subepochs. If the position of the delta was determined by conditions within the area to the north, this indicates that the subsidence is a result of the loading of the Gulf's margin by the sediments of the river. If, on the other hand, the load is thick because it filled an area that was subsiding mostly for reasons that are not understood, the case is not so interpreted. The evidence is not sufficiently conclusive that the proponents of either hypothesis have been converted to the other.

THE TIME OF ICE RETREAT

How long ago did the ice retreat? Geochemistry has in recent years given a fairly reliable answer to this. When the ice front last advanced into northern Illinois, it buried the edge of a forest, now exposed along the Wisconsin shore of Lake Michigan. The remains of these trees give a measure of age through the radioactivity of their carbon atoms. There are two isotopes of carbon (Fig. 23–8), having atomic weights of 14 and 12; the latter is common carbon. Carbon 14 is formed at the surface of the earth by the bombardment of nitrogen by neutrons released by cosmic rays. Carbon 14 is present in a rather constant proportion in the carbon dioxide of the air, and is used by living plants and animals. It is radioactive, so that its quantity is reduced to about one-half in about 5500 years, thus to one-fourth in 11,000 years, etc. By measuring the ratio of carbon 14 and 12, we can determine the time

Fig. 23–20. The successive courses of the ancestors of the Mississippi River during the past 5000 years (after H. N. Fisk). *A.* Maringouin course and delta some 5000 years ago.

B. Teche course some 3800 years ago.

C. St. Bernard delta some 2900 years ago.

D. Lafourche delta some 1500 years ago.

E. Plaquemines delta some 1200 years ago.

F. Modern course and delta of the Mississippi River.

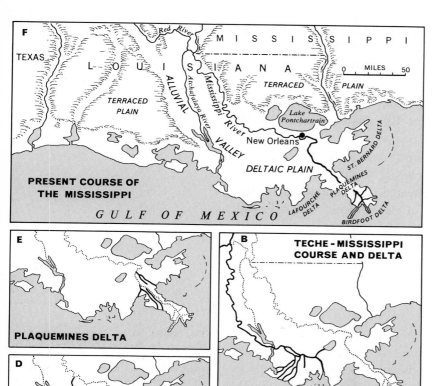

F

TEXAS

Red River

M I S S I S S I P P I

L O U I S I A N A

TERRACED PLAIN

0 MILES 50

ALLUVIAL

Atchafalaya River

Mississippi River

TERRACED PLAIN

Lake Pontchartrain

New Orleans

VALLEY

DELTAIC PLAIN

ST. BERNARD DELTA

PLAQUEMINES DELTA

PRESENT COURSE OF
THE MISSISSIPPI

LAFOURCHE DELTA

BIRDFOOT DELTA

G U L F   O F   M E X I C O

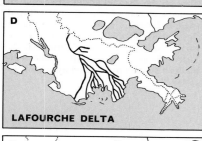

E

PLAQUEMINES DELTA

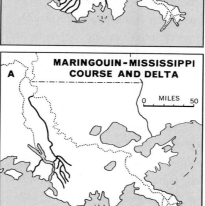

B

TECHE - MISSISSIPPI
COURSE AND DELTA

D

LAFOURCHE DELTA

A

MARINGOUIN - MISSISSIPPI
COURSE AND DELTA

0 MILES 50

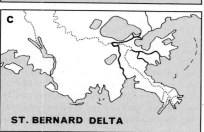

C

ST. BERNARD DELTA

595

since the substance containing the carbon was buried. The ratio becomes so small in material older than about 30,000 years that it is not practical to use carbon 14 measurements beyond that span. The method can be proved reliable by testing articles of known age, as in measuring the ratio in historic documents and by the growth rings of trees. The ice front stood in northern Illinois until about 10,000 years ago.

A somewhat longer time since the retreat of the ice had been judged by other methods. The Niagara River is cutting Niagara Falls at a known rate; the distance that the falls have retreated permits judgment of the time since they originated when freed of glacial cover. The river had several stages, reflecting the effects of crustal warping in adding and subtracting sources of water. Judgment had been reached that it was about twenty-five thousand years since the ice stood at the eastern end of Lake Erie, but this is likely to be too long an estimate.

VARVED CLAYS

The glacial lake deposits, particularly the smaller ones like those along the Connecticut River, have alternations of dark plastic clay and lighter colored silt and fine sand, and there are several such layers to an inch (Fig. 23–22). They are not of uniform thickness but variably thicker and thinner. These are varves, interpreted as the deposits of summer and winter seasons, the darker material having settled during the time when the lake was sealed by ice. Single sections of exposed sediment may have a few or a few score such varves. Within a drainage basin, storms that affected one lake should have affected others. Thus a hurricane might cause flooded streams that would bring in a thicker layer of silt in each lake in a region during a certain year, but might not recur for a number of succeeding years. The matching of the relative thicknesses in varves in lakes of the same drainage basin, the correlation of the laminae in the belief that the correlated layers are synchronous, yields a record of the number of years that water stood in each, and the duration of each with relation to those to the south, and the later lakes to the north. Because of the impossibility of gaining all varves in a single section, the total must be interpreted from the correlation of varves in many sections. Such correlations are difficult to demonstrate as precise, so they have been subject to criticism. In this way it was judged that the ice took several thousand years to retreat northward from Long Island Sound through New England. Thus, the ice must have stood over the northern part of the United States as recently as ten to fifteen thousand years ago.

Fig. 23–21. Cross-sections (*A* and *B*) through the channel of the Mississippi in its delta, showing *C*, the effect of a glacial stage in lowering water level resulting in the incision of the sediments laid in an earlier interglacial stage (after H. N. Fisk). Wood buried in the older sediments is too old to be dated by carbon 14 ratios.

Fig. 23–22. Pleistocene varved clay near Dorion, Ontario, on the northwest shore of Lake Superior (W. R. Farrand). Such clays have alternating coarser, silty bands and finer, clayey layers representing the deposits of summer seasons when streams carried sediment to the lake, and of winter seasons when the lake surface was frozen, the suspended, fine-textured material settling. Thus the duration of lakes can be determined from counts of the varves, and as their thicknesses are related to yearly climatic variations, changes in thickness introduced in one lake are likely to be recognizable in another, permitting correlation of the varve sequences over regions. In recent years, the introduction of carbon 14 age determinations has given a more accurate means of determining the age and correlation of sediments that contain carbonaceous matter in post-glacial lakes.

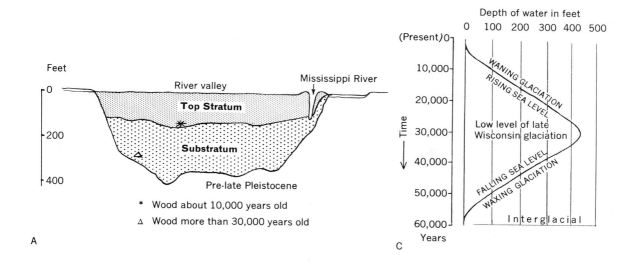

Feet

0

200

400

River valley　　　　　Mississippi River

Top Stratum

Substratum

Pre-late Pleistocene

* Wood about 10,000 years old
△ Wood more than 30,000 years old

A

Depth of water in feet

0　100　200　300　400　500

(Present) 0

10,000

20,000

Time

30,000

40,000

50,000

60,000

Years

WANING GLACIATION
RISING SEA LEVEL

Low level of late
Wisconsin glaciation

FALLING SEA LEVEL
WAXING GLACIATION

Interglacial

C

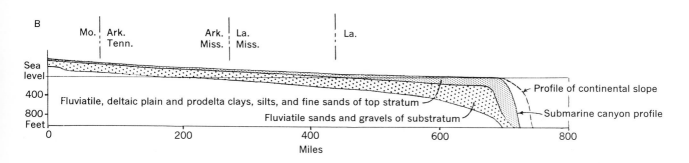

B

Mo. | Ark.
Tenn.

Ark. | La.
Miss. | Miss.

La.

Sea
level

400

800
Feet

0　　　　　200　　　　　400　　　　　600　　　　　800

Miles

Fluviatile, deltaic plain and prodelta clays, silts, and fine sands of top stratum

Fluviatile sands and gravels of substratum

Profile of continental slope

Submarine canyon profile

The ideal place to gain the continuous record of Pleistocene deposition should be in the marine deposits. Deposition proceeds very slowly in the bottom of the sea. Even if there were annual layers, they would be scarcely measurable. Lakes have varves because of the contrast produced by the influx of sediments in warmer months as compared to that in winter months when their surfaces and the source terrane are frozen; conditions differ from those in oceans. Moreover, the salts in marine water have ions that tend to flocculate or aggregate particles of many sizes into settling as clumps soon after they reach the sea. However, the marine cores of tens of feet length that have been obtained by oceanographic ships do have significant changing and alternating characters of other kinds.

It was shown in discussion of the Tertiary Period that certain microfossils, foraminifera, can be classified on their tolerance for waters of different depths and temperatures. The deep-sea cores can be divided on the basis of the foraminifera into bands laid in warm water and in cold. At a varying depth below the present surface (Fig. 23–23), each core passes from a zone having prevalence of warmer water species to

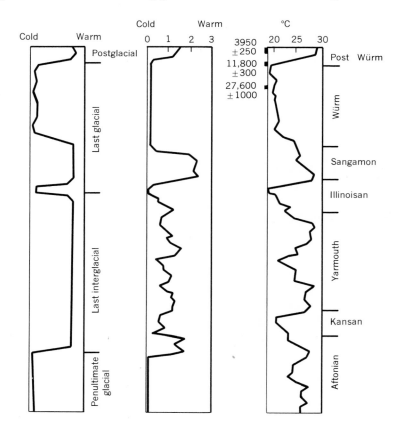

Fig. 23–23. Comparison of climatic curves based on variations in Foraminifera with curves based on the oxygen-isotope ratios in carbonates in the sediments. The curves on the left are based on the relative numbers of planktonic or free-floating Foraminifera that are believed to prefer warm water to those interpreted as cold-water habitants (after D. B. Ericson and G. Wollin); the sediments in the cores are brownish in the warm-water deposits and grayish in the colder water layers. The curves on the right show temperatures based on oxygen-isotope ratios measured in shells of the foraminifer *Globigerina* (after C. Emiliani). In addition, radiocarbon dates are recorded between the two columns on the right (after F. Suess). The curves are based on samples from cores taken from the bottoms of the Atlantic Ocean.

one having colder water forms. Moreover, the relative temperatures can be measured by determining the proportion of the isotopes of oxygen. By means of the study of the oxygen in the carbonate in the shells of the foraminifers, the change from cold has been determined as having been about eleven thousand years ago—thus corresponding rather closely with the determination of the age of wood at the last advance of the glacial ice. Deeper in the core, there are other concentrations of calcareous foraminifera evidencing the temperatures of the surface ocean waters from which they settled. The carbon radioisotope method does not permit age determination beyond a few tens of thousands of years, and these are too old, but the temperature minima can be determined from the oxygen isotopes. If the rate of deposition is assumed to have been rather constant in the core, we can use the distance to each cold-water horizon relative to the known age of the uppermost one to get a judgment of its age. The resultant curve is quite complex but it does have several prominent bands of cold which may be attributed to times of glaciation on the lands. Moreover, they have been correlated with peaks of solar radiation (Fig. 23–24) computed by astrophysicists. This correlation suggests that the glaciations of the past have been related to the varying radiation of heat by the sun and the changing energy received by the earth.

THE PRESENCE OF MAN

There is a suggestion that man may have inhabited the region while the glacier stood near, for in Minnesota, the skeleton of a child having the general skull form of an Indian was found in the excavation for a highway. The critical question was whether the skeleton was buried within the sediments as they were laid, or whether a grave had been dug to place the skeleton, and covered by very recent swamp and forest

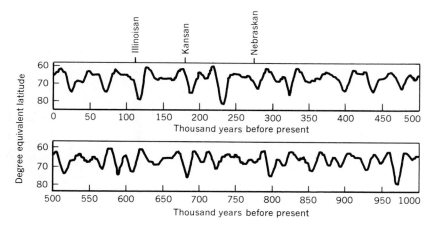

Fig. 23–24. Variations in the insolation at 65° N. Lat. during the past million years in terms of the amount at the present equivalent latitude (from D. Ericson after D. Brouwer and A. J. J. van Woerkom). The interpretations of the times of the main North American glaciations are shown on the right of the diagram; there seem to have been glaciations prior to the advance of the Nebraskan ice sheet.

layers. The direct evidence was not obtained unequivocally, for the bones were removed before they could be examined by experts. Whether or not the skeleton is one of a human who lived as the lake formed, there were men living in the southwestern states at that time, for artifacts, man-made stone tools (Fig. 23–25), have been found with skeletons of animals that did not survive glacial times.

ARTIFACTS

The evidence for the association of early man in North America with animals now extinct has become increasingly documented during the past two or three decades. There were glimmerings of such associations many years ago, but perhaps it can be said that the era of modern scientific proof for the contemporaneity of man and extinct mammals was ushered in during the nineteen twenties, the result of the discovery of an undoubted association of a flint point with the skeletons of extinct bison. This discovery was made near Folsom, New Mexico, and the flint found there, of a distinctive form, became the type of a class of flint points, called Folsom points, now known to be widely distributed in North America. These points are characterized by their fine crafts-manship, as exemplified particularly by a longitudinal groove down the middle of each side; they have since been found at various localities, frequently associated with the remains of extinct mammals. Among the mammals, now vanished, with which early man in America lived, were the several mammoths and the American mastodon, the giant ground sloths, the native horse, the camel, and the long-horned bison (Fig. 23–26). There were other mammals, too, no longer living. So it was that the early Americans lived in a richer and more spectacular world than the one that was unfolded to the wondering view of the early European explorers of this continent.

Fig. 23–25. A few examples of the fluted Folsom points, which have been found at various sites in North America in association with extinct mammals (E. H. Colbert, 1958).

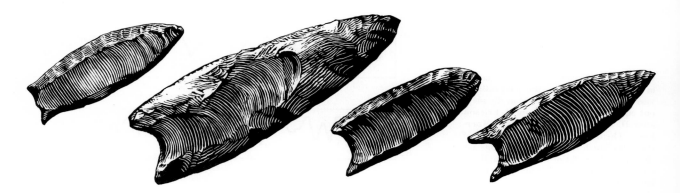

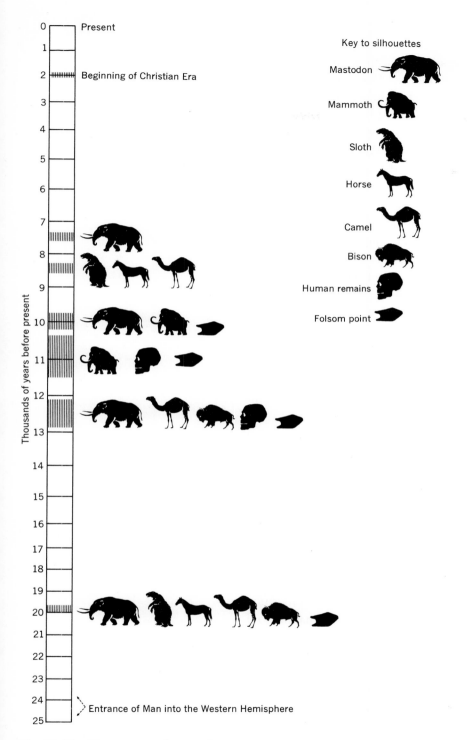

Fig. 23-26. The presence of man and certain mammals, now extinct, in North America, as established by carbon 14 dates.

These early Americans were the ancestors of the modern Indians. They were Asiatic mongoloids who entered the New World across the trans-Bering region, probably in a series of slow migrational waves that extended through some thousands of years. They were Neolithic men, with refined tools and weapons, with companion dogs, and with highly developed cultures. They came into the New World from the north, and made their way through the extent of the two Americas. Then, several thousand years after they came, many of the large mammals among which they had lived became extinct.

As a result of the refined techniques of radiocarbon dating, it has been possible in recent years to establish the time in years when the last of the mammoths and mastodons and other mammals which we think of as being typically "Pleistocene" shared the land with early men in North America. And generally speaking, the dates so derived accord nicely with dates similarly established for the beginning of the glacial retreat from the northern part of the United States. Briefly, man was living with mammals now extinct as late as ten thousand—even as late as eight thousand years ago. Consequently, if we wish to separate the Pleistocene from the Recent, we can set the boundary upon the basis of the retreat of the glaciers and the disappearance of various large mammals.

In a sense, however, we may still be living within the Pleistocene; we may now be enjoying the fourth interglacial stage, a phase of earth history that may well continue for many millenia before the fifth advance of continental glaciers from the north.

DURATION OF THE PLEISTOCENE

If the end of glacial ice in northern United States was somewhat more than ten thousand years ago, what was the duration of the whole epoch? The calcareous late Wisconsin drift has been leached of its lime to a depth of about two and one-half feet (Fig. 23–27); decomposition has not proceeded far. Early Wisconsin drift and the loess that blew off the moraine are leached about twice as much where not covered by later deposits; hence the end of early Wisconsin should be at least twice as long ago and probably somewhat longer, because the bottom foot would not weather as readily as that directly below the surface. Gumbotil still did not form. It has been found that where there are gumbotil and gravels, the gravels are leached about two or three times or more as deep as the gumbotil; so we judge that a foot of gumbotil corresponds to several feet of the leaching of gravels. Four to six feet of gumbotil on the third drift sheet (Illinoian) thus should represent ten feet or more of leaching. By such methods, the time since the first stage of the last glaciation has been estimated as more than twice as long as the time since the last stage, the interglacial stage preceding it as about five times as long, the interglacial time before that as about twelve times as long, and the first interglacial stage as eight times as long. Each figure is a minimum if the leaching of the bottom foot is at the same rate as the top. And in arriving at these figures, we assume that rates of weathering have been rather uniform through interglacial and recent times. The figures add to give the span since early Wisconsin and all the interglacial spans as about thirty times that since the retreat of the last glacial ice. The belief was that twenty-five thousand years had elapsed since the retreat of the ice, based on estimates of the retreat of waterfalls and of the annual layers of varves in post-glacial lakes. The twenty-five thousand years is nearly two and one-half times too great on the basis of the carbon ratio measures. This gives a duration of the intervals of non-glaciation within the Pleistocene as about three hundred thousand years. The multiplication factor is a minimum figure that should extend the duration. The results are in close agreement with those from deep-sea cores and solar radiation (Fig. 23–24).

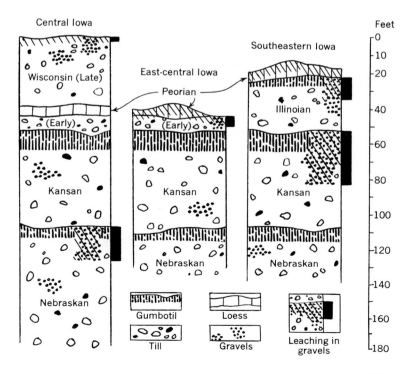

Fig. 23–27. Succession of drifts and associated glacial deposits in three areas in the upper Mississippi Valley, showing the depths of weathering in different materials in a succession of glacial and interglacial stages (after G. F. Kay).

24
Cenozoic Life

The Pleistocene horse, *Equus occidentalis*, of western North America (American Museum of Natural History).

24 *Cenozoic Life*

At the beginning of Cenozoic history, tropical seas circled the globe broadly on each side of the equator, and in these seas life was abundant. The fossil record of life in these waters is preserved in the sediments deposited in the long-persisting Tethys Sea that stretched across the Mediterranean region and covered parts of Europe, North Africa, northern India, and eastern Asia, and similarly in the Caribbean region. There are many likenesses among the fossils from these opposite sides of the world—an indication that the early Tertiary circumequatorial fauna spread widely around the world.

MARINE LIFE

Even though these early Tertiary seas offered extensive tropical environments in which life burgeoned, such life was markedly different from the life of the Cretaceous tropical oceans. Perhaps the most striking difference was the complete absence in the Cenozoic seas of ammonites, the invertebrate rulers of the Mesozoic seas. There were still shelled cephalopods, similar to the pearly nautilus of today's oceans, and undoubtedly many naked octopi and squids and their relatives, but in a large sense the climax of cephalopod evolution had been passed. In the Cenozoic seas the pelecypods and gastropods were the dominant molluscs, especially the former, which evolved along diverse lines of adaptive radiation, to occupy numerous ecological niches in the marine environment. Many of the seas were coral seas, in which an abundant fauna occupied the waters around the reefs. There were numerous echinoderms, for the most part free-swimming types among which the echinoids were particularly prolific. It was during this era of earth history, too, that the modern crustaceans evidently became established in the oceans far and wide, although the fossil record of these animals is admittedly a poor one. And everywhere in the marine waters of the Cenozoic were the foraminifers, protozoans with protective tests, which lived and were fossilized in incredible numbers. These are perhaps the most useful of Tertiary fossils for the correlation of Cenozoic marine sediments.

THE FISHES

The rich spectrum of invertebrate life in Cenozoic seas was enhanced by the swarms of teleost fishes which now swam through the oceans. From the beginning of Tertiary time to the present day the teleosts have been the most numerous and diverse among the vertebrates, and their evolutionary development during this time span has been a record

Fig. 24–1. The relationships of Cenozoic continents to each other, showing routes for and barriers to migrations of land-living animals.

Fig. 24–2. Cenozoic climatic zones of North America.

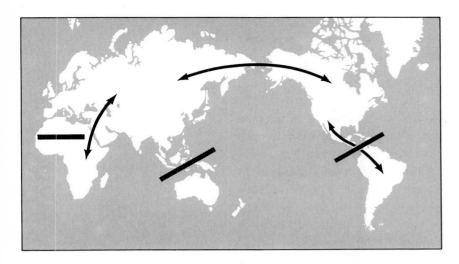

LATE PLIOCENE

TUNDRA
SUBARCTIC
TEMPERATE
SUBTROPICAL
TROPICAL

PRESENT
0 MILES 1000

TUNDRA ICE
SUBARCTIC
TEMPERATE
SUBTROPICAL
TROPICAL

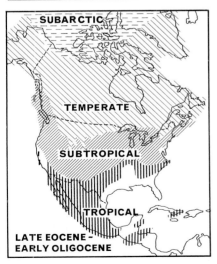

SUBARCTIC
TEMPERATE
SUBTROPICAL
TROPICAL

LATE EOCENE -
EARLY OLIGOCENE

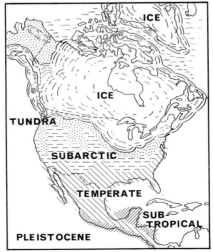

ICE
ICE
TUNDRA
SUBARCTIC
TEMPERATE
SUB
TROPICAL

PLEISTOCENE

607

of constant expansion, of ever more complex adaptations to ever-increasing environments. As has already been said, the Cenozoic was an Age of Teleost Fishes as fully as it was an Age of Mammals.

MARINE MAMMALS

The mammals, which were so ubiquitous and dominant on the land during the extent of the Cenozoic Era invaded the oceans at an early stage in the history of this great division of geologic time, just as in the early years of the Mesozoic the reptiles had invaded the seas. In short, the cetaceans or whales arose during the Eocene Epoch, probably from primitive carnivorous land mammals, and within a very short time, geologically speaking, became marvellously adapted for life in the open ocean. Many of the whales became the animal giants of all time, far exceeding the largest dinosaurs in size; many others, such as the numerous porpoises and their relatives, evolved as very fast-swimming predators of moderate size, in form remarkably similar to the ichthyosaurs of the Mesozoic. Here, as a result of convergent evolution, the small cetaceans, descended from completely different ancestors than those which gave rise to the ichthyosaurs, nevertheless imitated the reptiles in form and function because of the strict limitations invoked by the environment on animals living the kind of life that was lived by the ichthyosaurs and that is lived by the porpoises.

Fig. 24–3. The North American Eocene bird, *Diatryma*, as restored by Erwin S. Christman (American Museum of Natural History). This bird, as large as an ostrich, represents the evolutionary trend toward giantism and ground living that marked the development of birds during early Tertiary times, before carnivorous mammals had grown large.

It might be said at this place that other mammals also took to the sea during Tertiary times, some at an earlier, some at a later date than did the whales. These were the sea cows, of Paleocene origin, and the seals, sea lions, and walruses, first appearing in Miocene sediments.

MARINE TEMPERATURES

Thus we may envision the seas as broadly tropical and abundantly inhabited during the opening phases of Cenozoic history. The oceans continued to be abundantly inhabited during the extent of the Cenozoic, but as time went by the extent of tropical seas became continually restricted, for the Cenozoic Era was an age when climates evidently became zoned as they had never been zoned before. The oceanic waters of middle and high latitudes became progressively cooler as Cenozoic time continued and many marine faunas became adapted for cool and even for very cold waters.

Of course, the development of zonal belts in the oceans was not as simple as this; indeed it must have been extraordinarily complex through time. We know from our modern oceans that the several great oceanic currents, such as the Gulf Stream, cross wide belts of latitudes, altering the environments of sea and of land along the broad course of

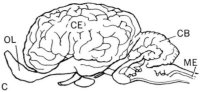

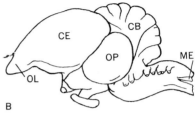

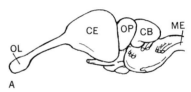

Fig. 24–4. Brains of a reptile, at the bottom, a bird in the middle, and a mammal at the top, as seen in lateral view. The comparison illustrates the great expansion of the cerebrum, the "thinking part" of the brain in the mammals, as contrasted with birds and reptiles. It emphasizes the crucial attribute of intelligence that was to lead to the dominance of the mammals during Cenozoic history. CE—cerebrum; CB—cerebellum; OP—optic lobe; ME—medulla; OL—Olfactory bulb.

their flow. And similar conditions may have held to a lesser degree during much of Cenozoic time, with consequent extensions and restrictions of faunas and floras in patterns sometimes transcending the climatic zones that were becoming ever more sharply defined.

GEOGRAPHIC ASPECTS AND MIGRATIONS

Certainly the Cenozoic era was the time when oceans became confined largely within their present limits. Likewise, it was the time when the continents gradually assumed their modern shapes and relationships each to the other. Of course, there were differences in continental outlines from those of the present time during the various Cenozoic epochs, but the differences decreased as the Cenozoic extended through the years. For example, the Gulf Coast and Florida were widely submerged during early Tertiary time but during the later years of the Tertiary the land kept emerging, with the sea correlatively retreating toward its present limits.

At various times within the Cenozoic Era, North America was connected to Asia across what is now the Bering Strait, so that various land animals could wander from the one continental land mass to the other, and at intervening intervals the connection between the continental masses was broken (Fig. 24–1). Eurasia was connected to Africa as it is today, although during the several Cenozoic epochs the location and extent of the connections differed from their present placement. After the Cretaceous, Australia was separate from the Asiatic continent, a separation that probably took place during the transition from Mesozoic to Cenozoic times; and throughout most of the Tertiary Period, South America was also an island continent, isolated from the rest of the world. It was linked to North America by an isthmian connection at the very beginning of Tertiary history, but the link was soon broken and remained broken until the end of Tertiary history.

These are relationships to be kept in mind in any review of Cenozoic life, because they determined not only the migration routes by which animals and plants spread from one continent to another, but also the courses of oceanic currents which had definitive effects upon the climates. Climates determined the distributions of many land-living organisms and, of course, the distributions of marine faunas as well. As for continental climates, some very important interpretations have been made within recent years as a result of the study of the distributions of Tertiary plants. It will be remembered that Mesozoic environments have been reconstructed to a large extent from continental reptilian distributions. Of course, the occurrences of large Cenozoic reptiles (this meaning, in effect, the crocodilians) can be of similar help in the

Fig. 24–5. The relationships of the orders of placental mammals (E. H. Colbert, 1955). The various orders were well defined by early Tertiary times, and most of them were carried through into late Cenozoic history. Several orders of primitive hoofed mammals became extinct in the early Tertiary, probably because of the competition from other, more highly adapted hoofed mammals.

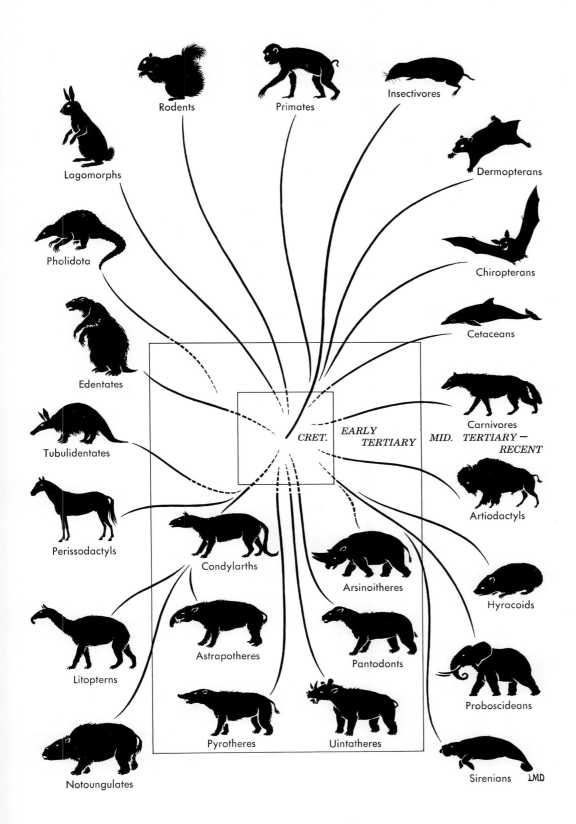

Lagomorphs

Rodents

Primates

Insectivores

Dermopterans

Chiropterans

Cetaceans

Pholidota

Edentates

Tubulidentates

Perissodactyls

CRET.   *EARLY*
          TERTIARY   *MID.*   *TERTIARY —*
                                 RECENT

Carnivores

Artiodactyls

Condylarths

Arsinoitheres

Hyracoids

Litopterns

Astrapotheres

Pantodonts

Proboscideans

Notoungulates

Pyrotheres

Uintatheres

Sirenians   LMD

611

interpretation of Cenozoic environments, but palaeoclimatological knowledge so gained is most usefully reinforced by what is to be learned from plant distributions during the past seventy million years. Fossil mammals which are the overwhelmingly dominant organic remains in most Tertiary continental beds are of limited use in the interpretations of past environments, because these warm-blooded animals (and likewise the warm-blooded birds) were in past ages, as in the present, not very strongly controlled by the external temperatures to which they were subjected.

CONTINENTAL CLIMATES AND FLORAS

Therefore, we return to the paleobotanical evidence concerning continental climates of Cenozoic times, and there can be no doubting the value of such evidence. Comparisons of the vegetation of modern forests and of other environments with the fossil record makes it possible to reconstruct with some feeling of confidence the topographic and climatic conditions that prevailed during the extent of Cenozoic history. The angiosperms that arose in early Cretaceous times and continued to evolve during the close of Mesozoic and the totality of Cenozoic times, are, like the reptiles, sensitive to the temperatures in which they live. If we

Fig. 24–6. *Barylambda*, a North American Paleocene amblypod or primitive hoofed mammal, as restored by Walter Ferguson (American Museum of Natural History). This was a large, herbivorous mammal, far more primitive than the hoofed mammals of the Eocene and later epochs. Note the low, generalized head, and the five-toed feet.

suppose that the oaks and willows and sassafras and other broad-leaved trees of the Tertiary Period were tolerant to much the same climates as are their modern relatives, and this is the only logical supposition that can be made, then we have in these fossils, as well as in the fossil seeds of grasses and other plants, temperature-sensitive organisms which by their distribution give us positive clues as to the nature of past climates.

The evidence indicates that after the beginning of Tertiary time there was for some millions of years a trend for floras to spread from the tropics toward the poles in the northern hemisphere, so that tropical and subtropical forests extended to latitudes of 50 degrees, in other words, to a region somewhat to the north of the United States–Canadian boundary in North America, and through Northern Europe in the Old World. The evidence for the southern hemisphere, which is not so clear, indicates that such forests were less extensive in that part of the world during early Tertiary time (Fig. 24–2).

Fig. 24–7. The replacement of archaic Eocene mammals by progressive Oligocene mammals (animals after W. B. Scott, 1937).

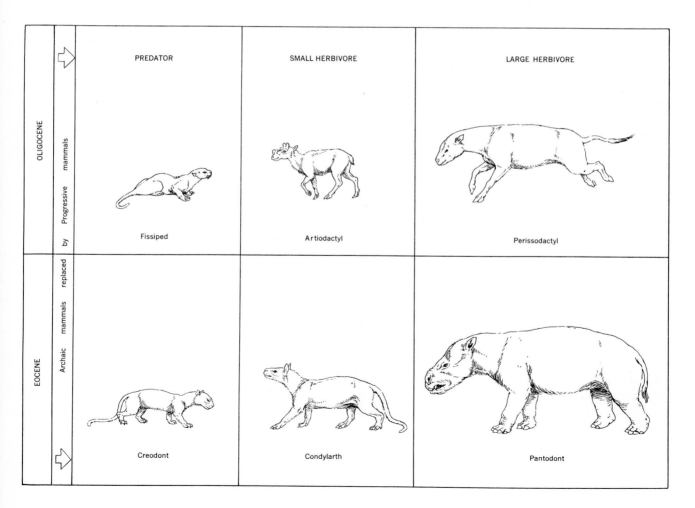

613

The gradual uplift of the continents during the Oligocene Epoch brought about the beginning of a restriction of tropical and subtropical climates, and through the remainder of Tertiary time the northern limits of the tropical-subtropical forests in North America moved southwardly, giving way to temperate floras coming down from Alaska, the southern borders of which also moved southwardly during the course of the years. Floras became ever more sharply zoned and ever more diverse, although it would seem that as late as the Miocene rather uniform forests covered much of the northern hemisphere. But during the Pliocene Epoch, climates became cooler and more varied, as is indicated by the distribution of the fossil plants.

The climax of this progressive change toward stronger zonation and greater diversity of climates was reached, of course, during the Pleistocene Epoch, the great Ice Age, when polar ice caps became extensive, and when in the northern hemisphere great continental glaciers advanced across the lands four successive times, each advance being followed by a glacial retreat and an interval of warm humid climates.

Thus the evolution of Cenozoic climates followed a trend from early uniformity and the wide extent of tropical and subtropical conditions—conditions that were frequently prevalent during the Paleozoic and Mesozoic eras—toward increasingly greater diversity, with the tropical-subtropical zone gradually restricted more and more to the equatorial and bordering regions, and with temperate climates moving southwardly to occupy the middle latitudes. By the end of the Cenozoic, climates had become strongly diversified, with truly arctic conditions characterizing the polar regions. It was in such evolving climates, as interpreted from the evidence of paleobotany, that the vertebrate faunas of the Cenozoic developed, with their constituent members showing progressive changes in adaptation to the changing environments.

THE BIRDS

It was mentioned in Chapter 22 that the birds were structurally highly advanced by the beginning of Cenozoic time, and that with the progress of Cenozoic history, evolution among these particular vertebrates was a matter of wide adaptations for varied environments. Very probably many of the complex inherited behavior patterns so characteristic of modern birds were also developed during the course of Cenozoic time.

Of course, the dominant trend in avian evolution was toward free flight, with food-gathering and reproductive activities spread over a wide range of environments—in trees and among low plants, on the ground and in the water. But at the beginning of the Tertiary Period there was established one trend of evolution among the birds that has been variously successful and frequently repeated during the span of Cenozoic time, namely the development of ground-living giants. One of the earliest of these terrestrial avian giants was *Diatryma,* found in the Eocene of Wyoming (Fig. 24–3). This robust bird possibly performed an ecological role that soon was to be taken over by large carnivorous mammals, and it represents an early adaptation among birds promising much but failing to continue. This large, predatory bird and others that may have similarly developed during early Tertiary time, was able to prosper only so long as the meat-eating mammals remained relatively small. As soon as the mammalian carnivores became large they became very potent predators within their faunas, and fulfilled their roles most thoroughly and successfully. Since the rise of large mammalian carnivores, giant ground birds have evolved time and again, frequently as independent lines of evolution, but commonly in habitats where they have been safe from the depredations of predators. Such were the gigantic Miocene phororhacids of South America, living in a fauna where the only carnivores were marsupials. Such was the Pleistocene elephant bird of Madagascar and the moas of New Zealand, living on islands devoid of large carnivores. Such today are the cassowary in Australia, again on an island devoid of any carnivores except the introduced dingo. And such, also today, are the rheas of South America and the ostriches of Africa, birds that have become so fleet of foot as to be able to escape their enemies.

Fig. 24–8. Adaptive radiation among the modern marsupials of Australia, the result of their isolation, and protection throughout Cenozoic times from competition with placental mammals (E. H. Colbert, 1955). Only a few of the many different adaptive types of marsupials are shown, but in this small sample we see a predator, the thylacine or Tasmanian "wolf," and several herbivores and insectivores, specialized for living in trees, for burrowing in the ground, and for living in various terrestrial habitats. Drawing by Lois Darling.

Marsupial "Mouse"

Tasmanian "Wolf"

Wombat

Phalanger

Koala

Bandicoot

Marsupial "Mole"

The mammals, in spite of their lesser abilities for locomotion than those possessed by the birds (many waterfowl are unique among animals by being simultaneously adept as flyers, as walkers and as swimmers and divers), benefited through the Cenozoic from certain evolutionary developments that were denied to the birds. The most striking of these was in the development of the brain. The high structural perfection attained by the birds at an early stage in their evolutionary history, thus enabling them to conquer the air, had the effect of putting an evolutionary straitjacket around the brain. The bird brain evidently was wonderfully developed for the control of balance and reflex actions by the beginning of Cenozoic time, and so it has remained to the present day, but in this brain there has been little development of the cerebrum, of the thinking part of the brain, in contrast to cerebral expansion in the mammals, especially in the placental mammals. Consequently the birds during Cenozoic times remained as primarily automatic animals, governing their actions and regulating their lives by inherited behavior patterns. In contrast, the mammals through the Cenozoic for the most part became increasingly independent animals, capable of meeting situations according to the circumstances, capable in greater or lesser degrees of solving "problems." In short, the evolution of the mammals during the Cenozoic Era was a story of the evolution of intelligence.

THE MAMMALS

This crucially important attribute, intelligence, shaped much of the evolutionary development of the mammals during seventy million years of Cenozoic history (Fig. 24–4). Of course, structural adaptations were of prime importance, too. It was the combination of wide structural changes to fit the mammals for life in diverse habitats, and the generally ever-expanding brain that were the paramount factors in the successful adaptive radiation of these animals. Theirs has been a history of diversity and perfection.

This history is much too involved for elucidation here, but a few outstanding events and principles involved in it may be briefly discussed. In the first place, with the disappearance of the dinosaurs and other ruling reptiles of the Mesozoic, great ecological lacunae were made available to land-living animals, to which the mammals responded. At the close of Cretaceous time the only mammals on the earth, so far as we can read the fossil record, were, as mentioned in Chapter 22, multituberculates, opossum-like marsupials and shrew-like and hedge-hog-like insectivores. Undoubtedly there were a few other groups, unrepresented as fossils, but the mammals were obviously limited in diversity.

At the opening of the Cenozoic, in Paleocene deposits we find, in addition to the three mammalian orders surviving from the Cretaceous fourteen other orders, all placental mammals (Fig. 24–5). These fourteen orders of placentals obviously arose from insectivore ancestors, and their initial evolutionary separation from an insectivore stem and early diversification must have taken place in a remarkably short span of geologic time, namely during the transition from the Cretaceous to the Paleocene. This primary diversification of Paleocene mammals established them in many habitats, as tree-living types, as burrowing animals, as herbivores of many kinds, ranging from those of small size to those of considerable size, and as predatory carnivores. In short, the initial evolution of Paleocene mammals resulted in an integrated fauna, with placental mammals adapted for many various roles.

SUDDEN APPEARANCE OF NEW FORMS

Here is an example of quantum evolution, of the sudden appearance within a short space of time of large taxonomic units—units in this instance at the level of orders. This phenomenon is encountered frequently in the stratigraphic record, and has been accorded much attention by paleontologists interested in evolutionary problems. One characteristic of quantum evolution, as preserved in the geologic record, is the lack of annectent forms. Thus new groups of high taxonomic rank suddenly appear, commonly without any fossils to show the transition between them and their putative ancestors. In such instances, we can only suppose that there were exceedingly rapid evolutionary trends in very definite directions to fill ecological vacancies, developments taking place at such speed that we are unable to find traces of them in the record of the rocks. In effect, there was an evolutionary explosion of mammals as a result of quantum evolution, with the advent of the Paleocene Epoch.

THE ARCHAIC FAUNAS OF THE PALEOCENE

This explosion established the archaic faunas so characteristic of Paleocene deposits (Fig. 24–6). The word "archaic" is used here to express the fact that these assemblages were, in terms of mammalian evolution, indeed composed of primitive mammalian types. They filled the available ecological niches on land, as has been said, but in many respects on a lower level of efficiency than is the case among modern-day mammals. There were numerous plant-eating herbivores, but they evidently did not transform the energy which they derived from the plants they ate into animal energy with the same degree of completeness as do their modern counterparts. Moreover, they were comparatively clumsy of foot and slow. Likewise the carnivores that preyed upon them were probably rather slow and less efficient as predators than modern carnivores. Yet strange and perhaps imperfect though these mammals might seem to our modern eyes, they were far advanced beyond reptiles in their adaptations for life. The tempo of life on the land had increased remarkably with the rise of the Paleocene mammalian faunas.

The archaic faunas thus established persisted as a matter of fact into and through much of the Eocene Epoch. These early faunas, which peopled the continents of the earth, contained animal groups of two categories. In the first place, there were the truly archaic mammals which in time were destined to become extinct. These were groups which during their heyday were quite successful, but lacked the potentialities for continued evolution in a changing world. Such were the persistent multituberculates, the strange-looking herbivorous taeniodonts and vaguely rodent-like (but large) tillodonts; such were the many large and even gigantic plant-eaters, among them the generalized condylarths and the heavy amblypods.

Secondly, there were within the archaic faunas of early Tertiary times groups of animals, the members of which though primitive, were destined to be ancestral to animals living today. Such were the marsupials, the insectivores which have continued through the years as primitive types, and their direct derivatives the primates, the early carnivores known as creodonts, the rodents, the lagomorphs or rabbits, the edentates, the sea cows and other groups.

MORE PROGRESSIVE LATE EOCENE MAMMALS

Although the archaic mammalian faunas, thus characterized, flourished through the early years of Tertiary history, they began to assume new appearances during Eocene time by the rise of certain truly progressive groups of mammals. By the late Eocene, mammalian faunas were of mixed composition, being made up of the very archaic elements soon destined to disappear, of the groups old in origin but strong in evolutionary potentialities, and in addition of new orders that evidently arose during Eocene times (Fig. 24–7). Among these latter were the proboscideans represented by the first mastodonts, the ancestral perissodactyls or odd-toed hoofed mammals and the ancestral artiodactyls or even-toed hoofed mammals. There should be mentioned also the appearance at this time of the first modernized carnivores, the fissipedes,

Fig. 24–9. The adaptive radiation of Tertiary mammals in South America (animals after W. B. Scott, 1937). Here we see varied autochthonous forms, ultimately derived from North American ancestors, but which evolved along independent lines during the greater part of Tertiary time when South America was an island continent.

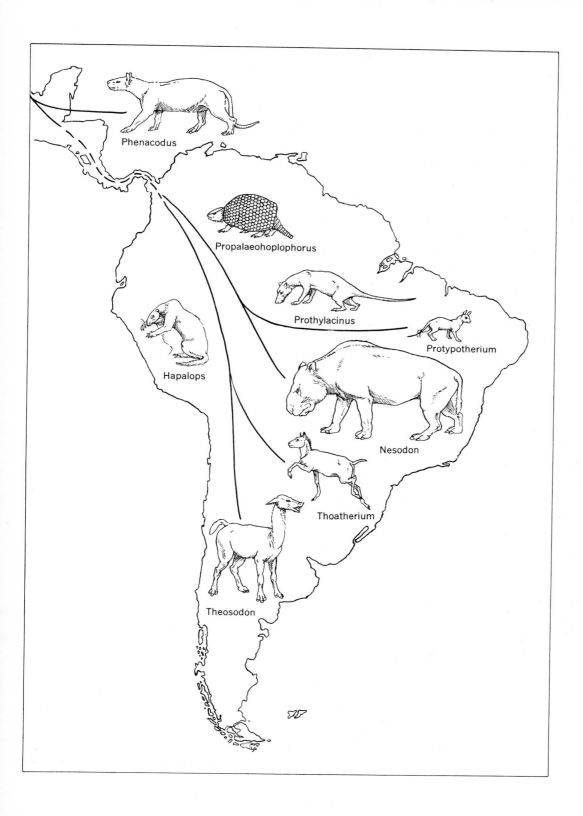

Phenacodus

Propalaeohoplophorus

Prothylacinus

Protypotherium

Hapalops

Nesodon

Thoatherium

Theosodon

which though of less than ordinal rank in most systems of classification, are equally important in an evolutionary sense to the groups just cited.

The appearance of these particular groups of Eocene mammals brought about the replacement of many members which had been so prominent in the archaic faunas of early Tertiary time. Once again we see an evolutionary phenomenon that has taken place frequently in the development of life through the ages. The perissodactyls, artiodactyls, and proboscideans were destined during the transition from Eocene to Oligocene times to replace the various ancient herbivores throughout most of the world. The fissipedes, among them the first of the dog group, the mustelids, the civets and the cats, were destined to replace the older and clumsier creodont carnivores, although this replacement was more gradual than that of the archaic herbivorous mammals. With such replacements taking place, and with the progressive evolution of various other mammalian orders of ancient lineage, the faunas of all the continents except Australia and South America had by the beginning of the Oligocene epoch begun to have rather "modern" characteristics. They were composed of animals that to us would seem more or less familiar.

FAUNAS IN AUSTRALIA AND SOUTH AMERICA

At this place, we may digress briefly from the consideration of the development of Tertiary mammalian faunas, so well exemplified by the fossils found in North America, to consider in very general terms the Cenozoic mammalian faunas of Australia and South America. It would seem that when Australia became isolated from the rest of the world at the close of Cretaceous time, the mammalian inhabitants of this great region were marsupials (and probably monotremes too) and nothing else (Fig. 24–8). Thus during the Cenozoic Era Australia

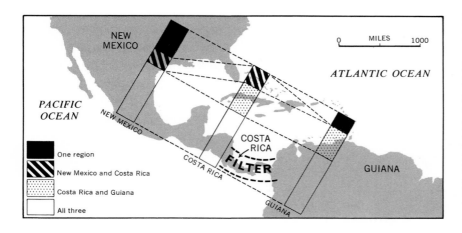

Fig. 24–10. The filter bridge of the Isthmus of Panama (after G. G. Simpson, C. S. Pittindrigh, and L. H. Tiffany, 1957). The bars indicate the relationships of modern mammalian faunas in New Mexico, Costa Rica, and Guiana. There has been a large interchange of mammals between southern North America and northern South America, as indicated by the blank segments of each bar. Nevertheless a considerable proportion of animals reach the isthmus from the north and from the south, but never get beyond this constricted land bridge. This is the filter effect. Some mammals never leave their homelands, as indicated by the solid segments of the bars representing the New Mexican and Guianan faunas.

became, after the extinction of the great reptiles, a theater for marsupial radiation, and these mammals assumed the many roles which in the rest of the world were performed by placentals. Australia became the home of marsupial herbivores and carnivores, marsupial "squirrels" and "moles" and so on. Only the rodents, which evidently entered the island continent as waifs on floating logs, and the bats, which could fly across the oceanic barriers, represented placental mammals on the isolated continent before the immigration of man.

South America likewise became isolated from the rest of the world as an island continent, but not until after the beginning of Tertiary time. Consequently, this region was inhabited by various archaic placentals before its separation from North America had occurred, by the drowning of the isthmian link. From that time until near the end of the Tertiary Period, South America was a detached theater for the evolution of mammals quite different from those which were evolving elsewhere. Yet although the Tertiary mammals of South America evolved in complete isolation from the mammalian faunas of other continents, their evolutionary development eventually was to lend considerable distinction to the very late Cenozoic faunas of North America, when the isthmian link was re-established.

The most notable feature of the successive Tertiary mammalian faunas in South America (Fig. 24–9) was the radiation of a whole complex of autochthonous herbivores, descended in the main from condylarths that had reached the southern land mass prior to the severance of the isthmian link. The bulk of these plant-eating mammals belongs to an order known as the notoungulates, within which there was great diversity. The early notoungulates were rather generalized hoofed mammals, but as Tertiary time extended through the years, the notoungulates spread into various habitats. Some of them, the toxodonts,

Fig. 24–11. A scene on what are now the high plains of North America during mid-Tertiary times, as restored by Charles Knight (American Museum of Natural History). In this warm temperate, savannah environment lived antelope-like artiodactyls (left foreground), long-necked and long-legged camels (left background), rhinoceroses (center) and three-toed horses (right).

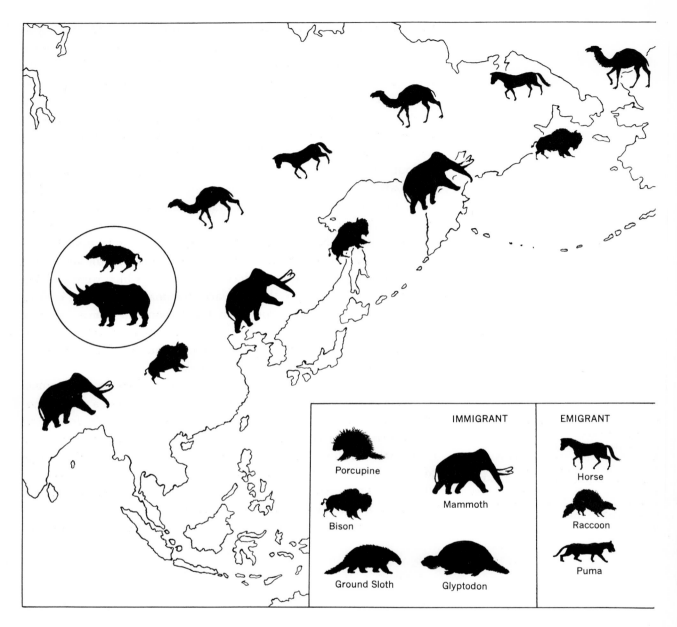

Fig. 24–12. The intercontinental migrations of certain mammals during the Pleistocene epoch. Many groups of mammals crossed back and forth between Asia and North America by way of the Bering Straits filter bridge, and between North and South America by way of the isthmian filter bridge. Horses and camels migrated from North America to Asia, mammoths and bison in the opposite direction. Pumas, raccoons, mastodonts,

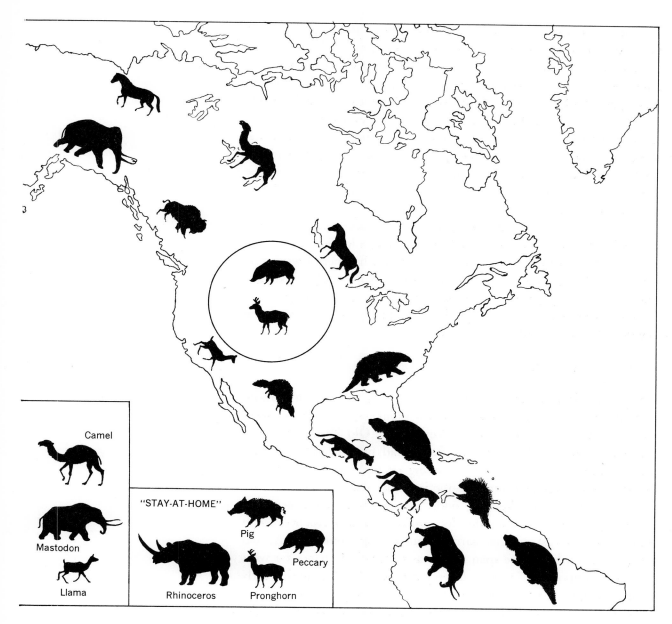

Camel

Mastodon

Llama

"STAY-AT-HOME"

Pig

Rhinoceros

Pronghorn

Peccary

horses, and llamas migrated from North to South America; ground sloths, glyptodonts, and porcupines in the opposite direction. Note that the rhinoceroses and pigs were confined by the Bering filter bridge to Asia, the pronghorns and peccaries to North America. These routes and barriers to intercontinental migrations were important in giving to the modern faunas of the several continents their distinctive characters.

were more or less like large rhinoceroses, some, the entelonychians, like the Cenozoic clawed chalicotheres of northern lands, and some, the typotheres, rather like large rabbits. Another order contains the large, heavy astrapotheres rather difficult to compare with herbivores of other lands. Still another South American order, the litopterns, contains the camel-like macrauchenids and the proterotheres, remarkable parallels to the horses that were simultaneously evolving in North America. Another order of mammals that evolved in the island continent of South America was the pyrotheres, gigantic mammals that may be compared in a general way with the mastodonts that were populating other continents.

The evolution of numerous herbivores was accompanied by the evolution of carnivores that preyed upon them, as has inevitably been the case through time. These carnivores in South America were marsupials, some of them more or less like foxes or wolves, others like cats. One of the last of the carnivorous marsupials of this region was *Thylacosmilus*, a saber-toothed predator uncannily similar to the great saber-toothed cats of the northern hemisphere.

PROBLEMS OF MIGRATION

One group of particular interest that had a long evolutionary history in South America is the monkeys. These monkeys, of common ancestry with but distinct from the monkeys of the Old World, evidently reached South America as waifs after it had become isolated as an island continent. Here is an outstanding example of what has been called sweep-stakes distribution—the invasion of a new area by some organism against overwhelming odds. On the face of the matter it would seem virtually impossible for monkeys to have reached South America across vast oceanic barriers, but geologic time is long, and through such a long time span there was a statistical chance for what seems the "impossible" to have taken place. At some time during early Tertiary history, it would seem that some monkeys were rafted to the island continent where they established themselves successfully. Then as time went on, they evolved in wide variety in this new and agreeable habitat to become almost as varied, but in distinct ways, as the monkeys of the Old World.

It is probable that the varied native rodents of South America evolved from ancestors that, like the monkeys, reached the continent as waifs.

The edentates, sloths and armadillos, ground sloths and glyptodonts, the ancestors of which had reached South America from the north at a very early stage in Tertiary history, evolved widely in this isolated region, and at the close of the Cenozoic invaded North America across the re-established isthmian link. And there was also an invasion from south to north of certain rodents.

The isthmian link along which these South American mammals traveled to the north was re-established near the close of Pliocene time (Fig. 24–10). The late Tertiary emergence of the isthmus had profound effects upon all faunas, both continental and marine. No longer was there a free passage from the Atlantic to the Pacific across the isthmian region, as there had been through much of Tertiary history. No longer was the land block to the north separated from that to the south.

Although there was a migration of ground sloths and glyptodonts, of porcupines and capybaras from South America into North America, by far the predominant migration was in the opposite direction. Various dogs and bears, mustelids, cats, horses, tapirs, peccaries, camels (llamas and their relatives), deer, and other animals crossed the isthmus from north to south, and these progressive and vigorous invaders spelled the doom of many mammals which for so long had been dominant on the southern continent. The South American herbivores could not hold their own against the North American plant-eaters. The carnivorous marsupials gave way before the influx of placental carnivores. The native rodents held on, as did the monkeys and edentates, and these mingled with the invading forms to constitute the South American faunas of late Pleistocene and Recent times.

The passage of various mammals in both directions along the Panamanian isthmus during late Cenozoic time illustrates very nicely the effect of what has been aptly termed a "filter bridge." The isthmus provided a link between the two continents and at the same time a sort of filter, that would seem to have allowed the intermigrations of some faunal elements but not of others. Thus in spite of the wanderings of many mammals back and forth along this bridge, the monkeys never made the trip all of the way (although they did go part of the way along the isthmus) from the south, and beavers, pronghorn antelopes and bison, among others, never made the trip from the north. It is not easy to deduce the reasons for such filtering effects in many cases; but we are certain of their impacts upon the faunas between which filtering migrations have taken place.

NORTH AMERICAN OLIGOCENE MAMMALS

This South American digression has taken us ahead of the story in North America. It will be recalled from page 620 that by the beginning of the Oligocene Epoch the mammalian faunas of North America and various other continents had begun to assume a rather modern look. There had been extensive replacements of archaic types by progressive types of mammals, and these were to be the ancestors of the mammals of today.

LATER CENOZOIC DIVERSION

During middle and late Cenozoic time the development of mammalian faunas was largely a story of divergent evolution among these progressive groups of mammals which had arisen during the Eocene-Oligocene transition. The evolutionary development of horses, briefly recounted in Chapter 21, is an outstanding example in the fossil record of such a phylogenetic history. But there were many others as well, many of them not as completely documented as the story of horse evolution, yet none the less with records sufficiently complete as to give clear ideas of their directions. Some of the phylogenies are known in great detail.

North America was a center for the continuation of the marsupials and the evolution of insectivores and bats, of dogs and bears, raccoons, mustelids, cats, rodents, lagomorphs or rabbits, mastodonts (which entered this region in the Miocene), horses, tapirs, rhinoceroses, chalicotheres, titanotheres (of Eocene and Oligocene age), peccaries, oreodonts, anthracotheres, camels, deer, and various pronghorns. Along the shores of the continent lived sea lions, seals and walruses, sea cows, porpoises, and whales. The development through time of all of these animals can be interpreted from the fossils that occur in the Cenozoic sediments of North America (Fig. 24–11). Some of these animals, such as the oreodonts and the pronghorns were strictly North American during the entire course of their evolutionary histories, some, such as the peccaries and representatives of various other groups, worked their way into South America at a late date, and some, such as horses and camels emigrated widely throughout the world.

PLEISTOCENE FAUNAS

With the advent of the Pleistocene Epoch, the North American mammalian fauna was augmented by incursions from other regions (as of course it had been at various times during Tertiary history) to enrich it and add variety to its complexion (Fig. 24–12). Porcupines, capybaras, armadillos, giant ground sloths, and armored glyptodonts came in from the south, as we have seen. From the Old World, by way of a Bering filter bridge, came elephants (the mammoths of the Ice Age), stags, and bison. Strangely, the Old World pigs and rhinoceroses (the native North American rhinoceroses had become extinct during Pliocene times) never made the crossing into North America (Fig. 24–13).

Fig. 24–13. A scene at the Rancho la Brea tar pits in southern California during very late Pleistocene time, as restored by Charles Knight (American Museum of Natural History). On the right are giant ground sloths, one of them mired in the tar. On the left are saber-toothed cats (*Smilodon*) and in the tree above them giant vultures. In the background are mammoths and wolves. The Rancho la Brea pits, now preserved in a park in the center of Los Angeles, have yielded vast quantities of bones of the animals that became entombed there during the late phases of Pleistocene history.

Mention of the Ice Age brings up a point that may be made in connection with the origins and intercontinental migrations of certain mammals. As has previously been said, the criteria of glacial advances as indications of the beginning of the Pleistocene Epoch, although applicable to a considerable degree in the northern hemisphere, do not hold over large areas of the earth. Therefore, reliance has been placed upon certain standard marine sections in Italy, and likewise upon the first appearances of modern horses, elephants, and cattle in the continental beds of this same region, as criteria for the beginning of Pleistocene time. The evidence of these mammals is especially valuable, because they are progressive animals of large size that, geologically speaking, spread instantaneously to all parts of the world. Horses of the genus *Equus,* originating in North America, spread to all of the continents except Australia and most of this migration took place with the advent of Pleistocene time. Elephants of the genus *Archidiskodon* and cattle of the genus *Bos* originating in the Old World, likewise spread widely although they did not, of course, reach Australia, nor did they populate South America.

Pleistocene conditions were at times harsh for animal and plant life throughout much of the northern hemisphere. The four advances of continental glaciers drove faunas and floras to the south, the retreats (and we are living in the fourth interglacial stage) allowed life on the continents to spread to the north again. Likewise, these alternating phases of refrigeration and warmth affected life in the oceans, so that at certain times walruses, animals of icy waters, lived far down our

southern coasts. Yet varied as were the environments of the Pleistocene, they none the less were suitable for rich mammalian faunas throughout the world. Above all, the Pleistocene was a time when large mammals dominated landscapes, when mastodonts and mammoths, ground sloths and glyptodonts, gigantic beaver, great bears, and saber-toothed cats, stags and giant bison roamed across the North American continent.

THE ADVENT OF MAN

But of particular significance is the fact that the Pleistocene was the time when man arose from ape ancestors and evolved into modern man. It was in the Old World that the evolution of man took place (Fig. 24–14). The first men would seem to have lived in Africa (perhaps elsewhere in the Old World, too, but it is Africa that yields the fossil record in cave deposits of probable early Pleistocene age). These were the creatures known as australopithecines, very ape-like in many respects but showing human characters in their teeth and in their upright posture. These early ape-men, if so they may be called, probably wandered in small groups and gathered food with the aid of sticks and even crudely shaped implements. Perhaps they had a language of sorts. Perhaps they had fire.

However we may regard the australopithecines, there can be no doubt as to the thoroughly human traits of the pithecanthropines, found in terrace deposits in Java, and in cave deposits near Peking, China. These are the so-called Java man and Peking man, of middle Pleistocene age. These were large-skulled men, with large brains but with low foreheads and massive jaws. They were rather short of stature. They were hunters who used flint implements and had fire. They undoubtedly had a culture.

Culture, man's crowning achievement, became well established and highly developed by late Pleistocene time. The men of the late Pleistocene were primarily of two types—Neanderthal man and modern or *sapiens* man. The Neanderthals, who lived during the last glacial stage and died out perhaps twenty-five thousand years ago, were short, heavy men with low foreheads and massive jaws. But they had very large brains and obviously were intelligent men. They made exquisite flints, and buried their dead with considerable ceremony. The modern men, our own ancestors, were essentially like ourselves and need no particular description. Among these men the Cro-Magnons of southern Europe were remarkable craftsmen and artists, who made lively and beautiful paintings deep in the recesses of caves through southern France and northern Spain. The pictures they made, together with sculptures, carved objects and tools of bone and ivory, tell us much

Fig. 24–14. Early man, after restorations by Lois Darling (E. H. Colbert, 1955). *Australopithecus* from the Pleistocene of Africa, is perhaps the most primitive of known men. *Pithecanthropus*, from the middle Pleistocene of Eurasia, knew the use of fire and had evolved a culture. Neanderthal and Cro-Magnon men inhabited Europe and other parts of the Old World in late Pleistocene times. Neanderthal man had a highly evolved culture and was skillful in the fashioning of flints. Cro-Magnon man of Europe made the beautiful cave paintings that are preserved in southern France and Spain.

Cro-Magnon

Pithecanthropus

Neanderthal

Australopithecus

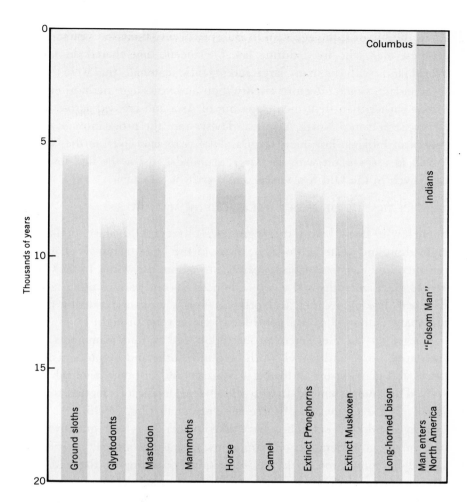

Fig. 24–15. The extinction of various mammals in North America. Most of the extinct large mammals that we think of as characteristically Pleistocene disappeared between five and ten thousand years ago—this has been established by careful radiocarbon dating. Camels may have persisted a little longer. Man entered North America from Asia by way of the Bering filter bridge probably as far back as twenty thousand years ago, possibly even earlier. The Folsom culture (see Fig. 23–25) can be dated as early as ten thousand years before the present.

about the life of stone-age man in Europe, twenty thousand years ago.

These men who lived during late Pleistocene time shared the Old World lands with the many large and gigantic mammals that were then so abundant. Some fifteen or twenty thousand years ago, perhaps even a little earlier than that, men came out of Asia and crossed across the Bering region into North America. These were the progenitors of the American Indians. For many thousands of years man lived in the New World in association with the large mammals, just as he had for so long lived in the Old World. He was a part of the fauna.

DISAPPEARANCE OF LARGE PLEISTOCENE MAMMALS

Some few millenia before the beginnings of recorded history, perhaps eight thousand years ago or less, many of the large mammals characteristic of the Pleistocene disappeared throughout the world. In the Old World the mammoths and woolly rhinoceroses and various other large mammals became extinct. In North America a wave of extinction that may have extended over a considerable time span brought to an end the hegemony of the American mastodon and various mammoths, of the native horse and camel, of the ground sloths and glyptodonts, of tapirs and giant bison and saber-tooth cats and of other mammals as well. The fauna changed from a Pleistocene fauna to a modern one, largely by the extinction of these large mammals (Fig. 24–15).

Why should this extinction have taken place? As is so frequently the case in any consideration of extinctions, and as has already been discussed in connection with the extinctions occurring at the end of Cretaceous time, this is not an easy question to answer. Perhaps man had something to do with the disappearance of the large Pleistocene mammals, but it is difficult to understand how primitive men, living in small populations, could have affected animals so large and well established and obviously so numerous as were the mammoths and mastodonts and horses in North America. We may only speculate. Probably the causes of extinction were complex and subtle and will never be well understood by us.

However that may be, the great mammals so typical of the Pleistocene did disappear. The glaciers retreated to the north. Man became successively a maker of polished stone implements, of bronze tools, and then of iron tools. He gradually abandoned hunting to become a pastoralist and then a farmer. He invented writing. He settled in large urban communities. The Pleistocene Epoch of earth history had, if you so wish to consider it, come to an end; now had come the dawn of Recent time.

25
The Progression of Events in the Continent and Its Origin

Folded Paleozoic sedimentary rocks in the Mackenzie Mountains, southwestern Northwest Territories, between the Mackenzie River and the border of Yukon (Royal Canadian Air Force).

25 *The Progression of Events in the Continent and Its Origin*

Historical geology gives an understanding of the principal features of the earth's crust in terms of their evolution. The continent of North America was described as separable into regions having exposures of several kinds of rocks. These can be re-examined in the light of the records that have been discussed. There are definite sequences of events that affected the parts of the continent at different times. We have then to discuss the manner of origin of the largest features, the continents and ocean basins. The earliest rocks that are preserved in each of many regions are great thicknesses of sedimentary and volcanic rocks; they were the oldest of the Precambrian rocks, and similarly form the cores of the oldest anticlines on the Atlantic and Pacific coasts, though of varying antiquity in the different places—frequently of rather indeterminate time of formation.

The associations of sediments and volcanic rocks along the continental margins in the early Paleozoic implied a geography with deeply subsiding belts that had linear tectonic, raised islands and many volcanoes, much as in the island archipelagoes of the present, the eugeosynclinal belts. These contrasted with adjoining non-volcanic subsiding belts, the miogeosynclinal belts, and with a relatively stable interior, a craton, having thinner sedimentary and volcanic Precambrian rocks. The paired geosynclinal belts were exemplified by the Paleozoic history through the earlier Devonian in the East and the Paleozoic and earlier Mesozoic record in the West. From time to time orogenies raised lands in the volcanic-sedimentary geosynclines so that some of the eroded detritus spread into the adjoining non-volcanic geosynclines; such was the case in the Ordovician along the Appalachian belt, and in the Carboniferous in the western states. The deposition of these terrigenous sediments permits dating of the spasms of orogeny, the rise of tectonic lands in the volcanic geosynclinal belts. At times and from place to place, the orogenies were associated with intrusions such as the linear belts of ultrabasic rocks, serpentine and peridotites, paralleling the geosynclines, formed at an early stage in orogenic history. Serpentine masses are scattered along the Atlantic Coast, as of the late Ordovician; and similar intrusives along the Pacific are as old as late Paleozoic in Alaska and as young as Jurassic in California and Oregon (Fig. 17–9). Intrusions of granitic rocks are much more widespread in the orogenic belts; their repeated presence is indicated by boulders and smaller fragments in the deposits in the volcanic-sedimentary geosynclines, as well as in the intrusions that invaded and altered them later. In time the terrigenous sediments spread far inland into geosynclines within the margin of the more stable craton (foredeeps or exogeosynclines). These geosynclines were of somewhat limited extent in the latest Ordovician and later Devonian of the East, and were greater and quite extensive in the Rocky Mountain and Great Plains regions in the Cretaceous. They resulted in downbowing of the margin of the previously stable craton, as well as the introduction of terrigenous sediments derived from highlands raised during orogenies in the bordering previously subsiding belts.

The volcanic-sedimentary geosynclines ultimately were severely intruded by granitic plutonic rocks, such as those that invaded the northern part of the Atlantic Coast during the Acadian Orogeny of the medial and late Devonian, and those of the Pacific Coast which began to intrude during the Nevadan Orogeny of the late Jurassic, continuing in places at least until the close of the Cretaceous. These plutonic events tend to close the more active phase of deformation in the belts, though limited geosynclines continue to subside, sometimes very deeply, in succeeding periods. Thus the Carboniferous troughs of New England and the Atlantic Provinces and the Cretaceous and Tertiary troughs of California are filled with miles of sediments.

Meanwhile, orogenic movements extended cratonward from the volcanic-sedimentary geosynclinal belts, producing the folds and faults in the non-volcanic geosynclinal belts and the geosynclines in the cratonal margin. The great thrusts and folds of the central and southern Appalachians originated in the latter part of the Paleozoic in the Appalachian Orogeny but those in western New England, along the St. Lawrence River and Gulf are mostly older. The structures in the Rocky Mountain region formed in the Laramian Orogeny at the close of the Mesozoic and beginning of the Cenozoic. Great thrusts were formed to the west in the Mesozoic, or perhaps late Paleozoic. Some of the most striking folds are those in the geosynclines in the margin of the craton, as in central Pennsylvania and central Wyoming and Colorado, in the areas that had subsided as they gained terrigenous detritus from uplifts on the outside in the late Ordovician, early Silurian and late Devonian in the East and the Cretaceous in the Rocky Mountain region.

A later phase involved great vertical movements, the rifting exemplified by the Triassic faulting along the Atlantic Coast with the associated subsidence of tilt blocks to depths of miles, and the fracturing that still continues from the Pacific Coast far inland. This rifting was accompanied by the extrusion of lavas, such as flowed into the subsiding blocks in the Triassic of the Atlantic Coast, and are particularly widespread in the Tertiary of the western interior.

Knowledge of the Arctic margin of North America is being acquired rapidly. The non-volcanic geosynclinal belt is well established; the volcanic-geosynclinal belt is of limited exposure, perhaps because it is buried beneath the sediments in later geosynclines. Much of the record on the southern part of the continent is buried, though the cratonal side of the orogenic belt has records similar to those on the east and west. In the same fashion, but to a more limited degree, the more oceanward record on the East Coast is concealed beneath the Atlantic Coastal Plain from New York southward. But to generalize from the limited example of the Gulf Coast, and from knowledge gained from geophysical surveys, the stage that follows crustal deformation, intrusion and fracturing in the volcanic-geosynclinal belts, is one of large-scale oceanward tilting, the formation of deeply subsiding coastal geosynclines.

In the first pages of this book there was a brief discussion of the distribution of the kinds of rocks that appear at the surface in North America. Their areas were divided into six categories. (1) Areas having a thin veneer of surficial rocks, disregarded on most maps. (2) Regions with lavas and other volcanic rocks that lie at the surface, as in large areas in the West. Sedimentary rocks underlie most of the United States and much of western Canada; the areas of distribution were separated as follows. (3) Those with little consolidated and undeformed sediments, such as extend along the Gulf and Atlantic coasts. (4) Those having consolidated but little deformed sediments, such as underlie most of the region south and west of the Great Lakes. (5) Those regions of folded and faulted sediments, as seen in the Rocky Mountains on the west, the Ouachita and Arbuckle mountains on the south, and the Appalachian Mountain System on the east. And (6) those great areas having a prevalence of igneous rocks intrusive into varying portions of metasedimentary and metavolcanic rocks, in the Canadian Shield, and in broad belts along the Atlantic and Pacific coasts, as well as smaller inliers within the other regions. We should now have a basis for understanding the origin of the rocks in these several regions.

Surficial sediments include: the recently formed gravels, sands, and clays deposited along present streams, in the beaches of lakes and oceans, and in the moving dunes; and the layers of weathered materials, such as those that accumulate with the alteration of the rocks, the residual clay soils of the humid regions, and the talus sliding from steep-walled mountain valleys. These are products of processes that operate in their present surroundings. Of greater interest in historical geology are the tills and water-carried drift, products of the multiple glaciation of the Pleistocene, and the silts and clays of the subsequently formed lakes. These sediments conceal the consolidated rocks in much of the northern part of the continent. The veneer of surficial rocks is shown generally only on large scale or special purpose maps.

The great regions of lava are within the Pacific Coast from the tip of Alaska to southern Mexico. The origin of the modern volcanoes, from those in the Aleutian chain to the scattered cones in southeastern Alaska and British Columbia, the stately peaks of the Pacific Northwest of the United States, and the many volcanoes of Mexico, as well as the associated and more extensive flows, ash-deposits, and cinder cones, is clear for they are little changed. Many other volcanic features are so fresh as to leave no doubt of their quite recent formation—the lavas of the interior northwestern states, as in the Craters of the Moon in Idaho, the flows in the valleys of northeastern New Mexico, and the scattered cones of northern Arizona. Other flows such as those along the Columbia in Washington and the Snake River in Idaho have been dissected and are older; and many of the lavas and volcanic sediments in the Basin and Range Region, as in Nevada and southern Oregon, have been tilted by movements along faults. Lava capped mesas in many localities have flows antedating the cutting of present relief. And there are isolated volcanic masses, such as the Devils Tower in Wyoming, Ship Rock in northwestern New Mexico, Spanish Peaks in Colorado, and the distant Monteregian Hills of southeastern Quebec, that evidence extensive volcanism, mostly in the Tertiary, associated with and presumably related to fractures in the rather rigid continental crust.

The unconsolidated sediments along the Atlantic and Gulf coasts were laid along margins of seas, much as present deposits form along the shores and floors of the same coasts today. This general geographic pattern was initiated about the Jurassic and has resulted in many miles of strata beneath the shores of the Gulf but only a few thousand feet within the Atlantic Coast, though greater thickness lies off shore. The gently seaward-dipping and thickening sediments fill geosynclines that conceal and bevel rocks severely deformed and altered in the contrasting orogenic earlier history of the belts. Little deformed Cenozoic sediments are exposed also along the coast of southwestern Washington and Oregon; Cenozoic and Mesozoic, too, are exposed in islands off the coast of British Columbia.

The interior of the continent has rather flat-lying sedimentary rocks that range from the Cambrian to the Tertiary System. The Mesozoic and Cenozoic systems surface the plains of the western interior states and provinces, thickening westward toward the mountains. They conceal varying thicknesses of Paleozoic rocks which underlie the region south and west of the Great Lakes. Geologic maps show the many systems in patterns representing gentle domes and basins, but the rocks in such broad structures show very little inclination. Some of the units thicken into areas of greater structural depression, whose deformation proceeded concurrently with the deposition through varying spans of time, but others thicken in patterns quite distinct from present structures. Thus the structural Michigan Basin as measured on the pre-Paleozoic surface shows much greater deformation than that visible on surface exposures and in near-surface wells, for the great subsidence was during the deposition of the middle Paleozoic sediments that accumulated in a span of only a few tens of millions of years. Earlier Paleozoic rocks indicate that the regional subsidence was in quite different patterns then. Gentle warping progressed at varying rates throughout the region, the rates and areas of subsidence changing from time to time.

The sedimentary rocks in the central part of the continent pass outward into belts where the rocks have been folded and faulted. Most conspicuous of these are the Appalachian Mountains in the east and the Rockies in the western interior. The deformation was concentrated in belts of thicker section, in geosynclinal areas. In the east, the Appalachian folds and faults are largely Paleozoic in the United States south of New England, but similar folds extend northeastward and are of earlier Paleozoic origin. The Appalachian folds in the south lie in a belt of thick late Paleozoic sediments, principally Mississippian and early Pennsylvanian, but in Pennsylvania they spread westward into the arcuate areas of thick later Devonian and later Ordovician. The major thrust faults lie within this belt, and extend in the south to about the flexure that bounded a main non-volcanic subsiding belt or geosyncline of the Ordovician.

The folds and faults in the West are most evident in the belt of thickest Cretaceous sediment that was laid within the margin of the earlier craton, but folding and faulting extends farther westward into the Paleozoic non-volcanic geosynclinal belt from eastern British Columbia to southern Nevada. There is a further thick section in the region of highest ranges and structure in central Colorado, where late Paleozoic sediments filled a two-mile deep subsiding trough. The great thrusts of the region are along the inner or cratonal side of the non-volcanic geosynclinal belt of the Paleozoic, but there are other great overthrusts to the western limit of that belt and beyond. The western structures on the margin of the interior plains are Laramian, formed in an orogeny in the latest Mesozoic and earliest Cenozoic. There are further limited areas of deformed sedimentary rocks along the southern margin of the interior, in the Ouachita and Arbuckle Mountains of Arkansas and Oklahoma, that have late Paleozoic sediments that were severely folded and faulted before the Jurassic sediments buried them. Sedimentary areas with varying degrees of folding are present along the Pacific Coast, as in the Valley and the Coast ranges of California, and in the East in the tilted downfaulted wedges of Triassic, and in the Carboniferous sedimentary basins of the Atlantic Provinces.

Finally, there are the regions of severe deformation, metamorphism and plutonic intrusion. The largest is the Canadian Shield, where the rocks are Precambrian, a time span that encompasses three-fourths of earth history. Precambrian is exposed in other smaller scattered inliers within the sediments to the south and west. The coastal belt extending from Newfoundland to New England, continuing within the Coastal Plain to Georgia, has Precambrian and early Paleozoic rocks deformed, metamorphosed, and intruded in several orogenic phases during the Paleozoic Era. A similar belt within the western continental margin has metamorphosed rocks deformed in several orogenies in late Paleozoic and Mesozoic, intruded most widely in the Nevadan Orogeny of mid-Mesozoic and somewhat later. The rocks of these sorts record history that has culminated in the invasion by extensive masses of granitic magma, but these events have occurred in the different belts at long-separated times. And the deformed, metamorphosed and intruded rocks, the schists, gneisses, and granites, are reached throughout the continent wherever the overlying rocks have been penetrated; they form the floor beneath all the sedimentary and volcanic rocks.

Thus the areas of different rock types on a geologic map of the continent also represent parts that have reached varying stages in a progression.

Fig. 25–1. Sections of the crust of the North American continent: the upper section drawn to scale, the lower with sea level as a plane, and the vertical distances on a scale ten times that of the horizontal distances.

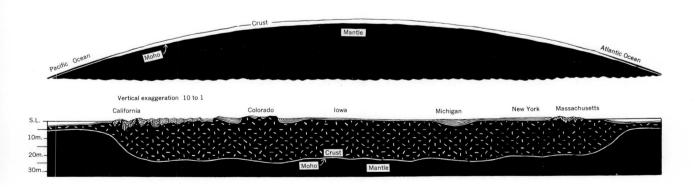

The study of historical geology has been concerned with the record of deposits formed on the continents, and with the seas that have invaded the continents or spread on their margins. The record has been reviewed from the earliest known rocks to the glacial times of the recent past. We shall now reverse the procedure, go back to what came before. Though the manner of origin of the continents and ocean basins is not subject to direct examination, we can profit from the knowledge of the behavior of the earth in the two billion years or so of recorded history. Hypotheses concerning the origin of continents are less speculative than those relating to the origin of the earth.

The continents not only have large areas with masses lying above the elevation of the sea but also other more fundamental contrasts with the ocean basins that constitute a much larger part of the earth's surface. The rocks that lie beneath the surface of the continents are of lighter density than those below ocean basins. Though there are dense, heavy and basic rocks on the continents, there are more extensive intrusions of quartz-rich, granitic rocks that are relatively lighter, and so rare beneath the oceans as to be virtually absent. The principle of isostasy explains the higher elevation of the lighter continental rocks, and the lower elevation of the denser rocks that underlie the oceans. The lighter rocks have been called "sial," in reference to their abundance of silica and alumina; the heavier "sima" has prevalent silica and magnesia. The terms are convenient for general reference. The sialic continental skins are conceived as floating on simatic material below them, which lies directly below the floors of the ocean basins. The thick sialic areas of higher elevation are supported by roots of sial extending down into the subcrust of sima. As was discussed in an early chapter, the Mohorovičić Discontinuity or Moho, the surface at the base of the crust, is deeper below the continents than the ocean basins (Fig. 25–1). The ocean basins are bounded by a line, the "andesite line," having belts of thin sial on the continental side; these are the belts of volcanic archipelagoes.

The geologic record presents good evidence bearing on the probable nature and the origin of the continents. The present lighter continents are conceived as of thick sial floating on a subcrust of sima, extending from below the ocean basins. The marginal parts of the continents, the volcanic island archipelagoes, are believed to be of thin sial. The geologic record shows a progression from belts like the island arcs to the stable continent—volcanic island arcs being deformed and narrowed by a succession of orogenic movements, invaded by ultrabasic rocks, soaked and intruded by great acidic bathylithic intrusives, then further warped and faulted before becoming an integral stable part of the continent. Wherever one goes on the continents, the oldest rocks have gone through such a history. The Precambrian granitic rocks of the Canadian Shield invaded thick sedimentary and volcanic sequences, much like those passing through similar stages at later times along the margins of the present continents. In fact, the fairly recent knowledge of ages of rocks by means of radiochemical techniques has shown that similar rocks, once thought to be of the same age, represent similar stages in orogenic belts formed at several times. There was not a great intrusion of large parts of the continent at a single time, but successive intrusions in narrower belts of the continent.

Geochemists are able to determine the age of plutonic rocks because they contain uranium-bearing minerals; the age of these rocks can also be found by means of potassium-argon and strontium-rubidium ratios. When the reported ages of the intrusions are compiled, they do not have wide and random scatter through time but are concentrated into rather short time spans, separated by times with rather few dated rocks (Fig. 25–2); orogenies were not continually in progress at an invariable rate but were interruptedly repeating.

On the Canadian Shield, the dates are not irregularly distributed geographically either, but those of about the same time span tend to be concentrated in belts. Ages are those of minerals at their formation. As the minerals will have developed in the last principal intrusion or deformation, ages are likely to be those of the last principal orogeny

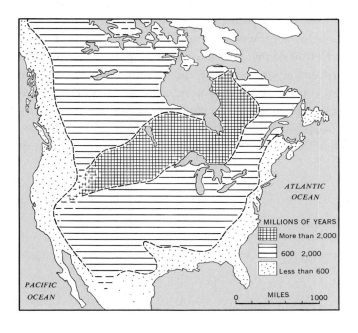

Fig. 25–2. Map of North America showing the ages of the rocks as determined from the ratios of radioactive isotopes (after R. G. Gastil). The ages are those of intrusion of plutonic rocks, or the metamorphism of rocks, such as are associated with orogenic periods. The map is simplified from others that separate the rocks into a larger number of areas. It shows the generally younger ages of the orogenic belts on the margins of the continents as compared to those in the central areas.

or of relics of older events from which mineral residues have not been altered. Moreover, it has been seen from the geologic evidence discussed in such a region as Minnesota that single belts have older orogenies as well as younger. So the record is not a simple addition of consolidated orogenic belts to a nucleus of a continent. The belts are rather orderly; the oldest ages being somewhat generally central, the successive younger ones more peripheral. The ages tend to fall quite systematically into cycles of about four hundred million years duration (Fig. 17–3). There are very ancient intrusives in some parts of the continent that suggest that the pattern, if orderly, is not simple. Old dates in central Texas, the Black Hills of South Dakota, and central Colorado show that the arrangement is not a regular progression of intrusions around a continental center. The older belts tend to be arranged with a northeasterly trend, the younger to parallel the Atlantic, Gulf, and Pacific. A reasonable conclusion is that the intrusions that accompanied North American orogenies are of progressive dates, culminating in cycles of four hundred million years or so, and that the progression was outward from within the Canadian Shield, but that there were other centers of growth that have coalesced with orogenies to form a single continent. Allusion has been made to the theories of drift-apart of continents. If two or more present continents were originally adjacent, orogenic belts of like age should connect; or the similar belts should have continued through present ocean basins. The data bearing on the solution of such problems are now accumulating.

Further judgments are more speculative. If orogenies take place in marginal oceanic belts having thin sialic crust, the island archipelagoes and associated troughs of present geography, all parts of the continent must have been at a time of that character, assuming that the intrusions in orogenic belts are a late stage in the development of volcanic geosynclinal belts. Following this reasoning, there should not have been any continent in the beginning—the continent would have grown by the addition of orogenic belts to those that preceded and had been consolidated. Then the earth must have been either all of thin sial or of thin sial and ocean basin. It can be demonstrated from the geologic record that the thin sial can become continent, but it has not been shown that the ocean basins can give rise to new archipelagoes along their margins. If they cannot, we must believe that the ocean basins have grown larger by the drift of thin sialic crust toward their margins to form thicker sial and the continents.

The sialic layer along the continental margin would have diminished in area and increased in average thickness as the land increased in area and average height. On the other hand, the formation of the earth without sialic crust is also compatible with the evidence. The oldest known rocks are volcanic-bearing sediment, not granitic basement; the sialic crust would have been built of differentiates from the simatic ocean basins. It has been written by Arthur Holmes: "Presumably continental submergence was complete, apart from volcanic islands, when the earth began her geological history. The first orogeny would produce long island arcs, and thereafter sediments would begin to be formed on a minor scale." The Precambrian history seems to have had progressive consolidation to one or more continental cores with volcanic geosynclinal belts along their margins. Whether the entire earth in the beginning of the rock record has a thin sial "skin" that has gradually been aggregated to form thick sialic continents, while leaving the void areas as basalt-floored ocean basins, or whether the whole was basaltic, the sialic thin crust being differentiated and subsequently gathered into thick crust, are problems that await solution through the ingenuity of earth scientists.

Fig. 25–3.   Histogram showing the relative abundance of age determinations in North America through geologic time (after R. G. Gastil). The fact that the results are not randomly distributed, but concentrated in peaks, suggests that there were times of principal orogeny separated by those having little orogeny, assuming that the sampling on which the data are based was not biased.

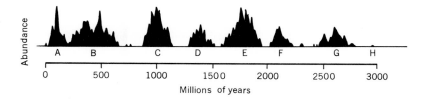

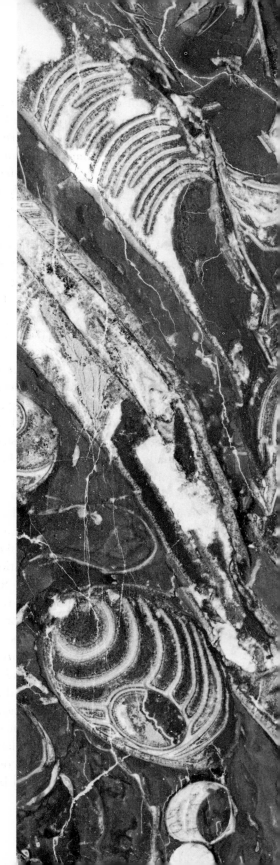

26
Fossil Organisms

Cephalopods sectioned in slab from Ordovician limestone, New York (The Smithsonian Institution).

Many sorts of organisms have been preserved partially as fossils. The characters of many of these have been considered in the discussion of the times when they were important. The present chapter assembles descriptions and examples of some of the sorts of fossils that are most prevalent and significant, and refers to others discussed on preceding pages. Paleontology is the study of the life of the geologic past—paleobotany concerning plants and paleozoology dealing with animals, the two kingdoms of the organic world. The successively smaller categories of classification are Kingdom, Phylum, Order, Class, Family, Genus, and Species. As previously stated (Fig. 6–5), the name given to a fossil has two parts, that of the genus and of the species, written as in Latin. The generic name, of the larger category, precedes that of the species, in accordance with the practice established by the Swedish scientist, Carl von Linné (later Linnaeus) in 1758. Thus *Olenellus thompsoni* is a trilobite of the genus *Olenellus* and is one of many species within that genus. It has been estimated that there are some one hundred thousand described species of fossils; a paleontologist can expect to know the distinguishing characters of only a small proportion of them. The soft parts of fossil organisms are directly known in an exceedingly small number of instances; only the more resistant parts are usually preserved as fossils. From the similarity of many fossils to living organisms, we can assume that they had similar soft structures. Thus the paleontologist gains a proper understanding of the fossils through the study of botany and zoology. In the following pages, the form and principal distinguishing characters of some of the fossil types will be described. The reader should consult a text on zoology or botany to gain better appreciation of the analogous living forms. Borradaile, Eastham, Potts, and Saunders' *The Invertebrata* is a text that discusses extinct animals as well as living ones. *Vertebrate Paleontology* by Romer treats the vertebrate animals.

The principal reference books on fossil invertebrates are the series of volumes, *Treatise on Invertebrate Paleontology,* published by The Geological Society of America and the University of Kansas Press. This chapter uses a somewhat simplified classification based on that authority. Conventionally, organisms are placed in two kingdoms, those of animals and plants. The *Treatise* places the simplest forms, the Phylum Protozoa and corresponding simplest "plants" in the Kingdom Protista; the thallophytes, also included, are considered to be in the plant kingdom in other standard works and will be so placed in this text. The classification of the plant kingdom is based on that in Andrews' *Studies in Paleobotany,* following Bold's *Morphology of Plants.*

Many of the fossil organisms have been described in the discussion of illustrations in the text. Others have received only fleeting reference. The former will be briefly noted and illustrated in this chapter, whereas the others will be given fuller description, insofar as their geologic significance warrants. The structures of fossils are more fully described in textbooks of paleontology, and typical index fossils of North America in a book of that title by Shimer and Shrock.

KINGDOM PROTISTA

The general practice has been to divide organisms into two kingdoms, the plants and animals. The simplest organisms do not always have the distinguishing characters of either plants or animals, so have been placed in a third kingdom, the *Protista.* The Protista can be considered as in two phyla:

      Phylum PROTOZOA
          Class *Foraminifera*—foraminifera
          Class *Radiolaria*—radiolarians
      Phylum PROTOPHYTA—including the bacteria

The phylum *Protozoa* contains those that commonly are placed in the animal kingdom. Two classes of protozoans, the *Foraminifera* and the *Radiolaria,* have shells that are preserved in such abundance in places as to be rock-forming, and the former are particularly valued as microfossils distinctive of successive stratigraphic units. The one-celled organisms sometimes called plants, the *Protophyta,* have left very little direct record, because they lack preservable parts. Fossil *Bacteria* are believed to be present in iron ore of the Animikian of Minnesota, nearly two billion years old, so are among the oldest known organisms. Bacteria have had incalculable influence on the formation of rocks and on the destruction and alteration of organic matter.

The *Protozoans* are represented today by many minute organisms of a single cell, some having internal skeletons, but many of which are free-moving individuals of soft-matter protoplasm. In the geologic record, two types of shell-secreting protozoans have been of great importance. The *Foraminifera* (Figs. 26–1 and 22–6) have internal skeletons, most of calcareous composition but some of siliceous and detrital character. Doubtfully present in the earliest Paleozoic, the foraminifera became abundant, rock-forming fossils in the Carboniferous. The fusulinids of the Carboniferous and Permian (Figs. 13–23 and 14–3) have generally spindle-shaped shells coiled spirally about an axis, with complexly partitioned interiors. The fossils are almost unbelievably intricate to have been formed within a single cell. This complexity and changing characters permit the discrimination of forms representative of successive strata. A second great burst of large shell-secreting foraminifera is represented by the *nummulitids,* most famed for their being the principal constituents of the early Tertiary limestones forming the Pyramids of Egypt. Both fusulines and nummulites were large enough to be readily seen with the unaided eye, reaching diameters of an inch and more. The vast majority of the foraminifera are microscopic, however, and of a great range of shapes—most of them coiled in one form or another (Fig. 22–6). Such foraminifera are appreciable constituents of some limestones.

The *Radiolaria* (Fig. 26–2) are a class of protozoans that secrete a delicate lattice-like open skeleton of hydrous silica or opal. Present radiolarians float within the sea and are preserved in such quantities as they settle to the bottom that in some areas they form radiolarian ooze. They were also abundant in many Paleozoic rocks, particularly associated in the volcanic geosynclinal belts, where they are preserved in cherts. When a block of radiolarian chert is polished and immersed for

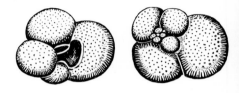

Fig. 26–1. Protozoa. Foraminifer *Globigerina* from Pliocene of California. This is a common genus in the present oceans.

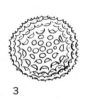

Fig. 26–2. Protozoa. Miocene Radiolaria from Maryland (after Maryland Geological Survey). (1) *Anthocyrtium,* (2) *Hexalonche;* (3) *Cenosphaera.*

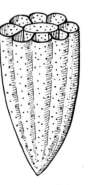

Fig. 26–3. Archaeocyatha; *Archaeocyathellus* sectioned to show the porous walls and compartments.

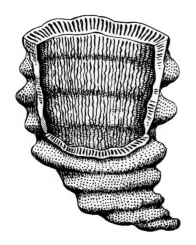

Fig. 26-4. Porifera. A cone-shaped sponge *Archaeoscyphia* from the Ordovician of Illinois.

a few minutes in hydrofluoric acid, the radiolarians differ enough from the surrounding matrix that they appear as minute oval or spheroidal objects, often forming a considerable part of the constitution of the rock. The frequent association with volcanic rocks probably reflects that the organisms secreting siliceous skeletons required marine water in which silica was available, as would be the case where volcanism was prevalent. Some radiolaria have been found in cherty limestones far away from volcanic regions; streams entering the sea carry an appreciable amount of silica in solution. Radiolaria have not been very exhaustively studied and have rarely been used in stratigraphic dating.

KINGDOM ANIMALIA

The animal kingdom contains a great variety of organisms, from those as simple and sedentary as corals to those as complex and mobile as the mammals. Classifications vary, but generally the animals are treated in two categories, those that lack backbones, the invertebrates, and those having vertebral columns, the vertebrates. The latter are within a single phylum. The following invertebrate animal groups are of geologic interest:

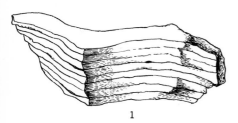

1

2a

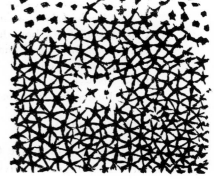

2b

Fig. 26-5. Coelenterata; (1) fragment of a specimen of *Stromatopora* from the Devonian of Germany; (2) vertical and tangential sections of *Actinostroma* from the Devonian of Belgium, enlarged about ten times.

Phylum ARCHAEOCYATHA—archaeocyathids
Phylum PORIFERA—sponges
Phylum COELENTERATA
    Class *Scyphozoa*—jellyfishes
    Class *Hydrozoa*—including stromatoporoids
    Class *Actinozoa* (or *Anthozoa*)—corals
Phylum BRYOZOA—bryozoans
Phylum BRACHIOPODA—brachiopoda
Phylum MOLLUSCA
    Class *Gastropoda*—snails
    Class *Cephalopoda*—cephalopods
    Class *Pelecypoda*—clams
Phylum ANNELIDA (or VERMES)—worms*
Phylum ARTHROPODA
    Class *Trilobita*—trilobites
    Class *Crustacea*—crustaceans
Phylum ECHINODERMATA
    Class *Asteroidea* (or *Stelleroidea*)—starfishes
    Class *Echinoidea*—sea urchins
    Class *Crinoidea*—crinoids
    Class *Blastoidea*—blastoids
    Class *Cystoidea*—cystids

Phylum STOMACHORDA
    Class *Graptolithina*—graptolites
Phylum CHORDATA
    Subphylum HEMICHORDATA—acorn worms
    Subphylum TUNICATA—tunicates
    Subphylum CEPHALOCHORDATA—lancelets
    Subphylum VERTEBRATA—vertebrates
      Class *Agnatha*—jawless vertebrates
      Class *Placodermi*—placoderms
      Class *Acanthodii*—acanthodian fishes
      Class *Chondrichthyes*—sharks
      Class *Osteichthyes*—bony fishes
      Class *Amphibia*—amphibians
      Class *Reptilia*—reptiles
      Class *Aves*—birds
      Class *Mammalia*—mammals

Fig. 26–6. Coelenterata. Anthozoa. Rugose corals. (1) *Homalophyllum* from the Devonian of Michigan, a simple coral; (2) *Billingsastrea*, a colonial form from the Devonian of Michigan.

Phylum ARCHAEOCYATHA

Representatives of a primitive type of organism, closely similar to sponges, the Pleospongia, or phylum Archaeocyatha are restricted to lower and middle Cambrian strata. They are cylindrical to long conical fossils an inch or so in diameter, having porous walls that surround a

Fig. 26–7. Coelenterata. Anthozoa. Tabulate coral *Strombodes* from the Silurian of Michigan.

* The worms are here placed in a single phylum because this is the most convenient manner to classify them for paleontological purposes. Our knowledge of modern worms shows that they belong in several phyla.

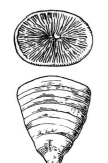

1

2

deep central cavity (Fig. 26–3), known in many parts of the world, as in Spain, Siberia, British Columbia, Labrador, and Australia. Since they are recognizable even when poorly preserved and restricted in stratigraphic range to the Cambrian, Pleospongia are very helpful in classifying the rocks that contain them.

Phylum PORIFERA

The Porifera or sponges are among the most primitive of animals, consisting essentially of a hollow attached vase-shaped structure of horny fibers penetrated by canals or pores that bring food to the many cells (Fig. 26–4). The skeleton has rods or needles made of calcium carbonate or of silica that are commonly preserved in stratified rocks. Sponges lived through Paleozoic and later times in great variety. Although rarely of much importance stratigraphically, they are locally quite varied and abundant, making considerable parts of the mass of some limestones.

Phylum COELENTERATA

The coelenterates include such marine animals as the jellyfish and corals, the former free-swimming and lacking hard parts, the latter sedentary and with a calcareous skeleton. The class *Scyphozoa,* includes jellyfishes such as represented among some of the oldest recorded fossils, their delicate structures preserved in the muds of the Cambrian (Fig. 15–2). An extinct order of the scyphozoans, the *Conulata* or conularids had pyramid-form shells of chitin and phosphate that are found occasionally in rocks from the Cambrian to the Triassic. The class *Hydrozoa* contains the extinct stromatoporoids that thrived in Ordovician to Devonian (Fig. 26–5) seas and were important constituents of the "coral" reefs (Fig. 11–21). Other types of hydroids were carbonate rock-forming in the Mesozoic and Cenozoic eras.

The class *Actinozoa,* the corals, are by far the most widespread of the coelenterates. In the Paleozoic there were two principal orders, the *Rugosa* and the *Tabulata.* The former (Fig. 26–6), which have an arrangement of septa that divides the calyx into four parts, include both the simple "horn" corals and colonial forms that are assemblages of corallites having comparable structure. The Tabulata include some of the most prolific colonial corals of the Paleozoic, having polygonal corallites with crenulate or septate borders and a succession of platforms or tabulae formed as floors of corallites at successive growth stages (Fig. 26–7). These Paleozoic corals were important constituents of reefs from the Ordovician to the Permian (Fig. 10–9).

It is doubtful that any of these corals survived the Paleozoic Era, unless their descendants were in the distinctive order *Scleractinia*—the hexacorals (Fig. 26–8) that appeared in the Triassic and are the corals of present reefs. Just as in the Rugosa, some Scleractinids are solitary individuals and others are colonial, growing into masses that form constituents of reefs in association with other carbonate-depositing animals and plants such as hydrocorallines and algae. The scleractinian corals or hexacorals are distinctive in that they have six more prominent septa or partitions radiating to their walls, whereas the Paleozoic rugose corals have but four; as will be evident in the illustrations, the six main septa are not particularly obvious. The Scleractinia very quickly took the place of the Rugosa, and formed reef structures in the Triassic Period and thereafter.

Though simple corals live today to high latitudes, reef-building forms are restricted to tropical and subtropical seas, to latitudes of 35 degrees or less. Hence corals have been thought to distinguish the warmer parts of ancient seas. The growth may be affected by seasonal changes in the elevation of the sun; some have tried to apply the principle to the determination of ancient latitudes.

Phylum BRYOZOA

The bryozoans are colonial animals, each small individual or zooid living in a minute cup or cell, a zoecium, within the solid, calcium carbonate zoarium or colonial structure. Fossil bryozoans (Fig. 26–9) are distinguished by the shapes of the colonies, by the internal structures in successive growth stages as revealed in thin-sections, and by the characters of the zoecial openings and the lids or opercula that cover some of them. The soft parts of living bryozoans differ from those of corals, but some fossil bryozoans resemble tabulate corals; the latter generally have corallites larger than bryozoan zoecia, and with different internal structures. The bryozoans range from minute, delicate, rod-like zoaria to forms building hemispherical mounds; many are branching, twig-like forms, and among the most beautiful are fan-like

Fig. 26–8. Coelenterata. Anthozoa. Scleractinid corals. (1) *Paracyathus*, **from the Miocene of Maryland; (2) the colonial** *Oulastrea* **from a modern reef, (a) a colony, and (b) enlargement of a few zoecia.**

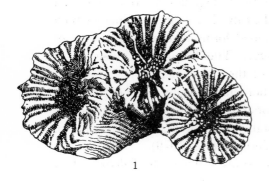

1

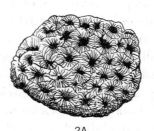

2A

2B

perforate fronds that are particularly prevalent in Carboniferous limestones. Some bryozoans were so massive, and built their colonies to such height as to form small reefs. They often contributed measurably to the calcareous matter in organic reefs or bedded biostromes (Fig. 26–10). Their skeletons have grown or accumulated to form significant parts of many limestones. Bryozoans have been rather common animals since the Ordovician, and they are frequent but not a conspicuous part of modern marine faunas. The changing assemblages permit recognition of rocks of successive ages, but they have not been generally used. Some classify the organisms in the phylum *Ectoprocta*.

Phylum BRACHIOPODA

Brachiopods are bivalved animals, each shell being bilaterally symmetrical, but of dissimilar shape (Fig. 26–11). They have lived from the Cambrian to the present, but were of greatest variety and abundance in the Paleozoic (Fig. 8–8). The phylum is separable into many contrasting types (Figs. 11–11, and 15–1). Although a few of the brachiopods have phosphatic and chitinous shells, the great majority are of calcium carbonate, frequently accumulating in such numbers as to build layers of limestone.

Fig. 26–9. Bryozoa. (1) Trepostome bryozoan *Mesotrypa*, (b) and (c) being enlarged tangential and radial cross-sections; (2) the trepostome *Phylloporina*; and (3) the cyclostome *Anolotichia*, all from the Ordovician of Minnesota. (4) The cheilostome *Coscinopleura*, from the Cretaceous of Germany. The figures are enlarged from ten to twenty times.

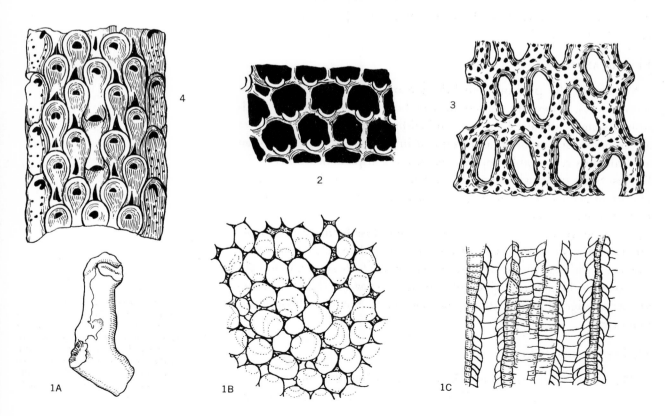

Phylum MOLLUSCA

The molluscs contain several classes of shell-bearing animals, the pelecypods, gastropods, and cephalopods being of greatest interest in historical geology. *Pelecypoda,* the clams, (Fig. 26–12) have been abundant from time to time from the Ordovician; they are bivalved animals having two similar valves to the left and right of the plane of junction (Figs. 15–4; 22–2A). *Gastropoda* (Fig. 26–13) or snails have lived since the Cambrian, becoming particularly common in the Tertiary (Fig. 22–2B).

The class *Cephalopoda* is divided into two orders, the *tetrabranchs* and *dibranchs.* Of the former, the *nautiloids,* having simple septa and sutures (Fig. 26–14), lived in greatest number and variety in the early Paleozoic (Fig. 15–5). The *ammonoids,* tetrabranchs, which evolved rapidly in the late Paleozoic (Figs. 15–7 and 15–8), became the most significant invertebrate fossils of the Mesozoic (Figs. 16–24, 17–4, and 18–4). The ammonoids are separable into several distinctive types, such as the *goniatites* (Fig. 26–15), distinctive of the later Paleozoic, the *ceratites* (Fig. 26–16) of the Triassic and the *ammonites* (Fig. 26–17) of the later Mesozoic. The dibranch cephalopods are represented today by the octopus, squid, and cuttlefish, having internal skeletons; in the late Paleozoic and Mesozoic there were frequent *belemnites* (Figs. 26–18 and 17–5).

Phylum ANNELIDA

Worms, of the phylum Annelida, are of many sorts preserved in several ways. They are represented in borings that penetrate sandstones, some with raised castings on bedding surfaces; such structures are common in Cambrian and frequent in later rocks. The Cambrian also contains remarkable impressions of the soft parts of worms in the Burgess Shale in British Columbia. Probably worms have been extremely important agents in the mixing of sedimentary layers and the destruction of primary structures.

Paleozoic rocks contain some microscopic fossils that can be separated, because they are insoluble residues when the enclosing rocks are dissolved with the proper acid reagents. *Scolecodonts* are minute fossils that seem to be chitinous jaws of annelids. *Conodonts* (Fig. 26–19), of greater stratigraphic importance, are phosphatic microfossils of uncertain origin.

Phylum ARTHROPODA

This phylum includes many classes of organisms that are characterized by having jointed legs. Few classes have been of stratigraphic

Fig. 26-10. Bryozoans forming much of the mass of a bed in the Silurian Clinton shales at Rochester, New York (The Smithsonian Institution).

Fig. 26-11. Brachiopoda; (1, 2) exterior brachial and side views of terebratulid *Oleneothyris* from Paleocene of New Jersey; and (3, 4) interior views of brachial and pedical valves of terebratulid *Megallania* from the present seas; c—cardinal process, etc., h—hinge line, l—loop, po—pedicle opening, br—brachial valve, pe—pedicle valve.

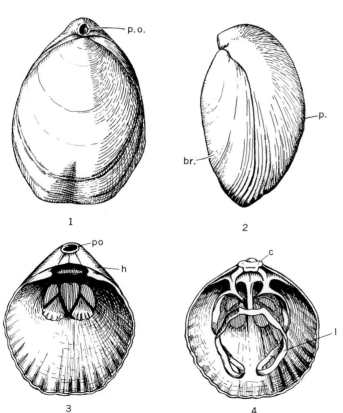

1

2

3

4

p.o.

p.

br.

po

h

c

l

value, but others have great biologic interest. The *trilobites* (Fig. 26–20) were present throughout the Paleozoic, their relative importance decreasing from the Cambrian (Fig. 6–7), where they are the pre-eminent means of carrying time-correlations. The fact that these highly organized animals are the first abundant fossils suggests that we lack knowledge of a wealth of life that preceded the Cambrian without leaving records (Fig. 26–21). *Ostracods* (Fig. 26–22) are known throughout the Paleozoic and later eras, are abundant in many rocks, and are quite useful in chronology (Fig. 10–17), particularly because their small size saves them from destruction in the drilling of wells. Several other types of arthropods are of interest at certain times and places.

Arthropods moult as they grow, shedding successively larger carapaces or shells. A single animal thus leaves a succession of growth-stage records. It is possible to reconstruct the life-histories or ontogenies of the fossil trilobites and ostracods in this manner.

Insects are rarely found but are known in great numbers from several localities and systems, particularly as fossils impressed in fine-textured silts or ash-beds (Fig. 26–23) and occasionally in the peculiar external molds in amber, a hardened extrusion from trees. The great number of insects at some of these fossil localities gives significant information on the biology of the class, even though they have little stratigraphic use because of their infrequency. Among scorpion-like animals, the *euryp-terids* (Fig. 26–24) were common locally in Ordovician and Silurian rocks, and were the giants of the arthropods, the largest approaching ten feet in length (Fig. 26–25). The horseshoe *crabs* of the present beaches, *Limulus*, have ancestors that look very much like them in rocks as old as Devonian (Fig. 26–26). And there were many sorts of arthropods with soft parts recorded, in the Cambrian Burgess Shale

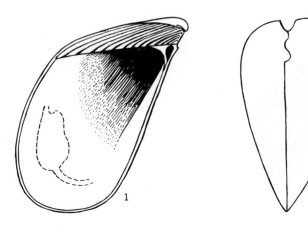

Fig. 26–12. Pelecypoda; Carboniferous pelecypod *Liebia* from Kansas, interior of left valve and anterior view, enlarged three times. The two valves are similar, but each is not symmetrical.

A

B

Fig. 26–13. Gastropoda. *A.* Eocene gastroped *Tornatellaea* from Virginia.

B. Gastropods of the genus *Turritella* from Eocene sands in cliffs along the Potomac River near Aquia, Stafford County, Virginia; the fossils are external and internal molds of specimens that were buried and have been dissolved away (The Smithsonian Institution).

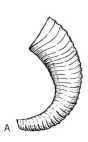

A

Fig. 26–14. Nautiloid cephalopods. *A.* *Plectoceras* from the Ordovician of Illinois. Nautoloid cephalopods have regular sutures, and vary in shape from straight to forms coiled tightly in a plane.

B. Nautiloid cephalopods shown in section in a polished surface of limestone from the Ordovician Chazyan along Lake Champlain in northeastern New York; see Fig. 9–2*B* (The Smithsonian Institution).

B

fauna (Fig. 15–2) that give us further evidence of the substantial constancy of some of the forms through the past half-billion years.

Phylum ECHINODERMATA

Echinoderms include several classes that are geologically quite important and others that are rare. Some remain attached to the sea floor by stems through life. The classes *Crinoidea, Blastoidea,* and *Cystoidea* all are common at some times. Crinoids (Fig. 26–27), known from the Cambrian to the present, are particularly prevalent in the Carboniferous (Fig. 12–5). Blastoidea (Fig. 26–28) were contemporaries in the Paleozoic (Fig. 12–5C) but disappeared in the Permian. Cystoids (Fig. 26–29), closely resembling crinoids, are quite varied in the arrangement of their plates. They usually have distinctive perforated plates, pore-rhombs, and some have small arms. Cystoidea lived from the Cambrian to the Permian.

Fig. 26–15. Goniatite cephalopod *Koenenites* from Michigan, showing sutures exposed by removing shell.

The principal examples of the echinodermata that are free to move along the sea bottom or in the water, are the *Echinoidea* (Fig. 26–30), which became important organisms in the late Mesozoic, though first appearing in the early Paleozoic. Another class of free echinoderms is the *Asteroidea,* and the common starfish is an example. Starfishes (Fig. 26–31) are known from the Cambrian but they rarely have been preserved and are only occasionally abundant. A third class, *Holothuroidea,* differs from other echinoderms in having leathery skin containing imbedded distinctive calcareous parts. The microscopic plates of geometric pattern and hooked structures are seen occasionally in rocks. Holothurians are known among the animals from the remarkable impressions in Cambrian Shales in British Columbia (Fig. 15–2). The class may have been common in the past but rarely evidenced by fossils.

Phylum STOMACHORDA

This assemblage of organisms contains one class of geologic importance.

Class Graptolithina: the classification of graptolites is uncertain. The order Graptolithina is now placed in the phylum Stomachorda, because graptolites extracted from cherts in Poland have remarkably well-preserved structures that resemble those in modern hemichordates. Graptolites are colonial organisms made of chitin (Fig. 26–32), a very resistant substance, in which the individuals live in small cups, theca, distributed along rod-like stripes that, when joined, form the rhabdosome. The graptolites formed colonies that were suspended from a structure that floated at the surface of the sea; they are often preserved

Fig. 26–17. Ammonite cephalopod *Hantkiniceras* from the Cretaceous of Germany. Note the serrate lobes and saddles of the sutures.

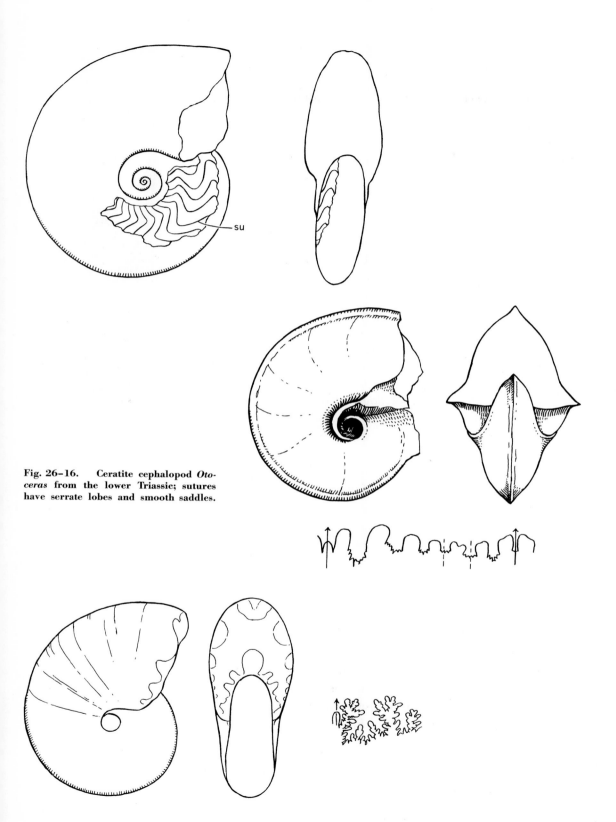

Fig. 26–16. Ceratite cephalopod *Otoceras* from the lower Triassic; sutures have serrate lobes and smooth saddles.

su

as a carbonaceous film (Fig. 26–33). Thus they are very widely distributed in the Cambrian to Mississippian rocks that contain them, being among the most important fossils in classifying marine sediments of the Ordovician (Fig. 8–2) and Silurian. Biologically they are of great interest because the growth of each individual, the ontogeny preserved in its rhabdosome, can be compared with that of other genera and the interrelationships or phylogeny can be established.

Phylum CHORDATA

These are the animals with some sort of a supporting structure along the dorsal side of the body, animals with "backbones" composed in the more primitive members of a flexible notochord, in the majority of forms of a series of interlocking vertebrae. The main trunk of the nervous system is located above the notochord or vertebral column; the circulatory system below. There is a head end, containing an enlargement of the nerve cord, the brain, and in the circulatory system a heart that pumps the blood. These animals are typically bilaterally symmetrical with the axis of the body in a horizontal position. Representative forms were described in Chapters 15, 20, and 24.

The hemichordates, tunicates, and cephalochordates are small, modern animals that seem to give us some idea of the basic chordates. *Amphioxus,* the lancelet, is much used in the classroom to demonstrate a chordate more primitive than any of the vertebrates.

The oldest known vertebrates are the agnathans, in which there are no lower jaws. These are the ostracoderms of lower and middle Paleozoic sediments. Certain specialized agnathans, the lampreys and hagfishes, survive today. The placoderms and acanthodians are early evolutionary developments of jawed vertebrates, both destined for early extinction. These vertebrates, of fish-like form, have the paired fins variously developed, as is the case in the agnathans. The sharks and the bony fishes are the successful fishes from the Devonian to the present. In these jawed fishes there are two pairs of fins, and various median fins, one, the caudal, being specialized as a propulsive tail.

The agnathans, placoderms, acanthodians, sharks, and bony fishes— all aquatic, all breathing by gills, may be grouped as fishes in the large sense of the word, the *Pisces,* and set off against the tetrapods, *Tetrapoda,* vertebrates with limbs, which (except for those reverting to the water) live on the land, and invariably breathe air (*see* Fig. 26–38, p. 668).

The most primitive of the tetrapods, the amphibians, arose from fish ancestors at the end of the Devonian Period. In these animals the fish fins had become transformed into limbs. But the amphibians from their inception to the present day have, during individual growth, gone

Fig. 26–18. Belemnite cephalopod *Pachyteuthis*, a dibranch cephalopod from the Jurassic of western South Dakota (The Smithsonian Institution).

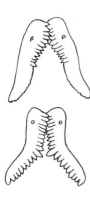

Fig. 26–19. *Conodonts. Scottognathus,* an assemblage in shale in the upper Carboniferous Pennsylvanian at Oglesby, Illinois; and a restoration showing the paired similar parts each form having a name that was applied before they were found to be parts of a single animal (F. H. T. Rhoades).

The conodonts are minute fossils that can be described as serrate or tooth-like. Although each shaped form is treated as of a particular kind or species, the fact that several kinds are consistently found in association indicates that they are of an organism that has several differing parts; as there are parts that differ only as mirror images, the animals must have been bilaterally symmetrical. Beyond this, the nature of the source of conodont parts is speculative. Their composition resembles that of vertebrate bones. Their cosmopolitan distribution and presence in rocks of varied facies shows that their source was free-swimming and adaptable. Conodonts appear from late Cambrian into the Triassic, being most useful in the stratigraphy of the Ordovician, late Devonian, early Carboniferous, and earlier Triassic rocks. Although biologically enigmatic, they are stratigraphically quite significant.

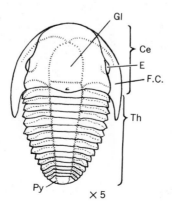

Fig. 26–20. *Trilobita.* Cambrian trilobite *Peltura,* from Sweden; Ce—cephalon; E—eye; F.C.—free cheek outside the facial suture; Gl—glabella with glabellar furrows; Py—pygidium; Th—thorax with segments.

659

through a fish-like tadpole stage, in which the newly hatched individual is legless and breathes by gills. As it grows to an adult the gills characteristically are lost, legs grow out from the body, and the animal becomes an air-breather.

The reptiles are a step beyond the amphibians because they lay a protected egg on the land (or hatch the young within the oviduct). Development is direct; there is no intermediate stage.

All of the vertebrates described so far are and were throughout their history ectothermic; their temperature varying more or less as the temperature of the environment varies. But two other classes of vertebrates evolved during Mesozoic times from reptilian ancestors—these being the endothermic or "warm-blooded" birds and mammals. Endothermism in these two classes was independently achieved. The birds are very reptilian in structure but have the advantages of a constant body temperature and the powers of flight (except in those that have lost such powers). Flight is made possible in part by the evolution of the reptilian scales into feathers, these forming wing surfaces. They also insulate the body.

The mammals are the most progressive of the vertebrates, not only by virtue of their constant body temperatures but also because of the growth of the cerebral part of the brain and the great increase of intelligence. These animals have hair for insulation. The young are

Fig. 26–21. Trilobites of the genus *Phacops* in Devonian shale from Arkona, near Sarnia, Ontario. Carapaces of trilobites usually became dismembered on death, and many specimens are the fragments of moults; thus an assemblage of whole specimens such as in the illustration was an exceptional find (The Smithsonian Institution).

born alive (except in the reptile-like monotremes, which hatch the young from eggs) and are nurtured during the early stages of their life upon milk produced by special glands in the female. Protracted care of the young by the parents has been a touchstone of success among the mammals during the course of their evolutionary history.

KINGDOM PLANTAE

The student of fossil plants is a paleobotanist. He is concerned with plants as the principal or recognizable constituents of rocks and as the means toward understanding the biology of plants. Some plants that are excessively rare or even unique have disparate importance, because they contribute significantly to our knowledge of the nature of ancient life.

Classifications of plants differ. Some plant groups that are of greatest importance to the paleobotanist have little interest to the student of modern plants, the coal-forming Carboniferous plants, for instance. On the other hand, the simplest plants, algae and fungi, placed in the thallophytes by some constitute a dozen or more phyla in other classifications, only two or three having appreciable importance as rock-formers. Others of these phyla are barely or rarely known because, having unstable compositions, they left little or no record of their former existence. Some of them did exist and some are even recorded—

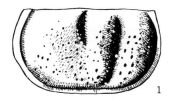

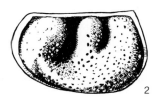

Fig. 26–22. *Ostracoda.* Silurian ostracod *Zygobeyrichia* from West Virginia; male left valve and female left valve with brood pouch; enlarged about ten times.

Fig. 26–23. Arthropoda. Fossil insects are found in several localities in rocks of late Paleozoic age and younger (Frank Carpenter, from specimen in University of Texas collections). Among many kinds in the Permian System in eastern Kansas are huge insects related to modern dragon flies. The specimens are of the genus *Typus,* having a wing spread of nearly ten inches.

for example, in the evidences of fungal damage to the wood of ancient plants. Additional organisms such as the bacteria that have been called protophytes are now placed in the separate kingdom Protista.

The following classification includes only the types of plants that are of interest to the geologist. The principal groups of plants are the divisions resembling the phyla among the animals. The divisions that have little importance in the geologic record are omitted. The classification is based on that of H. C. Bold, *Morphology of Plants* (1957), with revisions proposed by H. N. Andrews, Jr., *Studies in Paleobotany* (1961).

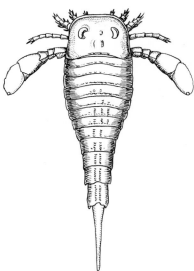

Fig. 26–24. Arthropoda. An eurypterid, *Eurypterus*, from the Silurian of New York; specimens reached a length of a foot or more.

Fig. 26–25. The arthropod *Eurypterus*, an eurypterid, diorama of the Silurian of New York; commonly a foot or so long, some eurypterids reached a length of as much as ten feet (Chicago Natural History Museum).

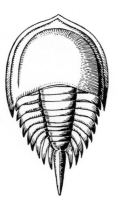

Fig. 26–26. Arthropoda. A primitive horseshoe crab, *Pseudoniscus*, from the Silurian of New York. Specimens of crabs very like those living today are found in middle Paleozoic rocks.

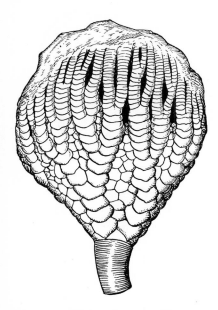

Fig. 26-27. Echinodermata. The crinoid *Forbesiocrinus*, a flexibiliate from the lower Carboniferous, Mississippian of Kentucky.

The Thallophytes—algae and fungi:
    Division CYANOPHYTA—blue-green algae
    Division CHLOROPHYTA—green algae
    Division PHAEOPHYTA—brown algae
    Division RHODOPHYTA—red algae
    Division CHRYSOPHYTA—diatoms
    Division PYRROPHYTA—dinoflagellates
The Bryophytes—mosses—are little known
The Pteridophytes—ferns and associated plants:
    Division PSILOPHYTA—the psilophytes
    Division MICROPHYLLOPHYTA—club mosses
    Division ARTHROPHYTA—horsetails
    Division PTEROPHYTA—ferns
The Spermatophytes—seed-bearing plants:
    Division PTERIDOSPERMOPHYTA—seed-ferns
    Division CYCADOPHYTA—cycads
    Division GINKGOPHYTA—ginkgos
    Division CONIFEROPHYTA—conifers
    Division ANTHOPHYTA—flowering plants

The simplest plants are the algae, organisms that produce chlorophyll, enabling them to convert light to food through photosynthesis. The algae have many divisions, corresponding to phyla in the animal kingdom. Because of their simple structures and poor preservation, many are difficult to classify with assurance. Though there are repre-

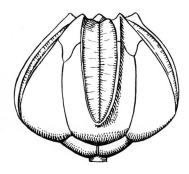

Fig. 26-28. Echinodermata. A blastoid from the Carboniferous, Mississippian of Kentucky; a—ambulacra or food grooves; b—basal plates.

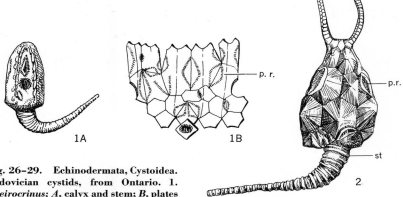

Fig. 26-29. Echinodermata, Cystoidea. Ordovician cystids, from Ontario. 1. *Cheirocrinus*; *A*, calyx and stem; *B*, plates of calyx as though laid out on a flat surface. 2. *Pleurocystites*: p.r.—pore rhomb; st—stem.

sentatives of many of the plant divisions, only a few are important rock builders, particularly the green algae and the diatoms.

The Precambrian rocks in many places have mound structures of limestone that resemble those built by modern blue-green algae, *Cyanophyta*. These *stromatolites* have been shown from Montana (Fig. 5–15), and Ontario (Fig. 5–16); similar forms, *Cryptozoon,* (Fig. 26–34), are displayed in the Cambrian at Saratoga Springs, New York. Ordovician rocks have green algae *Chlorophyta,* such as *Mastopora* (Fig. 26–35), a type of alga known through most succeeding systems. Red algae, *Rhodophyta,* such as *Solenopora* (Fig. 26–36), appear in such abundance in a few Ordovician limestones as to form principal constituents. Similar structures are in the rocks of succeeding systems.

Algae of unknown sort, perhaps brown algae, *Phaeophyta,* are thought to have been the source of hydrocarbons in some bituminous Ordovician "oil rock," and similar pre-Carboniferous sediments. The *diatoms, Chrysophyta,* were discussed in connection with the formation of porcellaneous cherts from diatomaceous earth in the Tertiary (Fig. 21–29). They form thick sequences of siliceous rocks of Cretaceous and Cenozoic age. *Coccoliths,* disk-shaped calcareous bodies of extremely small size, form a large part of some chalks such as those of the Cretaceous cliffs of Dover, England (Fig. 18–2). They belong in the *Pyrrophyta* or dinoflagellate division, which have come to be of particular interest because of their utility in stratigraphic micropaleontology.

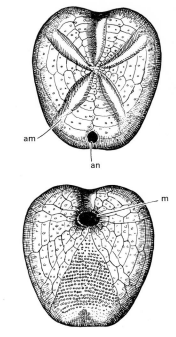

Fig. 26–30. Echinodermata, Echinoidea. A Cretaceous echinoid *Epiaster,* from Texas, (a) top and (b) bottom views. a—anal opening; m—mouth; the radiating rays of plates on the top side are the ambulacra. The living animal had many movable spines, like the sea urchin of the present seas.

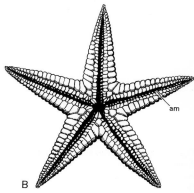

Fig. 26–31. An asteroid, or starfish, one of the Echinodermata; *A,* a cast of the base of *Devonaster,* from the middle Devonian on the campus of Colgate University at Hamilton, New York; there are also casts of brachiopods and of another starfish (The Smithsonian Institution); *B,* restoration of base of Devonaster.

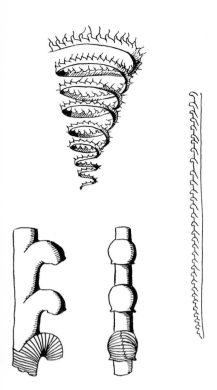

Fig. 26–32. Graptolithina: representatives of the Silurian genus *Monograptus*,
the stipes being enlarged about twice in
the contrasting stipes, and about twenty
times in the details of the thecae.

Fig. 26–33 (above). Graptolithina: a
graptolite of the genus *Tetragraptus* from
lower Ordovician shales at Levis, Quebec, opposite the city of Quebec; see Fig.
8–4 (The Smithsonian Institution).

Fig. 26–34. Reef of the alga *Cryptozoon* in the Petrified Gardens, Saratoga
Springs, New York, in upper Cambrian
dolomite (New York State Museum and
Science Service). The structures resemble those built by modern blue-green
algae.

The liverworts and mosses (*Hepatophyta* and *Bryophyta*) are such fragile plants that they are rarely preserved. A few examples are known from coal beds of the Carboniferous, and scattered specimens are reported from the succeeding systems. They did not form rocks, nor add significantly to our understanding of plant history.

The pteridophytes include the psilophytes, the club mosses, the horsetails, and the true ferns. We have read that the psilophytes formed plant layers in Devonian swamps, excellently preserved in the cherts at Rhynie, Scotland (Fig. 11–2). Similar plants and club mosses are well known from rocks as old as Silurian and very doubtfully from the Cambrian. These earliest known land plants were succeeded by the descendants much like them, and by the great outburst of tree plants in the medial Devonian, such as the forests of true ferns or pterophytes that grew on the stream plains of the eastern New York (Figs. 11–24 and 11–25). Trees of this sort and others of the arthrophyte or horsetail division, thrived in the later Devonian, and were among the principal plants of the Carboniferous (Fig. 12–14) forests from which such great coal resources have come.

The spermatophytes or seed-bearing plants came into being in great

2

1

Fig. 26–35. *Mastopora* or *Nidulites*, a rock-forming plant classified as a green alga or chlorophyte. Both specimens are from the middle Ordovician. The surface of the weathered limestone slab (1) is from Strasburg, Virginia (The Smithsonian Institution); the silicified specimen (2) was etched from limestone from near Clifton Forge, Virginia. The fossils are about an inch long.

profusion in the Carboniferous *Pteridospermophyta,* the seed ferns (Fig. 12–14). Though it is assumed from their advanced development that ancestors lived in the Devonian, none is now recognized. Some of the tree ferns of the Devonian were long thought to belong in this phylum. The pteridosperms seem to have lived into the early Mesozoic Era. The distinctive and paleogeographically important *Glossopteris* (Fig. 19–5) flora of the Permian and Triassic of the southern hemisphere is dominated by seed-bearing plants that are generally classed as pteridosperms. In the Jurassic and Cretaceous, the cycads, *Cycadophyta* (Fig. 18–27), are known from many places, and the *Ginkgophyta* or maidenhair trees were widespread in early Mesozoic (Fig. 17–10), though doubtfully known also from latest Paleozoic rocks.

The remaining plant division, the *Anthophyta* or angiosperms, contains the flowering plants with which we are most familiar from their prevalence today. As discussed in the Cretaceous chapters, the flowering plants burst forth in great numbers in the middle of the Mesozoic, though a few seem to have been living in late Triassic. In the Tertiary, the grasses appeared in profusion to give the continents the floras much as we see them today.

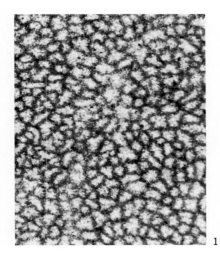

1

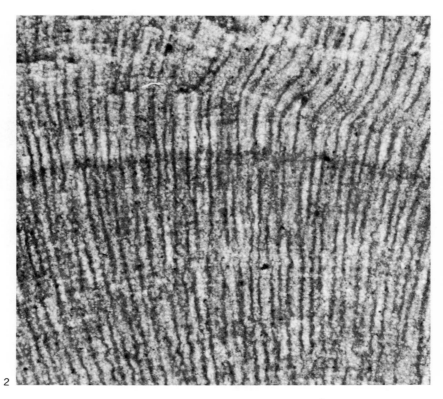

2

Fig. 26–36. *Solenopora,* a rock-forming plant thought to be a red alga, or rhodophyte. The fossils are globular, about like marbles or small golf balls, found widely distributed in Ordovician rocks; the specimens are from Chazyan limestone at Joliette, Quebec, and are tangential (1) and longitudinal (2) sections enlarged about forty times.

667

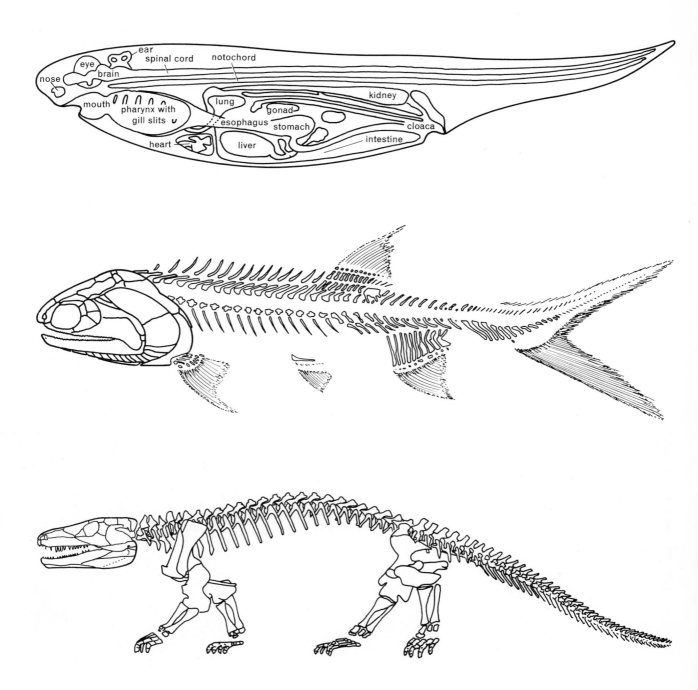

Fig. 26-37. *Top*—A generalized longitudinal section of a vertebrate, showing the relationships of some of the major organs to each other. This plan illustrates the horizontal axis of the body, with its main support, the notochord, beneath the nerve cord.

Middle—A primitive fish, showing the basic vertebrate adaptations for life in the water; with a persistent notochord, which passes up, posteriorly, into the upper lobe of the caudal fin, or tail. On the back is the dorsal fin. Beneath the body, from front to back, are the paired pectoral and pelvic fins, and the single anal fin.

Bottom—A primitive tetrapod, showing the basic vertebrate adaptations for life on the land, with well-developed, interlocking vertebrae, to form a strong support for the body. This vertebral column is slung within the anterior pectoral girdle and the posterior pelvic girdle, to which the limbs are attached.

27
The Origin of the Earth

by DEAN B. MCLAUGHLIN

The Horsehead Nebula in Orion, an opaque cloud of dust-size particles, the type of material from which it is considered possible the solar system developed (Mount Wilson and Palomar Observatories).

27 *The Origin of the Earth*

BY DEAN B. McLAUGHLIN

When we attempt to reconstruct the history of the formation of the earth and the other members of the solar system, we are following a trail that has become very "cold" with the passage of time. The evidence from radioactivity indicates that the earth became a planet some four and one half billion (4.5×10^9) years ago. Just as erosion has destroyed parts of the record of geologic history, so to a much greater degree have later events in the solar system erased many of the clues to its origin. But many still remain and they indicate boundaries within which our hypotheses must be restricted.

Dimensions of the planets' orbits, and physical data and dimensions of the planets themselves, are given in Table IV.

TABLE IV

Body	Distance from Sun in Astronomical Units[1]	Diameter		Mass	Density
		Miles	Earth = 1	Earth = 1[2]	gm/cm^3
Sun		864000	109	332500	1.4
Mercury	0.387	3100	0.39	0.056	5.1
Venus	0.723	7570	0.95	0.82	5.2
Earth	1.000	7927	1.00	1.00	5.52
Moon		2160	0.27	0.012	3.3
Mars	1.524	4200	0.53	0.108	4.0
Jupiter	5.20	88700	11.2	318	1.33
Saturn	9.54	75100	9.5	95	0.68
Uranus	19.18	29000	3.7	14.6	1.7
Neptune	30.06	28000	3.5	17.3	2.2
Pluto	39.52	(?)	<1	<1	>6?

[1] One Astronomical Unit = 92.9×10^6 miles = 150×10^6 km (approximately).
[2] Mass of Earth = 6×10^{27} gm = 6×10^{21} tons (approximately).

REGULARITIES IN THE SOLAR SYSTEM

The broad features of the solar system emphasize that it is indeed a *system*. The planets are not merely a haphazard collection of satellite bodies in randomly oriented orbits around a star; they are arranged and move in a very orderly manner. Their orbits are nearly circular, and all lie nearly in the same plane. The motions of all the planets around the sun are in the same direction: counterclockwise as viewed from above the north pole of the earth (Fig. 27–1). The sun itself rotates in that same direction around an axis inclined only a few degrees from a perpendicular to the average plane of the planetary orbits. All of the planets, so far as their axial rotations are known, also rotate in the same sense, save for Uranus whose axis lies almost in its orbital plane. Such regularity could not have arisen from chance capture; it indicates a common origin from a single mass of matter with a well-defined rotational motion from its very beginning.

All of the planets except Mercury, Venus, and Pluto have satellites revolving about them. The satellite systems of Jupiter, Saturn, and Uranus have the appearance of small-scale models of the solar system (Fig. 27–2), with nearly circular, coplanar orbits, and motions in the same direction as the axial rotations of their central planets. This similarity of the satellite systems to the larger system of planets tributary to the sun is of the very greatest significance. It shows that the processes of planet formation, *whatever they were,* took place on two different scales. Any hypothesis that assigns to the satellites a mechanism of origin different from that for the planets is foredoomed by its own artificiality.

A closer look at the system shows departures from perfect uniformity. A few satellites in the outer parts of the Jupiter and Saturn systems have strongly inclined and rather eccentric orbits, and some of these revolve in the reverse direction. The comets contrast rather strikingly with the regularity of the planets: comet orbits are oriented nearly at random in space, with all angles of inclination, and with motion in either direction. The outstanding characteristic of comet orbits is their extremely great eccentricity.

GENERAL NATURE OF COSMOGONIC HYPOTHESES

Nearly all hypotheses of the origin of planets fall into two main groups. According to one group, the sun and planets all were formed together from a primordial mass of nebulous material. According to the other,

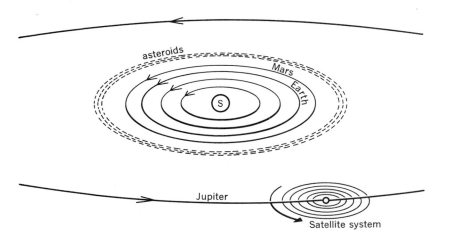

Fig. 27–1.   The orbits of the planets.

671

the sun as a star came into existence first, and the formation of the planets was an independent later event—usually the result of an encounter between the sun and another body.

The first two hypotheses of the modern scientific period, both proposed near the middle of the eighteenth century, represent the two contrasting points of view. Georges Louis Buffon (1707–1788) envisioned a collision of a comet with the sun; he supposed that the material that was spattered into space collected to form the planets. At that time, comets were thought to be very massive bodies; we now know this to be incorrect. In more modern form, this encounter hypothesis employs a star instead of a comet. Immanuel Kant (1724–1804), on the other hand, proposed that the sun and planets originated simultaneously from a rotating nebula.

It will not be possible here to review all the hypotheses in detail. Historically, the nebular hypothesis of Pierre Simon de Laplace (1749–1827) (which has considerable resemblance to Kant's earlier speculation) and the planetesimal hypothesis of T. C. Chamberlin (1843–1928) and F. R. Moulton (1872–1952) (an encounter mechanism) each held the field longer than any others. Currently the process that seems to give the best account of the system is the one developed by Kuiper from one by C. F. von Weizsacker. It has some very general resemblances to Laplace's nebular hypothesis. Variations on the same general theme have been presented by H. C. Urey, H. Alfvén, and Fred Hoyle.

THE NEBULAR HYPOTHESIS

Laplace knew only the dynamical aspects of the solar system: the masses of the bodies, the character and arrangement of their orbits. In his time nothing could be known of the chemical compositions of the sun and planets. Further, while it was recognized that the sun was just one of many stars in a large galaxy, this seemed to have little relevance to the immediate local problem of the earth's beginnings.

Laplace (about 1820) pictured the primordial state of the solar system as a great spherical cloud, or nebula, of hot gases in slow rotation. As it was pulled together by its own gravitational attraction, the rotation speeded up, in accordance with the principle of conservation of angular momentum, and the nebula consequently became more and more flattened toward a plane perpendicular to its axis of rotation (Fig. 27–3). Eventually (according to Laplace's speculation), the centrifugal force at the rim would just balance gravitation, and an outer ring of matter would then become detached while the main mass shrank inside it. Undoubtedly, this type of structure was suggested by the

Fig. 27–2. The orbits of Saturn's satellites, from the American Ephemeris for 1961 (U.S. Nautical Almanac Office).

visible example of Saturn with its rings (Fig. 27–4). This process was supposed to have happened several times. The remaining central mass developed into a dense stellar body, the sun. Each detached ring accumulated into a single cloud revolving about the sun, and these clouds in their turn went through an evolution very similar to that of the main one, contracting, rotating faster, and dropping off rings at their rims. Each central mass became a planet, and the rings accumulated to form the satellites. Laplace saw clearly that the basic history of the whole solar system and of the satellite systems must have been the same, with only a difference of scale.

In its original form, Laplace's hypothesis contains fatal errors. The most important weakness is its failure to account for the present distribution of angular momentum in the system. The sun is over-whelmingly the most massive body in the solar system, but it is rotating so slowly that it has only about 2 per cent of the total angular momentum. The planets, though they comprise much less than 1 per cent of the mass, have 98 per cent of the angular momentum. If the sun had rotated rapidly enough to drop off a ring where Mercury now revolves, it should now be rotating more than one hundred times as rapidly as it

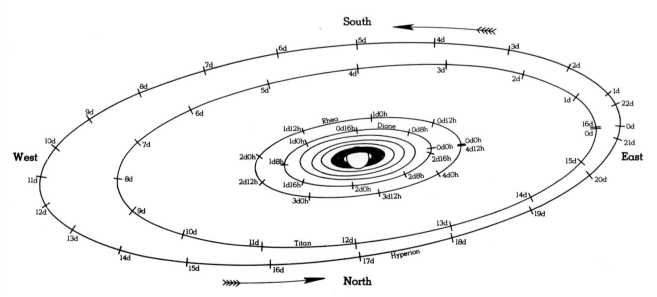

APPARENT ORBITS OF THE SEVEN INNER SATELLITES, AT DATE OF OPPOSITION, JULY 19

	Name	Mean Synodic Period			Name	Mean Synodic Period	
		d	h			d	h
I	Mimas	0	22.6	VI	Titan	15	23.3
II	Enceladus	1	08.9	VII	Hyperion	21	07.6
III	Tethys	1	21.3	VIII	Iapetus	79	22.1
IV	Dione	2	17.7	IX	Phœbe	523	15.6
V	Rhea	4	12.5				

does. Once the planets were separate from the sun and fully formed, there seems to have been no way of changing the distribution of angular momentum by any large amount.

THE PLANETESIMAL HYPOTHESIS

Recognition of this difficulty led to virtual abandonment of the nebular hypothesis. In its place arose the planetesimal hypothesis of Chamberlin and Moulton (ca. 1900) and the somewhat similar Jeans-Jeffreys tidal hypothesis, both based on an encounter of the sun with another star. Viewed in retrospect, these encounter hypotheses appear as measures of desperation, adopted in the attempt to inject into the solar system the large angular momentum now possessed by the planets, without giving the sun a rapid rotation.

In the planetesimal hypothesis, the sun was already formed as a normal star before the birth of the planets resulted from the accident of the close passage of another star. As the two stars passed one another, it was supposed that tides were raised on each, and the tidal forces, *aided by supposed violent eruptive activity of the sun,* caused large amounts of gas to burst out from the sun's interior and range out to great distances from it (as far as the outermost planet). This material was thought to have taken the shape of spiral arms (Fig. 27–5). It was dragged along in the direction of motion of the passing star. By this process, it was conjectured, the matter that was to become the planets acquired the large angular momentum that the planets now have. The hot ejected gas cooled quickly and condensed into liquid drops and solid particles. Then began a slow process of accumulation into larger bodies and eventually into the small number of planets. Although each particle was at first moving in a rather eccentric orbit, collisions and combination of particles canceled out much of the radial motion and left the resulting planets moving in nearly circular orbits.

The thinking of Chamberlin and Moulton was strongly influenced by the observed prevalence of spiral "nebulae" (Fig. 27–6) in the sky. The real nature of these objects was as yet unknown, and though it was recognized that they must be much larger than our planetary system, it was supposed that the spiral form originated through tidal action and that they might be taken as legitimate models of the early stage of our solar system. Not until a quarter century later did it become clear that these "nebulae" were galaxies of stars many millions of times as large as a planetary system. The origin of their two-armed spiral form is still an unsolved problem, but it seems practically certain that tidal action is not the cause.

When the planetesimal hypothesis was proposed in 1900, it was

Fig. 27–3. Development of the solar system, according to the nebular hypothesis of Laplace.

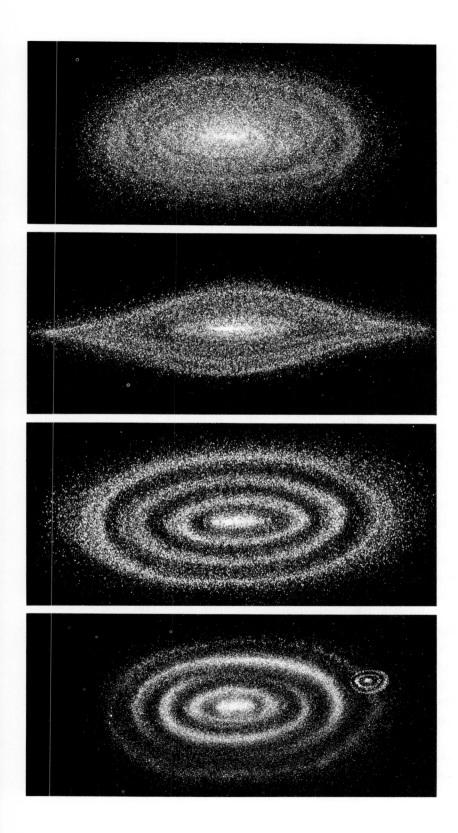

675

thought that the solar eruptive prominences were violently expelled *from below the sun's surface.* We now know that this is incorrect. Prominences simply form in certain regions of the solar corona and may remain relatively quiescent for a considerable period before they are rapidly accelerated outward. This means that the sun could not be counted on to help the tidal forces in moving solar material out to great distances.

THE TIDAL HYPOTHESIS

James H. Jeans and Harold Jefferys in about 1925 attempted to account for the detachment of the planetary material through the tidal action of the passing star alone. Their calculations showed that the unaided tidal forces would not have been competent to pull out masses large enough to form the planets to great distances unless the encounter were a very close one—practically a collision of the stars. The material was supposed to have been drawn out of the sun as a long filament of hot gas which broke up into several parts, each of which cooled off to become a planet. But in such a close encounter the material would be pulled out almost radially from the sun (Fig. 27–7), so that it would either escape entirely or fall back very close to the sun again, traveling in highly eccentric orbits. In order to account for the nearly circular orbits of the planets, it was necessary to introduce a resistance to their motion. A nebula of diffuse matter could produce the effect, but the orbits would become too small in the process. Later it was shown by Spitzer that the hot filament, due to its small mass and high temperature, would have so little gravitational cohesion that it would simply diffuse into space and never collect to form planets.

Both the planetesimal and tidal hypotheses met especially great difficulties in connection with the origin of satellite systems. To suppose that the planets also underwent tidal disruption meant piling one improbable assumption upon another. All in all, the encounter hypotheses failed very badly, in spite of all attempts to resuscitate them.

An important objection to encounter hypotheses in principle is the extremely low probability of the required close approach. The stars in our galaxy are so far apart that only a few such encounters would occur in the whole history of the galaxy (of the order of 10^{10} years). Only one star in millions would have a close tidal encounter. If the planets originated in that way, our solar system must indeed be a rare specimen, a cosmic accident or freak.

As a variant of the encounter hypothesis, Lyttleton proposed that the sun was originally a double star and that the companion was disrupted tidally by another passing star. The angular momentum of the compan-

Fig. 27–4. Saturn and its ring system, photographed with the 100-inch telescope (Mount Wilson and Palomar Observatories).

Fig. 27–5. Tidal disruption of the sun, according to the planetesimal hypothesis of Chamberlin and Moulton.

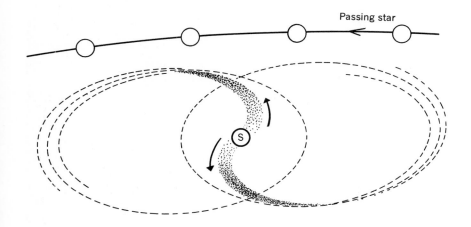

Passing star

S

ion was thus the source of the present angular momentum of the planets. However, getting rid of the companion star posed serious difficulties and required very special assumptions.

A very different and unique hypothesis, suggested by Hoyle, also makes use of the double star idea, but the companion was supposed to have undergone a supernova explosion which expelled a large amount of material, some of which was "captured" by the sun and revolved about it. This material then supposedly collected to form the planets. This hypothesis has the same weakness as all mechanisms that involve the ejection of hot gases: the matter would simply diffuse, because its gravitational cohesion was too weak. Furthermore, the velocity of ejection in a supernova explosion is so enormous that there seems to be no possibility that any of the gases could have been caught and retained by the sun.

THE SOLAR SYSTEM IN THE GALAXY

Since we have reviewed what we may call the historically unsuccessful attempts to formulate a theory of origin, we should now take a closer

Fig. 27–6. The spiral galaxy Messier 81, in Ursa Major, photographed with the 200-inch telescope. Formerly thought to be a model for the embryonic solar system, these spiral "nebulae" are now known to be galaxies thousands of light years in diameter, and composed of billions of stars (Mount Wilson and Palomar Observatories).

look at the relevant evidence as it appears today. The entire stage-setting, that is, the place of the planets and sun in the larger universe, needs to be viewed. Possible clues to the birth of the stars in general, and of the solar system in particular, have been uncovered only recently in the study of nebulae and star clusters. Notice must be taken of the general properties of stars (for the sun is just one of them) and especially of the chemical composition of stellar and interstellar matter. Whatever the mechanism by which they originated, the sun, earth, planets, and comets were built out of the available chemical elements mixed in proportions that were dictated by the composition of the universe as a whole and by the physical and chemical processes that operated either to separate or to bring together certain elements in preference to others. The chemical compositions of the various bodies may well differ considerably, depending on their different histories. The chemical evidence was largely ignored in all the hypotheses reviewed above, chiefly because it was unavailable, but partly because its relevance simply was not fully appreciated by astronomers until recently.

The sun is only one member of a vast system of probably one hundred billion stars that we call the galaxy. This system, 100,000 light years in diameter, is strongly flattened and circular in outline, with a thickness less than a tenth its diameter. A denser nucleus is shaped like an oblate spheroid or a very thick lens, forming a thick bulge in the center. Within the disk-like system are spiral arms in which the stars are more numerous, or at least more luminous, than in the regions between. The whole is in rotation in its plane and the spiral arms trail behind the rotation of the central nucleus. The sun is located in a rather faint spiral arm in the outer regions of the flattened system which, from our point of view, makes the luminous band of the Milky Way across our skies. In the sun's neighborhood, the average distances between stars are a few light years.*

*One light year is about 63,000 times the distance from the sun to the earth, or about 6 $\times$ 10^{12} miles.

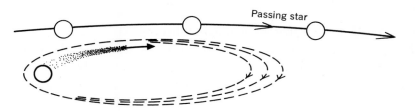

Fig. 27–7. A very close tidal encounter of the sun with a star, showing how material would be drawn out almost radially.

THE NATURE OF THE SUN AND STARS

The sun is a fairly typical average star. In dimensions, mass, and temperature it lies fairly midway in the range covered by the known stars. Its diameter is 864,000 miles, its mass over 300,000 times that of the earth, and its superficial temperature about 6000 degrees absolute. Many stars are known whose characteristics are closely similar to these. Stars that are hotter than the sun are generally larger, more massive, and much more luminous. Stars that are appreciably cooler than the sun fall into two groups. On the one hand are the dwarfs which are smaller, less massive, and dimmer than the sun. On the other are the giants, ten to one hundred times the sun's diameter and one hundred times its brightness. The hottest stars, the solar type stars, and the dwarfs show a steady gradation of characteristics from high to low luminosity, diameter, and mass. Taken all together, they are known as the *main sequence*. The giants form a group apart; they are stars that were once members of the main sequence but, owing to changes of structure that resulted from partial exhaustion of their nuclear energy sources, they now have very different and distinctive features.

INTERSTELLAR MATTER

Gases and dust-like material are associated throughout large regions of interstellar space in our galaxy. This material is especially abundant in the regions of the spiral arms. Luminous gaseous nebulae have been known ever since the discovery of the Orion nebula (Fig. 27–8) early in the seventeenth century. They are composed primarily of hydrogen and helium, with oxygen, nitrogen, neon, and other gases in smaller amounts. These gas clouds, many light years in diameter, are rendered luminous by excitation of their atoms by absorption of the ultraviolet radiation from hot stars imbedded in them. These illuminating stars are recognized as "young" objects (see below). The fainter and cooler stars that are associated with them in these nebulae also show characteristics that stamp them as young. Star formation apparently is either going on before our eyes in these gaseous nebulae, or it has occurred only very recently.

Besides the visible luminous nebulae, diffuse clouds of cold non-luminous hydrogen occur in extensive regions of the spiral arms of the Milky Way. These are revealed through the characteristic radio microwaves (length 21 cm) that are emitted by hydrogen.

In some of the gaseous nebulae, and in many other parts of the Milky Way, are seen evidences of non-luminous solid material. Dark clouds, some of them nearly opaque, others partially transparent, appear silhouetted against dense star fields or luminous nebulae (Fig. 27–9 and

Fig. 27–8. The great nebula in Orion, a cloud of gas several light years in diameter, rendered luminous by stars imbedded in it. Photographed with the 100-inch telescope (Mount Wilson and Palomar Observatories).

chapter opening). These dark nebulae must be composed of tiny particles, as shown by the way in which they dim light of different wavelengths, and they are called "dust clouds." However, it is not likely that they are chiefly dust in the usual sense. The term "dust" refers to size rather than to composition. Though tiny grains or crystals of mineral matter and metals may be present, it is likely that most of the so-called dust is really *snow*—not only the snow of water but the solid form of many simple compounds that are gaseous at room temperature but solid at the very low temperatures of interstellar space. Among these are carbon dioxide, methane, and ammonia, as well as elementary permanent gases.

ORIGIN AND EVOLUTION OF STARS

Out of this sort of stuff it seems likely the stars can be formed. Certain very small, round, and highly opaque dark nebulae are especially intriguing in this connection. They are known as *globules* (Fig. 27–10). Their dimensions can be roughly estimated as several thousand times that of the earth's orbit. Their opacity tells something of the amount of matter in them: it is of the order of the mass of a star. Calculation shows that *cold* clouds of this size and mass are capable of contracting gravitationally. These globules are, therefore, regarded as very possibly stars in the process of formation out of the interstellar gas and dust. It seems possible that the sun and planets originated from the gravitational contraction of such a globule or "dust cloud," and that the formation of planets may have been a fairly normal accompaniment of the creation of stars. The "dust cloud hypothesis," proposed in 1948 by Fred Whipple, was the first to connect the origin of the solar system explicitly with a dark globule.

Detailed calculations of the contracting phases of stars at the beginnings of their lives have been carried out. As the cloud that is to become a star contracts under its own attraction, the material is compressed until its internal temperature is raised to several million degrees. At this temperature atomic nuclear reactions can begin in the deep interior, and the star is then self-sustaining without any further contraction. The nuclear reactions in its core keep up a steady supply of energy that is radiated to space. The size, temperature, and total brightness of the star (that is, its place in the main sequence) are fixed by its mass, and it remains about the same for most of its life, that is, until the nuclear "fuel" in its core is almost exhausted. A cloud of matter with a mass less than about one twentieth of the sun cannot become a star, since gravitationally it does not have the power to compress itself until its

Fig. 27–9. A star cloud in the Milky Way in Sagittarius, partially obscured by foreground clouds of dust (dark nebulae). Photographed with the 48-inch Schmidt camera (Mount Wilson and Palomar Observatories).

central temperature is raised to the "kindling point" for the nuclear reactions.

The principal nuclear reaction is the building of helium nuclei from hydrogen, which releases the energy that keeps the stars shining. In small stars the reaction goes relatively slowly, but in stars of large mass (ten times the sun's mass and greater) it goes very rapidly. The luminosity of a star is roughly proportional to the cube of its mass, so that a star with ten times the sun's mass gives out one thousand times the sun's light. Since the available "fuel" is about the same percentage of any star, this means that a massive star will burn itself out in a much shorter time than one of small mass. Specifically, the star with a mass ten times the sun's would have a life only about one hundredth as long as the sun.

It is not our purpose to follow all the possible paths of the evolution of stars. The chief point of interest here is the influence of the massive

Fig. 27–10. Numerous dark "globules," possibly destined to develop into stars, in the Rosette Nebula in Monoceros. Photographed with the 48-inch Schmidt camera (Mount Wilson and Palomar Observatories).

stars upon the over-all chemical composition of the universe. Not only is helium built up out of hydrogen in their interiors, but near the ends of their lives when they change from main sequence stars to giants, their cores reach extremely high densities and temperatures. Then heavier elements can be built up out of helium. Of these heavy elements, oxygen is one of the most abundant, and iron is the chief one among those still heavier. Finally, the star may have a tremendous supernova explosion which scatters the material of its interior into space, where it becomes mixed with the interstellar gas and "dust." Since the very massive stars evolve rapidly (their lifetimes are only a few tens of million of years), there may have been many generations of these heavy element factories during the life of our galaxy. In the course of billions of years the interstellar matter has been steadily enriched in the heavy elements.

Spectroscopic analyses of stellar atmospheres and gaseous nebulae show that the composition of the observable part of our universe is about 75 per cent hydrogen, 24 per cent helium, and 1 per cent all heavier elements. These proportions are quite compatible with the hypothesis that the original composition of the universe was pure hydrogen, and that the now-existing helium and heavier elements were all manufactured in the interiors of generations of massive stars that have

Fig. 27-11. An edgewise view of a spiral galaxy, showing the "globe-and-disk" form (central nucleus and flat spiral arm system). NGC 4565 in Coma Berenices, photographed with the 200-inch telescope (Mount Wilson and Palomar Observatories).

long since disappeared. Stars that now form out of the interstellar matter are of somewhat different composition from those that were born in the very early history of our galaxy. If the cloud from which our galaxy formed was originally pure hydrogen, then the first generation of stars could not have had planets like those in our solar system.

CHEMISTRY AND STRUCTURE OF THE PLANETS

The earth's chemical composition is almost entirely in the 1 per cent of heavy elements. It contains hardly any helium except that which results from radioactive decay of heavy atoms. Its hydrogen is almost solely that which is locked in the water of the oceans. But with few exceptions, the composition of the earth is fairly well in line with the verdict of the stars as to the composition of the 1 per cent of heavy elements. The most abundant of the elements heavier than helium is oxygen, both in the earth and in stars and nebulae. Silicon, the next most abundant terrestrial element, is very abundant in stellar atmospheres. But although neon is conspicuous in nebulae, it is very scarce on earth, for well-understood physical and chemical reasons: it escaped from the earth because it was not locked in compounds, and the earth's attraction was too feeble to hold it at the temperatures that existed during the earth's formation.

Compositions of the planets are certainly not all alike. They separate into two broad groups. The four inner planets, Mercury, Venus, Earth, and Mars, are called the terrestrial planets, because there is every reason to consider the earth fairly typical of them. They all have densities between the limits 4 and 6 times that of water. (See Table IV.) The asteroids probably have similar composition. The meteorites, the only direct samples we have from the universe outside, are interpreted as fragments from a collision of asteroids, and their average compositions are essentially earth-like. Even the separation into iron and stony meteorites corresponds roughly to the core and mantle of the earth.

The large outer planets, Jupiter, Saturn, Uranus, and Neptune, form the Jovian group. Their densities are much lower than that of the earth, and they must be composed predominantly of material that is nearly lacking in the earth's composition. Jupiter, with a mass over 300 times that of the earth and about a thousandth that of the sun, is the largest of them. Its density is only 1.4 times that of water, and it is inferred that it is composed mainly of hydrogen and helium.

Perhaps it is no mere coincidence that the heating effect of the sun is considerable throughout the region occupied by the terrestrial planets, but extremely feeble at Jupiter and beyond. Is this correlation of density with distance a cause-and-effect relation? The low density of

Jupiter must be due to a large amount of volatile matter which would, nevertheless, be frozen at the low temperature existing at such distance from the sun but which would have been vaporized at the distance of the earth or Mars. This material then could have escaped from the inner region, especially if the gravitational attraction of the terrestrial planets was too feeble to hold it. The fact that Jupiter is as dense as it is must be due largely to the compression of its interior by the weight of the overlying layers. Saturn, with less than a third of the mass of Jupiter, is much less dense (only 0.7 as dense as water), but the difference is probably due entirely to lesser compression of the same material as composes Jupiter.

Uranus and Neptune, though less massive than Jupiter and Saturn, are nevertheless appreciably denser (Table IV) and must contain a greater percentage of non-volatile and probably terrestrial-type material. They are intermediate in both mass and density between Jupiter and the earth, and are probably intermediate in composition also. During their formative stages the gravitational attraction of Uranus and Neptune was sufficient to hold only a fraction of the volatile materials that could be retained by the most massive planets. The inferred structure of Uranus and Neptune consists of a core of earth-like composition surrounded by outer layers of the more volatile material that composes Jupiter and Saturn.

The oblatenesses of the various planets tell something of the internal distribution of density. Gravitational theory gives a relation between the degree of polar flattening and the amount of central condensation. If the mass is concentrated at the center and the outer portion highly rarefied, the planet is less oblate, for a given rate of rotation, than a planet whose density is uniform throughout. Comparison of theory and observation shows that Jupiter and Saturn must be much denser at their centers, and that the increase of density is close to that expected for hydrogen and helium under the compression of the overlying layers. The earth's oblateness agrees with the conclusion that it has a much denser core, probably of nickel-iron. Mars, on the other hand, seems to be more nearly of uniform density throughout. If it contains the same percentage of iron as the earth, the iron must be almost uniformly distributed, and not concentrated in a core. It is likely that Mars never became sufficiently molten in its interior to bring about the differentiation of core and mantle that we find in the earth's structure.

The comets belong to the outermost reaches of the solar system, and except for some that have been "captured" into short-period orbits by the attraction of Jupiter, they only briefly visit the region of the planets. From their spectra and their association with meteor swarms, it seems most likely that a comet is primarily a mass of ices with considerable solid non-volatile material imbedded in it. This is described as the "icy conglomerate" or "dirty iceberg" model. On nearing the sun, the outer ices are vaporized and the molecules partly dissociated. The vapors fluoresce under the sun's radiation, and the solid particles are expelled along with the gases to form the coma and tail. Carbon molecules, hydrocarbons, carbon monoxide, cyanogen, and the radicals NH and OH are present in the gaseous envelope. The composition is probably similar in many ways to that of the Jovian planets.

The numerous details of composition and structure described above make the task of accounting for the solar system much more complicated than the merely dynamical problem that was discussed by Laplace, Chamberlin and Moulton, Jeans, and Jeffreys. But they also narrow the range of permissible hypotheses. It is clear that the planets, whether they came from the sun or were formed separately from it, are not merely masses of solar-type material. They represent the products of processes of separation or concentration that operated differently in various regions of space.

PREVALENCE OF ROTATION IN THE UNIVERSE

One more important fact concerning stars and star systems generally needs to be noticed. Rotation is extremely prevalent among the young, massive, hot stars. Most of those that are apparently single have enormously broadened spectrum lines, for which the only satisfactory explanation is rapid axial rotation. A large percentage of stars are double. The *visibly* double ones are fewer than the spectroscopic binaries, pairs of stars so close together that no telescope can separate them, but whose duplicity is shown by the presence of two superimposed spectra that shift periodically as the stars revolve, or single spectra that shift to and fro, indicating motion as a member of a pair with another star too faint to record. Both axial rotation and orbital motion indicate large angular momentum associated with these stars. It is especially noteworthy that stars of the same type as the sun, as well as cooler stars, do not usually show rapid rotation, though doubles are frequent. Granted that large angular momentum is characteristic of stars generally, the slow rotation of solar-type stars as a class is difficult to explain. Whatever the reason, the slow rotation of the sun may be the result of a process that operates upon all stars like it, and we may ask the question whether it was related in some way to the origin of the planets which have the greater part of the angular momentum of the system.

The failures of the tidal hypothesis, and new knowledge of the nature of interstellar material, led to a re-examination of the whole

Fig. 27-12. Development of the solar system according to the protoplanet hypothesis.

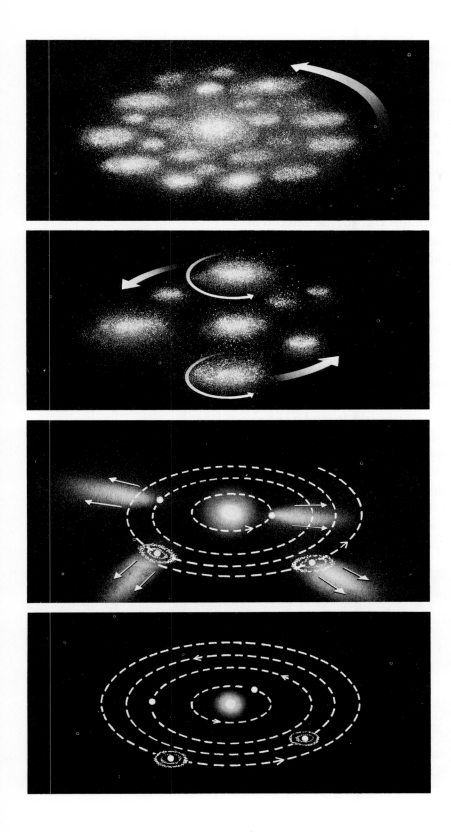

problem of solar system origin. Even though the distribution of angular momentum seemed to rule out the Laplacian hypothesis in its original form, it appeared, nevertheless, that the idea of a single extensive rotating cloud of matter must be basically correct. Von Weizsacker (1945) was particularly impressed by the general resemblance between the solar system, the satellite systems, Saturn's rings, and even the forms of whole galaxies. Each of these has the same basic "globe and disk" form: the planet with its coplanar satellites, Saturn with its rings, the sun with its coplanar planets, and the galaxy with its oblate spherical nucleus and the wide-spreading flat disk that contains the spiral arms (Fig. 27–11). These are all forms that suggest rotation of a formerly connected mass of material. As Von Weizsacker expressed it: "The form of the system must be understood historically. It is a relic of the time when the planets were still physically interconnected."

TURBULENT MOTION IN CLOUDS

Von Weizsacker introduced a new point of view in recognizing the importance of *turbulence* in the rotating nebula. Formerly astronomers had thought either of each atom pursuing its own regular orbit, or of the cloud rotating as a continuous mass. Neither of these conditions is physically realistic.

Turbulence in a cloud of gas or particles takes the form of a chaotic swirling and churning motion superimposed on whatever main rotation the cloud may have. Its effects are similar to those of friction in a fluid. The motion of rapidly moving parts can be communicated to slower ones, and fast-moving parts can be slowed down by friction with the slower. In a rotating nebula that has a strong gravitational attraction, the material in any zone lying within a limited range of distance from

Fig. 27–13. Halley's comet at its return in 1910, showing the tail streaming away from the sun (Mount Wilson and Palomar Observatories).

the center will tend to revolve about the center in a definite period. Zones nearest the center will have the most rapid motion and shortest periods of revolution, but within each zone and between adjacent zones, a great deal of turbulent motion will occur, forming a complicated system of eddies. Large and fairly systematic eddies contain smaller and more chaotic ones. Von Weizsacker supposed that the rapid rotation of the solar mass at the center would be communicated outward and that the sun's rotation would thereby be slowed down, while the speeded-up material of the disk would spread out farther and farther as its angular momentum was increased. Qualitatively this is correct, but quantitatively this mechanism alone is not adequate to account for the extreme slowness of the sun's rotation. The operation of turbulence does explain why the thin and wide-spreading disk is formed, and this is a real step forward.

The spiral galaxies, our solar system, the satellite systems, and Saturn's rings, as we have mentioned, have the same basic form. If we could fully understand how any one of them originated, we should understand them all. Turbulence is not the complete explanation, but it is an essential part of it.

THE PROTOPLANET HYPOTHESIS

G. Kuiper (in 1950) modified and extended Von Weizsacker's hypothesis. With minor departures our discussion from here on will follow Kuiper's proposed mechanism, which has been named the *protoplanet hypothesis* (Fig. 27–12).

The sun is much younger than our galaxy as a whole. Stages of history of the entire universe and of our galaxy before the birth of the solar system have been briefly touched on and need not further concern us

here. At the time when the solar system formed, some five billion years ago, our galaxy was similar in composition and organization to what we observe today. Somewhere in interstellar space, in a nebula of gas and "dust," denser condensations formed, producing dark globules like those we see now in the Lagoon and Rosette nebulae. One of these globules was destined to become the sun and its entourage of planets. We recall that the material was very cold, only a few degrees above absolute zero. Most of the solid particles were various ices, and the cold gases were held gravitationally by the attraction of the whole mass.

The rotation of a globule is practically inevitable. The material composing it was originally spread over a volume of space a large fraction of a light year in diameter. Due to the rotation of the galaxy, the differences of velocity of parts of the material at different distances from the galactic center would be appreciable. A globule several thousand times as large as the earth's orbit and with a mass a little greater than the sun's would be able to contract under its own gravitational attraction. Originally its slow rotation was probably rather chaotic, with large components transverse to the "average plane." As it contracted the rotation speeded up, due to conservation of angular momentum, and the cloud began to become oblate. Already the density was too great to permit each particle to revolve in an independent orbit around the center of mass. Mixing and collision of clumps of particles and molecules resulted in canceling out the motions transverse to the equatorial plane, and the whole cloud collapsed into a flat disk, all parts of which were caught up into a systematic whirl. Turbulent friction also tended to cancel out radial motions of material, so that the main motion of each part of the mass was in a nearly circular orbit around the center. The greater part of the mass was concentrated at the center, but a good-sized fraction, estimated at one-tenth the mass of the sun, was in the disk region and destined to take part in the formation of the planets, while the main central condensation contracted to form the sun. The large mass of material in the planet-forming disk is the distinctive feature of the Von Weizsacker–Kuiper hypothesis. The present planets altogether contain only about 2 per cent of this original disk. The rest of the mass has been lost. In the formative stages the light-weight and fast-moving hydrogen molecules and helium atoms were able to escape from the attraction of the planetary material. Even so, Jupiter and Saturn are still mainly composed of those elements.

FORMATION OF PROTOPLANETS

When the cloud had shrunk to dimensions somewhat larger than the present solar system, a balance between gravitation and centrifugal force was reached and the disk had become extremely flat, only 1 or 2 per cent as thick as its diameter. At this stage, the density was high enough to allow condensations to form. For a condensation to persist, its gravitational attraction must exceed the tidal action of the central mass (the future sun). With a mass of one tenth of the sun in the disk, this critical density (called the Roche density) would be exceeded in the regions of the terrestrial and Jovian planets. Saturn's rings illustrate what happens when the density is insufficient or the tidal action too great for the density of the material. No satellite can form so close to Saturn; any collection of ring particles would be quickly broken up. Similarly, no planet could be formed much closer to the sun than Mercury.

The process of gravitational break-up of the disk was gradual. Numerous condensations, small swirling clouds of particles and molecules, formed at first. Many of these collided and merged to form larger clouds, and particles coalesced to build larger masses, up to yards in diameter. Gradually the largest of the swarms of particles and larger chunks swept up and incorporated most of the material in several different rings extending over limited ranges of distance from the central mass. Each of these swarms was what we call a protoplanet, and was rotating and considerably flattened just as the main solar system cloud had been. Each had swept more or less clear a zone within its range of action and they revolved around the central mass without interfering with one another. They were enormously larger than the present planets. The earth protoplanet was at least a few million miles in diameter; Jupiter's protoplanet may have exceeded a hundred million miles in diameter.

The time scale for development from the original globule to the protoplanet stage has been estimated as some ten million years. The protoplanets were still very cold material, mainly hydrogen and helium, with large amounts of ices (chiefly compounds of hydrogen) and other solids such as metals and silicates. The over-all chemical composition was similar to that of the sun. From the outer fringes of the protoplanets, however, hydrogen molecules and helium atoms could escape their rather weak attractions. The most massive protoplanets, Jupiter and Saturn, lost the least material; small ones like the earth suffered enormous losses.

The most important cause of mass loss was the birth of the sun as a luminous star. Gravitational self-compression of the central mass raised its internal temperature to several million degrees. Even before nuclear processes were "turned on" the sun became very bright as violent convection brought hot gases from the interior to the surface. The transition from a dimly glowing cloud to a bright star was rather "sudden" in terms of cosmic time. A great flood of radiation spread through the space around the sun, raising the temperature in the region of the terrestrial planets to its present level, and for a time perhaps even higher. The greater part of the ices in that region were then vaporized, and much of the vapor escaped. Solar radiation pressure and a stream of particles ejected from the sun (the so-called "solar wind") swept most of the gases away from the terrestrial protoplanets and even caused considerable depletion of the Jovian planets, though their great masses enabled them to hold much hydrogen and helium. The temperature remained relatively low at their distances from the sun, and this also favored the retention of gases.

The repulsive action due to the sun is well illustrated by the behavior of comets' tails today (Fig. 27–13). As a comet approaches the sun, the tail forms and streams away at a rate that indicates the action of forces many times as great as the sun's gravitational attraction. These forces are effective only on atoms, molecules, and very fine particles. As a result of the sun's repulsive action the terrestrial planets are now composed almost exclusively of the heavy residue of silicates and metals, having lost practically all of their free gases. Jupiter and Saturn remain more nearly of "cosmic" composition. The earth's present mass is estimated to be about one five hundredth the mass of its protoplanet. Even Jupiter's mass is considered to have been reduced to a tenth or a twentieth that of its protoplanet.

DEVELOPMENT OF PLANETS AND SATELLITES
FROM PROTOPLANETS

During the protoplanet stage each disk was passing through a process similar to the development of the whole solar nebula. In the flattened disks condensations formed, to grow into the satellites, while the central masses became more compact, developing toward the final planetary condition. In each case the tidal action of the sun set a limit to the size of the system of satellites that could develop. As the protoplanets lost mass, some outer satellites escaped. The asteroids of the "Trojan group" are thought to have originated in this manner. Occasionally an escaped satellite would again encounter the protoplanet and be recaptured; this is considered to account for those outlying satellites that have highly inclined or eccentric orbits or that revolve in the direction opposite to that of the "regular" satellites.

A systematic direction of rotation was impressed on each protoplanet by the sun's tidal action. As the protoplanets shrank gravitationally, their rotation speeded up and the satellite systems that formed, as well

Fig. 27–14. The Copernicus region of the moon, showing probable impact craters on a lava plain. The large central crater, Copernicus, is 56 miles in diameter. It is surrounded by a system of rays, interpreted as material splashed over the surface by the impact of a large meteor. At upper left from Copernicus, a chain of small craters probably represents local volcanism (Mount Wilson and Palomar Observatories).

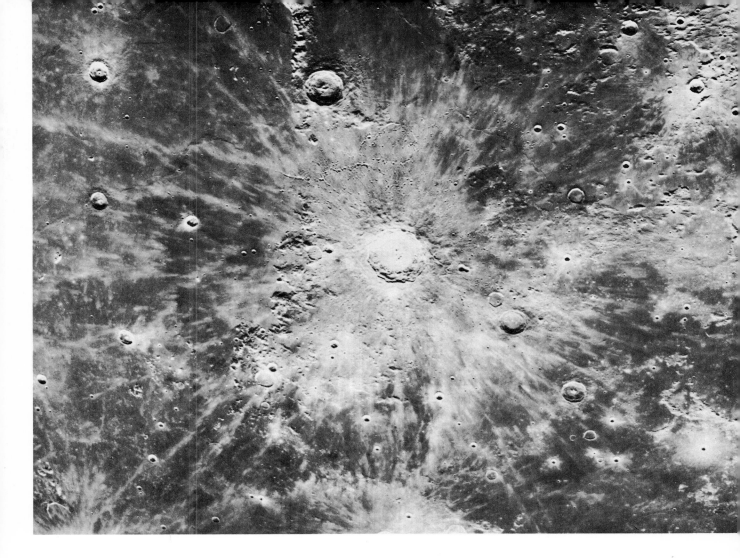

as the main planets at their centers, now rotate in the same sense as their orbital motion around the sun. Uranus is an outstanding exception.

Each of the protoplanets just before its final coalescence was a great gravitating swarm of particles and larger masses that had grown from clumping of particles. The presence of ices probably aided greatly in the process of accretion; Urey has suggested that the Jovian planets accumulated as tremendous "snowstorms" in space. The terrestrial planets, on the other hand, could have had only small amounts of ices. They must have been rotating swarms of material of various sizes that simply lost energy through numerous collisions until the whole mass collapsed upon itself.

The moon and earth are believed to have developed separately but in close proximity, possibly as two condensations in a single proto-planet, or perhaps the moon was "captured" by the earth during the protoplanet stage. A hypothesis formerly held, that the moon was formed by fission of the earth, now appears highly improbable. Incidentally, the moon may be a fairly good example of the appearance of any of the terrestrial planets at the end of the accumulation. Since it has no water or atmosphere, its surface has undergone no profound modification through erosion, and the scars made by the last infalling masses remain visible today as the numerous craters that pit its surface (Fig. 27–14).

The comets are explained as accumulations of ices and mineral particles that formed in the outermost cold parts of the nebula, where the density was simply too low for any sizable planet to form. The apparently chaotic arrangement of their orbits is speculatively attributed to disturbances by encounters with stars, which may be rather frequent at the great distances to which the comets range.

MAGNETIC BRAKING OF THE SUN'S ROTATION

When the sun first became a star it must have had a very rapid rotation, for its losses of angular momentum by turbulent friction could not have been great. The sun's slow rotation at present was a major stumbling block of the original Laplacian nebular hypothesis. It now appears that the slowing of the sun's rotation may have occurred during its early history through the operation of a magnetic braking action. The sun today has a weak general magnetic field, and very strong fields occur in sunspots. Numerous other stars are known to have powerful magnetic fields. If the sun in its early stages had a strong magnetic field, and if it ejected ionized (electrically charged) particles as it now does, then we have a means of braking. Ionized particles do not readily cross magnetic lines of force. The lines of force rotate with the sun, and any ionized particle that leaves the sun is swept forward by the magnetic field so as to rotate at more nearly the same angular rate as the sun, instead of lagging behind. But this means that the particle is given an angular momentum much greater than it had when at the sun's surface. For every gain of momentum by the ejected matter, there must be an equal loss of angular momentum from the rotating main body of the sun. Only a small percentage loss of mass from the sun would be required to account for the slowing of its rotation by this mechanism. While this is still somewhat speculative, there is some hope of overcoming the old angular momentum difficulty.

EARLY HISTORY OF THE EARTH AND ITS ATMOSPHERE

The earth, once it was a single compact mass, had already been considerably heated by the infall of the material. Gravitational compression further heated the interior, and the low conductivity of the silicate components prevented the rapid escape of internal heat. The structure of the earth, with its well-defined separation of core and mantle, is most easily explained by melting of at least the greater part of it. When the earth was newly formed, uranium 235 and potassium 40 were much more abundant than now. Radioactive decay of these atoms released a great amount of heat. This, added to the heat of infall and compression, probably is adequate to explain the melting, but according to some calculations the time since the earth's origin is insufficient for *both* the melting *and* the resolidification of the entire rocky mantle. The recent discovery of rifts in the midocean ridges raises the question whether the earth is now actually expanding, perhaps because its interior is still heating up.

The newly formed earth had very little atmosphere at first. Some ices in the interior of the protoplanet would have been shielded from the sun and were therefore trapped when the earth became a single mass. The resultant gases leaked from the interior to form a primitive atmosphere. At first, it is believed, this envelope was composed of methane, ammonia, and water, like the atmospheres of the Jovian planets (except that they also contain much free hydrogen, and their water is all frozen out). Solar radiation dissociated water vapor, and the freed oxygen reacted with methane to yield carbon dioxide and more water. Dissociation of ammonia liberated nitrogen which accumulated in the atmosphere. Thus the earth's atmosphere developed into one like that of Venus, rich in carbon dioxide and probably nitrogen. The present oxygen content of our air may have been largely produced by the action of plant life which fixed carbon and liberated oxygen. Part of it may have come directly from dissociation of water, the excess hydrogen having escaped to space. The oceans are generally thought to have originated from within the earth, accumulating during geologic time from the emanations of volcanoes, though the new additions now may be less voluminous than those of the earliest periods of earth history.

SUMMARY

To recapitulate, the solar system probably originated from a cloud of gases and dust like many that we can observe in space. The cloud, at first rotating very slowly, developed a rapid rotation as it contracted under its own gravitational attraction, and a part of it developed into a flat disk which was to become the planets, while the greater fraction at the center contracted to form the sun. The general resemblance between the satellite systems and the solar system convinces us that these bodies originated through some process broadly similar to that first outlined by Kant and Laplace, and later modified by Von Weizsacker and Kuiper. Instead of Laplace's hot gaseous nebula, we now adopt a cold cloud of gas and dust. Tidal encounter between the sun and another star failed to account satisfactorily for the present structure of the system. However, the process of accumulation of the planets according to the current hypothesis does resemble that pictured by the authors of the planetesimal hypothesis. The formation of the solar system appears to have been a fairly normal event in the history of the universe, and not a cosmic accident. In some young star clusters that contain nebulosity, dark "globules," and peculiar variable stars, it seems possible that formation of planetary systems like our own may now be going on. There is no reason to believe that the solar system is unique; it may well be a fairly common type in the population of our galaxy.

28
Stratigraphic Principles and Graphic Methods

Stratified Cambrian sedimentary rocks, Mt. Eisenhower, Alberta.

The principles that can be applied in determining the events that have transpired in the history of the earth have been considered in the preceding chapters. They will now be summarized in a more systematic manner. The reader may recall examples that have been discussed. As references to the specific illustrations are not given within this chapter, the subjects should be consulted in the index. The basic materials that concern us are rocks and fossils, and the skill with which each can be described bears on the knowledge that is gained of its history.

THE PETROLOGY OF SEDIMENTARY ROCKS

Petrology is the field that concerns the nature and origin of rocks; conventionally, there are treatments of igneous, metamorphic, and sedimentary rocks. The stratigrapher is concerned not only with the origin of the rocks, but particularly and peculiarly with their succession in time. Among sedimentary rocks, this is expressed in the general "law of superposition," in that the younger rocks will be laid over the older ones. To distinguish among the rocks in a succession, the sedimentary petrologist compares them with respect to colors, textures, compositions, and structures. The color, size, shape, and arrangement of the particles, and their chemical and mineral compositions and structures are also pertinent, particularly those features that are initial or primary. The methods of description of particles in unconsolidated and readily disintegrated rocks have come to be subject to elaborate statistical analysis, permitting not only their differentiation and classification but also judgment as to origin from correlation of the characters with those of known origin. Rocks studied in thin-section offer somewhat greater problems in analysis, but the theory of conversion of these characters for comparison with the unconsolidated counterparts has advanced. Primary structures, such as grain orientation, arrangement of cross-stratification and directions of sole-markings are gained from laboratory study of specimens whose original orientation is known, or from the examination of the rocks in the field. The geochemist, from the study of radioisotopes, can make determinations bearing on the age of the rocks; some of these determinations are particularly applicable to igneous and metamorphic rocks, others also to sediments, some to ancient rocks, and still others to those rather recent. The study of isotopes of oxygen, is applicable to determination of temperature of formation. There are other geochemical studies that bear on the salinity of the waters from which certain sediments were deposited. Such observations and determinations are made directly on rocks that are observable.

SUBSURFACE EVIDENCE

In the past few years, additional information has been gained from indirect, geophysical observations. When a well is drilled, the penetrated rocks can be retrieved in rock cores, though this is a time-consuming and costly procedure. Bits of the rocks, including their microfossils, can be gathered from the drilling fluids that rise from the well. And electronic devices can be dropped down the well, recording what are broadly called electric logs—probes that record a range of data such as resistivity and radioactivity—that permit interpretation of the physical character of the rocks and their contained fluids. Even more indirect information is gained by seismic methods, which record the density of subsurface materials through the speed with which they transmit waves, and the position and depth of layers through reflections from interfaces having contrasting properties; such exploration has progressed until quite continuous sections can be compiled from a progression of readings along a track.

BIOLOGIC INFORMATION

The fossil content of beds can be considered as part of the rock descriptions, quite aside from the biological aspects. Surface exposures have considerable advantages in permitting examination of whole specimens of larger fossils that will be preserved only as fragments in cores from wells, and in giving access to large areas on single richly populated surfaces. The refinement in taxonomy, the examination and classification of organisms, has become so great that few individuals are competent to presume authority on more than a limited group of organic taxa and generally those of a limited span of geologic time. With the advent of the digital and integral computers, analyses of the variations in fossil associations can be much more sophisticated than was humanly possible without them. Thus the rocks that are found in a succession of strata can be described and compared in their many attributes in veritably limitless detail, and interpretations derived from the data of such study can be of great interest.

THE DETERMINATION OF ORIGINAL SUCCESSION

The succession of events as expressed in the rocks is established by several variants of the law of superposition, which only in its simplest case involves the laying of a younger horizontal layer on an older one.

Rocks that have been tilted or turned to vertical or beyond must have determination of the top surface of layers, as by aspects of sedimentary structures of several sorts; structural criteria, such as the relations of bedding planes to cleavage related to the forming of the folds, can be very useful. Intrusive rocks cut through or affect rocks that are older, so that criteria have been devised that permit recognition of the relative ages of the rocks. The radioactive measurements based on isotope disintegration are rarely applicable to determination of local succession, because the spans to be compared in the local succession are rarely as great as the limitations of reliability in the geochemical methods; measurements of ages within the past few tens of thousands of years by carbon disintegration methods are an exception.

CORRELATION

After he has established the succession of rocks and the interpreted events in a single succession, the stratigrapher is concerned with extending his knowledge laterally. The comparison of two or more sections constitutes correlation which in its most sophisticated sense relates likenesses and differences quantitatively. Correlations are of two sorts, those pertaining to present or original continuity of lithology—homotaxial correlation—and those representing interpretation of contemporaneity or synchroneity; the latter may be of the general time equivalence of units in the same or different depositional areas, or of the synchroneity of depositional surfaces, relative or absolute correlation. In stratigraphy, the determination of synchrony is essential to the reconstruction of stratigraphic sections in vertical charts, or of maps presenting events of a time. Thus correlation tables are concerned principally with the placement of rocks of the same age at the same level on a graphic plot. The continuity of lithic units can be demonstrated often, but the precise contemporaneity of surfaces is rarely established. It has become known that lithic boundaries cross horizons of identical age in intertonguing facies, but techniques for the recognition of the planes of synchrony are barely known. Classification is a form of correlation, the probable general equivalence to a unit in a standard section of reference, often in a geographically distant place.

LITHOLOGY IN CORRELATION

Correlation of units in two sections can be demonstrated in simplest form by tracing or "walking out"; mapping on aerial photographs has contributed greatly to the accuracy of such work in many areas. The matching of electric logs is but an application of the method to subsurface studies; modern reflection seismic sections present data that show vertical continuity of subsurface strata remarkably well. Comparisons may be based on similarity in lithology of one unit in each of two sections; this implies that the rocks were deposited under conditions so similar that methods of study do not permit differentiation, that there is belief in their continuity, or both. Whether they are of the same age depends on whether the conditions progressed in time from one section to the other. Thus the basal unit succeeding an unconformity has probability of continuity but also the probability of varying age in all but one trend; ordinarily, continuity is the greatest practical concern. The distinction between lithologic and time-stratigraphic units has received much discussion; it relates to terminology that will be summarized in a later page.

Determinations of precise contemporaneity in separated sections is rarely possible, for few events produced broad, immediate, and yet distinctive effects on sediments. Sudden uplift may have spread peculiar and essential synchronous initial detrital sediments over great areas. Rarely, falls of volcanic ash have great utility. Exceptional storms produced intraformational conglomerates, thin shales in offshore deposits, and peculiar primary structures. Possibly, earthquakes have caused submarine flow breccias or some clastic dikes. These are but extremes in normal depositional history, the products of catastrophic events that produced lithologies that may be more than ordinarily useful in precise correlation.

Probability of correct correlation should be increased if a sequence of similar lithologies is represented in each section. Experience has shown that the seeming greater probability can be misleading when correlation is intended to imply synchroneity, for there are normal progressive lithologic sequences. Thus there are progressions indicating deepening of water and recession from shore in marine overlap, others showing derivation from rising source lands or those gradually reduced by erosion; most confusing are the many sorts of cyclothems that have repetitions even in details. The establishment of correlation between or among sections having contrasting lithologies is commonly established by tracing the intertonguing facies. Time planes lie within such tongues; a single depositional surface does not cross both upper and lower boundaries of a tongue.

FOSSILS IN CORRELATION

There are inherent advantages in correlation by paleontology that do not pertain to lithology. There has not been such change in processes from early to recent geologic time as to assure determination of age by lithology; conditions have been so similar locally at many times as to form sediments that are relatively identical. On the other hand, there have been progressive changes in the organisms, and the differences can be classified significantly.

The advantages in paleontology decrease as the scale of correlation becomes refined. In classifying units into systems, series, or even stages, changes normally are sufficient to permit some assurance of validity. Successive faunas may be so similar in smaller units as to be essentially indistinguishable, and those in restricted areas and lithologies may have greater resemblance to others of different age than to contemporaneous faunas from different provinces and habitats. At the smallest scale, variations in populations in synchronous beds are likely to be as great as contrasts in successive beds. Thus paleontology can practically never result in precise correlation of depositional horizons, whereas there is possibility of such correlation in limited areas by lithic criteria.

Correlation by fossils or faunas involves relative identity of forms. Though useful in judging synchrony, identity does not prove it. Relative age is established by superposition in single sections. Forms present in any population represent the influence of the time, but also of habitat reflected in lithology (lithofacies) and of paleogeographic factors controlling migration. Comparisons are commonly made with faunas assembled from several localities; insofar as any faunule deviates in age from another, or the assemblage adds fossils from beds mistakenly correlated, errors are compounded and correlations become less precise.

In early studies, geologists were concerned with classification into the largest categories, systems and series. Correlations were made on the general aspect of the fauna, which usually meant the presence of a few or many genera that seemed significant. As studies progressed with demand for refinement, the preferred correlation of a fauna with one of several in another section on the percentage of common species developed a quasi-mathematical method having an unwarranted impression of validity. The unit is deemed correlative with that in the compared section having the largest percentage of its species. There may be in fact no equivalent, in which case there is, nevertheless, a preferred correlation. Two halves of a single sample may have a low percentage of common forms. The sections compared should be of similar lithofacies and of biofacies as suggested by the presence of similar organisms. There must be comparable discrimination in the identifications; if the paleontologist tends to separate species on minor differences, he will gain smaller percentage than if he does not. There should be comparable study of successive units in the reference section, and the units should have faunules of similar size. As far as practical, control should be quite continuous. It has been written that "for establishing the sequence of faunas within a given region, the percentage method is advantageous. It may be useful in affording a clue to proper allocation of a given formation. But in the face of the numerous qualifying factors now recognized, one must conclude that the evidence which the method supplies is relative rather than absolute" (M. Keen, 1940). Comparing percentages of fossil and recent forms has been a basis for classifying Tertiary rocks since its introduction by Lyell more than a century ago.

Other methods of similar type, such as the correlation of forms in one unit with the median of midpoints of ranges in another, have disadvantages additional to those of the percentage method. The long ranges of some forms may make it impossible to avoid correlation of many units in one section with those near the middle of that compared.

Correlations and classifications are based on one or more of several criteria: (1) the presence of one species or genus that in experience has a limited range; (2) the presence of forms whose upper known range overlaps the lower range of others; (3) sequences of forms or faunas; (4) the presence of an evolutionary stage in a series, a corollary of the preceding; and (5) abundance of forms elsewhere abundant in limited range. The judgment of an expert is based on consideration of these kinds of evidence in varying degrees.

THE RANGES OF FOSSIL ORGANISMS

Fossils most characteristic of a restricted range frequently are known in older and younger beds. Maximum range can never be known, inasmuch as it requires finding the initial and final representatives of the form. Since all must have had ancestors, and since some left descendants, the range for each fossil depends on the discrimination between these two factors. Abrupt changes imply a gap in stratigraphic record in the section, a change in habitat admitting a new biota, or a paleogeographic change allowing introduction of previously foreign elements. Undue stratigraphic emphasis is sometimes given to absence of forms as evidence of lack of equivalents, most stratigraphers are familiar with inexplicable absence of species in rocks that seem continuous with beds in which they abound. Observation of faunas on present shores indicates that populations vary locally. Collections never give the complete fauna, even of durable organisms. Differences in temperature, salinity, and depth that are not apparent in lithology may have affected the marine life appreciably. And there are factors of chance and human inadequacy that defy analysis.

THE PRECISION OF CORRELATION

The accuracy of correlation depends on the rapidity with which faunas evolved, so that the descendent forms are distinguishable, and on the presence of linear descendents in successive strata. Though the discrimination of species may be refined, there is the probability that the range of variation in an ancestor will overlap that in the descendents. The initial appearance of derived forms is never exactly known and ancestral forms persist; frequently there is not progressive and exclusive presence of successive forms, but replacement of older by younger approaching continuous progressive variation and mutation. In a single sedimentary basin having free communication and abundant specimens, correlation may be made fairly confidently on proportions of successive forms, though each will be a probable rather than a precise correlation. Restriction of ingress of forms from locus of origin leads to seeming retardation, with apparent correlation with units older than proper. And if the fossils were differently controlled by ecology, the possibilities of close correlation diminish. Phylogenies rarely are known among fossils that are used as guides in correlation. Thus correlation is based usually not on progressive evolutionary stages, but on forms having scattered places in biological sequence and whose similar ancestors lived elsewhere.

Though developing techniques constantly improve the precision of correlations by fossils, conclusions must remain as among many alternatives. Collection and description continually increase knowledge of forms and their ranges. Thorough studies of large collections are needed, so that a range of variations can be established and comparisons made of synchronous populations. Most descriptions of species are based principally on an original reference specimen, the holotype, with addition from a few other specimens in a subjective manner; frequently a species is named from very few specimens that fall within the known variation of other named species, though differing from their holotypes. Designations of horizon and locality of types are normally not exact enough to enable further collection of variants; species are type specimens to some and related variant groups to others. Systematic studies between fossils and lithologies are revealing many interesting relationships; such are particularly informative when applied to rocks whose continuity can be established confidently by other means. There is promising research in the distribution of organisms in modern sediments with respect to lithofacies, and to other conditions of habitat that are not directly evidenced in consolidated sediments. Paleontologists are becoming well informed on studies of modern organisms. These and other projects are in the field of paleoecology. Quantitative statistical studies of the measurable characters of fossils—their morphological variations and of the associations in populations—are being increasingly applied.

LIMITATIONS IN THE UTILITY OF FOSSILS

Those who use fossils must be cognizant not only of the possibilities of the methods but also of the limitations. Age is proved only by superposition; identical conditions produce identical deposits, not necessarily laid at the same time, encouraging similar life; ranges are limited by experience and each case is a possible exception. Relations often are determined by restoration of sections rather than by methods that can be applied empirically.

CLASSIFICATION OF STRATIGRAPHIC UNITS

The categories of stratigraphic units have been divided into those that concern the lithologic types of rocks, rock-stratigraphic units; those having similar fossil assemblages, biostratigraphic units; and those having rocks formed in the same time span, time-stratigraphic units that designate the time of formation applied to successively smaller intervals or spans. The principal classes of rock-stratigraphic units are formations, their subdivisions called members, and the association of two or more formations, a group. The units are designated by such terms as Niobrara Limestone or Tellico Formation. Those of biostratigraphic units are zones, bearing the names of fossils that are representative, as Zone of *Exogyra costata*. The subdivisions are successively subzones and zonules. The time-stratigraphic units in succession from greatest time span to smallest are the system, series, stage, and substage, the corresponding time terms are period, epoch, and age. And spans of time containing a succession of periods are known as eras, there being no corresponding time-stratigraphic term. The time and time-stratigraphic units are conventionally recognizable from their "-an" or "-ian" endings.

Generally, rock-stratigraphic and time-stratigraphic units have geographic names, and each has a designated section of rocks that is typical. The definitions of zones are more complex because there are biozones of several sorts, based on assemblages of fossils, on ranges of specified fossil categories and other factors. The procedures in establishing and using stratigraphic terms are presented in a code of stratigraphic nomenclature that is under the supervision of the American Commission of Stratigraphic Nomenclature, an international body having representatives of the principal geologic organizations in Canada, United States, and Mexico. A list is kept of all of the formal units that have been used and recorded in order to guard against the application of identical names in conflicting ways.

CLASSIFICATION OF DEFORMATIONAL EVENTS

In addition to stratigraphic units, times of deformation gain geographic designations. The time span through which the deformation is judged to have endured is varyingly called a revolution, disturbance, or orogeny, the terms being subjective as they reflect the attitude of the sponsor; generally, the geographic names have "-an" or "-ian" endings, as Nevadan Revolution or the Caledonian Disturbance. The structures formed as a result of the deformation are known by the same root with the ending "-ide" as the Nevadides or the Caledonides. It is the practice in some instances to consider that a revolution spans time that contains several or many pulses or phases, each of which may be designated as a disturbance or as a phase of the revolution.

STRATIGRAPHIC SECTIONS

A number of graphic methods and conventions present the stratigraphic data in condensed and readily appreciated form. The stratigraphic succession in a locality is represented in a columnar section, a strip having representation of the thickness of each unit to uniform scale with the lithology represented by conventional symbols, such as dots for sand or sandstone, brickwork for limestone and short horizontal dashes for shales or argillites; or the rocks can be sketched with their relative resistance shown by the prominences and re-entrants on the side of the column; or the rocks can be represented with several closely parallel columns, with the thickness of each rock type represented in the appropriate column. Another type of columnar section is that having the record of geophysical effects, electric logs, showing the resistivity through the column or other properties. The columnar section is a linear description as though from a well.

The data can be treated in two dimensions in vertical section or horizontal plan. In the simplest form, two dimensions can be represented by a succession of columns distributed laterally relative to their horizontal separation on the ground; lines can connect points of correlation, lithic or time-stratigraphic. A structure section is a restoration of the present conditions as they are interpreted as though on the wall of a vertical slice; conventionally, sections have north or west on the left side. Horizontal scale is generally much less than the vertical scale; this can produce apparent thickening and thinning in folded beds, so usually the sections are drawn with approximation of scale. Restored sections are those that represent the stratigraphy and structure as it is interpreted in the past, as at the close of deposition of the stratigraphic interval; they show in profile the elevations of source lands, the slopes of depositional environments, and depths of marine waters. Lithologies are represented by conventional patterns commonly separated by lines of equal time (isochrons) and by lines of similar lithic contrast (isoliths); relations of the latter to the former give evidence of facies patterns, reflecting the physical and biological conditions that developed. Sections can express three dimensions on a flat surface by placing columns as though one were examining a perspective model having columns distributed on a map, a stereographic projection; the lines representing correlations among such columns can be quite confusing. A more lucid method is to present restored sections stereographically, a diagram resembling a set of intersecting walls or fences on which the stratigraphic relations are portrayed. The effort to portray three dimensions is effected also in maps.

MAP PROJECTIONS

A map of a part of the surface of the earth must show on a flat sheet the projection of data that are at elevations varying with respect to sea level. The effect of the spheroidal form of the earth is not significant on large scale maps of small areas. But the portrayal of large areas such as continents must introduce distortion of some form. Thus the many map projections are concerned with portrayal having the least distortion in such properties as relative areas in different parts of the map, or constancy of directions, or variations in shapes. Maps of ancient geographies become involved in still another problem, that of relating the present rocks to their relative original positions. This arises when changes have produced alterations in relative distances and directions of the placement on the rocks from their initial locations; these changes might be of large magnitude involving presumed drift of continents or those of smaller magnitude, such as the displacements on thrust faults or on strike-slip faults and the distortion of positions brought about by folding. Palinspastics concerns the placement of rocks in their relative original positions, and maps that make these corrections are palinspastic maps; their preparation is most important in regions that have been the sites of mountain-making.

PALEOGEOGRAPHIC MAPS

Maps are truly geographic if they show conditions at a moment of time, stratigraphic if they represent the cumulative information through a span of time.

The simplest paleogeographic maps are those separating seas and lands. Hypsometric maps that have contours or hachures are used in illustrating present topography, the common "topographic maps." As past lands are interpreted chiefly from the deposits gained from their denudation, evidence is rarely sufficient to warrant their representation by contours; they have been used occasionally in local areas on unconformities having high relief. Increasing numbers of smaller scale maps employ shading or physiographic symbols to portray concepts of lands. Shore lines should have appropriate symbolic shore features rather than smooth curves.

Bathymetric maps show the form of sea floors by lines of equal depth, isobaths. Rocks preserve many pertinent facts bearing on the depths of waters. Paleolithologic maps have lines, isoliths, connecting points of similar lithology and separating rocks of differing nature, designations of color, texture, or composition; isoliths include such lines of equal abundance or magnitude (isopleths) as those of equal average sediment grain sizes. Paleolithologic maps carry tectonic implications; rocks of coarsest texture are deposited from fluids that have the greatest energy. Thus stream-laid sediments will be coarsest toward the source where gradients are highest, and marine sediments toward shores where agitation is greatest and the coarsest sediments lag as the finer are carried away in suspension. Turbid flow to submarine depths will also produce aprons of deposit of decreasing average texture. Character of sediment cannot be converted empirically to depth, but the nature and distribution of textures, characters of bedding, cross-lamination, and ripples aid in judgment. Maps of fossil distribution, with evaluation of the habitat of the organisms, add evidence. Thus from the lithology and fauna, it is possible to construct paleobathymetric maps. Paleolithologic and paleobathymetric maps are useful not only as concise summaries of regional records but also in interpreting stratigraphy and structure.

An atlas of geography contains maps showing the courses of streams, ocean current, and wind directions. In the rocks, currents in air and water have oriented organisms, ripple marks, flute-casts and cross-laminations, as well as distributed lithofacies. Faunal and floral distributions have climatic correlation, as have paleolithologic maps recording pertinent data.

Studies in the orientation of fossils have been few and those on the orientation of ripples are not common. However, these ripple studies bear on climate if they are the effects of prevailing winds, on coastal trends if they are the results of offshore currents. Cross-lamination in several aeolian sandstones shows distinctive patterns of orientation that have significance in interpreting climatic belts on the earth; these patterns have contributed to the discussions of continental drift. The many studies made in fluviatile deposits reveal the prevailing directions of past stream flow. Another type of primary structure is that formed by fragments carried by submarine density currents; the slopes of past marine bottom surfaces have been plotted, giving valuable paleogeographic bathymetry. Orientation studies offer a field of investigation that has been neglected until quite recently and is giving important paleogeographic results.

ISOPACH MAPS

Lines connecting all points of equal thickness are isopachs. Inasmuch as isopach maps cover a span of time, they are correctly stratigraphic rather than paleogeographic. They are objective only in areas having full preservation of units, but their construction permits better judgment of the former extent of sediment, for the solid geometric forms are truncated by present and ancient erosion surfaces. Ultimate knowledge may become greater of older than younger sediments, for their very burial results in broader preservation, though it reduces probability of penetration in wells. Isopach maps, first prepared in any numbers in the nineteen thirties, have become the most common kind of stratigraphic maps. Structure contour maps are those showing the elevation relative to sea level of a stratigraphic datum; for example, they can show the present configuration of the surface of the top of Precambrian rocks. If the surface of reference was originally level, structure contours give evidence of warping—epeirogenic movements that have affected the surface of reference. Many factors determine the present thicknesses of sediments. However, isopachs are not simply structure contours of a past time, though in most cases no serious errors result from considering them so; isopachs are not isobases, lines of equal epeirogenic deformation. These exceptions are of such significance that they bear discussion.

Compaction will have reduced thicknesses, perhaps most where thickness is greatest, but varyingly with sediments of differing compositions and textures. Though a well-sorted quartz sandstone will be reduced to less than two-thirds the original volume, a clay may be compacted to only one-tenth its original thickness through loss of water in its porosity. Changes in sea level, eustatic movements, add to or subtract from effects of differential deformation. To analyze the remaining factors, the original forms of lower and upper surfaces of reference must be considered. Maps of units bounded by unconformities or disconformities give an approximate measure of epeirogeny if the surfaces were peneplanes. If the lower and upper surfaces differed in slope or relief, isopachs may give little information about the movement that took place during the time of deposition. The character of the lower and upper surfaces might be expressed on paleobathymetric maps, but these are never available. The comparison of paleolithologic maps at different levels in the stratigraphic interval will enable better recognition of contrasts.

Thus isopach maps must be interpreted cautiously. But generally they are in large part a record of epeirogeny. Contrasts in successive isopach maps are among the most lucid means of revealing changing trends of deformation and thus of tectonic development.

LITHOFACIES MAPS

Maps showing the changing proportions of sediment types within a time-stratigraphic span in an area are lithofacies maps. They differ from paleolithologic maps in that they are concerned with the average through a thickness of rocks rather than the distribution on a depositional surface at a stratigraphic horizon; thus they are three dimensional and related to isopach maps. In the simplest form, lithofacies maps portray the proportions between two end constituents averaged from columns distributed over the area of the map, such as the ratio of sand to shale (sand-shale ratio) or of terrigenous or insoluble to non-terrigenous, soluble constituents ("clastic ratio"); such maps have lines connecting points of equal proportions. More complex are comparisons of three end constituents, such as the proportions of limestone-sandstone-shale at each point on the map within the stratigraphic unit. A triangle is drawn, with 100 per cent of each constituent at one point; the interior is divided into areas having differing proportions of the constituent, each shaded with a distinctive pattern; the areas on the map are separated into regions having similar proportions shaded by the same patterns. They carry tectonic implications; as rocks of coarsest texture are generally deposited from the fluids having greatest energy, and agitation is related to water depth, lithofacies maps like paleolithologic maps present evidence suggestive of conditions of deposition. If conditions changed during the span represented by the mapped unit, or if the stratigraphic interval is more fully represented in some areas than others, averages can fail to represent the conditions from time to time. Lithofacies maps have been prepared for a range of attributes or parameters of sedimentary rocks, for the correlation of the variables, enhanced by the application of computing techniques, bears on the exploration for petroleum as well as leading to better understanding of the geologic history.

Lithofacies maps are widely used with isopach maps to portray and interpret past conditions of deposition and deformation. From their study, judgments are expressed on the relative stability of different parts of an area in a span of time, the interpretation expressed in paleotectonic maps. Locally, the maps may suggest the directions in which changes can be anticipated. Maps can portray the larger tectonic elements, such as the mobile volcanic and non-volcanic geosynclinal belts and the more stable cratonic areas, as well as the relatively more and less active basins and platforms within the latter. Just as lithofacies maps integrate the sediment types through a span of time, biofacies maps can present the dominant biologic aspects of a span of rocks; certain sorts of organisms have had sharply limited belts or areas of distribution in the past, reflecting the areas of favorable conditions for living and preservation.

PALEOGEOLOGIC MAPS

The term geologic map is applied to a class of maps showing the distribution on the present surface of rocks of each lithic kind—the formations of the large scale map—or of rocks of successive ages, systems, and series on smaller scale maps. Paleogeologic maps show the areal geology of an ancient surface. They permit better understanding of regional structure beneath unconformities. Comparison of structural trends evidenced in paleogeologic maps with those shown in properly evaluated isopach maps of the preceding and succeeding deposits facilitates understanding of the regional deformation. A complementary type of map is that showing the age of rocks progressively overlapping an unconformity, as though one lifted the blanket of rocks covering an unconformity and looked at the bottom of the blanket.

TECTONIC PATTERNS

The term tectonic concerns deformation, the events following it, and their products. Thus a tectonic map is one showing the distribution of the larger structural features as well as the pattern of the smaller ones. The physical record of the earth's crust is dependent on deformation. Particles eroded or salts dissolved from parts of the crust that have been relatively raised, flow toward parts of the earth that are relatively depressed. Large parts of the crust are relatively stable. These are the cratons—the low ocean basins and the high continental platforms. Between and within them are the more mobile belts, bands on the earth that subsided deeply, such as the oceanic troughs, or that rose to form the mountain ranges. These are relative distinctions, but there are fundamental differences in the crust between the oceanic and continental cratons, quite aside from their elevations. Deformation that is in the form of warping or block faulting, dominated by vertical or lateral-slip movements, is epeirogenic. Though the term orogenic strictly means mountain building, it has been transferred to deformation such as produces folds and low-angle faults, as well as accompanying invading intrusions; the mechanics are not fully understood.

Tectonics and sedimentology are subjects in themselves, though closely related to stratigraphy. The larger implications of stratigraphy, the determination of structural events that produced the sediments and permitted their deposition, involve tectonics and sedimentation. The thickness of deposited rocks depends on there having been a place where sediments could be retained; volcanic rocks can accumulate independently. A deep-sea trench has abyssal depth because the narrow belt subsided deeply, more rapidly than sediment and volcanic materials accumulated. Thus the measure of deformation is not alone thickness of sediment and associated volcanic rocks but also evidence of subsidence without filling; the surfaces beneath the empty and the filled troughs may have similar cause. The base of the elongate filled trough is invariably called a geosyncline, the oceanic-deep trough somewhat reluctantly so (a leptogeosyncline—from "thin," in reference to the sparse deposits). Subsidence and available material determine the thickness of geosynclinal sediments. Rates of subsidence are quite variable, rarely reaching as much as a thousand feet in one million years through any great span of time; if such rates were to continue for any great time, they would exceed the thickness of the crust!

The more mobile belts of the earth, those having folded and thrust-faulted stratified rocks, have been known for a century to be in the beds in which rocks are relatively thickest; thus the more mobile belts were previously the belts of greatest subsidence; with recognition of the latter, they have become known as orthogeosynclinal ("straight") belts.

CLASSIFICATION OF TECTONIC BELTS

In formal classification, it is useful to have distinctive terms that represent discrete entities, as in some categories of biological taxonomy or end members of continual series as in isomorphic mineral groups. So-named, the types of geosynclines can be compared, and each name expresses information that would require repeated descriptive phrases. The "clams and snails" of the nursery rhyme are pelecypods and gastropods, terms that are part of a formal classification that removes the inadequacies of vernacular words. The distinctive aspect of an orthogeosyncline is that it is a great earth downfold formed through time, an elongate trough in a long belt. It is not defined by the contained rocks, though the proportions of several rock types in orthogeosynclines will be found to differ from those in other geosynclinal classes.

The rocks commonly have been severely deformed, and often were metamorphosed and intruded in accompanying and following orogenic phases; the determination of sequences and thicknesses in mobile belts is difficult. The kinds of rocks that are found in the deeply subsiding belts depend on the sources that yielded them. Great thickness may be the effect of long-continued slow subsidence or short-time rapid sinking. But the character of sediments will be a reflection of the rigor of erosion of the source. Rocks gained from a broad, stable region where streams and currents slowly sort sediment constituents may reach considerable thickness if carried into areas of great sinking; the same depression can receive poorly sorted materials from other sources in unstable lands nearby. Thus the sediments in a geosyncline reflect relative rates of sinking and filling, the nature of source areas, and the environment through which the sediments have been transported.

From the description and correlation of rocks of successive systems, it has been found that geosynclines differ in character—some are in belts having prevalent or frequent volcanism and others in belts lacking volcanic rocks; the geosynclines are the "earth downfolds," the rocks are their contents. Thus there are volcanic and non-volcanic orthogeosynclinal belts: eugeosynclines, "true" ones, those that are volcanic; and miogeosynclines, "lesser" ones, not volcanic. Eugeosynclines and miogeosynclines were not simply paired troughs that extended continuously along the borders of the continents. Their patterns in the belts were complex. Some volcanic geosynclines merged laterally into non-volcanic ones, others had intervening tectonic barriers. The volcanic geosynclinal belts commonly had a succession of elongate troughs separated by tectonic welts, some forming shoals that subsided little, others rising to highlands that yielded sediment to adjoining troughs; and associated were submarine flows or volcanic archipelagoes that contributed appreciably to the filling of the troughs. The mobility and instability encouraged the distribution of poorly sorted sediments, but some were products transported across the more stable non-volcanic miogeosynclinal belt from erosion there or from the craton. In these regions of crustal instability the structural geologist and student of igneous and metamorphic petrology meet their most challenging problems.

Stability is a relative thing. Geosynclines are conventionally elongate surfaces below thick sedimentary and volcanic-rock successions. The great troughs are within these mobile belts. But toward the continental interior there are other elliptical and oval subsiding troughs and basins that formed through time, as deposits filled them in the craton. Some were deeps formed in the border of the continental interior, filled by detritus carried from tectonic lands raised in the more mobile belts; they have been called foredeeps, or exogeosynclines, the deposits coming "from without" the continent. Farther within the continental interior, some regions subsided for a time rather rapidly, the oval or elongate basins acquiring substantial volumes of sediments, the intra-cratonal basins that are independent of complementing source lands, or autogeosynclines. "Self-sinking," and isolated, they are generally called "basins"; their resemblance to more typical geosynclines is only in that they subsided considerably through time. Basin is a non-technical term that is applied in geology to the structural form evidenced by a structure contour map on a surface of reference, to a physiographic or bathymetric surface such as an intermontane basin or a lake basin, and to the form developed by subsidence through time represented in the autogeosyncline or "autobasin." A few similar subsiding basins had complementing interior source lands; these zeugogeosynclines are "yoked" to highlands. Such classifications are made to create categories that are to be compared to others, to exemplify concepts rather than to impose terminology. Technical words are often the economical ones, for they say a great deal in a few syllables. One should not fail to emphasize ideas rather than semantics.

After orogenies have consolidated mobile orthogeosynclinal belts, deep, linear basins have formed and filled with sediments and with some volcanic rocks. When fault bounded, they have been called taphrogeosynclines—from taphrogeny, block breaking. Epieugeosyn-clines, "lying above" eugeosynclines, seemed to be relatively very narrow belts sinking deeply through time; experience has shown that such deformation is commonly associated with fracturing, so some that were interpreted as deep down folds may actually be taphrogeo-synclines. Paraliageosyncline, "coastal," is a term that has been applied to the downfold that developed in late geologic time along the southern margin of North America opposite the Gulf of Mexico; such a linear, non-volcanic belt resembles a miogeosyncline, but the identity is questionable, so the specific terms permit comparisons. Thus the classification of these larger earth features, based on the correlation and interpretation of their contained rocks, enables the stratigrapher to learn more of the contrasting development of the earth through time.

The aspiration of stratigraphers is to describe the succession of rocks in the earth's crust, and to interpret them so as to understand the events that have transpired. An individual geologist can become familiar with the record in a limited area, or within a limited time-stratigraphic span. The collation of the vast fund of information that has been obtained involves the application of principles, and subjective analysis. Many of the conclusions are those that seem most probable; new observations and reinterpretations of old ones lead to constant refinements and occasionally to drastic changes in view. The study of the organisms within the rocks by the paleontologists is carried forward by many scientists, each especially informed and qualified by his training and experience within a limited domain.

Although words can convey a picture, much of historical geology is expressed concisely and vividly by the graphic methods of sections and maps. The details of the history of a continent and of the earth have barely been introduced. The principles that develop the history from the interpretations of rocks have been brought out. Few books endeavor to treat the record of the whole earth, and these must do so in a very limited way. Very few have endeavored to consider the record of a single system, and these have been directed toward some aspect of the system, such as the sequence of faunas. In North America, atlases are being prepared that give some of the basic data relating to each of several systems. Other publications have brought together correlations within the continent. Knowledge of the surface and the subsurface increases continually, contributing to the revision of the history. The earth offers such a great fund of knowledge to be gained that the prospects of learning are nearly without limits.

Though no man can draw a stroke between the confines of day and night, yet light and darkness are upon the whole tolerably distinguishable.

Edmund Burke
Thoughts on the Cause of the Present Discontent, 1770

Index

Isostasy (cont.):
    adjustment, post-glacial, 584; origin
        of continents, 638
Isotherms, 392
Isotopes: analysis by mass spectrograph,
    71; of potassium, 71; oxygen isotopes
    and temperature scale, 393; age
    determination by carbon[14], 584–85
Isthmus of Panama: "filter bridge," 620;
    effect on continental and marine
    faunal migrations, 625

Jackson Dome, Mississippi: Cretaceous
    reefs, 420, 422, 566; volcanics, 420,
    422; structure section, 430–31;
    Paleocene surface contour map,
    430–31; Cenozoic relations, 558
Jasper: origin of, 37; in Precambrian
    Soudan formation, 37
Java Man, Pleistocene, 628
Jeans-Jeffreys tidal hypothesis: solar
    system origin, 674
Joggins, Nova Scotia: Carboniferous tree
    casts, 259; amphibians, 261; 267
Jovian planets: names of, 686; chemical
    composition and density, 686
Juniata, 172
Jupiter, 686
Jura, Mount, 401, 403
Jurassic:
    Alaska Peninsula: 398–99, 404
    Arctic coast, 413
    Atlantic coast, 413
    Black Hills, South Dakota, 408
    California: Yosemite Valley, 389; Sac-
        ramento Valley, 397
    climates of, 414, 492
    conglomerates, 398–99
    continental development, 398
    England, 412
    fauna of:
        ammonites, 390, 392, 401, 403, 488
        as climate indicators, 392–94
        belemnites, 401
        birds, 483, 492
        dinosaurs: migratory routes, 412; in
            Morrison formation, Como Bluff,
            Wyoming, 408, 413–14; at Red
            Rocks, Colorado, 414; 490, 491,
            492
        faunal zones, 392
        fish, 489, 490
        parallelism in evolution, 489
        restoration of sea bottom life, 413
        vertebrates, 413–14
    flora of: 397, 403, 667
    geomagnetic poles during, 394–95
    Great Plains states, 408
    Gulf Coast: subsurface thickness and
        distribution, 408–11; depositional
        pattern, 413
    Idaho, 408
    in Arizona, 404
    in Rocky Mountains, 404
    in Sierra Nevadas, 403
    Jura Mountains, France, 390–91
    Mexico: paleogeographic map, 406–07;
        408–09, 408, 410
    Morrison formation: maps of, 404–05;
        stratigraphic succession, 404; isopach
        map, 404–05; Como Bluff, Wyoming,
        406–07, 408, 414; uranium ore, 406;
        reptiles, reconstruction of, 410–11;
        394, 403, 412

Jurassic (cont.):
    naming of system, 388
    Nevadan Orogeny in, 398, 404
    Oregon: flora, 397; marine sequences,
        403
    Pacific Coast, 401–04
    Rock of Gibraltar, 388
    section, Idaho to western South Dakota,
        402
    section, Nevada, Arizona, New Mexico,
        402
    southwestern United States, 410
    stages of, 390
    Texas, 408
    Utah, 408
    wind direction, 398, 400
    Yukon, 404
    Zion National Park, Utah: the Great
        White Throne, 387, 404–05; Navajo
        sandstone, cross-bedding in, 400,
        404–05, 406
Juresania, 318–19
Juvavites, 378–79

Kaibab: fauna of, 304; 289
Kaibab Plateau, 300–01
Kansan till, 580–82
Kansas: Central Platform of, 280; cyclo-
    thems, 293; Permian salt mines,
    294–95
Kant, I.: solar system, origin of, 672
Karoo System, South Africa: fauna of,
    484–85
Keewatin, 50, 56, 60, 61, 74
Kentrosaurus, 492
Kentucky: Mississippian cyclothems in
    Illinois Basin, 246
Kettleman Hills, California: anticline,
    528, 530; petroleum, 535
Kettle Point, Lake Erie, Ontario, 217
Keweenawan: lavas in, 39, 40, 46, 50;
    sedimentary rocks of, 46, 50; copper
    mining, 50; intrusions, 56; compared
    with Belt Group, 80; 52, 60, 74, 80
Killarnian, 83
Kimmeridgean: dinosaurs in, 412; fauna,
    compared with Morrison, 414; 390,
    394
Kinderhookian, 238, 244–47
Kingston, Ontario: Ordovician-Precam-
    brian contact, 158–59
Klamath Mountains California: Paleozoic
    eugeosynclinal deposition, 230–31;
    fossil plants of, 403
Knife Lake Group: age of, 33, 34; con-
    glomerates of, 36; correlation of, 58;
    34, 40, 56
Koenenites, 330, 331
Kronosaurus, 496
Kuiper, G.: solar system origin, nebular
    hypothesis, 672, protoplanet hypoth-
    esis, 691

Labrador: metamorphic and igneous rocks
    of, 18–19; iron formations, 64
Labyrinthodonts, 300, 384
Lagomorphs, 618
Lahontan, Lake, 593
Lake terraces: post-glacial in New York,
    588–89
Lampreys, 234–35, 658
Laplace, P.: solar system, nebular hypoth-
    esis of, 672–74
Lapworth, Charles, 88, 128

Laramian Orogeny: 283, 416, 440. Laramian
    Revolution:
    dating of, 509, 520, 522–24
    epeirogenic movements, 509
    extent of, 520, 522
    faults of: Chief Mountain fault, Mon-
        tana, 79; in Idaho, 450–51; in Wy-
        oming, 450–51; in Rocky Mountain
        region, 522, 528
    in Rocky Mountains, 448, 509, 522
    in western United States, 434, 441, 504,
        508–10, 521
    in zeugogeosynclines, 523
    overthrusts and miogeosynclinal belts,
        522–24
    subsidence in: Denver Basin, 510; Pow-
        der River Basin, 510; Laramie Basin,
        510
    taphrogenic movements, 509
Laramides, 436, 509, 520
Laramie Basin: Tertiary subsidence in,
    510
Laramie Peak: monadnock, 512
Laramie Range: peneplanes in Eocene
    and Oligocene, 512; monadnocks of,
    512, 515
Latimeria, 382
Laurentide ice sheet, 578–79
Lava flows: See also pillow lava
    amygdules in, 38
    Coahuila, Mexico, 294
    Columbia Plateau, 5
    Idaho, Snake River canyon, 541
    Triassic, Newark Group, 356–61
    Utah, 521
    Washington, 5
    Wyoming, Absaroka Mountains, 526
Lead. See also geochemical dating: isotopes
    of, 68; and uranium disintegration,
    68
Lepidocyclina, 565
Lepidodendron, 252
Lepidophloios, 253
Leptogeosyncline, 714
Levis, Quebec: graptolites, 132–33; con-
    glomerate, 134, 139; geologic map,
    161
Libya, 565
Liebia, 654
Life: origin of, 340; expansion through
    time, 341
Limulus, 654
Lingula: Cambrian, related form, 94; mod-
    ern, 136
Linnaeus, Carl, 94, 644
Lipalian interval, 74
Liparoceras, 392
Lithic tongues, 110–11
Lithic zones, 164
Lithofacies: on paleographic maps, 114;
    222
Lithofacies maps:
    clastic ratios, 712
    Cretaceous, Rocky Mountains, 440–41
    Jurassic: Morrison, 404–05
    Ordovician: Adirondack-New England,
        143; southwestern Virginia, 170–71
    sand-shale ratio, 712
    Silurian: in Illinois, 193; Cayugan saline
        deposits, Great Lakes region, 200–01
    three end constituents, 712
Lithological changes, 304
Lithologic map, Bahama Banks, 566–67
Lithology, in correlations, 701–03

727

Mohorovičić discontinuity, Moho: at Sigsbee Deep, 558; Gulf of Mexico, 559; 23, 638
Mollusca, 549, 652
Monadnocks: Laramie Peak, 512; Pikes Peak, 512; in Medicine Bow Range, 512; in Laramie Range, 515
Monograptus, 665
Monotremes, 661
Monroe Arch: reefs on, 566
Monteregian Hills: volcanic necks, 428; isotope age determination, 434; in St. Lawrence Lowlands, 434–35; 635
Montmorency Falls, Quebec: Ordovician-Precambrian relationships, 158, 161
Montreal, Quebec: map of vicinity, 433
Moon: impact craters, 694–95; origin, 696
Moraines: origin of, 578; Pleistocene, California, 591
Morrison, Mount, 403
Mosasaurs, 496
Mosses, 666
Moulton, F. R. *See also* Chamberlin, T. C., 674
Mountain peaks, volcanic, 22
Mountain ranges: Wegner's theory of origin, 472
Mucrospirifer, 220
Muirwoodia, 318, 321
Multituberculates, 617–18
Murchison, R.: Silurian, 86, 88, 182; Devonian, 208, 210; Permian, 290
Myalina: evolutionary trend, Carboniferous and Permian, 326–27

Nautiloids. *See* Cephalopods, 328–29
Neanderthal man. *See* Man
Nebulae: luminous, gaseous, 680; dark, 682; "globules," 682
Negaunee iron formation. *See* Marquette, Mich.
Nemagraptus, 130, 131
Neocomian: Gulf Coast, 418; British Columbia, 442; California, 442; *See also* Cretaceous
Neospirifer, 318–19
Neotremata, 136
Nevada:
  Antlerian Orogeny, 272–73
  Basin and Range Province, 514
  Cambrian, 114–15
  graptolites, 130, 184
  interior drainage, 515
  Jurassic, thrusts, 401
  Ordovician, palinspastic map of, 142
  Paleozoic: restored section, 142–43; overthrust faulting of, 271
  semi-arid basins, 515
  Tertiary orogeny, 526
Nevadan Orogeny: 398, 400
  age of, 403
  Jurassic: Pacific Coast, 398, 404, 440, 442; North America, western interior, 401, 408, 445, 448, 493, 633
  ultrabasic intrusive belts, 400–01
  Yosemite Valley, 388; geochemical age determination, 389
Nevadides, 440, 442, 448
Newark Group. *See* Triassic
New Brunswick:
  geologic map of, 94–95
  graptolites, 128
  oil fields, 264
  Ordovician intrusions, 174

New Brunswick (*cont.*):
  volcanism, 264
New England: Taconic Mountains, 174; Devonian, 212; 18, 21. *See also* Atlantic Coast
Newfoundland:
  Cambrian: fossils, relation to Wales, 94, 100; 85, 94–95
  folded and faulted sediments, 16
  fossils, eastern and western compared, 100
  graptolites, 128
  Hampden Fault, 465
  iron formations, 64
  Lukes Arm Fault, 465, 466
  metamorphic rocks, 18
  Ordovician, 130, 134
  Paleozoic of British Isles compared, 465
  pillow lava, 135
  Silurian: submarine troughs, current direction, 176; flute-casts, 176; sole markings, 176–77
  Taconian Orogeny in, 175
Niagara Falls: columnar section of, 182–83; Whirlpool Gorge, rocks of, 183; Pleistocene dating, 596
Niagaran escarpment: trend of, 186; 183, 185
Niagaran Series: fauna of, 184; stratigraphy of, 184, 186; 182, 184, 186, 200
Niagara River: Cayugan Series, 184
Nickel-copper mining, 54, 57, 58. *See also* Sudbury region
Nidulites, 666
Niobe, 92–94
Nipissing, Lake: Pleistocene glaciation, 584
North America:
  age determination through geologic time, histogram, 641
  Cambrian: western geosynclinal belt, 108–110; sections on eastern and western margins, 108–09; age of rocks at base of, 116
  eastern coastal belt, metamorphosed and intruded rocks, 637
  graptolites, 128
  mammals, 601, 621, 629
  maps of: geographic, 7; elevations, 8; physiographic, 9; distribution of principal rock types, 11; ages of rocks, inside book jacket; elevation of basement of metamorphic and igneous rocks, 13; Paleozoic volcanic rocks: distribution, 306–07; rocks of comparable age, distribution, 639
  Ordovician, paleogeographic map of, 147
  Pleistocene, extent of ice sheet, 578–79
  post-glacial uplift, 586–87
  Silurian, deformation of interior, 199
  surficial rocks, distribution of, 634–35
  volcanic rocks, 12
Northwest Territory: Devonian, 226; Mackenzie Mountains, 631; 15–16
Norway, graptolites, 128
Nothosaurs, 371, 484
Nuée ardente, 62, 526
Nummulites, 646

Oceans: crustal rocks under, 23, 470; origin of, 632
Ochoan: occurrence of, 290–91; gypsum, potash and salt in, 296
Ocoee, 121–122
Ohio: glacial sediments, 12, 13; Carboniferous, geologic map of, 248

Oil: Devonian, 230, 236; Mississippian, New Brunswick, 264; Carboniferous, 336, 337; Paleozoic, 336–37; Cenozoic, 337; Mesozoic, 337; Tertiary, 337, 547; source of, 337; Cretaceous, Mexico, 439
Oil shale: Eocene, Utah and Wyoming, 514–19
Oistoceras, 392
Oklahoma: graptolites, 184; Pennsylvanian paleographic map, 278–79
Old Red Sandstone. *See also* Devonian occurrence of, 208–11; Miller, Hugh, 208; Rhynie chert, 210; 86, 88
Olenellus, 91, 92, 99, 100, 106, 108, 110, 122
Oleneothyris, 652–53
Olenus, 92–94, 96, 99
Oligocene: 502
  Absaroka Mountains, Wyoming, lavas, 526–27
  Badlands, South Dakota, 510, 514
  mammals, 613, 620
  monadnocks, Laramie Peak, Pikes Peak, 512
  Nebraska, 510
  peneplanes, Laramie and Medicine Bow Ranges, 512
  reef limestone on salt domes, 565
Omphalocirrus, 226–27
Onaping: rhyolite breccia in, 56–57; 60–61
Ontario: iron ranges, 26. *See also* Ordovician, Precambrian, Silurian
Ontogeny: of trilobites, 90–91; definition of, 330; in goniatites, 332–33
Oolina, 554
Oolitic limestone. *See* Indiana limestone, 241
Oppel, Albert, 392
Orbigny, Alcide d', 46, 392
Ordovician:
  Adirondack line, 162
  Appalachian deposits resulting from uplifts, 177
  Arbuckle Mountains, Oklahoma, 280
  Bolarian, 140, 170
  Canadian, 140, 158
  Chazyan, 140, 147, 156–57, 158
  Cincinnatian, 140, 170, 172–73
  correlation tables of, 146, 148, 150, 151
  current flow: direction by fluke marks, grooves, sole markings, 176
  deformation during:
    Adirondack Mountains, deformation in, 163
    paleotectonic elements, 158
    Taconic Revolution, folding, 168, 172
    tectonic land: 172–73; Blountia, Taconica, Vermontia, 177
    thrust faults: Nevada, 139; Ouachita Mountains, Oklahoma, 139; Taconic Range, New York, 139; Vermont, 139
  epeirogenic movement, 148
  eustatic change, 148, 152, 153
  facies: "shaly," 134, 138; "shelly," 134; carbonate rock facies, classification, 140; changing facies, 134, 138, 141, 142, 144–45
  fauna, invertebrate:
    Brachiopoda, 136–37
    graptolites, 128, 130, 134; graptolite shales of North America, 132–33
    restoration, sea bottom fauna, Indiana, 153

Rhine River, Germany: Devonian, 211
Rhodophyta, 664, 667
Rhynie chert: Devonian plants, 210, 211
Rhyolite breccia, 57
Richardsonoceras, 328–29
Rifts: Pacific Coast, 633; Atlantic Coast faulting, Triassic, 633–34; lava flows, 634
Ripple marks: Precambrian, 34; sedimentary structures, 38; wave direction, 51; orientation diagrams indicating current flow and wind direction, 165
Robson, Mt., 20, 106, 107
Robulus, 554
Roche density, 693
Rock of Gibraltar, Jurassic, 388
Rocks:
    age of: geochemical dating, 2, 639; North America, map of distribution, 639; 20, 22, 32, 34, 82
    epeirogeny, 20, 574
    geographic controls of, 192
    kinds of: 10–21; North America, map of distribution, 11; and eustatic levels, relation to, 574
    origin, Wernerian doctrine, 48
    stratigraphic relations: 2; unconformable contacts, 30
    units: Table III, 90; rock-stratigraphic, 707
Rock Springs, Wyoming, 443
Rocky Mountains:
    Cretaceous: in western Canada and United States, 438–39; paleolithologic maps, Albian and Campanian, 440; lithofacies map, 440–41; Front Range, development in Wyoming, 510
    Laramian Orogeny, folding and faulting, 448, 522, 528, 636, 637
    of Colorado, 18, 264
    origin of, 436
    Tertiary, mountain areas and basins, 512
Rodents, 618
Rosette Nebula, 684
Rostricellula, 136
Rotation: of planets and sun, 670, 671; in universe, 688, 690; and solar tidal action, 694
Royal, Mount: volcanic neck, 426, 428; structure section, 433; intrusives of, 433; age determination by potassium-argon ratios, 433
Rubidium-strontium ratios: geochemical dating, 70, 72
Rugosa, 649
Russia: Devonian fishes, 234; Permian, 290–91

Sabine Arch, East Texas Oil field, 427, 566
Sacramento Valley, California: Jurassic, 397; Cretaceous, 397, 442
Saganaga granite, 36, 38, 40
Saganaga Lake, Minnesota: pillow lava, 37
Saganagan Orogeny, 74, 80
Saguenay River, Quebec: graben, 428, 435
St. Croixan: distribution in North America, 116, 121; cross bedding, 118–19; 110, 121
St. Elias Mountains, Yukon, 21

St. Francis Mountains, Missouri, 18
St. Helen's, Island, Quebec, volcanic vent, 428
St. Hilaire, Mount, Quebec. *See* Monteregian Hills, 434
St. John, Lake, Quebec: graben structure of, 435
St. Lawrence Lowlands, map of, 434–35
St. Mary's Bay, Nova Scotia, 356
St. Peter Sandstone, 152
St. Pierre and Miquelon, 94–95
Salem limestone. *See* "Indiana" limestone, 241
Salt: Silurian in Great Lakes area, 200–04; Devonian, Alberta, 226, Michigan, 226, 230; Mississippian in Michigan, 244; Permian, Kansas, 296, in Ochoan, 296; Gulf Coast, Mississippi, Vera Cruz, 559–66
Salt basin, Pennsylvanian, 280
Salt domes: Eagle Mills salt, 410–11; Carribean Sea, 412; and petroleum, 559, 564, 566; and sulphur, 559, 562, 564; origin of, 560, 564; geophysical examination of, 562, 564; faults around, 563; and reef distribution, 565; Vera Cruz, 566; Mississippi, 566
    Tertiary Gulf Coast: Avery Island Dome, Louisiana, 560; High Island Dome, Texas, 560; Grand Saline Dome, Texas, 562; Weeks Island Dome, Texas, 563; Spindle Top, Texas, 564
Salt water precipitation, 202, 296
San Andreas Fault, California: and Denali Fault, 530; lateral displacement in, 530; 536–38
Sands, shore: changing age of, diagram, 45; effect of sorting 556; marine regression, 557; sand-shale ratio, 712
Sandstone: stream flow trend, map of, 250–51; Pennsylvanian sources, 254, 256
San Juan River, Utah: entrenched river, 263, 524–25
San Marcos Arch, 424–25
Santa Catarina System, South America, 485
Satellites: origin, 671; revolution around planets, 671; development of, 694; "Trojan group," 694
Saturn: satellites, orbits of, 672–73; density, 686–87; composition, 687; 676–77
Sauropods, 482
Scandinavian ice sheet, 579
Scaumenac, Escuminac Bay, Quebec: Devonian amphibians, 236; 224–25
Schizotreta, 136
Schwagerina, 292
Scleractinia: reef builders, 649–50; distribution through time, 649; 480
Scolecodonts, 652
Scorpion, 345, 347
Scotland, Old Red Sandstone unconformity, 211
Scottognathus, 659
Scouring rushes. *See* Calamites
Scyphozoa, 649
Seamounts: origin of, 574; Atlantic, 574; age of, 574–75; Hawaiian Islands, 574; Bermuda, 575
Sea urchins, 480

Secondary system, 88
Sedgwick, Adam: 52, 86, 88, 182, 210
Sedimentary petrology, 2, 256, 700
Sedimentary rocks:
    Alaska, 15, 16
    Atlantic Coast, 16
    Canada, 15, 16
    continental interior, 635–36
    folded and faulted sediments, 16
    Gulf Coast, thickness, 14, 16
    in basins, 16
    Michigan Basin, 636
    New Mexico, 14, 16
    North America: distribution by age, map of, inside book jacket
    on domes, 16
    Texas, 14, 16
    thickness determination, geophysical methods, 12, 16; North America, distribution, 12, 14, 16
    tops of beds, criteria for determining, 33, 34
Sedimentology, 2, 714
Sediments: British Columbia, 15; Hudson Bay, 18; Pacific Belt, 18; Atlantic Belt, 18; deposition, time of, 22; distribution, interpretation of, 114, 116, 118–20; epeirogenic movement, 118–19; eustatic movement, 118–19
Seismic waves: velocity of, 23; and geophysical methods, 432
Seneca Rocks, West Virginia, 200–01
Series, definition of, 90
Serpentine: North American belts, age, composition, distribution, relation to tectonic belts, 396; "verde antique marble," 398
Serpentinite: eugeosynclines, 400; 401
Shales, laminated, 515, 518–19
Sharks, 234–35; 489–90; 658
Shasta, Mount, 12, 231, 533, 542–43, 577
Sherman Fall, New York, 140
Shickshockian Orogeny, Acadian Orogeny, 218, 223–24
Shinarump: Petrified Forest, 368
Ship Rock, New Mexico, 635
Shishaldin, Mount, Alaska, 530–31
Shoreham Limestone: depth related to texture and thickness, 164–65
Sial: and continental drift, 456, 470; character of, 638; and continental development, 640–41; 23
Sierra de Parras, Mexico, 408–09
Sierra Nevada: rocks of, 18; bathylith, age of, 389; fossil plants, 403; upland surface, Yosemite National Park, 590–91; 400, 526
Sigillaria, 252–54
Sigsbee Deep: geophysical studies of, 556, 558–59; Mohorovičić discontinuity, 558
Silurian:
    Alexandrian, 192, 194, 196
    Alexandrian Sea, 194–95
    Anticostian Sea, 194–95
    Caledonian Orogeny, 208, 210–11
    California: Sierra Nevadas, 230; Klamath Mountains, 231
    Cayugan: 184, 198–204; Great Lakes region, isopach and lithofacies maps, 200–01; saline deposits, 200–01; paleogeographic map, Great Lakes, 200–01
    Clintonian, 186, 190, 200–01